Using
Literature

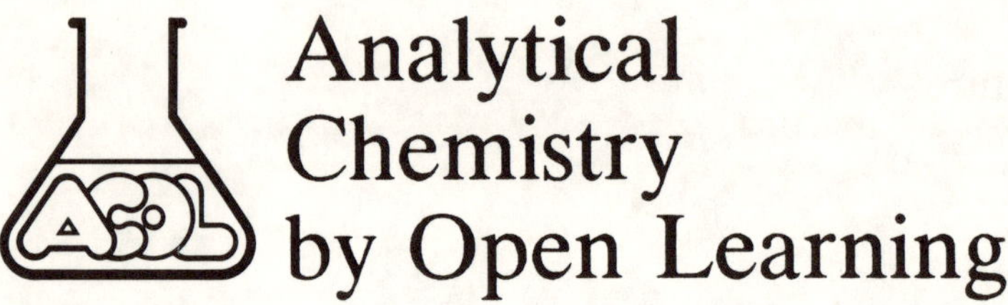
Analytical
Chemistry
by Open Learning

**Titles in Series:**

Samples and Standards
Sample Pretreatment and Separation
Classical Methods
Measurement, Statistics and Computation
Using Literature
Instrumentation
Chromatographic Separations
Gas Chromatography
High Performance Liquid Chromatography
Electrophoresis
Thin Layer Chromatography
Visible and Ultraviolet Spectroscopy
Fluorescence and Phosphorescence Spectroscopy
Infra Red Spectroscopy
Atomic Absorption and Emission Spectroscopy
Nuclear Magnetic Resonance Spectroscopy
X-Ray Methods
Mass Spectrometry
Scanning Electron Microscopy and X-Ray Microanalysis
Principles of Electroanalytical Methods
Potentiometry and Ion Selective Electrodes
Polarography and Other Voltammetric Methods
Radiochemical Methods
Clinical Specimens
Diagnostic Enzymology
Quantitative Bioassay
Assessment and Control of Biochemical Methods
Thermal Methods
Microprocessor Applications

# Using Literature

Analytical Chemistry by Open Learning

Author:
STUART JAMES
*Paisley College of Technology*

Editor:
NORMAN B. CHAPMAN

*on behalf of ACOL*

Published on behalf of ACOL, Thames Polytechnic, London
by
JOHN WILEY & SONS
Chichester • New York • Brisbane • Toronto • Singapore

Published by permission of the Controller of
Her Majesty's Stationery Office



***Library of Congress Cataloging in Publication Data:***

Analytical chemistry by open learning.
Includes index.
Contents: –[5]. Analytical literature / author, Stuart James.
1. Chemistry, Analytic–Programmed instruction.
I. Chapman, N.B. (Norman Bellamy), 1916–
QAD75.2.A519 1986 543'007'7 86–15684

ISBN 0 471 91220 4 (v. 5)
ISBN 0 471 91221 2 (pbk. : v. 5)

***British Library Cataloguing in Publication Data:***

James, Stuart
Analytical literature.—(Analytical chemistry by open learning)
1. Chemistry, Analytic—Information services 2. Chemistry, Analytic—Bibliography.
I. Title II. Chapman, Norman III. Series

ISBN 0 471 91220 4 (Cloth)
ISBN 0 471 91221 2 (Paper)

Printed and bound in Great Britain

# Analytical Chemistry

This series of texts is a result of an initiative by the Committee of Heads of Polytechnic Chemistry Departments in the United Kingdom. A project team based at Thames Polytechnic using funds available from the Manpower Services Commission 'Open Tech' Project has organised and managed the development of the material suitable for use by 'Distance Learners'. The contents of the various units have been identified, planned and written almost exclusively by groups of polytechnic staff, who are both expert in the subject area and are currently teaching in analytical chemistry.

The texts are for those interested in the basics of analytical chemistry and instrumental techniques who wish to study in a more flexible way than traditional institute attendance or to augment such attendance. A series of these units may be used by those undertaking courses leading to BTEC (levels IV and V), Royal Society of Chemistry (Certificates of Applied Chemistry) or other qualifications. The level is thus that of Senior Technician.

It is emphasised however that whilst the theoretical aspects of analytical chemistry can be studied in this way there is no substitute for the laboratory to learn the associated practical skills. In the U.K. there are nominated Polytechnics, Colleges and other Institutions who offer tutorial and practical support to achieve the practical objectives identified within each text. It is expected that many institutions worldwide will also provide such support.

The project will continue at Thames Polytechnic to support these 'Open Learning Texts', to continually refresh and update the material and to extend its coverage.

Further information about nominated support centres, the material or open learning techniques may be obtained from the project office at Thames Polytechnic, ACOL, Wellington St., Woolwich, London, SE18 6PF.

# How to Use an Open Learning Text

Open learning texts are designed as a convenient and flexible way of studying for people who, for a variety of reasons cannot use conventional education courses. You will learn from this text the principles of one subject in Analytical Chemistry, but only by putting this knowledge into practice, under professional supervision, will you gain a full understanding of the analytical techniques described.

To achieve the full benefit from an open learning text you need to plan your place and time of study.

- Find the most suitable place to study where you can work without disturbance.

- If you have a tutor supervising your study discuss with him, or her, the date by which you should have completed this text.

- Some people study perfectly well in irregular bursts, however most students find that setting aside a certain number of hours each day is the most satisfactory method. It is for you to decide which pattern of study suits you best.

- If you decide to study for several hours at once, take short breaks of five or ten minutes every half hour or so. You will find that this method maintains a higher overall level of concentration.

Before you begin a detailed reading of the text, familiarise yourself with the general layout of the material. Have a look at the course contents list at the front of the book and flip through the pages to get a general impression of the way the subject is dealt with. You will find that there is space on the pages to make comments alongside the text as you study—your own notes for highlighting points that you feel are particularly important. Indicate in the margin the points you would like to discuss further with a tutor or fellow student. When you come to revise, these personal study notes will be very useful.

Π When you find a paragraph in the text marked with a symbol such as is shown here, this is where you get involved. At this point you are directed to do things: draw graphs, answer questions, perform calculations, etc. Do make an attempt at these activities. If necessary cover the succeeding response with a piece of paper until you are ready to read on. This is an opportunity for you to learn by participating in the subject and although the text continues by discussing your response, there is no better way to learn than by working things out for yourself.

We have introduced self assessment questions (SAQ) at appropriate places in the text. These SAQs provide for you a way of finding out if you understand what you have just been studying. There is space on the page for your answer and for any comments you want to add after reading the author's response. You will find the author's response to each SAQ at the end of the text. Compare what you have written with the response provided and read the discussion and advice.

At intervals in the text you will find a Summary and List of Objectives. The Summary will emphasise the important points covered by the material you have just read and the Objectives will give you a checklist of tasks you should then be able to achieve.

You can revise the Unit, perhaps for a formal examination, by re-reading the Summary and the Objectives, and by working through some of the SAQs. This should quickly alert you to areas of the text that need further study.

At the end of the book you will find for reference lists of commonly used scientific symbols and values, units of measurement and also a periodic table.

# Contents

# Acknowledgements

Fig. 1.2a is taken from a Chemical Abstracts Services Report and is reproduced by permission of the American Chemical Society

Pages 8–9 are extracted from *Chemical Methods of Rock Analysis*, 3rd Ed., Jeffery and Hutchison, 1981, and are reproduced by permission of Pergamon Press

Pages 10–13 are extracted from R B Lucke *et al*, *Analytical Chemistry*, **57**, 1985, 633 and 639 and are reproduced by permission of American Chemical Society

Pages 19–37 are extracted from the BSI Catalogue 1985, and are reproduced by permission of the British Standards Institute

Pages 42–63 are extracted from the *Annual Book of ASTM Standards*, 1984 and are reprinted, with permission, from the *Annual Book of ASTM Standards*, copyright, American Society for Testing and Materials, 1916 Race Street, Philadelphia, PA 19103, USA

Pages 65–79 are reproduced from *Official Standardised and Recommended Methods of Analysis*, N W Hanson, Society for Analytical Chemistry, 1973 and are reproduced by permission of the Royal Society of Chemistry

Pages 81–90 are extracted from the *Official Methods of Analysis*, 14th ed., 1984 and are reproduced by permission of the Association of Official Analytical Chemists

Pages 92–109 are extracted from British Pharmacopoeia, 1980, Vols 1 and 2 and are reproduced by permission of the Controller of Her Majesty's Stationery Office

Pages 118–137 are reproduced from *Comprehensive Analytical Chemistry*, **Vol XIII**, 1981, Wilson and Wilson, Elsevier Scientific Publishing Company. Permission has been requested

Pages 140–157 are extracted from the *Encyclopedia of Industrial Chemical Analysis*, F D Snell and C L Hilton, by permission of John Wiley & Sons Inc.

Pages 159–202 are reproduced from the *Encyclopedia of Chemical Technology*, Kirk-Othmer, by permission of John Wiley & Sons Inc.

Pages 205–224 are extracted from *Handbook of Analytical Chemistry*, 1st ed., L. Meites, and are reproduced by permission of McGraw-Hill Book Company, USA

Pages 226–243 are extracted from *The Merck Index*, 10th Ed., 1983, and are reproduced by permission of Merck & Co., Inc., USA

Pages 253–273 are extracted from *Handbook of Reactive Chemical Hazards*, 2nd ed., L Bretherick, 1979, Butterworth's, London and are reproduced by permission of L Bretherick and Butterworth's, London

Pages 275–305 are extracted from *Adhesives Handbook*, 3rd Ed., J Shields and are reproduced by permission of the publishers, Butterworths, London.

Pages 346–382 are extracted from the *Dewey Decimal Classification and Relative Index*, 1979, Vols 2 and 3, and are reproduced by permission of Forest Press, USA

Pages 384–387 are extracted from the International Medium Edition of the *Universal Decimal Classification* and are reproduced by permission of the Internation Federation for Documentation and the British Standards Institution

Pages 398–404 are extracted from *Current Technology Index*, 1983 and are reproduced by permission of Library Association Publishing Ltd.

Pages 409–438 are extracted from *Chemical Abstracts* and are reproduced with permission from the American Chemical Society

Pages 441–450 are extracted from *Analytical Abstracts*, 1984 and are reproduced by permission of the Royal Society of Chemistry

Pages 453–454 are extracted from SCI-Sample Searches in Chemistry leaflet and are reproduced with permission from and copyright owned by the Institute for Scientific Information ®, Philadelphia, PA, USA

Pages 455–459 are extracted from *Current Contents* and are reproduced with permission from and copyright owned by the Institute for Scientific Information ®, Philadelphia, PA, USA.

Pages 475–478 are reproduced as facsimile print out from *Food Science and Technology Abstracts* and are reproduced by permission of International Food Information Service.

Pages 469, 492–495, 566–570 are reproduced as facsimile output from the CAS computer-readable files with the permission of the American Chemical Society

Pages 479–480 are reproduced by permission of ESA-IRS.

Pages 481–482 are reproduced by permission of Datastar Marketing Ltd.

Pages 483–484 are reproduced by permssion of Pergamon Press Ltd.

Pages 505–507, 582–586 are reproduced as facsimile print out from the *Encyclopedia of Chemical Technology*, Kirk-Othmer. Permission has been requested from John Wiley & Sons Inc.

Pages 510, 588–589, 591–593 are reproduced as facsimile print outs. Permission has been requested from Merck & Co., Inc., USA

Page 513 is reproduced as a facsimile print out from the Martindale Online database. Permission has been requested from the Pharmaceutical Society of Great Britain

Pages 572–574 are reproduced as facsimile print out from *International Pharmaceutical Abstracts* and are reprinted with permission from American Society of Hospital Pharmacists.

Pages 576–577 are reproduced as facsimile print out from the *Laboratory Hazards Bulletin* and are reproduced with permission from the Royal Society of Chemistry

Pages 578–580 are reproduced as facsimile print out from *Chemical Engineering Abstracts* and are reproduced with permission from the Royal Society of Chemistry

# 1. Introduction

## 1.1. THE NATURE AND IMPORTANCE OF INFORMATION

Why should you need to undertake a course in analytical literature? And what am I doing talking about information instead of literature? The answer to the first question will become obvious during the course of this Introduction (and even more so as you work through the rest of the Unit). As to the second question, well, stop and think for a moment: when you look something up in a textbook or a reference book, what are you really looking for? Information, of course. The analytical literature is the vehicle through which you acquire information on analytical chemistry.

Scientific and technical information has always been of great importance, but in recent years that importance has been understood, and formalised, to a greater degree than ever before. Information is now considered by many economists as an important resource; certainly, increasing numbers of firms and consultants are proving information to be a marketable and profitable commodity. But, by using library services and publications such as I am going to describe to you in this Unit, you can acquire much of the information you are ever likely to need (if not all of it) at the expense of just a little time and effort on your part.

Π You are looking for information on a chemical topic. How do you expect that information to be presented to you?

A lot of information is presented in words of course, eg the text of this Unit. But chemical and other scientific and technical information can be presented usefully in other formats. These formats, among others, might be: numerical data in tabular form; numerical and alphabetical data in tabular form; alphabetical information in tabular form; mathematical expressions; chemical formulae and equations; graphical information (diagrams, photographs, etc).

One last point here: to be of any value or relevance, information *must be communicated* in some form or other. If it is not communicated it can have no impact and, therefore, no value.

## 1.2. THE LITERATURE 'EXPLOSION'

One of the major problems concerning scientific and technical literature (and thus information) is that there is simply an awful lot of it—far too much for anybody to contemplate trying to read all the new papers being published within even some very specialised field. The growth of chemical literature is aptly illustrated by *Chemical Abstracts* (we shall be examining this service in detail in Parts 3 and 4). In 1907 just a few thousand publications were abstracted, but in 1984 alone almost 500,000 were abstracted; on the 19th March 1984 *Chemical Abstracts* published its ten millionth abstract since 1907. Just to illustrate this growth and emphasise what you are up against, I have included here as an illustration an advertisement for *Chemical Abstracts* announcing and celebrating their ten millionth abstract, Fig. 1.2a. You will notice from their graph how dramatic the growth has been since 1945: although many people consider the literature/information explosion to be a purely post-war phenomenon, it does in fact represent approximately the steep end of an exponential growth-curve which actually began in the late seventeenth century. In practice an exponential growth is often finite: but this curve shows no signs of reaching a limit and few signs even of decreasing in slope; and that despite recent economic and other trends in the industrialised nations which have inhibited scientific research and its publication. In other words, you face an already immense amount of literature, the growth of which is unlikely to diminish in the foreseeable future; it might even continue to grow at an increasing rate. A sobering thought, isn't it?

**Fig. 1.2a** *Growth in number of abstracts published by Chemical Abstracts*

## 1.3. DEFINITION OF INFORMATION REQUIREMENTS

So you begin to see the scale of your problem. Information (as embodied in the analytical literature) is of vital importance to you, the analytical chemist (just as it is to anybody working in any field in science and technology or elsewhere); and there is an awful lot of it. Not only that but, as we shall see in Part 2, the literature is produced in different forms, all with differing purposes and functions. So here is one most important word of advice at this early stage, and one to bear in mind throughout your study of this unit and whenever you are seeking information. Before you ever start looking for information of any kind on any topic, *stop and think*: funny how that expression keeps cropping up—don't think you have heard the last of it! Define exactly what information you think you need and why you need it. If you just want the formula of a particular compound or the value of a particular parameter of that compound in order to check a reaction or test, then don't go combing *Chemical Abstracts* or some multi-volume handbook, or reading ten or twenty pages of a textbook: you can look that kind of information up much more conveniently in some of the reference books I shall discuss in Part 2. If you come across a specialist chemical subject for the first time, you don't want an advanced research paper or detailed post-graduate level book to introduce that subject to you: an article in a chemical or scientific encyclopaedia (or even a dictionary) is more likely to give you the comprehensible introduction you need. Once you have digested that kind of information, if you still need more details, that will be the time to move on to handbooks, and eventually research papers. Always decide exactly what kind of information you need and for what purpose you need it, before starting to look things up. That way you will save yourself a lot of time and energy and avoid a lot of frustration. And of course, when you have worked through this Unit the knowledge and experience you have gained will give you an even greater likelihood of success in seeking information.

## 1.4. GROWTH OF INFORMATION TECHNOLOGY

Obviously the computer has affected, and will continue to affect, incalculably the production and dissemination of scientific and technical information. After the turn of the twentieth century the scientific and technical information field will have gone through a complete revolution and will look quite different from today's scene. At present, we are partially into that revolution, and this presents some problems for this Unit. I have covered much of the ground through traditional printed materials and confined electronic services in their own Part. That is how the information scene for most analytical chemists still looks today: a reliance on printed reference books and other printed documents, but an increasing awareness of the significance of computerised services (many of the printed materials we shall examine are in fact compiled and produced from computer data bases). The pattern is very uneven between different organisations and groups of information users, and between different information services; some of each group are for various reasons pushing ahead into electronic services faster than others. The pattern is likely to remain uneven for quite some time yet, although if I were asked to produce a new edition of this Unit in, say, five years time, I should place an even greater emphasis on electronic services in some instances than I do today. I shall deliberately avoid mentioning computers at all (except occasionally in passing) until we reach Part 4. If you are already familiar with printed sources, you will find the switch to electronic information retrieval much more comprehensible and logical when we do come to examine it later.

## 1.5. FURTHER READING

I have set out to make this Unit completely self-contained for you. But in order to contract the material into the time allocated and to avoid introducing complications which might confuse you, I have had to over-simplify some aspects and to omit detailed discussion of others. Perhaps you want to know more about chemical literature, or some of the matters I raise here but do not pursue. The first book I would recommend to you is:

> Mellon, M. G. *Chemical publications: their nature and use* 5th edn. New York. McGraw-Hill, 1982. 419pp.

This covers for the whole of chemistry that which I restrict to analytical chemistry. It discusses all the aspects I cover in this Unit (including computerised information retrieval) and lists useful chemical publications of all kinds. It is an American publication and so can usefully counteract any British bias you might find in this Unit. The second book I recommend is a British one:

> Bottle, R. T. (ed) *Use of chemical literature* 3rd edn. London. Butterworth, 1979. 306pp.

This has a series of narrative chapters (with works referred to listed in notes) discussing in greater detail many of the topics we shall cover in this Unit. A third book on the same subject which you might find useful is:

> Maizell, R. E. *How to find chemical information: a guide for practising chemists, teachers and students.* New York: Wiley, 1979. 261pp.

I have also mentioned a couple of other items in the text of the Unit. One caution though: I recommend any (or all if you are so interested) of these books as useful supplements, but I do suggest that you complete your study of this Unit before consulting any of them. I have designed this Unit, *Using Literature*, as an integrated whole, and you will need to study all of it to cover the subject adequately, before supplementing or widening your experience. Also, these books don't cover the subject in the order that I have adopted.

## 1.6. LAYOUT OF THIS UNIT

I have constructed the Unit as an integrated whole. That means that I expect you to start here at the beginning and work through the material in the order in which I present it. When you have completed one section I shall assume that the knowledge and experience you have gained there can be applied in later sections where necessary. I advise you not to try and dip in at the middle of the Unit or some later stage: you will certainly find it irritating when I refer you back to something you might have skipped or if you don't in fact know something that I assume you do know.

There are three main elements in the Unit: the text, in-text questions, and self-assessment questions. *All these are completely interdependent and help to build the unit into a whole.* Don't think you can miss out the self-assessment questions. They are designed to test what you should have learned from the text; but, because this is an intensely practical subject and I cannot know what access you might or might not have to the publications I describe, these questions are often also designed both to give you practical experience and to amplify my text. If you try to ignore the self-assessment questions you will miss some vital points in the questions and responses, which are *not* to be found in the text. I hope that you find the text itself completely self-explanatory; I have tried to keep it as simple as I can (even to the point of over-simplification on occasions) and to explain any terms you might not understand. In one or two places I have had to compromise between the demands of keeping the text interesting while giving a lot of factual information, and not turning parts of the Unit into a glorified book list. There are various in-text questions throughout the Unit. These are not just rhetorical questions: do (that magic advice again!) stop and think for a moment when you have read the question before you go on to read my replies.

I have made no prior assumptions about you, the learner, except that you are a practising or aspiring analytical chemist who can read English. My questions don't require much chemical knowledge—indeed one of the virtues I would claim for this Unit is that it will show you how to find information on subjects of which you might be quite ignorant! So I do hope that you will achieve all the aims I set out for you throughout the Unit, and that, at the end of it, you will be able to appreciate the usefulness of analytical literature to you and, above all, to find and use that literature with confidence and to your own ultimate benefit.

## 1.7. BIBLIOGRAPHIC REFERENCES

Don't think this Introduction is simply a quick (but essential) preamble and that you can just read it and get away without learning and doing something practical! Obviously we shall be talking for most of our time about different publications. So, our first problem is: whenever we encounter a publication for the first time, how do we describe it concisely, consistently, and accurately enough for anyone else to trace it should they wish? Or, to put it more technically, how do we write a bibliographic citation? Whatever the publication you wish to refer to, there are certain items you must include in your description. These are as follows.

(*a*) *The author's name*, usually with the surname first (eg Smith, John).

Where there are two authors you give the names of both (eg Smith, John and Jones, Michael). But where there are three or more authors you give only the first named in the list and use the Latin abbreviation *et al* for the rest (Smith, John *et al*). If there is no author named at all then you enter the work by its title.

(*b*) *The title* of the book or journal paper. Always give the main title in full, and the sub-title separated by a colon eg (*Analytical chemistry by open learning: using literature*). In fact for books you can omit the sub-title if you want to save space, but the main title is essential and must never be altered or abbreviated.

(*c*) *The Publication details*. For books this means: the edition, only if it is not the first edition (abbreviated thus: 2nd ed, Rev ed, etc); the place of publication and the name of the publisher are advisable but not absolutely essential: give only the main place of publication (many publishers list several towns or cities on the title page), and cut out terms such as 'Inc', 'and Co.', 'Ltd' from the publisher's name; the date of publication is always absolutely essential; the number of pages is optional. For journal papers the publication details must comprise the title of the journal in which the paper is published; the volume number; the date of the issue; and the pages on which the paper appears.

Let's just see how these will look and how they are punctuated. A typical entry for a book, quoting all the data I have indicated, would look like this:

> Loebel, Arnold B. *Chemical problem-solving by dimensional analysis: a self-instructional progam.* 2nd ed. Boston: Houghton Mifflin, 1978. 423pp.

If, on the other hand, you wanted just a very brief bibliographic citation, the following is the *minimum permissible* information for this same book:

> Loebel, A. B. *Chemical problem-solving by dimensional analysis*. 2nd ed. 1978.

A typical entry for a journal paper would look like this:

> Nollins, Ann and Cullen, W. R. Determination of organotin compounds contained in aqueous samples using capillary gas chromatography. *Analyst*, v.109 (Dec 1984), 1527–1529.

A couple of points arise from this, although I am trying to keep this as simple as I can for you. Titles of journals (but *never* of the individual papers) may be abbreviated, and many journals in fact indicate a standard abbreviation you can use; but do be very wary of abbreviating a journal title on your own initiative and, if in doubt, always quote it in full. Usually but not always you need quote only a year for the date, but to be on the safe side it is as well to give the precise date of issue. Alternatively, you can quote the issue number after the volume, number, eg v.12(4), (1984).

There are many complications which may arise in trying to write bibliographic citations to these and other kinds of documents. But the advice I have given you here will cover most of your requirements and you will find that further practice turns this into a fairly routine operation. One last word of advice: whenever you are recording any publication you come across, if in doubt, include more information than you are likely to need. It is always easier subsequently to edit material out, than to have to go back and find extra information you didn't bother to record the first time.

So let's finish here with a practical exercise for you. Then, when you have finished that, we will start out into the main contents of the Unit. Remember, this question isn't just a practical or revision exercise: I shall be making one or two points in my response.

**SAQ 1.7a**

I have given you, on the next few pages, a copy of the title page of a book on rock analysis, and one of the first and the last page of a journal paper on polycyclic aromatic hydrocarbon separation (with the title pages of the journal in which it was published). The book has 379 pages. Write out a bibliographic citation for each by using the full form of citation I have indicated in my text.

# Chemical Methods of Rock Analysis

by

P G JEFFERY, *Deputy Director (Resources)*
*Laboratory of the Government Chemist, London, UK*

and

D HUTCHISON, *Geochemistry and Petrology Division, Institute of Geological Sciences (N.E.R.C.), London, UK*

**THIRD EDITION**

PERGAMON PRESS

OXFORD · NEW YORK · TORONTO · SYDNEY · PARIS · FRANKFURT

| | |
|---|---|
| U.K. | Pergamon Press Ltd., Headington Hill Hall, Oxford OX3 0BW, England |
| U.S.A. | Pergamon Press Inc., Maxwell House, Fairview Park, Elmsford, New York 10523, U.S.A. |
| CANADA | Pergamon Press Canada Ltd., Suite 104, 150 Consumers Road, Willowdale, Ontario M2J 1P9, Canada |
| AUSTRALIA | Pergamon Press (Aust.) Pty. Ltd., P.O. Box 544, Potts Point, N.S.W. 2011, Australia |
| FRANCE | Pergamon Press SARL, 24 rue des Ecoles, 75240 Paris, Cedex 05, France |
| FEDERAL REPUBLIC OF GERMANY | Pergamon Press GmbH, Hammerweg 6, D-6242 Kronberg-Taunus, Federal Republic of Germany |



First edition 1970
Second edition 1975
Peprinted (with corrections and additions) 1978
Third edition 1981
Reprinted (with corrections) 1983

**British Library in Cataloguing Data**

Jeffery, Paul Geoffrey
Chemical methods of rock analysis. - 3rd ed.
(Pergamon series in analytical chemistry)
1. Rocks - Analysis - Laboratory manuals
I. Title II. Hutchison, D.
552'.06 QE438

ISBN 0-08-023806-8

**Library of Congress Catalog Card no.**: 81-81234

*In order to make this volume available as economically and as rapidly as possible the authors' typescripts have been reproduced in their original forms. This method unfortunately has its typographical limitations but it is hoped that they in no way distract the reader.*

*Printed in Great Britain by A. Wheaton & Co. Ltd., Exeter*

analytical chemistry®

**MARCH 1985**

**COVER FEATURE**

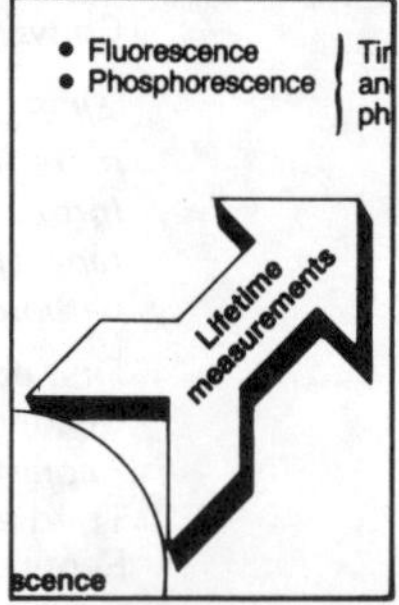




Volume 57, No. 3
March 1985
ANCHAM
57(3) 365A–488A
577–784 (1985)
ISSN 0003 2700



ANALYTICAL CHEMISTRY (ISSN 0003-2700) is published monthly except semimonthly in April and August by the American Chemical Society at 1155 16th St., N.W., Washington, D.C. 20036. Editorial offices are located at the same ACS address (202-872-4600; TDD 202-872-8733). Second-class postage paid at Washington, D.C., and additional mailing offices. Postmaster: Send address changes to Membership & Subscription Services, P.O. Box 3337, Columbus, Ohio 43210.

**Claims for missing numbers** will not be allowed if loss was due to failure of notice of change of address to be received in the time specified; if claim is dated (a) North America: more than 90 days beyond issue date, (b) all other foreign: more than one year beyond issue date, or if the reason given is "missing from files."

**NEWS**

**The ACS Spring National Meeting** will be held April 28–May 3 in Miami Beach, Fla. The Analytical Division program will be highlighted by the presentation of five ACS awards

**408 A**

**ANALYTICAL APPROACH**

**Gunpowders.** Using the wrong type of powder can injure the shooter and result in lawsuits against the manufacturer. Peter Lott and Darwin Dahl describe their work to determine the powder involved in such cases

**446 A**

**BOOKS**

**Critical reviews.** Books on immunoassay, food analysis, and FT–NMR are reviewed by G. A. Hudson, J. F. Lawrence, and T. C. Farrar

**455 A**

**EDITORIAL**

**The Center for Analytical Chemistry** at NBS serves as the nation's reference laboratory for chemical composition measurements. With its broad base of expertise it can be considered our national analytical laboratory

**577**

**1985 subscription rates include air delivery outside the U.S., Canada, and Mexico**

| | 1 yr | 2 yr |
|---|---|---|
| **Members** | | |
| Domestic | 20 | 34 |
| Canada and Mexico | 40 | 74 |
| Europe | 57 | 108 |
| All Other Countries | 80 | 154 |
| **Nonmembers** | | |
| Domestic | 30 | 51 |
| Canada and Mexico | 50 | 91 |
| Europe | 99 | 179 |
| All Other Countries | 122 | 225 |

Three-year and other rates contact: Membership & Subscription Services, ACS, P.O. Box 3337, Columbus, Ohio 43210, (614) 421-3776.

**Subscription orders by phone** may be charged to Visa, MasterCard, Barclay card, Access, or American Express. Call toll free at (800) 424-6747 from anywhere in the continental U.S.; from Washington, D.C., call 872-8065. Mail orders for new and renewal subscriptions should be sent with payment to the Treasurer's Office at the Washington address.

**Subscription service inquiries and changes of address** (include both old and new addresses with ZIP code and *recent mailing label*) should be directed to the ACS Columbus address noted above. Please allow six weeks for change of address to become effective.

**ACS membership information:** Ann Donahue, Washington address.

**Single issues,** current year, $7.50 except review issue and LabGuide, $8.50; **back issues and volumes and microform editions** available by single volume or back issue collection. For information or to order, call (800) 424-6747 or write the Sales Department at the Washington address.

**Nonmember rates in Japan:** Rates at left do not apply to nonmember subscribers in Japan, who must enter subscription orders with Maruzen Company Ltd., 3–10 Nihonbashi 2-chome, Chuo-ku, Tokyo 103, Japan. Tel: (03) 272-7211.

*Anal. Chem.* **1985**, *57*, 633–639 **633**

# Integrated, Multiple-Stage Chromatographic Method for the Separation and Identification of Polycyclic Aromatic Hydrocarbons in Complex Coal Liquids

**Richard B. Lucke,* Douglas W. Later, Cherylyn W. Wright, Edward K. Chess, and Walter C. Weimer**

*Chemical Methods and Kinetics Section, Pacific Northwest Laboratory, Operated by Battelle Memorial Institute, Richland, Washington 99352*

**A multiple-stage methodology is described for the separation and identification of neutral polycyclic aromatic hydrocarbons (PAH) in complex coal liquids. Fractional distillation followed by adsorption chromatography on neutral alumina with subsequent reverse-phase high-performance liquid chromatography were used for the isolation of neutral PAH subfractions. Capillary column gas chromatography, gas chromatography–mass spectrometry, and direct-probe mass spectrometry were used for the identification of sample components. This method results in chemically less complex subfractions which enable more detailed characterization of both trace level and major neutral PAH components. The applicability of this multiple-stage separation method is demonstrated with a solvent refined coal (SRC) II material.**

In recent years, synfuels have become of increased interest because of their potential as an alternate energy source. The need to identify potentially toxic and carcinogenic compounds which may be present in alternate fuel materials has resulted in an increased effort to obtain detailed information about their chemical composition as it relates to biological response observed in laboratory test systems. However, synfuels, such as coal-derived liquids, are extremely complex, and the relationships between chemical composition and biological response are not always straightforward. The molecular weight distribution, degree of alkylation and hydrogenation, chemical functionality, and heteroatom content of coal liquids are all variables that contribute to the increased complexity of these materials. This complexity has led to the development of a wide variety of analytical methods aimed at achieving improved separations for chemical and biological characterization of synfuels.

Coal liquid fractionation procedures have included fractional distillation (*1–4*), size-exclusion chromatography (*5–9*), solvent extraction (*10–15*), adsorption chromatography with alumina and/or silica (*16–20*), normal-phase high-performance liquid chromatography (HPLC) (*21–24*), and reverse-phase HPLC (*25–28*). Analytical methods using combinations of these separation techniques are often required for complete chemical characterization and biological testing of these highly complex mixtures. For example, multidimensional coupled-column HPLC techniques that produce discrete fractions have been used to analyze polycyclic aromatic hydrocarbons (PAH) in coal liquids (*28, 29*). However, coupled-column HPLC requires modified instrumentation and is impractical for the preparative-scale fractionation of material that is often required for biological testing.

In this laboratory, a four-step analytical methodology for the separation, identification, and quantitation of neutral PAH and their alkylated analogues in coal liquids has been developed and applied (Figure 1). In general, the neutral PAH are the most abundant compound class in coal-derived synfuels and have been shown to be primarily responsible for the tumorigenic activity of these materials (*30, 31*). The integrated multiple-stage chromatographic method described in this paper employs fractional distillation followed by adsorption chromatography on neutral alumina, with subsequent reverse-phase HPLC. The combination of these separation techniques produces fractions which are chemically much less complex than the starting crude material and which are well defined in terms of chemical class and subclass. Furthermore, correlation of biological response to chemical constituency is facilitated. Capillary column GC, GC/MS, and direct-probe MS are used for the identification and quantitation of individual PAH components in each fraction.

## EXPERIMENTAL SECTION

**Materials.** This analytical approach was demonstrated using a solvent refined coal II (SRC II) material from the Fort Lewis, WA, pilot plant. Fractional distillation of the full boiling range material (stage one of the separation methodology) was performed by Gulf Research and Development Co. (*32*). A distillate cut with a boiling point range of 800–850 °F was used in this investigation.

**Alumina Adsorption Column Chromatography.** The adsorption method used for separating the neutral PAH fraction from coal liquids has been described by Later et al. (*16*). Briefly, 150–200 mg of the 800–850 °F distillate cut was adsorbed on 3 g of neutral aluminum oxide and packed on top of 6 g of neutral aluminum oxide (Brockman activity I, 80–200 mesh, Fisher No. A 950, Pittsburgh, PA). The alumina was stored in an oven at 160 °C prior to use in the adsorption column chromatographic step. This oven temperature produced an alumina with ~1.5% water content (by weight). The details of this alumina standardization technique are available elsewhere (*33*). Sequential elution with hexane, benzene, and chloroform with 1% ethanol and methanol (spectral grade, Burdick & Jackson, Muskegon, MI) was done to obtain four fractions: aliphatic hydrocarbons (AH), neutral polycyclic aromatic hydrocarbons (PAH), nitrogen-containing polycyclic aromatics compounds (N-PAC), and hydroxy PAH (HPAH), respectively (see Figure 1). The PAH fraction was dried under nitrogen for gravimetric analysis and redissolved in methylene chloride (spectral grade, Burdick & Jackson) for further separation by HPLC.

**Reverse-Phase High-Performance Liquid Chromatography.** The PAH fraction was further separated and analyzed by using a Spectra Physics 8100 liquid chromatograph equipped with a Spectra Physics 8110 autosampler and a Waters Assoc. 440 UV absorbance detector fixed at a wavelength of 254 nm. A Spectra Physics 4100 computing integrator was used for data readout. Samples (subfractions) were collected with a modified Hewlett-Packard 79825 fraction collector.

A Supelcosil RP-8-DB, 5 μm, 25 cm × 10 mm i.d. reverse-phase column was used for the HPLC separation. The mobile-phase profile consisted of 69% methanol (spectral grade, Burdick & Jackson) in water (doubly distilled, deionized, Milli Q grade) for 40 min followed by a 20-min linear gradient to 79% methanol and a 10-min linear gradient to 100% methanol which was then held for 10 min. Finally, a 12-min period was used to return and equilibrate to initial conditions for repetitive injections. The column temperature was held at 50 °C, with an eluent flow rate of 6 mL/min. The injection volume was 50 μL. Twenty repetitive

0003-2700/85/0357-0633$01.50/0 

ANALYTICAL CHEMISTRY, VOL. 57, NO. 3, MARCH 1985 • 639

are now directly identified without undergoing these procedures. The benzonaphthothiophenes were well resolved in HPLC subfraction 1, and the methylbenzonaphthothiophenes were identified in HPLC subfraction 4.

Since only selected components were identified and quantified, the sum total of the concentrations of components in Table II does not account for the total weight found in the HPLC subfractions. When comparing the quantitative composition of the PAH fraction before and after HPLC separation (last two columns in Table II), quantitative agreement was observed in some cases, although discrepancies in concentration are noticeable. The differences can be partly explained by errors in integration due to differences in the base-line definition of unresolved components in the more complex neutral PAH fraction vs. the less complex HPLC subfractions. Furthermore, the combined HPLC subfraction component concentrations were only corrected by using the 85% recovery obtained from gravimetric analysis of the HPLC subfractions and not from the recovery of individual compounds spiked into the complex mixture. It is also possible that coeluting components were present in the GC analysis of the neutral PAH fraction prior to HPLC separation.

This separation method is of interest from the biological testing standpoint for two reasons. First, it isolates discrete chemical fractions in which the parent PAH are separated from their alkylated species. Also, it isolates many of the putative carcinogens such as benzo[*a*]pyrene, 5-methylchrysene, and 7,12-dimethylbenz[*a*]anthracene. The biological responses of the individual HPLC subfractions compared to that of the unfractionated neutral PAH fraction may give valuable information in understanding the effect of the complex sample matrix on the biological expression of PAH components in a coal-derived process material.

The integrated multiple-stage chromatographic methodology presented and discussed in this paper has been applied to the separation and analyses of a wide variety of coal liquefaction process materials including SRC II 800–850 °F and 850 °F plus distillate cuts, 650–700, 700–750, and >800 °F distillate cuts from the EDS process, and an integrated two-stage liquefaction (ITSL) 850 °F plus distillate cut. Results from the detailed chemical analyses of the SRC II 800–850 °F distillate were presented in this paper. Results of the biological testing of the HPLC subfractions will appear in a subsequent paper. To date, this multiple-stage separation scheme has only been applied for analysis of the PAH fraction but could easily be extended for analysis of the AH, N-PAC, and HPAC fractions from the alumina column chromatographic separation.

**Registry No.** 1-Methylpyrene, 2381-21-7; methylbenzo[*b*]naphthofuran, 93755-97-6; benzo[*b*]naphtho[2,1-*d*]thiophene, 239-35-0; benzo[*b*]naphtho[1,2-*d*]thiophene, 205-43-6; benzo[*b*]naphtho[2,3-*d*]thiophene, 243-46-9; benz[*a*]anthracene, 56-55-3; benz[*b*]anthracene, 92-24-0; methylbenzo[*b*]naphthothiophene, 67526-85-6; 3-methylchrysene, 3351-31-3; 5-methylchrysene, 3697-24-3; methylchrysene, 41637-90-5; benzofluoranthene, 56832-73-6; benzo[*k*]fluoranthene, 207-08-9; benzo[*e*]pyrene, 192-97-2; benzo[*a*]pyrene, 50-32-8; perylene, 198-55-0; benzo[*e*]pyrene, 192-97-2; benzo[*a*]pyrene, 50-32-8; perylene, 198-55-0.

## LITERATURE CITED


(1) Wilson, B. W.; Willey, C.; Later, D. W.; Lee, M. L. *Fuel* **1982**, *61*, 473–477.
(2) Wilson, B. W.; Pelroy, R. A.; Mahlum, D. D. In "Chemical Characterization and Genotoxic Potential Related to Boiling Point for Fractionally Distilled SRC-I Coal Liquids"; NTIS: Springfield, VA, 1982; PNL-4277.
(3) Later, D. W.; Wright, C. W.; Wilson, B. W. *Prepr. Pap.—Am. Chem. Soc., Div. Fuel Chem.* **1983**, *28*, 273–284.
(4) Pelroy, R. A.; Wilson, B. W. In "Fractional Distillation as a Strategy for Reducing the Genotoxic Potential of SRC II Coal Liquids"; NTIS: Springfield, VA, 1981; PNL-3787.
(5) Streuli, C. A. *J. Chromatogr.* **1971**, *56*, 225–229.
(6) Jones, A. R.; Geurin, M. R.; Clark, B. R. *Anal. Chem.* **1977**, *49*, 1766–1771.
(7) Rao, T. K.; Allen, B. E.; Ramey, D. W.; Epler, J. L.; Rubin, I. B.; Guerin, R. M.; Clark, B. R. *Mutat. Res.* **1981**, *85*, 29–39.
(8) Ho, C.-h.; Ma, C. Y.; Clark, B. R.; Guerin, M. R.; Rao, T. K.; Epler, J. L. *Environ. Res.* **1980**, *22*, 412–422.
(9) Toste, A. P.; Sklarew, D. S.; Pelroy, R. A. *J. Chromatogr.* **1982**, *249*, 267–282.
(10) Mima, M. J.; Schultz, H.; McKinstry, W. E. In. "Analytical Methods for Coal and Coal Products"; Karr, C, Ed.; Academic Press: New york, 1978; Vol. 1, Chapter 19, p 557.
(11) Burke, F. P.; Winschel, R. A.; Wooton, D. L. *Fuel* **1979**, *58*, 539–541.
(12) Schwager, I.; Yen, T. F. *Fuel* **1978**, *57*, 100–104.
(13) Steffgen, F. W.; Schroeder, K. T.; Bockrath, B. C. *Anal. Chem.* **1979**, *51*, 1164–1167.
(14) Boduszynski, M. M.; Hurtubise, R. J.; Silver, H. F. *Anal. Chem.* **1982**, *54*, 372–375.
(15) Schweighardt, F. K.; Thames, B. M. *Anal. Chem.* **1978**, *50*, 372–375.
(16) Later, D. W.; Lee, M. L.; Bartle, K. D.; Kong, R. C.; Vassilaros, D. L. *Anal. Chem.* **1981**, *53*, 1612–1620.
(17) Callen, R. B.; Simpson, C. A.; Bendoraitis, J. G.; Votz, S. E. *Ind. Eng. Chem. Prod. Res. Dev.* **1976**, *15*, 222–233.
(18) Boduszynski, R. J.; Hurtubise, R. J.; Silver, H. F. *Anal. Chem.* **1982**, *54*, 375–381.
(19) Leary, J. A.; Lafluer, A. L.; Longwell, J. P.; Peters, W. A.; Kruzel, E. L.; Biemann, K. In "Polynuclear Aromatic Hydrocarbons: Formation, Metabolism, and Measurement"; Cook, M., Dennis, A. J., Eds.; Battelle Press: Columbus, OH, 1983; pp 799–808.
(20) Farcasiu, M. *Fuel* **1977**, *56*, 9–14.
(21) Toste, A. P.; Sklarew, D. S.; Pelroy, R. A. In "Advance Techniques in Synthetic Fuels Analysis"; Wright, C. W., Weimer, W. C., Felix, D. W., Eds.; NTIS: Springfield, VA, 1983; Chapter 5, pp 74–94.
(22) Wise, S. A.; Chesler, S. N.; Hertz, H. S.; Hilpert, L. R.; May, W. E. *Anal. Chem.* **1977**, *49*, 2306–2310.
(23) Hertz, H. S.; Brown, J. M.; Chesler, S. N.; Guenther, F. R.; Hilpert, L. R.; May, W. E.; Parris, R. M.; Wise, S. A. *Anal. Chem.* **1980**, *52*, 1650–1657.
(24) Tomkins, B. A.; Griest, W. H.; Caton, J. E.; Reagan, R. R. In "Polynuclear Aromatic Hydrocarbons: Physical and Biological Chemistry"; Cooke, M., Dennis, A. J., Fisher, G. L., Eds.; Battelle Press: Columbus, OH, 1982; pp 813–824.
(25) Burke, F. P.; Winschel, R. A.; Pochapsky, T. C. *Prepr. Pap.-Am. Chem. Soc., Div. Fuel Chem.* **1981**, *60*, 562–572.
(26) Dark, W. A.; McFadden, W. H. *J. Chromatogr. Sci.* **1978**, *16*, 289–229.
(27) Chmielowiec, J. *Anal. Chem.* **1983**, *55*, 2367–2374.
(28) Ogan, K.; Katz, E. *Anal. Chem.* **1982**, *54*, 169–173.
(29) Sonnefeld, W. J.; Zoller, W. H.; May, W. E.; Wise, S. A. *Anal. Chem.* **1982**, *54*, 723–727.
(30) Mahlum, D. D. *J. Appl. Toxicol.* **1983**, *3*, 1, 31–33.
(31) Later, D. W.; Pelroy, R. A.; Mahlum, D. D.; Wright, C. W.; Lee, M. L.; Weimer, W. C.; Wilson, B. W. In "Polynuclear Aromatic Hydrocarbons: Seventh International Symposium on Formation, Metabolism, and Measurement"; Cooke, M. W., Dennis, A. J., Eds.; Battelle Press: Columbus, OH, 1982; pp 771–783.
(32) Gary, J. A. "Solvent Refined Coal (SRC) Process: Selected Physical, Chemical and Thermodynamic Properties of Narrow Boiling Range Coal Liquids for the SRC II Process"; NTIS: Springfield, VA, 1981.
(33) Later, D. W.; Lee, M. L. In "Advanced Techniques for Synthetic Fuels Analysis"; Wright, C. W., Weimer, W. C., Felix, W. D., Eds.; NTIS: Springfield, VA, 1983; Chapter 4, pp 44–73.
(34) Wilson, B. W.; Pelroy, R. A.; Mahlum, D. D.; Frazier, M. E.; Later, D. W.; Wright, C. W. *Fuel* **1984**, *63*, 46–55.
(35) Wright, C. W. *HRC CC, J. High Resolut. Chromatogr. Chromatogr. Commun.* **1984**, *7*, 83–88.
(36) Vassilaros, D. L.; Kong, R. C.; Later, D. W.; Lee, M. L. *J. Chromatogr.* **1982**, 252, 1–20.
(37) Later, D. W.; Lee, M. L.; Wilson, B. W. *Anal. Chem.* **1982**, *54*, 117–123.
(38) Mahlum, D. D.; Wright, C. W.; Chess, E. K.; Wilson, B. W. *Cancer Res.* **1984**, *44*, 5176–5181.
(39) May, W. E.; Wise, S. A. *Anal. Chem.* **1984**, *56*, 225–232.
(40) Willey, C.; Iwao, M.; Castle, R. N.; Lee, M. L. *Anal. Chem.* **1981**, *53*, 400–407.


RECEIVED for review July 2, 1984. Accepted October 10, 1984. This work was supported by the Department of Energy under Contract DE-AC06-76RLO.

**Objectives**

When you have worked through this Introduction you should be able to:

- recognise the nature and importance of scientific and technical information;
- appreciate the potential usefulness of analytical literature in providing such information;
- understand the problems you face because of the amount of chemical information already published and still being published;
- realise the general implications for analytical information and literature of the growth of information technology;
- proceed through this Unit with confidence and gain the maximum benefit from it;
- write accurate and consistent bibliographic citations for most of the publications you are likely to come across.

# 2. Main Types of Literature

## 2.1. PUBLISHED STANDARDS

### 2.1.1. Introduction

After working through the Introduction to this Unit you should now have a reasonable idea of what we are going to do to investigate the analytical literature and why we need to do it. If you look at the list of contents you will see that Part 2 is dominated by reference books. Instead of examining in detail all the types of literature you might (or might not) encounter, we shall concentrate on those which are likely to prove of most practical value to you; hence the concentration on reference books.

Let's start with published standards. First of all, what are standards and why have them? Well, standards are basically rules: as far as analytical chemists are concerned, standards tell you how to carry out particular tests or analyses—what chemicals and reagents to use and in what quantities, and how to make a determination and record your results; in the industrial and commercial world they not only tell you how to do things, but also specify how particular products of all kinds are to be composed. You will see examples of all this later in the Part. To emphasise the importance of standards, just think what might happen if they did not exist: what use would forensic evidence be if there were no agreed methods for obtaining and presenting results? What protection would the public have if there were no specifications for the preparation of drugs and pharmaceuticals and no agreed methods of testing their safety? I think the general point about the need for standards is obvious, don't you?

Standards have, of course, either to be agreed by the interested parties, or imposed by some regulatory agency (usually a government department). When we look at published standards in a moment, you will see the various kinds of bodies which issue standards, and the way they compile them. Most industrial countries have a national standardisation body of some kind, although of course the standards issued from major countries such as the USA, Japan, Great Britain, or West Germany, are often adopted in other countries as well, either formally or under economic and commercial necessity. Also, there is the *International Organisation for Standardisation*: I will tell you more about that organisation in a moment. Normally standards are compiled by agreement in committees, between representatives of three types of interested parties: producers, consumers, and regulatory or research agencies. In addition to the national bodies, some learned institutions or associations of practitioners also produced standards for the guidance of their members. Again, you will soon see examples of these.

I have emphasised the regulatory origin and nature of standards, but before we look at practical examples I have another important point to make. Over and above their regulatory or advisory function, standards are also of great use as reference sources for technical information. You can use them not only to check on rules and procedures for particular analyses, tests, or compounds, but also to find answers to particular questions: How do I do this test? What constituents must I use for a particular compound or product? What reagents in what quantities do I use in a specific determination? Other examples should occur to you as we work through the SAQs. But don't forget the usefulness of standards as excellent sources of authoritative technical data which may not be readily available in other publications.

So, with this general introduction made, let's now have a look at various standard publications in more detail. I shall give you a brief introduction to each organisation or publication, and then take a closer look at them, as well as giving you practical experience of looking things up, in the form of a SAQ in each sub-section.

### 2.1.2. *British Standards Institution, and the International Organisation for Standardisation*

The British Standards Institution (BSI) is an independent, non-profit-making body founded in 1901, and is the United Kingdom's nationally recognised organisation for the preparation of standards for all industries and technologies. There are at present over 9,000 standard publications listed in its catalogue, and some 600 new or revised standards are issued each year. The standards are compiled by committees representing all those with a particular interest in the subject under discussion: manufacturers, users, research organisations, government departments, professional bodies, individual experts, and, where appropriate, consumers. Drafts of new standards are made public for comment before their final publication.

Two regular publications help us to find and identify British Standards:

(*a*) **BSI Catalogue**: annual. This lists all the current standards and other BSI publications in their serial-number order. It has a subject index.

(*b*) **BSI News**: monthly. This updates the *Catalogue* with, for example, lists of new, revised, or withdrawn standards. It carries news of the progress of committees and the standards on which they are working and of general standardisation matters; it also has regular feature articles on technical or commercial matters involving particular standards.

The BSI is one of about 90 national standard bodies comprising the *International Organisation for Standardisation* (ISO) whose headquarters are in Geneva, Switzerland. Its object is to promote the development of standardisation and related activities in the whole world with a view to facilitating international exchange of goods and services, and to developing international scientific, technological, and economic co-operation. It has established certain technical committees to develop standards in various priority fields, with the agreement of national bodies. All ISO standards are reviewed at least every five years. It also issues a catalogue (with text in English and French) to help to identify its standards.

**ISO Catalogue**: annual. This comprises a numerical list of standards and other ISO publications. It has a subject index and a Universal Decimal Classification Index (this is a classification scheme used in some scientific and technical libraries: more of this in Section 3 of this Unit).

Now, let's have a go at the first SAQ in this Section. This will give you practice in looking up British Standards and show you more details of them.

**SAQ 2.1a**

This is the first SAQ I have set you in this Section, so let me make a very important general point right at the start. I cannot be certain that you will have convenient access to all, or any, of the books I refer to in this Unit; so, you are being given, in this book, copies of at least a few of the pages of each major work to enable you to answer the SAQ if you do not have access to the necessary books. But, if you do have convenient access in a local public, college, or works library to any of the books on which I set you questions, then *ignore the copies and use the books themselves* to answer the questions; you will derive much more benefit from this Unit if you are able to do so. And please bear with me when I remind you briefly of this advice at every relevant question. The important thing with these questions is to make you look things up for yourself in the material concerned, and you will do that more profitably in the books themselves; the copies will be useful substitutes, *but only if you cannot get access to the books.*

So, let's start by having you look up some British Standards. For these questions you will need a copy of the *BSI Catalogue*: use the copies only if you cannot get hold of a copy.

(*a*) Imagine you have to select a technique and apparatus for a gas chromatographic analysis to a British Standard.

⟶

**SAQ 2.1a (cont.)**

What two standards would you need to refer to? Just write down their numbers and dates of publication.

(*b*) Imagine you are preparing chemical samples for distribution to and use in chemical laboratories in schools in your area. You need to check the preparation of potassium nitrate. As your work has always to be to the British standard, what standard specification do you need to check? Again just write down the serial number and date of publication.

(*c*) Suppose you are going to carry out an analysis for lead by anodic stripping voltammetry to British standards, by using hydrogen peroxide as a reagent. What British standard would you need to check for a specification for the reagent and method of test? Just give the serial number and date of the specification. I must warn you that unless you are already familiar with the standard concerned you might find this particular question very tricky. If you cannot find the right standard after three or four attempts, don't get upset but go straight to my response: there is a moral to this question!

(Information complete to 30 September 1984)

# BSI Catalogue 1985

British Standards Institution
Incorporated by Royal Charter

2 Park Street
LONDON W1A 2BS

Telephone: 01-629 9000
Telex: 266933 BSILON G

# Contents

BS 3274 General Series

**BS 3274 : 1960**
**Tubular heat exchangers for general purposes**
Design, construction, inspection, testing of cylindrical shell and plain tube heat exchangers. Shell diameters 6 in to 42 in, tube lengths 16 ft to 16 ft, tube diameters ½ in to 1½ in. Type 1: fixed tube-plate (non-removable tube bundle); Type 2: U-tube (removable tube bundle); Type 3: floating head (removable tube bundle).
*PD 3998, December 1960* ®
*PD 4147, April 1961* ®
*PD 5937, November 1966* ®
*PD 6144, May 1967* ®
48 page Gr 8 PVE/1

**BS 3275 : 1960**
**Glass for signs and recommendations on glazing for signs**
Quality, marking of glass for box and panel signs, including toughened and laminated glass. Use of glass in these signs, including glazing, fixing, strength of supporting materials, ventilation.
8 page A5 size Gr 2 ECB/4

BS 3276 : 1960 (withdrawn)
Thermometers for measuring air cooling power
(August 1980, obsolete)

BS 3277 : 1960 (withdrawn)
Thigh corsets and cuff tops for orthopaedic calipers
(December 1977, superseded by BS 2574 : Part 1)

BS 3278 : —— (withdrawn)
General recommendations for the sampling of imported iron ores
(superseded by BS 4103 : Part 1)

**BS 3279 : 1960 (1979)** ≠ ISO/R 55
**Seedlac**
Six grades. Description, sample, volatile matter, matter insoluble in hot alcohol, expressing analytical results. Optional clauses: colour index, matter soluble in water, non-volatile matter soluble in cold alcohol, wax. Supersedes relevant parts of BS 954.
16 page Gr 4 PVC/4

**BS 3280 : 1960 (1979)** ≠ ISO 56/1
**Hand-made shellac**
Five grades. Appearance and colour, sample, matter insoluble in hot alcohol, orpiment, rosin, expressing analytical results. Optional clauses: colour index, volatile matter, matter soluble in water, non-volatile matter soluble in cold alcohol, ash, grit, wax, flow, life under heat, acid value, iodine value, arsenic, lead, adhesion to mica. Supersedes relevant parts of BS 954.
*PD 5499, May 1965*
40 page A5 size Gr 7 PVC/4

**BS 3281 : 1960**
**Rectangular metal boxes for use in high vacuum steam sterilizers**
Stainless steel or brass for holding dressings, surgical rubber gloves, etc. Two sizes allow loading of sterilizing chamber to make best use of space in pressure steam sterilizers to BS 3220.
*PD 4322, October 1961*
8 page A5 size Gr 2 SGC/14 12

**BS 3282 : 1969**
**Glossary of terms relating to gas chromatography**
20 page A5 size Gr 4 C/40

**BS 3283 : ——**
**Non-reversible connectors and appliance inlets for portable electrical appliances (for circuits up to 250 volts)**

**BS 3283 : Part 1 : 1980**
**Specification for 13 A connector and appliance inlets**
Requirements and tests for both re-wirable and non-rewirable types. Suitable for use with appliances where the temperature of the contact pins does not exceed 120 °C.
*AMD 3964, July 1982 (Gr 0)*
12 page Gr 6 PEL/4

**BS 3283 : Part 2 : 1974**
**6 A connector and appliance inlet**
Dimensions, materials, tests. Connectors for appliances where temperature of contact pins up to 120 °C. Requirements, tests similar to Part 1. Cannot be used interchangeably with plugs or other connectors complying with other British Standards. Gauges to show interchangeability and contact-making illustrated.
*AMD 1758, July 1975 (Gr 1)*
16 page Gr 7 PEL/4

BS 3283 : Part 3 : 1966 (withdrawn)
6 A non-reversible connector and appliance inlets for Class II appliances

**BS 3284 : 1967**
**Polythene pipe (Type 50) for cold water services**
Requirements for extrusion compound, pipe material, classification and dimensions of pipes, physical and mechanical characteristics, sampling, marking, and stocking and transport. Test methods for toluene extract of carbon black and antioxidant content in appendices.
*AMD 1237, September 1973 (Gr 2)* ®
*AMD 1615, November 1974* ®
*AMD 2235, February 1977 (Gr 1)*
*AMD 4461, November 1983 (Gr 0)*
20 page A5 size Gr 4 PLM/9

**BS 3285 : 1960**
**Method of sampling superheated steam from steam generating units**
Principles, apparatus, including sample probe, piping; procedure.
28 page A5 size Gr 5 EPC/37

**BS 3286 : 1960**
**Method for laboratory evaluation of disinfectant activity of quaternary ammonium compounds by suspension test procedure**
Test conditions may be selected according to purpose of test. Test organisms, preparation of cultures, cleaning of glassware, choice of range of dilutions, temperature, time of contact, and inactivator, addition of organic matter. Examples of application of procedure.
20 page A5 size Gr 4 DIC/12

BS 3287 : 1960 (withdrawn)
Domestic electric clothes-washing machines
(superseded by BS 3456 : Section 2.11)

**BS 3288 : ——**
**Insulator and conductor fittings for overhead power lines**

**BS 3288 : Part 1 : 1973 (1979)**
**Performance and general requirements**
Definitions, general requirements, protection against corrosion. Mechanical and electrical type, sample and routine tests, as appropriate, for insulator pins, insulator set fittings, earth conductor fittings, tension joints, anchor clamps, non-tension joints, suspension clamps, electrical control fittings and mechanical protective fittings of any material or design.
12 page Gr 6 PEL/70

**BS 3288 : Part 2 : 1977** ≠ IEC 120
**Dimensions**
Insulator pins, line post insulator studs, insulator set fittings, anchor and suspension clamps, ball and socket couplings, sag adjusters.
*AMD 2512, March 1978 (Gr 1)*
*AMD 3611, April 1981 (Gr 0)*
64 page Gr 9 PEL/70

**BS 3289 : 1982** ≠ ISO 283, = ISO 284, ≠ ISO/R 340
**Specification for conveyor belting primarily for use underground (including fire performance)**
Covers details of requirements for textile reinforced conveyor belting together with methods of test.
24 page A5 size Gr 7 RUM/16

**BS 3290 : 1960**
**Toughened polystyrene extruded sheet**
Covers two types of sheet, one plain and one faced on one side with a thermoplastic foil. Includes details of composition, dimensions and physical properties.
8 page A5 size Gr 2 PLM/40

BS 3291 : 1960 (withdrawn)
10-gallon alloy milk can and lid
(September 1979)

**BS 3292 : 1960**
**Direct reading hygrometers**
Two grades of non-recording direct reading hygrometers are specified, both calibrated for use over the range 20 to 100 per cent relative humidity. Permissible error, temperature correction, drift, response time and constructional requirements are specified and methods of test described in appendices.
8 page A5 size Gr 2 LBC/4

**BS 3293 : 1960**
**Carbon steel pipe flanges (over 24 in nominal size) for the petroleum industry**
Specifies materials, dimensions and marking for forged carbon steel slip-on welding and welding neck flanges, of nominal sizes 26 in to 48 in for Classes 150, 300, 400 and 600.
*AMD 1004, August 1972 (Gr 0)* ®
20 page Gr 7 PVE/14

**BS 3294 : ——**
**The use of high strength friction grip bolts in structural steelwork**

**BS 3294 : Part 1 : 1960**
**General grade bolts**
Use, in structural steelwork to BS 449, of bolts to BS 3139 : Part 1. Rules for design, based on 'slip factor' defined as ratio of load required to produce slip in pure shear joint to shank tension induced in bolt by tightening. Methods of tightening to produce necessary shank tension. Inspection procedure.
*PD 4429, January 1962* ®
*PD 5262, June 1964* ®
*PD 5963, December 1966* ®
*AMD 183, January 1969* ®
*AMD 678, January 1971* ®
*AMD 1046, December 1972 (Gr 0)* ®
16 page A5 size Gr 4 CSB/29

BS 3295 : 1960 (withdrawn)
Unit heads (slide type)
(superseded by BS 3884)

BS 3296 : 1960 (withdrawn)
Safety requirements for domestic electric hair dryers
(superseded by BS 3456 : Section 2.31)

**BS 3297 : ——**
**Post insulators of ceramic material or glass for nominal voltages greater than 1000 V**

**BS 3297 : Part 1 : 1982** ≠ IEC 168
**Methods of test**
Specifies tests for post insulators and insulator units for indoor and outdoor service in electrical installions or equipment operating on alternating current with a nominal voltage greater than 1000 V and a frequency not greater than 100 Hz.
16 page Gr 7 PEL/70

**BS 3906** General Series

**BS 3906**: 1976
**Electrolytic compressed hydrogen**
Specifications and methods of test for hydrogen type 1 (for general industrial use) and type 2 (reduced oxygen and moisture content).
**12 page Gr 6** CIC/19

**BS 3907**: ——
**Methods for the analysis of magnesium and magnesium alloys**

**BS 3907: Part 1**: 1965 = ISO 791
**Determination of aluminium in magnesium alloys (gravimetric method)**
Aluminium content 1.5 to 12 per cent. Reagents, sampling methods, test procedure.
**8 page A5 size Gr 2** NFM/35

**BS 3907: Part 2**: 1966 = ISO 792
**Determination of iron in magnesium and magnesium alloys (photometric-1, 10-phenanthroline method)**
Describes the reagents required, recommended methods of sampling and test procedure for the photometric determination of iron in magnesium and magnesium alloys in the range 0.002 per cent to 0.05 per cent.
**8 page A5 size Gr 2** NFM/35

**BS 3907: Part 3**: 1966 = ISO 794
**Determination of copper in magnesium and magnesium alloys (photometric method)**
Describes the reagents required, recommended methods of sampling and test procedure for the photometric determination of copper in magnesium and magnesium alloys in the range 0.002 per cent to 0.4 per cent
**8 page A5 size Gr 2** NFM/35

**BS 3907: Part 4**: 1966 = ISO 809, ISO 810
**Determination of manganese in magnesium and magnesium alloys photometric-periodate method)**
Describes the reagents required, recommended methods of sampling and test procedure for the photometric determination of manganese in magnesium and magnesium alloys having a manganese content in the range 0.01 per cent to 0.8 per cent.
**8 page A5 size Gr 2** NFM/35

**BS 3907: Part 5**: 1966 = ISO 810
**Determination of manganese in magnesium and magnesium alloys photometric-periodate method) (low contents)**
Describes the reagents required, recommended methods of sampling and test procedure for manganese contents less than 0.01 per cent.
**8 page A5 size Gr 2** NFM/35

**BS 3907: Part 6**: 1969 = ISO 2354
**Zirconium in magnesium and magnesium alloys (photometric method)**
Soluble zirconium contents 0.1 per cent to 1.0 per cent and insoluble zirconium contents 0.02 per cent to 0.3 per cent. Reagents, sampling methods, test procedure.
**12 page A5 size Gr 3** NFM/35

**BS 3907: Part 7**: 1969 = ISO 1178
**Nickel in magnesium alloys (photometric method)**
Nickel content 0.001 to 0.2 per cent. Reagents, sampling methods, test procedure.
**8 page A5 size Gr 2** NFM/35

**BS 3907: Part 8**: 1970 = ISO 2355
**Total rare earths in magnesium alloys (gravimetric method)**
Gravimetric method for total rare earth contents 0.2 per cent to 10 per cent. Reagents, sampling methods, test procedure.
**8 page A5 size Gr 2** NFM/35

**BS 3907: Part 9**: 1969 = ISO 1784
**Zinc in magnesium alloys (ion-exchange-volumetric EDTA method)**
Zinc contents 0.10 per cent to 8.0 per cent. Reagents, sampling methods, test procedure.
**8 page A5 size Gr 2** NFM/35

**BS 3907: Part 10**: 1969 = ISO 2353
**Manganese in magnesium alloys containing zirconium rare earths, thorium and/or silver (photometric method)**
Manganese contents less than 0.2 per cent. Reagents, sampling methods, test procedure.
**8 page A5 size Gr 2** NFM/35

**BS 3907: Part 11**: 1970 = ISO 1975
**Silicon in magnesium and magnesium alloys (photometric method)**
Silicon contents between 0.01 per cent and 0.6 per cent. Reagents, sampling methods, test procedure.
**8 page A5 size Gr 2** NFM/35

**BS 3907: Part 12**: 1971 = ISO 3255
**Aluminium in magnesium and magnesium alloys (photometric method)**
Photometric method for acid-soluble aluminium contents 0.005 per cent to 0.20 per cent. Reagents, sampling methods, test procedures.
**8 page A5 size Gr 2** NFM/35

**BS 3907: Part 13**: 1972
**Tin in magnesium and magnesium alloys (photometric method)**
Photometric method for tin contents up to 0.08 per cent. Reagents, sampling methods, test procedure.
**8 page A5 size Gr 2** NFM/35

BS 3907: Part 14: 1976 (withdrawn)
Zirconium and thorium in magnesium and magnesium alloys (volumetric method)
(September 1980, obsolete)

**BS 3907: Part 15**: 1976
**Lead in magnesium and magnesium alloys (atomic absorption method)**
Lead contents up to 0.3 per cent. Reagents, sampling methods, test procedure.
**4 page Gr 3** NFM/35

**BS 3908**: ——
**Methods for the sampling and analysis of lead and lead alloys**

**BS 3908: Part 1**: 1965
**Sampling of ingot lead, lead alloy ingots, sheet pipe, and cable sheathing alloys**
Selection of ingots, preparation of surfaces, method of taking samples by sawing, precautionary methods against contamination of the sawings.
*AMD 371, November 1969* ®
**8 page A5 size Gr 2** NFM/22

**BS 3908: Part 2**: 1967
**Arsenic in lead and lead alloys (photometric method)**
Reagents required, recommended method of sampling and analytical procedure for the photometric determination of arsenic in lead and lead alloys having an arsenic content in the range 0.001 per cent to 0.01 per cent.
**8 page A5 size Gr 2** NFM/22

**BS 3908: Part 3**: 1967
**Bismuth in lead and lead alloys (photometric method)**
Reagents required, recommended method of sampling and analytical procedure for the photometric determination of bismuth in lead and lead alloys having a bismuth content in the range 0.001 per cent to 0.1 per cent.
**8 page A5 size Gr 2** NFM/22

**BS 3908: Part 4**: 1967
**Copper in lead and lead alloys (photometric method)**
Reagents required, recommended method of sampling and analytical procedure for the photometric determination of copper in lead and lead alloys having a copper content in the range 0.003 per cent to 0.10 per cent.
**8 page A5 size Gr 2** NFM/22

**BS 3908: Part 5**: 1968
**Nickel in lead alloys (photometric method)**
Nickel content 0.0001 per cent up to 0.01 per cent. Reagents, sampling methods, test procedure.
**8 page A5 size Gr 2** NFM/22

**BS 3908: Part 6**: 1971
**Tellurium in lead and lead alloys (photometric method)**
Photometric method for tellurium contents 0.005 per cent to 0.1 per cent. Reagents, sampling methods and test procedure.
**8 page A5 size Gr 2** NFM/22

**BS 3908: Part 10**: 1967
**Antimony in lead and lead alloys (volumetric method)**
Reagents required, recommended method of sampling and analytical procedure for the volumetric determination of antimony in lead and lead alloys having an antimony content in the range 0.05 per cent to 12 per cent.
*AMD 372, November 1969*
**8 page A5 size Gr 2** NFM/22

**BS 3908: Part 11**: 1968
**Tin in lead and lead alloys (volumetric method)**
Tin content 0.1 per cent up to 0.5 per cent. Reagents, sampling methods, test procedure.
**8 page A5 size Gr 2** NFM/22

**BS 3908: Part 13**: 1970
**Antimony in lead and lead alloys (low contents) (photometric method)**
Antimony content 0.002-0.05 per cent. Reagents, sampling methods, test procedure.
**8 page A5 size Gr 2** NFM/22

**BS 3908: Part 15**: 1972
**Iron in lead and lead alloys (Photometric method)**
Photometric method for iron contents 0.001 per cent to 0.005 per cent. Reagents, sampling methods, test procedure.
**8 page A5 size Gr 2** NFM/22

**BS 3909**: 1965 (1982)
**Ingot lead for radiation shielding**
Three grades. Test sample selection, inspection and testing. Guidance on the use of each grade appended.
**12 page A5 size Gr 3** NFM/22

BS 3910: 1965 (withdrawn)
Cone and cheese trays for filament yarns

**BS 3911**: ——
**Lead-acid starter batteries for internal combustion engines**

**BS 3911: Part 1**: 1982 = IEC 951, ≠ IEC 95-2, 95-3
**Specification for batteries requiring regular maintenance**
Standard parameters, methods of test, and test requirements for lead-acid secondary batteries with a nominal voltage of 6 V or 12 V, used primarily for starting and ignition of internal combustion engines and also for accessories of internal combustion engines.
**16 page Gr 7** LEL/12

BS 3912: 1965 (withdrawn)
Polythene filament ropes (hawser laid) (superseded by BS 4928: Part 2)

BS 4584 : Part 14 : 1977
**Phenolic cellulose paper copper-clad laminated sheet of medium electrical quality and defined flammability : PF-CP-Cu-14**
Specifies electrical and non-electrical properties and gives a method of test for flammability.
*AMD 2448, February 1978 (Gr 0)*
**4 page** Gr 3 ECL/19

BS 4584 : Part 15 : 1978
**Adhesive coated polymeric films: PETP-F-15 and PL-F-15**
Requirements for properties of polymeric films of polyester and polyimide used as cover sheets for flexible printed circuits.
**12 page** Gr 6 ECL/19

BS 4584 : Part 16 : 1978
**Epoxide glass-reinforced copper-clad laminated sheet of defined flammability: YP-GCA-Cu-16**
Requirements for properties of sheet whose reinforcement includes non-woven glass filaments in addition to woven glass fabric.
*AMD 3906, October 1982 (Gr 0)*
**4 page** Gr 3 ECL/19

BS 4585 : ——
**Methods of test for spices and condiments**

BS 4585 : Part 1 : 1983 ≡ ISO 927
**Determination of extraneous matter**
Method involves manual separation of extraneous matter and is, therefore, applicable mainly to whole and broken spices.
**4 page** Gr 3 FAC/7

BS 4585 : Part 2 : 1982
**Determination of moisture content (entrainment method)**
States definitions, reagent, apparatus, procedure and test report.
**4 page** Gr 4 FAC/7

BS 4585 : Part 3 : 1981 ≡ ISO 928
**Determination of total ash**
Definition, principle, reagent, apparatus, sampling, procedure, expression of results and test report.
**4 page** Gr 2 FAC/7

BS 4585 : Part 4 : 1970 = ISO/R 940
**Determination of alcohol-soluble extract**
Definition, principle, reagent, apparatus, sampling, procedure, expression of results and test report.
**8 page** A5 size Gr 2 FAC/7

BS 4585 : Part 5 : 1980 ≡ ISO 941
**Determination of cold water-soluble extract**
Definition, principle, apparatus, sampling, procedure, expression of results, and test report.
**4 page** Gr 2 FAC/7

BS 4585 : Part 6 : 1981 ≡ ISO 1108
**Determination of non-volatile ether extract**
Definition, principle, apparatus, sampling, procedure, expression of results and test report.
**4 page** Gr 2 FAC/7

BS 4585 : Part 7 : 1977 ≡ ISO 3513
**Determination of Scoville index of chillies**
Investigation of greatest dilution at which the stimulus threshold of the pungent sensation can be detected.
**4 page** Gr 2 FAC/7

BS 4585 : Part 8 : 1977 ≡ ISO 3588
**Determination of degree of fineness of grinding—Hand sieving method (Reference method)**
Apparatus, procedure, properties of ground spices. Read in conjunction with BS 410 and BS 1796.
**4 page** Gr 2 FAC/7

BS 4585 : Part 9 : 1981 ≡ ISO 930
**Determination of acid-insoluble ash**
Definitions, principle, reagents, apparatus, procedure, expression of results and test report.
**4 page** Gr 2 FAC/7

BS 4585 : Part 10 : 1981 ≡ ISO 929
**Determination of water-insoluble ash**
Definitions, principle, reagents, apparatus, procedure, expression of results and test report.
**4 page** Gr 2 FAC/7

BS 4585 : Part 11 : 1983 ≡ ISO 5567
**Determination of volatile organic sulphur compounds in dehydrated garlic**
Principle, reagents, apparatus, procedure, expression of results and test report.
**4 page** Gr 4 FAC/7

BS 4585 : Part 12 : 1983 ≡ ISO 5564
**Determination of piperine content of pepper**
Spectrophotometric method. Applies to black and white pepper, whole or ground.
**4 page** Gr 3 FAC/7

BS 4585 : Part 13 : 1983 ≡ ISO 5566
**Determination of colouring power of turmeric**
Spectrophotometric method. Colouring power is reported as curcuminoids content expressed as curcumin as percentage by mass.
**4 page** Gr 2

BS 4585 : Part 14 : 1983 ≡ ISO 1208
**Determination of filth**
Quantitative determination of filth (mineral matter plus matter of animal origin).
**8 page** Gr 4 FAC/7

BS 4586 : 1984 ≡ ISO 3862
**Specification for spiral wire reinforced rubber covered hydraulic hoses and hose assemblies**
Four types of spiral wire reinforced rubber hoses and hose assemblies are specified according to their design working pressure for use with petroleum and water-based fluids within a temperature range of —40 °C to +100 °C.
**8 page** Gr 6 RUM/9

BS 4587 : 1970
**Recommendations for the selection of apparatus and techniques for the analysis of gases by gas chromatography**
Analysis of gases by gas chromatography, taking each step in the procedure and indicating the means by which the most suitable apparatus and technique can be selected.
**48 page** A5 size Gr 7 EPC/17

BS 4588 : 1970 (1981)
**Methods of test for determining the registration obtained on duplicating machines**
Methods of determining the registration obtained on offset litho, stencil and spirit duplicating machines. Tables of limits to classify the registration achieved for each type of machine.
**12 page** A5 size Gr 3 OIS/3

BS 4589 : 1970
**Abbreviations for rubber and plastics compounding materials**
Abbreviations for plasticizers, accelerators, antioxidants and antiozonants, and stabilizers.
**12 page** A5 size Gr 3 PLM/12 & RUM/11

BS 4590 : —— (withdrawn)
⅓ litre (12 fl oz non-returnable soft drink glass bottles (Part 1 : 1970 withdrawn October 1980, obsolete)

BS 4591 : 1971 = ISO/R 918
**Method for the determination of distillation characteristics**
Apparatus and method for the determination of distillation characteristics (distillation yield and distillation range) of liquids.
**16 page** A5 size Gr 4 CIC/4

BS 4592 : 1970
**Industrial open type metal flooring and stair treads**
Design (including dimensions), construction, testing, installation, fixing methods and erection. Appendix gives details of information required by manufacturer from purchaser. Does not apply to 'pressed metal' type flooring.
*AMD 2346, September 1977 (Gr 0)* ®
**16 page** A5 size Gr 4 ECB/2

BS 4593 : 1970 = ISO/R 1003
**Ginger (whole, in pieces and ground)**
Description, odour, flavour, freedom from moulds, insects, etc., extraneous matter, freedom from coarse particles, chemical requirements for ginger whole and in pieces and for ground ginger, packaging and marking. Recommendations on storage, transport, grinding sample for test, determination of calcium.
**12 page** A5 size Gr 3 FAC/7

BS 4594 : 1970 = ISO/R 973
**Pimento (allspice) (whole and ground)**
Description, odour, flavour, freedom from moulds, insects, etc., extraneous matter, freedom from coarse particles (ground pimento), grades, chemical requirements for whole pimento and ground pimento, packaging, marking. Recommendations on storage, transport, grinding sample for test.
**12 page** A5 size Gr 3 FAC/7

BS 4595 : 1970 = ISO/R 959
**Black pepper and white pepper (whole and ground)**
Description, odour, flavour, freedom from mould, insects, etc., extraneous matter, light berries, pinheads, total defects, freedom from coarse particles (ground), grades (whole), chemical requirements for whole and for ground pepper, packaging, marking. Recommendations for storage, transport, grinding sample for test, determination of light berries.
*AMD 959, May 1972*
**12 page** A5 size Gr 3 FAC/7

BS 4596 : 1981 ≡ ISO 882
**Specification for cardamoms**
Physical and chemical requirements for cardamoms as whole capsules or as separated seeds. Also packing, and marking. An Annex gives recommendations on storage and transport.
**4 page** Gr 3 FAC/7

BS 4597 : 1970 (withdrawn)
General requirements and methods of test for multilayer printed wiring boards using plated through holes
(September 1982, superseded by BS 6221 : Part 2)

BS 4598 : 1970
**Dental impression plaster**
Various performance requirements including the mechanical properties; includes details of packaging, and information to be provided by the supplier.
**16 page** A5 size Gr 4 DNC/4

BS 4599 : 1970 (withdrawn)
Dimensional features of optical sound recording on 8 mm (Type S) motion picture film with picture
(January 1982, superseded by BS 5550 : Subsection 1.4.5)

BS 5437 : 1977 ≈ EN 49
**Wood preservatives. Determination of toxic values against *Anobium punctatum* (De Geer) by egg-laying and larval survival (laboratory method)**
Exposure of preservative-treated test specimens of wood to female *Anobium punctatum* and observation of egg-laying, hatching of eggs and survival of larvae. Supersedes BS 3652.
*AMD 3291, June 1980 (Gr 0)*
**8 page Gr 5** WPC/10

BS 5438 : 1976
**Methods of test for flammability of vertically oriented textile fabrics and fabric assemblies subjected to a small igniting flame**
Three methods for observing and measuring aspects of flammability relevant to apparel fabrics and those fabrics that will be held loosely in an essentially vertical position, for example curtains and drapes. Appendices give notes for users and proposed criteria of acceptance.
*AMD 3595, April 1981 (Gr 1)* ®
**16 page Gr 7** FBM/19

PD 2777 : 1977
**Fabric flammability burning accidents and the relevance of BS 5438**
Reaction of textiles to fire. Statistics of burning accidents involving textiles. Relevance of BS 5438.
**12 page Gr 6** FBM/19

BS 5439 : 1977 (1983)
**Specification for aluminium foil catering containers**
Design, materials and dimensions of a range of sizes of containers complying with the modular dimensions specified in BS 4874.
**4 page Gr 3** FHM/26

BS 5440 : ——
**Code of practice for flues and air supply for gas appliances of rated input not exceeding 60 kW (1st and 2nd family gases)**

BS 5440 : Part 1 1978
**Flues**
Choice and installation of flues forming parts of installations for domestic or commercial purposes but excluding industrial specialist applications. Complete flue equipment, from point of issue of the combustion products from the appliance to their discharge to outside air. Materials, components and appliances, design considerations, installation constructional work, inspection and testing. Supersedes CP 337.
*AMD 4639, September 1984 (Gr 0)*
**40 page Gr 8** GSE/30

BS 5440 : Part 2 : 1976
**Air supply**
Air supply requirements for domestic and commercial gas appliances installed in rooms and other internal spaces and in purpose-designed compartments. Air vent areas are specified where appropriate.
**8 page Gr 5** GSE/30

BS 5441 : 1977 (1982)
**Methods of test for knitted fabrics**
Tests for barriness, fabric construction, wales and courses per centimetre, linear density of component yarns of knitted fabrics; stitch and course length of weft knitted fabrics and run-in of warp knitted fabrics.
**16 page Gr 7** FBM/16

BS 5442 : ——
**Classification of adhesives for construction**

BS 5442 : Part 1 : 1977
**Adhesives for use with flooring materials**
Classifies the preferred adhesives for laying the different types of flooring material used in building construction.
**4 page Gr 2** ADC/10

BS 5442 : Part 2 : 1978
**Adhesives for use with interior wall and ceiling coverings (excluding decorative flexible materials in roll form)**
**4 page Gr 3** ADC/10

BS 5442 : Part 3 : 1979
**Adhesives for use with wood**
Classifies preferred adhesives for bonding wood and non-wood adherends as used in the construction industry.
**8 page Gr 5** ADC/10

BS 5443 : 1977 ≠ ISO 2718
**Recommendations for a standard layout for methods of chemical analysis by gas chromatography**
**12 page Gr 6** C/40

BS 5444 : 1977 (1983)
**Recommendations for preparation of copy for microcopying**
Technical quality of copy to be reproduced on black-and-white microfilm or microfiche and layout required for microfiche.
**4 page Gr 3** DOS/16

BS 5445 : ——
**Specification for components of automatic fire detection systems**

BS 5445 : Part 1 : 1977 ≡ EN 54 Part 1
**Introduction**
Gives scopes of subsequent parts of the standard, with definitions for components of fire detection systems.
**8 page Gr 5** FSM/12

BS 5445 : Part 5 : 1977 ≡ EN : Part 5
**Heat sensitive detectors—point detectors containing a static element**
Requirements, test methods and performance criteria for three response grades of heat sensitive (point) detectors containing a static element. English language version of EN 54 Part 5. Supersedes BS 3116 : Part 1.
**16 page Gr 7** FSM/12

BS 5445 : Part 7 : 1984 ≡ EN 54 Part 7
**Specification for point-type smoke detectors using scattered light, transmitted light or ionization**
Requirements, test methods and performance criteria for re-settable smoke detectors. English language version of EN 54 : Part 7 : 1982.
**20 page Gr 7** FSM/12

BS 5445 : Part 8 : 1984 ≡ EN 54 Part 8
**Specification for high temperature heat detectors**
Requirements, test methods and performance criteria for heat sensitive (point) detectors containing a static element, that have high response temperatures. English language version of EN 54 : Part 8 : 1982.
**16 page Gr 7** FSM/12

BS 5445 : Part 9 : 1984
**Methods of test of sensitivity to fire**
Describes the six test fires used in other Parts of BS 5445 to determine the sensitivity of fire detectors. English version of EN 54 : Part 9 : 1982.
**12 page Gr 6** FSM/12

BS 5446 : ——
**Specification for components of automatic fire alarm systems for residential premises**

BS 5446 : Part 1 : 1977
**Point-type smoke detectors**
Performance requirements and test methods for smoke detectors for protecting life in residential applications where provision is made for people to sleep.
*AMD 4097, February 1983 (Gr 3)*
**8 page Gr 5** FSM/12

BS 5447 : 1977
**Methods of test for plane strain fracture toughness ($K_{1c}$) of metallic materials**
Definitions and symbols. Bend and tensile test pieces and testing equipment. Procedures for determining the plane strain fracture toughness, for analysing the data, and for recording the results. Supersedes DD 3.
**12 page Gr 6** ISM/NFM/4

BS 5448 : 1977 ≡ ISO 3791
**Specification for keyboard layouts for numeric applications on office machines and data processing equipment**
Basic layout of numerals and symbols on numeric and alphanumeric keyboards for applications where the data are generally numeric.
**4 page Gr 3**

BS 5449 : ——
**Code of practice for central heating for domestic premises**

BS 5449 : Part 1 : 1977
**Forced circulation hot water systems**
Includes sealed systems and microbore systems. Unless otherwise stated, all types of forced circulation hot water heating systems are covered. Supersedes CP 3006 : Part 1.
*AMD 3516, April 1981 (Gr 2)*
**16 page Gr 7** RHE/24

BS 5450 : 1977 (1984)
**Specification for sizes of hardwoods and methods of measurement**
Range of basic sizes of sawn hardwood at 15 per cent moisture content. Measurement of moisture content and sizes. Includes a table for reductions by manufacturing processes from the basic sawn hardwood sizes, for some end uses and products.
**4 page Gr 3** TIB/1

BS 5451 : 1977 ≠ ISO 2024, ISO 2251
**Specification for electrically conducting and antistatic rubber footwear**
Construction, thickness of sole, test requirements (particularly limits for conducting and antistatic properties), marking and labelling. Supersedes BS 2506 and BS 3825.
**4 page Gr 3** RUM/7

BS 5452 : 1977 (1984)
**Specification for hospital hollow-ware made of plastics material**
Range of sizes, material requirements, design and dimensions for rectangular and compartmented instrument trays, kidney dishes, wash bowls, lotion bowls, gallipots and measuring jugs made of plastics. Drawings with working tolerances. Sterilization and load tests.
**8 page Gr 5** SGC/7

BS 5453 : 1977 ≡ ISO 3776
**Specification for anchorages for seat belts in protective cabs and frames on agricultural tractors**
Positioning and testing of anchorages for pelvic restraint belts for operators of agricultural tractors fitted with protective cabs or frames.
**4 page Gr 3** AGE/6

BS 5454 : 1977
**Recommendations for the storage and exhibition of archival documents**
Site, structure and security of repositories; equipment and climatic conditions for storing paper, parchment, photographic material, gramophone records and magnetic tape.
**12 page Gr 6** DOS/9

## BS 5918 General Series

**BS 5918**: 1980
**Code of practice for solar heating systems for domestic hot water**
Deals with solar heating systems having flat plate collectors with liquid heat transfer media for heating water for domestic purposes in single family dwellings. Design considerations, manufacture, handling, installation, operation and maintenance. Excludes applications to space heating and swimming pools.
*AMD 4446, March 1984 (Gr 4)*
**20 page Gr 7** RHE/25

**BS 5919**: 1980
**Specification for children's anoraks**
Specifies make-up, design and performance requirements.
*AMD 3667, June 1981 (Gr 0)*
**4 page Gr 3** CLM/5

**BS 5920**: 1980
**Method for determination of dimensions of textile floor covering test specimens**
Specifies a method for use with the dimensional stability test methods described in BS 4682 and is applicable to textile floor coverings of all types with maximum thickness 15 mm.
**4 page Gr 2** FBM/16

**BS 5921**: 1980
**Method for determination of size, squareness and straightness of edge of textile floor covering tiles**
Applicable to tiles of all types of construction, maximum thickness 15 mm, and for all methods of installation, e.g. loose laid or adhered.
*AMD 3501, January 1981*
**4 page Gr 3** FBM/16

**BS 5922**: 1980 ≡ ISO 4799
**Specification for glass condensers for laboratory use**
Specifies five types of glass condensers in different sizes with or without conical or spherical glass joints suitable for general use in laboratories. Gives detailed drawings and nominal dimensions. Supersedes BS 1848 and 3787.
**8 page Gr 5** LBC/25

**BS 5923**: — —
**Methods for chemical analysis of rubber**

**BS 5923: Part 1**: 1980 ≡ ISO 247
**Determination of ash**
Describes three methods: heating over a gas burner and then transferring to a muffle furnace; closed furnace method and a sulphation method. Specifies apparatus, procedure and reporting of results. With Part 2 supersedes BS 903 : Part B13.
**4 page Gr 3** RUM/37

**BS 5923: Part 2**: 1980 ≠ ISO 2454
**EDTA titrimetric method for determination of zinc content of rubber products**
Specifies reagents, apparatus, procedure and reporting of results. With Part 1 supersedes BS 903 : Part B13.
**4 page Gr 2** RUM/37

**BS 5923: Part 3**: 1981 ≠ ISO 248
**Determination of volatile matter content of new rubbers**
Describes a hot-mill method and an oven method. Specifies apparatus, procedure and reporting of results. Together with BS 5923 : Part 1 : 1980 Supersedes BS 1673 : Part 8 : 1970.
**4 page Gr 2** RUM/37

**BS 5924**: 1980
**Specification for safety requirements for electrical equipment of machines for resistance welding and allied processes**
Requirements for safety rules for construction and installation of electrical equipments for resistance welding and allied processes particularly with regard to risk of direct or indirect contact with live parts and also with regard to mechanical hazards. ≠ HD 389
**36 page Gr 8** WEE/26

**BS 5925**: 1980
**Code of practice for design of buildings: ventilation principles and designing for natural ventilation**
Recommendations for the supply of outside air and the processes by which these recommendations may be fulfilled. The basis for the choice between natural and mechanical ventilation and guidance on the design of natural ventilation systems. Supersedes CP 3 : Chapter I(C).
**24 page Gr 7** BDB/2

**BS 5926**: 1980 ≡ ISO 3687
**Method for determination of air resistance (Gurley) of paper and board**
Applicable to papers and boards that (under the conditions specified) permit the passage of 100 ml of air in five to 1800 seconds, but not suitable for rough-surface papers such as creped and corrugated papers which cannot be securely clamped to avoid leakage.
**4 page Gr 3** PAM/11

**BS 5927**: 1980 ≡ ISO 4482
**Guide for laying asbestos-cement pipelines**
Recommended procedure for the installation of pipelines for both pressure and non-pressure application. Covers the types of laying conditions most commonly encountered in practice.
**8 page Gr 5** CSB/10

**BS 5928**: 1980 ≡ ISO 4796
**Specification for bottles for laboratory use**
Specifies an internationally acceptable series of bottles suitable for the storage of liquid chemicals and reagents in general laboratory use.
**4 page Gr 2** LBC/23

**BS 5929**: — —
**Methods for sensory analysis of food**

**BS 5929: Part 1**: 1980
**Introduction and general guide to methodology**
Background for users of sensory methods, general requirements, procedures and interpretation of results of methods, guidance on choice of procedure appropriate to a particular problem.
*AMD 4286, May 1983 (Gr 0)*
**12 page Gr 6** FAC/20

**BS 5929: Part 2**: 1982 = ISO 5495
**Paired comparison test**
Gives details of assessors and procedure with expression and interpretation of results for one- and two-sided tests, whether or not 'forced choice' technique is chosen. This Part supplements Part 1.
**8 page Gr 5** FAC/20

**BS 5929: Part 3**: 1984 ≡ ISO 4120
**Triangular test**
Describes a method for detecting differences between two products, particularly when differences are slight.
**8 page Gr 6** FAC/20

**BS 5930**: 1981
**Code of practice for site investigations (formerly CP 2001)**
Covers investigation of sites when assessing their suitability for the construction of civil engineering and building works and acquiring knowledge of the characteristics of a site, that affect the design and construction of such work and the security of neighbouring land and property. Supersedes CP 2001 : 1957.
**148 page Gr 11** CSB/1

**BS 5931**: 1980
**Code of practice for machine laid in-situ edge details for paved areas**
Information and recommendations on use of extruded asphalt or concrete together with slip-formed concrete for road edge features such as kerbs or channels. Details of asphalt and concrete mixes that are suitable for the process, and construction details.
**8 page Gr 5** RDB/31

**BS 5932**: 1980 ≠ IEC 285-1, -1A & -2
**Specification for sealed nickel-cadmium cylindrical rechargeable single cells**
Performance requirements, cell designation, terminations, marking, dimensions, charging requirements and methods of test for low and high rate of discharge at normal and low temperature, charge retention, overcharge, life cycle, mechanical and after storage
*AMD 3418, May 1980*
**4 page Gr 3** LEL/12

**BS 5933**: 1980 ≡ ISO 6055
**Specification for overhead guards for high-lift rider trucks**
Requirements and testing for overhead guards designed to protect the operator from falling objects.
**4 page Gr 2** MEE/112

**BS 5934**: 1980 ≡ ISO 5048
**Method for calculation of operating power and tensile forces in belt conveyors with carrying idlers on continuous mechanical handling equipment**
Sets out methods for the calculation of the operating power requirements on the driving pulley of a belt conveyor, and of the tensile forces exerted on the belt.
**16 page Gr 7** MEE/128

**BS 5935**: 1980 ≡ IEC 635
**Specification for toroidal strip-wound cores made of magnetically soft material**
Specifies dimensional and constructional characteristics for toroidal cores wound from strips of magnetically soft steel or alloy over the thickness range 0.003 mm to 0.35 mm. Establishes standard ranges of main dimensions and gives guidance on strip thickness, core protection, marking and packing.
**8 page Gr 5** ECL/20

**BS 5936**: 1981
**Specification for chemicals used in schools**
Specifications and associated test methods for 84 commonly used reagents (6 acids, 3 alkalis, 59 inorganic and 16 organic reagents) to meet particular requirements of school laboratories and to assist in their purchase.
**60 page Gr 9** CIC/26

**BS 5937**: — —
**Pressboard and presspaper for electrical purposes**

**BS 5937: Part 1**: 1980 ≡ IEC 641-1
**Definitions and general requirements**
Covers solid pressboard and presspaper, dyed or natural coloured, having a calendered or non-calendered finish and supplied in an unimpregnated condition. Includes pre-compressed pressboard but excludes laminated material and strawboard.
**4 page Gr 3** GEL/16

BS 6330 : 1983
**Code of practice for the reception of sound and television broadcasting**
Gives guidance for obtaining good reception of sound and television broadcasting in buildings. Includes recommendations for the provision of aerial systems and cabled distribution systems, accommodation for equipment and cables and protective measures against certain hazards. Supersedes CP 1020 : 1973.
**44 page Gr 8** EEL/25

BS 6331 : — —
**Mounting dimensions of single rod double acting hydraulic cylinders**

BS 6331 : Part 1 : 1983 ≡ ISO 6020/1
**Specification for 160 bar medium series**
Specifies interchangeability dimensions for eleven mounting styles of cylinder with round or square heads and with nominal bores in the range 25 mm to 500 mm.
**8 page Gr 5** MEE/186

BS 6331 : Part 2 : 1983 ≡ ISO 6020/2
**Specification for 160 bar compact series**
Specifies interchangeability dimensions for twelve mounting styles of cylinder with nominal bores in the range 25 mm to 200 mm.
**16 page Gr 7** MEE/186

BS 6331 : Part 3 : 1983 ≡ ISO 6022
**Specification for 250 bar series**
Specifies interchangeability dimensions for seven mounting styles of cylinder with nominal bores in the range 50 mm to 500 mm.
**8 page Gr 5** MEE/186

BS 6332 : — —
**Thermal performance of domestic gas appliances**

BS 6332 : Part 1 : 1983
**Specification for thermal performance of central heating boilers and circulators**
Thermal efficiency requirements and associated methods of tests for boilers and for circulators of rated heat input up to and including 60 kW and 8 kW respectively burning 1st, 2nd and 3rd family gases. Partially supersedes BS 1250 : Part 3 : 1963 which is to be withdrawn.
**8 page Gr 5** GSE/17

BS 6332 : Part 2 : 1983
**Specification for thermal performance of gas fires**
Specifies thermal performance requirements and associated methods of test for open flued gas fires burning 1st, 2nd and 3rd family gases. Partially supersedes BS 1250 : Part 4.
**12 page Gr 6** GSE/17

BS 6332 : Part 3 : 1984
**Specification for thermal performance of combined appliances: gas fire/back boiler**
Specifies thermal efficiency requirements and associated methods of test for combined appliances comprising a boiler or circulator and a gas fire and burning 1st, 2nd and 3rd family gases.
**12 page Gr 6** GSE/17

BS 6332 : Part 4 : 1983 GSE/17
**Specification for thermal performance of independent convector heaters**
Specifies thermal efficiency requirements and associated methods of test for independent convector heaters operating under natural draught and burning 1st, 2nd and 3rd family gases. Partially supersedes BS 1250 : Part 4.
**8 page Gr 5**

BS 6333 : 1983
**Adjustable wrenches**
Gives dimensions and test requirements for adjustable wrenches with maximum jaw opening in a range from 13 mm to 52 mm. Covers design, finish, marking and torque tests.
**4 page Gr 2** MEE/120

BS 6334 : 1983 ≡ IEC 707
**Methods of test for the determination of the flammability of solid electrical insulating materials when exposed to an igniting source**
Describes methods for checking the consistency of the characteristics of a material for comparison and classification purposes. Also standardizes the test chamber and conditioning of the specimen.
**8 page Gr 6** GEL/16

BS 6335 : — —
**Methods for determining pressure ripple levels generated in hydraulic fluid power systems and components**

BS 6335 : Part 1 : 1983
**High impedance method for pumps**
Describes procedures for the determination of pressure ripple levels generated by positive displacement hydraulic pumps over a frequency range of 125 Hz to 8000 Hz under controlled conditions and operation which are suitable for providing a basis of comparison in terms of ten individual harmonics of pumping frequency and the overall r.m.s. rating.
**8 page Gr 5** MEE/186

BS 6336 : 1982
**Guide for the development and presentation of fire tests and for their use in hazard assessment**
Describes the theory of fire behaviour, of fire hazard and its assessment and the associated need for fire tests. Gives guidance on the principles to be followed in the development of fire test methods, the expression of results and how to convey information on fire characteristics. Recommends an objective and unambiguous system of terminology. Supersedes DD 64 : 1979.
**8 page Gr 5** M/3

BS 6337 : — —
**General methods of chemical analysis**

BS 6337 : Part 1 : 1983 ≡ ISO 6228
**Method for determination of traces of sulphur compounds by reduction and titrimetry**
Applicable to amounts of sulphate between 4.5 and 450 g either in solution or in the test portion. The sulphate can be present as such or can be produced by suitable pre-treatment.
**8 page Gr 6** CIC/-/1

BS 6337 : Part 2 : 1983 ≡ ISO 6382
**Method for determination of silicon content (reduced molybdosilicate spectrophotometric method)**
Applicable to the determination of quantities of silicon, expressed as $SiO_2$, between 2 and 200 g in the aliquot portion of the test solution taken for the determination. Reference should be made, for the preparation of the test solution to the standard relating to the product being analyzed, which should also give any modifications.
**8 page Gr 6** CIC/-/1

BS 6337 : Part 3 : 1983 ≡ ISO 6685
**Method for determination of iron content (1,10-phenanthroline spectrophotometric method)**
Applicable to test solutions from which an aliquot portion can be taken containing between 10 and 500 g of Fe in not more than 60 ml. Reference should be made, for the preparation of the test solution, to the standard relating to the product being analyzed, which should also give any modification.
**4 page Gr 4** CIC/-/1

BS 6337 : Part 4 : 1984 ≡ ISO 6227
**Method of determination of chloride ions by potentiometry**
Covers reagents, apparatus, procedure, expression of result and test report.
**8 page Gr 6** CIC/-

BS 6338 : 1982 ≡ ISO 4520
**Specification for chromate conversion coatings on electroplated zinc and cadmium coatings**
Covers the method of application, classification (into two classes and four types) and requirements (general, adhesion of coloured coatings and corrosion resistance).
**4 page Gr 3** SRC/9

BS 6339 : 1983 ≡ ISO 6580
**Specification for dimensions of circular flanges for general purpose industrial fans**
Specifies the dimensions of circular flanges for general purpose fans defined as fans for handling air which is non-toxic, non-saturated, non-corrosive, non-flammable, free from abrasive particles and not exceeding a temperature of 80 °C or 40 °C if the motor or the fan bearings are in the air stream.
**4 page Gr 4** MEE/185

BS 6340 : — —
**Shower units**

BS 6340 : Part 1 : 1983
**Guide on choice of shower units and their components for use in private dwellings**
Provides details of minimum requirements to ensure that the facility provided is adequate and safe. Does not cover shower units in boats and caravans nor for the use of disabled persons.
**8 page Gr 3** SEB/37

BS 6340 : Part 2 : 1983
**Specification for the installation of shower units**
Minimum performance requirements, in addition to the performance details given in Part 1, to ensure the facility is adequate and safe.
**4 page Gr 2** SEB/37

BS 6340 : Part 4 : 1984
**Specification for shower heads and related equipment**
Materials, dimensions, testing of domestic shower heads and related equipment used in conjunction with shower controls complying with the requirements of BS 1415 : Part 1, BS 1010 : Part 2 and BS 5412/3.
**8 page Gr 5** SEB/37

BS 6340 : Part 5 : 1983
**Specification for prefabricated shower trays made from acrylic material**
Covers finish, tolerances, functional characteristics and tests.
**8 page Gr 5** SEB/3

BS 6340 : Part 6 : 1983
**Specification for fabricated shower trays made from porcelain enamelled cast iron**
Covers finish, tolerances, functional characteristics and tests.
**8 page Gr 5** SEB/3

**BS 6360** : 1981 ≠ IEC 228
**Specification for conductors in insulated cables and cords**
Requirements for aluminium, copper and copper-clad aluminium conductors in insulated wires and cords. Supersedes BS 6791 : 1969 and BS 4990 : 1973.
*AMD 4378, September 1983*
**8 page Gr 5** GEL/3

**BS 6361** : 1983 ≡ ISO 4400
**Specification for three-pin plug connectors for electrically controlled fluid power equipment**
Characteristics and requirements for a general purpose three-pin electrical connector with earth contact for use with single solenoid operators for fluid power control valves. These connectors are intended for use in conditions where they are protected from physical damage.
**4 page Gr 5** MEE/186

**BS 6362** : 1984
**Specification for stainless steel tubes suitable for screwing in accordance with BS 21 'Pipe threads for tubes and fittings where pressure-tight joints are made on the thread'**
**4 page Gr 3** MEE/196

**BS 6363** : 1983 ≠ ISO 4019
**Specification for welded cold formed steel structural hollow sections**
Specifies the dimensions, tolerances and properties of longitudinally welded cold formed steel hollow sections for structural application.
**20 page Gr 8** MEE/196

**BS 6364** : 1984
**Specification for valves for cryogenic service**
Specifies requirements supplementary to those in the applicable valve product standards when the valves are used for cryogenic service. Applies to the size range DN15 to the maximum size in the applicable product standards and in the temperature range of - 15 °C to - 196 °C.
**8 page Gr 5** MEE/191

**BS 6365** : 1983
**Specification for precision vernier depth gauges**
Gauges having ranges of measurement in metric units up to 600 mm and in inch units up to 24 inches. Nomenclature, construction, scales, accuracy, protection Recommendations for beam sections and methods of test are appended.
**4 page Gr 4** MEE/59

**BS 6366** : 1983
**Specification for studs for rugby football boots**
Specifies design, performance and marking requirements for replaceable (screw-in) studs used in rugby football boots.
**8 page Gr 5** PSM/31

**BS 6367** : 1983
**Code of practice for drainage of roofs and paved areas**
Recommends design methods for roof and paved area drainage based on modern hydraulics and meteorological knowledge. Also deals with the choice of materials and with site-work, including inspection, testing and maintenance. Supersedes CP 308 : 1974.
*AMD 4444, January 1984*
**56 page Gr 9** SEB/14

**BS 6368** : 1983 ≡ IEC 701
**Specification for axial lead cores made of magnetic oxides or iron powder**
Specifies dimensions, mechanical characteristics, solderability and resistance to soldering heat, and the appropriate test methods for axial lead cores made primarily of magnetic oxides (ferrites) or iron powder, for use in small wound inductors intended primarily for use in telecommunication and allied electronic equipment.
**8 page Gr 6** ECL/20

**BS 6369** : 1983
**Specification for battery-operated electric fence controllers suitable for connection to the supply mains**
Specifies definitions, construction and performance requirements and methods of test.
**36 page Gr 8** LEL/105

**BS 6370** : 1983 ≡ IEC 712
**Specification for helical-scan video-tape cassette system using 19 mm (¾ in) magnetic tape, known as U-format**
Gives dimensional and other characteristics necessary to permit the interchangeability of cassettes for both 50 Hz/625-line and 60 Hz/525 line television signals when using portable U-matic recorders for such purposes as electronic news gathering, programme interchange and archive purposes.
**24 page Gr 8** EEL/22

**BS 6371** : 1983
**Recommendations for citation of unpublished documents**
Concise guidance for authors and editors on methods of setting out references in books and journal articles. Applies to references to documents existing only as unique originals or in a very limited number of copies. Includes methods of making attribution within the text.
**8 page Gr 5** DOS/1

**BS 6372** : 1983 ≡ ISO 6939
**Method for determination of breaking strength of yarn from packages: Skein method**
**8 page Gr 4** FBM/16

**BS 6373** : ——
**Burning behaviour of textiles and textile products**

**BS 6373 : Part 1** : 1983 ≡ ISO 4880/1
**Glossary of terms: series 1**
Gives 48 terms and definitions in English and French.
**8 page Gr 5** FBM/19

**BS 6374 : Part 3** : 1984
**Specification for lining with stoved thermosetting resins**
Requirements for lining equipment for process industries including choice and application of lining, rectification of faults, storage, handling, transportation and installation of lined equipment. Applies to thermosetting resin-based linings applied as thin coatings and stoved. Prerequisites necessary for successful lining of equipment (design, construction, etc.) are also specified.
**12 page Gr 6** PVE/1

**BS 6375** : ——
**Performance of windows**

**BS 6375 : Part 1** : 1983
**Classification for weathertightness**
Gives a classification for watertightness in terms of test pressure levels for air permeability, watertightness and wind resistance. Will be supplemented by a BS Guide on the selection of windows. Supersedes DD 4 : 1971.
**8 page Gr 5** ECB/15

**BS 6376** : ——
**Reagents for chemical analysis**

**BS 6376 : Part 1** : 1983 ≡ ISO 6353/1
**Methods of test**
Test methods for verifying compliance of reagents for chemical analysis with specifications to be published in further parts of this standard.
**16 page Gr 8** CIC/26

**BS 6376 : Part 2** : 1984 ≡ ISO 6353/2
**Specifications (first series)**
Specification for acetic acid; acetone; ammonia solution; ammonium acetate and chloride; barium chloride dihydrate; chloroform; citric acid monohydrate; copper (II) sulphate pentahydrate; cyclohexane; ethanol; (ethylenedinitrilo) tetracetic acid disodium salt dihydrate; hydrochloric acid; hydrogen peroxide; hydroxylammonium chloride; magnesium chloride hexahydrate; magnesium oxide; methanol; nitric acid; oxalic acid dihydrate; perchloric and phosphoric acids; potassium dichromate, hyroxide, iodide and permanganate; potassium sodium tartrate tetrahydrate; silver nitrate; sodium acetate trihydrate; sodium carbonate anhydrous; sodium carbonate decahydrate; sodium chloride; disodium hydrogenphosphate dodecahydrate, sodium hyroxide; sodium sulphate anhydrous; sodium thiosulphate pentahydrate; sulphuric acid; tin (II) chloride dihydrate; toluene and zinc.
**60 page Gr 12** CIC/26

**BS 6377** : 1983 ≡ ISO 6575
**Specification for fenugreek**
Applies to whole and ground seeds of Trigonella foenum-graecum Linnaeus. An Annex relates to storage and transport conditions.
**4 page Gr 3** FAC/7

**BS 6378** : 1983 ≡ ISO 94
**Specification for dimensions of spindle gauges for ring-spinning and ring-doubling frames**
Specifies the spindle gauge dimensions. Supersedes BS 2707.
**4 page Gr 2** FBM/22

**BS 6379** : ——
**Sampling of coffee and coffee products**

**BS 6379 : Part 1** : 1983 ≡ ISO 4072
**Method of sampling green coffee in bags**
Method primarily intended for sampling consignments shipped in ten bags or more.
**4 page Gr 4** FAC/15

**BS 6379 : Part 2** : 1984 ≡ ISO 6670
**Methods of sampling instant coffee in cases with liners**
Describes a general method, and an alternative method for use with fragile particles, for examination of bulk density and particle size.
**4 page Gr 4** FAC/15

**BS 6379 : Part 3** : 1983 ≡ ISO 6666
**Specification for a coffee trier for sampling green coffee**
Details of materials, manufacture and dimensions.
**4 page Gr 3** FAC/15

**BS 6380** : 1983
**Guide to low temperature properties and cold weather use of diesel fuels and gas oils (classes A1, A2 and D of BS 2869)**
Provides basic guidance to help users overcome problems which may be encountered in very severe winters.
**12 page Gr 6** PTC/2

# Subject Index

*ALPHABETICAL SUBJECT INDEX TO BRITISH STANDARDS AND RELATED PUBLICATIONS*

*The following is primarily an alphabetical subject index to the publications listed in the BSI Catalogue. Except where otherwise indicated (e.g. DD, PP etc.), the numbers against each entry are BS numbers.*

## A

**Anhydrous Subject Index**



## B

**Channels Subject Index**

**Gas Subject Index**

## L

**Rampley's Subject Index**

**Vacuum Subject Index**

### 2.1.3. American Society for Testing and Materials

Of a few US standardisation bodies you might come across, the *American Society for Testing and Materials* (*ASTM*) is likely to be one of the most significant for you. Founded in 1898 for 'the development of standards on characteristics and performance of materials, products, systems and services, and the promotion of related knowledge', this is now one of the world's leading standardisation bodies. It has some 140 main technical committees representing producers, users, and general-interest participants. It publishes at present over 7,900 standards, all under constant review by the technical committees. The ASTM issues five basic types of publication: (*a*) *Specifications*—precise statements of a set of requirements to be satisfied by a material, product, system, or service; (*b*) *Test methods*; (*c*) *Classifications*—defining a systematic arrangement or division of materials, products, systems, or services divided into groups based on similar characteristics; (*d*) *Definitions of terms*; (*e*) *Practices*—a guide to procedures, which may or may not be auxiliary to a test method or specification.

Every year the ASTM republishes its entire range of standards in:

*Annual Book of ASTM Standards.* I have reproduced the 1985 contents list here, showing 66 volumes in 15 subject areas. About 30% of the standards are revised each year. There are indexes in each volume and a comprehensive index volume (Vol. 00)

01 IRON AND STEEL PRODUCTS

01.01 Steel piping, tubing, and fittings
01.02 Ferrous castings, ferroalloys, shipbuilding
01.03 Steel plate, sheet, strip, and wire
01.04 Structural steel, concrete-reinforcing steel, pressure-vessel plate and forgings, steel rails, wheels, and tyres
01.05 Steel bars, chain, and springs; bearing steel, steel forgings
01.06 Coated steel products

02 NON-FERROUS METAL PRODUCTS

02.01 Copper and copper alloys
02.02 Die-cast metals, light metals and alloys
02.03 Electrical conductors
02.04 Non-ferrous metals—nickel, lead, and tin alloys; precious metals, primary metals, reactive metals
02.05 Metallic and inorganic coatings, metal powders, sintered P/M structural parts

03 METALS TEST METHODS AND ANALYTICAL PROCEDURES

03.01 Metals—mechanical testing, elevated and low-temperature tests
03.02 Metal corrosion, erosion, and wear

03.03 Metallography, non-destructive tests
03.04 Magnetic properties and magnetic materials; metallic materials for thermostats, and for electrical resistance, heating, and contacts
03.05 Chemical analysis of metals, sampling and analysis of metal bearing ores
03.06 Emission spectroscopy, surface analysis

04 CONSTRUCTION

04.01 Cement, lime, gypsum
04.02 Concrete and mineral aggregates
04.03 Road and paving materials
04.04 Roofing, waterproofing and bituminous materials
04.05 Chemical-resistant non-metallic materials, vitrified clay and concrete pipes and tiles, masonry mortars and units, fibre-cement products, precast concrete products
04.06 Thermal insulation, environmental acoustics
04.07 Building seals and sealants, fire standards, building constructions
04.08 Natural building stones, soil and rock
04.09 Wood

05 PETROLEUM PRODUCTS, LUBRICANTS AND FOSSIL FUELS

05.01 Petroleum products and lubricants (1)
05.02 Petroleum products and lubricants (2)
05.03 Petroleum products and lubricants (3), catalysts
05.04 Test methods for rating motor, diesel, and aviation fuels
05.05 Gaseous fuels, coal and coke

06 PAINTS, RELATED COATINGS AND AROMATICS

06.01 Paint: tests for formulated products and applied coatings
06.02 Paint: pigments, resins, and polymers
06.03 Paint: fatty oils and acids, solvents, miscellaneous; aromatic hydrocarbons, naval stores

07 TEXTILES

07.01 Textiles: yarns, fabrics, and general test methods
07.02 Textiles: fibres, zippers

08 PLASTICS

08.01 Plastics (1)
08.02 Plastics (2)
08.03 Plastics (3)
08.04 Plastic pipe and building products

09 RUBBER

09.01 Rubber, natural and synthetic, general test methods, carbon black
09.02 Rubber products, industrial: specifications and related test methods, gaskets, tyres

10 ELECTRICAL INSULATION AND ELECTRONICS

10.01 Electrical insulation (1)
10.02 Electrical insulation (2)
10.03 Electrical insulating liquids and gases, electrical protective equipment
10.04 Electronics (1)
10.05 Electronics (2)

11 WATER AND ENVIRONMENTAL TECHNOLOGY

11.01 Water (1)
11.02 Water (2)
11.03 Atmospheric analysis, occupational health and safety
11.04 Pesticides, resource recovery, hazardous substances and oil-spill response, waste disposal, biological effects and environmental fate

12 NUCLEAR, SOLAR, AND GEOTHERMAL ENERGY

12.01 Nuclear energy (1)
12.02 Nuclear (2), solar, and geothermal energy

13 MEDICAL DEVICES

13.01 Medical devices

14 GENERAL METHODS AND INSTRUMENTATION

14.01 Molecular and mass spectroscopy, chromatography, resinography, temperature measurement, microscopy, computerised systems
14.02 General test methods, non-metals, statistical methods, appearance of materials, hazard potential of chemicals, particle size measurement, thermal measurement, laboratory apparatus, metric practice, durability of non-metallic materials

15 GENERAL PRODUCTS, CHEMICAL SPECIALITIES AND END—USE PRODUCTS

15.01 Refractories, manufactured carbon and graphite products, activated carbon
15.02 Glass, ceramic whitewares, porcelain enamels
15.03 Space simulation, aerospace materials, high modulus fibres and their composites
15.04 Soap, polishes, cellulose, leather, resilient floor coverings
15.05 Engine coolants, halogenated organic solvents, industrial chemicals

15.06 Adhesives
15.07 End-use products
15.08 Fasteners
15.09 Paper, packaging, flexible barrier materials, business copy products

00 INDEX

So now turn to the next SAQ for further practice and detail.

**SAQ 2.1b**

By the time you have worked through my SAQs you will have found yourself working, in imagination at least, in an extraordinary variety of laboratories and industries! But apart from giving you a wide range of questions to try, and making them as interesting as I can for you, there is a more serious purpose too: I want you to realise just how easily you can find information on subjects which might be completely unfamiliar to you.

Here a couple of straightforward questions from the *Annual Book of ASTM Standards* find you concerned with metals analysis and gas chromatography (nothing very esoteric here, but wait for some later questions!).

(*a*) You find yourself working in the metal industry. You have to determine the amount of silicon in a bronze sand-casting ingot. How many methods are recommended by ASTM? What weight of sample is needed?

(*b*) You are going to use a flame-photometric detector in a gas chromatography analysis. Find an ASTM standard which will tell you how to go about your task.

Use the copies only if you have to: try and consult copies of the books if you can.

VOLUME **00.01** **Subject Index; Alphanumeric List**

Publication Code Number (PCN): 01-000184-42

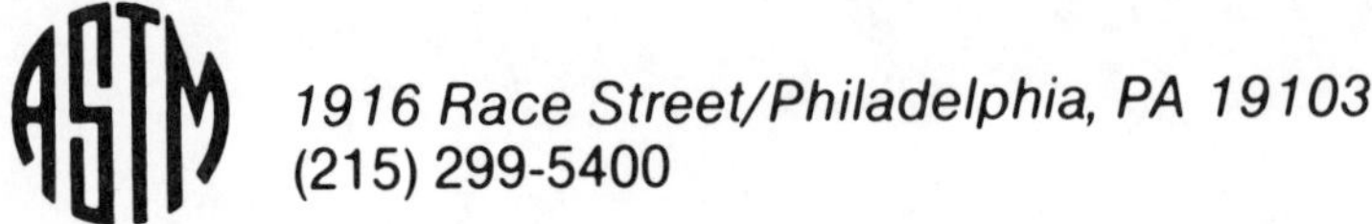


*1916 Race Street/Philadelphia, PA 19103*
(215) 299-5400

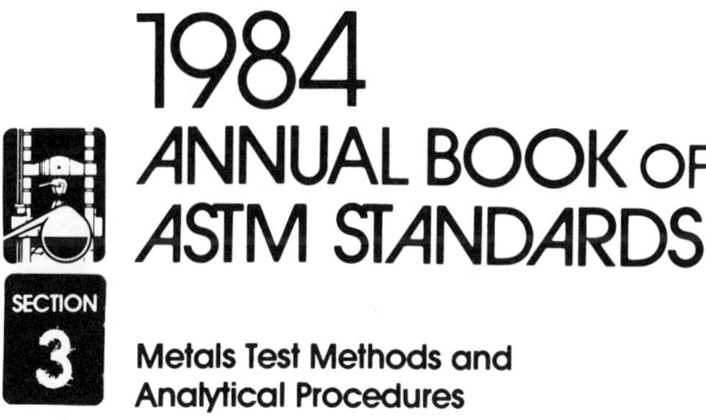


VOLUME
**03.05** **Chemical Analysis of Metals and Metal-Bearing Ores**

***Includes standards of the following committees:***

E-3 on Chemical Analysis of Metals
E-16 on Sampling and Analysis of Metal-Bearing Ores and Related Materials

Publication Code Number (PCN): 01-030584-24

**Designation: E 54 – 80**

An American National Standard

## Standard Methods for
# CHEMICAL ANALYSIS OF SPECIAL BRASSES AND BRONZES[1]

This standard is issued under the fixed designation E 54; the number immediately following the designation indicates the year of original adoption or, in the case of revision, the year of last revision. A number in parentheses indicates the year of last reapproval. A superscript epsilon (ε) indicates an editorial change since the last revision or reapproval.

### 1. Scope

1.1 These methods cover procedures for the chemical analysis[2] of the commercial alloys known as copper-base alloy ingots for sand castings, forging rods, bars, and shapes; aluminum brass; manganese bronze; phosphor bronze; copper-silicon alloys; and similar alloys having chemical compositions within the following limits.[2]

| | |
|---|---|
| Copper, percent | 50 and over |
| Lead, percent | 0.0 to 27 |
| Tin, percent | 0.0 to 20 |
| Silicon, percent | 0.0 to 5 |
| Aluminum, percent | 0.0 to 12 |
| Nickel, percent | 0.0 to 5 |
| Iron, percent | 0.0 to 5 |
| Manganese, percent | 0.0 to 6 |
| Sulfur, percent | 0.0 to 0.1 |
| Phosphorus, percent | 0.0 to 1.0 |
| Arsenic, percent | 0.0 to 1.0 |
| Antimony, percent | 0.0 to 1.0 |
| Zinc[a] | 0.0 to 50 |

[a] In the case of copper-base alloys containing 5.0 percent and over of zinc, the zinc is usually calculated by difference.

Whenever possible the technique and procedures for analysis should be checked against a National Bureau of Standards standard sample having a composition comparable to the material being analyzed.

1.2 The analytical procedures appear in the following order:

| | Sections |
|---|---|
| Copper, or Copper and Lead Simultaneously, by the Electrolytic Method | [2a] |
| Lead: | |
| Electrolytic Method | [2c] |
| Sulfate Method | [2b] |
| Tin by the Iodimetric Titration Method | [2b] |
| Silicon: | |
| Sulfuric Acid Dehydration Method | 17 |
| Perchloric Acid Dehydration Method | 18 and 19 |
| Aluminum by the Gravimetric Method | 20 to 22 |
| Nickel by the Dimethylglyoxime Method | [2a] |
| Iron by the Dichromate Method | 26 to 28 |
| Manganese: | |
| Bismuthate Method | [2b] |
| Persulfate Method | [2b] |
| Bromate Method | [2d] |
| Phosphorus by the Alkalimetric Method | [2b] |
| Arsenic by the Distillation-Bromate (Moffat) Method | 41 to 43 |
| Arsenic and Antimony by the Distillation - Iodometric Method | [2d] |
| Antimony and Tin by the Manganese Coprecipitation Method | [2d] |
| Sulfur: | |
| Direct Combustion - Iodate Method | 50 to 52 |
| Evolution Method | 53 to 55 |

[1] These methods are under the jurisdiction of ASTM Committee E-3 on Chemical Analysis of Metals and are the direct responsibility of Subcommittee E03.05 on Nonferrous Metals.

Current edition approved May 30, 1980. Published July 1980. Originally published as E 54 – 46, replacing former B 27, B 28, B 45, and B 46. Last previous edition E 54 – 79a.

[2] For procedures for sampling wrought products, see ASTM Methods E 55, Sampling Wrought Nonferrous Metals and Alloys for Determination of Chemical Composition (*Annual Book of ASTM Standards*, Vol 03.05). For procedures for sampling cast products, see ASTM Method E 88, Sampling Nonferrous Metals and Alloys in Cast Form for Determination of Chemical Composition (*Annual Book of ASTM Standards*, Vol 03.05). For procedures for the chemical analysis of other brasses, see ASTM Method E 36, for Chemical Analysis of Brasses (Discontinued; Replaced by Methods E 478, *Annual Book of ASTM Standards*, Vol 03.05). For photometric procedures applicable to special brasses and bronzes, see ASTM Method E 62, Photometric Methods for Chemical Analysis of Copper and Copper Alloys (*Annual Book of ASTM Standards*, Vol 03.05).

[2a] Discontinued as of June 30, 1975.

[2b] Discontinued as of Aug. 27, 1976.

[2c] Discontinued as of April 27, 1979.

[2d] Discontinued as of May 30, 1980.

E 54



**2. Apparatus and Reagents**

2.1 Apparatus and reagents required for each determination are listed in separate sections preceding the procedure. The apparatus, standard solutions, and certain other reagents used in more than one procedure are referred to by number and shall conform to the requirements prescribed in ASTM Recommended Practices E 50, for Apparatus, Reagents, and Safety Precautions for Chemical Analysis of Metals.[3]

**3. Precautions**

3.1 For precautions to be observed in these methods, reference shall be made to Recommended Practices E 50.

### COPPER, OR COPPER AND LEAD SIMULTANEOUSLY, BY THE ELECTROLYTIC METHOD

(This method, which consisted of Sections 4 through 7 of this standard, was discontinued in 1975.)

### LEAD BY THE ELECTROLYTIC METHOD

(This method, which consisted of Sections 8 and 9 of this standard, was discontinued in 1979.)

### LEAD BY THE SULFATE METHOD

(This method, which consisted of Sections 4 through 7 of this standard was discontinued in 1976.)

### TIN BY THE IODIMETRIC TITRATION METHOD

(This method, which consisted of Sections 12 through 14, was discontinued in 1976.)

### ZINC BY THE OXIDE OR FERROCYANIDE METHOD

(This method, which consisted of Sections 15 and 16 of this standard, was discontinued in 1975.)

### SILICON BY THE SULFURIC ACID DEHYDRATION METHOD

**17. Procedure**

17.1 *Solution of Samples Containing Under 1.0 Percent of Silicon*—Transfer 5.0 g of the sample to a 340-ml porcelain casserole, cover, and dissolve in 10 ml of HCl and 20 ml of $HNO_3$. When solution of the sample is complete, add 30 ml of $H_2SO_4$.

17.2 *Solution of Samples Containing 1.0 to 5.0 Percent of Silicon*—Transfer 1.00 g of the sample to a 340-ml porcelain casserole, cover and dissolve in 5 ml of HCl and 10 ml of $HNO_3$. When solution of the sample is complete, add 15 ml of $H_2SO_4$.

17.3 Place the casserole on a hot plate and with a cover glass placed slightly to one side evaporate until the HCl and $HNO_3$ have been expelled. Then, with the casserole well covered heat strongly for several minutes while dense white fumes are being driven off. Allow to cool add carefully 100 ml of HCl (1+5), and bring to a boil. If lead is present in the alloy, add g of $NH_4Cl$ and continue the boiling until all $PbSO_4$ is in solution. Filter the solution immediately through an 11-cm close-texture, ashless paper, scrubbing the casserole with a policeman and washing with HCl (1+99) to completely transfer the residue to the paper. Wash the paper and residue (Note 1) thoroughly with HCl (1+99) and reserve.

NOTE 1—If black silicides, such as those of manganese or iron, are present in the residue, ignite in accordance with 17.3. Fuse the residue with $Na_2CO$ and a crystal of $NaNO_3$ or $KNO_3$. Dissolve the melt in water and $H_2SO_4$ and continue in accordance with 17.3. Combine the filtrate with the original filtrate from 17.3, and complete the determination in accordance with 17.4 to 17.8.

17.4 Return the filtrate to the casserole and treat in accordance with 17.3.

17.5 Transfer the two papers and residues to a platinum crucible. Ignite at a low temperature until the paper has been consumed, and then at 1100 to 1150 C to constant weight (see Note 1). Cool in a desiccator and weigh.

17.6 Add 1 or 2 drops of $H_2SO_4$ (1+1) and sufficient HF (2 to 5 ml) to dissolve the residue and evaporate the solution slowly to dense white fumes. Continue to heat the crucible until all of the free $H_2SO_4$ has been expelled, and

[3] *Annual Book of ASTM Standards*, Vol 03.05.

then ignite at 1000 C for 5 min. Cool in a desiccator and weigh. The loss in weight represents $SiO_2$.

17.7 *Blank*—Make a blank determination, following the same procedure and using the same amounts of all reagents.

17.8 *Calculation*—Calculate the percentage of silicon as follows:[4]

$$\text{Silicon, percent} = [((A - B) \times 0.4675)/C] \times 100$$

where:

A = grams of $SiO_2$,
B = correction for blank, g, and
C = grams of sample used.

## SILICON BY THE PERCHLORIC ACID DEHYDRATION METHOD

### 18. Reagents

18.1 *Hydrogen Peroxide (3 percent).*

### 19. Procedure

19.1 *Solution of Samples Containing Under 1.0 Percent of Silicon*—Transfer 5.0 g of the sample to a 340-ml porcelain casserole, cover, and dissolve in 10 ml of HCl and 20 ml of $HNO_3$. When solution of the sample is complete, add 40 ml of $HClO_4$.

19.2 *Solution of Samples Containing 1.0 to 5 Percent of Silicon*—Transfer 1.00 g of the sample to a 340-ml porcelain casserole, cover, and dissolve in 5 ml of HCl and 10 ml of $HNO_3$. When solution of the sample is complete, add 25 ml of $HClO_4$.

19.3 If appreciable amounts of tin or antimony, or both, are present, add, while stirring, 10 to 30 ml of HBr to volatilize these constituents. Place the casserole on a hot plate and evaporate to white fumes. Heat the covered casserole strongly for 15 min while copious white fumes are being driven off. Allow to cool, but not to become cold, add 100 to 150 ml of hot water, stir well, and heat until the salts are dissolved. If manganese dioxide is present, bring it into solution with a few drops of $H_2O_2$ (3 percent). Filter the solution immediately through an 11-cm, close-texture, ashless paper, scrubbing the casserole with a policeman and washing with HCl (1+99) to completely transfer the residue to the paper. Wash the paper and residue (see Note 1) thoroughly with HCl (1+99) and reserve.

19.4 Return the filtrate to the casserole and treat in accordance with 19.3.

19.5 Complete the determination of silicon as described in 17.5 to 17.8.

## ALUMINUM BY THE GRAVIMETRIC METHOD

(Aluminum Content 0.10 Percent and Over)

### 20. Apparatus

20.1 *Mercury Cathode Cell*—Apparatus No. 10B.

20.2 *Filtering Crucible*—A 35-ml fritted-glass crucible of fine porosity (Apparatus No. 2).

### 21. Reagents

21.1 *Ferric Chloride Solution (100 g/liter)*—Dissolve 100 g of $FeCl_3 \cdot 6H_2O$ in water and dilute to 1 liter.

21.2 *Methyl Red Indicator Solution (0.2 g/liter)*—Dissolve 0.02 g of methyl red indicator in 100 ml of hot water, cool, and filter.

21.3 *Bromcresol Purple Indicator Solution (0.4 g/liter)*—Dissolve 0.04 g of bromcresol purple indicator in 100 ml of hot water, cool, and filter.

21.4 *8-Hydroxyquinoline Solution (25 g/liter)*—Add 50 ml of acetic acid to 25 g of 8-hydroxyquinoline and warm gently to effect solution. Pour the resulting solution into 900 ml of water at 60 C. Cool, filter if necessary, and dilute to 1 liter.

### 22. Procedure

22.1 For samples containing 0.5 percent and over of aluminum, the electrolyte reserved from the determination of copper on a 2-g sample (6.4) may be used, or a separate sample may be taken. If the electrolyte is used, continue in accordance with 22.10. If a separate sample (Note 2) is taken, proceed in accordance with 22.2 or 22.3.

NOTE 2—The preliminary separation of aluminum from interfering elements avoids too great contamination of mercury during the subsequent electrolysis. In certain cases, this may not be objectionable if the sample is not larger than 1 g, and a direct separation may be carried out in the mercury cathode cell as follows: Dissolve 1.000 g of the sample as described in 22.3. Continue as described in 22.4, 22.5,

[4] For the recommended procedure for rounding calculated values, see 3.4 and 3.5 of ASTM Recommended Practice E 29, Indicating Which Places of Figures Are to Be Considered Significant in Specified Limiting Values, *Annual Book of ASTM Standards*, Vol 14.02.

VOLUME
**14.01** **Analytical Methods — Spectroscopy; Chromatography; Temperature Measurement; Computerized Systems**

***Includes standards of the following committees:***

E-13 on Molecular Spectroscopy
E-14 on Mass Spectrometry
E-19 on Chromatography
E-20 on Temperature Measurement
E-23 on Resinography
E-25 on Microscopy
E-31 on Computerized Systems

Publication Code Number (PCN): 01-140184-39

*1916 Race Street/Philadelphia, PA 19103*
(215) 299-5400

**Designation: E 840 - 81**

## Standard Practice for
# USING FLAME PHOTOMETRIC DETECTORS IN GAS CHROMATOGRAPHY[1]

This standard is issued under the fixed designation E 840; the number immediately following the designation indicates th year of original adoption or, in the case of revision, the year of last revision. A number in parentheses indicates the year of la reapproval. A superscript epsilon (ε) indicates an editorial change since the last revision or reapproval.

## 1. Scope

1.1 This practice is intended as a guide for the use of a flame photometric detector (FPD) as the detection component of a gas chromatographic system.

1.2 This practice is directly applicable to an FPD that employs a hydrogen-air flame burner, an optical filter for selective spectral viewing of light emitted by the flame, and a photomultiplier tube for measuring the intensity of light emitted.

1.3 This practice describes the most frequent use of the FPD which is as an element-specific detector for compounds containing sulfur (S) or phosphorus (P) atoms. However, nomenclature described in this practice are also applicable to uses of the FPD other than sulfur or phosphorus specific detection.

1.4 This practice is intended to describe the operation and performance of the FPD itself independently of the chromatographic column. However, the performance of the detector is described in terms which the analyst can use to predict overall system performance when the detector is coupled to the column and other chromatographic system components.

1.5 For general gas chromatographic procedures, Practice E 260 should be followed except where specific changes are recommended herein for use of an FPD.

## 2. Applicable Documents

2.1 *ASTM Standards:*

E 260 Recommended Practice for General Gas Chromatography Procedures[2]

E 355 Recommended Practice for Gas Chromatography Terms and Relationships[2]

## 3. Terminology

3.1 *Definitions*—For definitions relating to gas chromatography, refer to Practice E 355.

3.2 *Descriptions of Terms*—Descriptions of terms used in this practice are included in Sections 6 through 16.

3.3 *Symbols*—A list of symbols and associated units of measurement is included in Annex A1.

## 4. Principles of Flame Photometric Detectors

4.1 The FPD detects compounds by burning those compounds in a flame and sensing the increase of light emission from the flame during that combustion process. Therefore, the FPD is a flame optical emission detector comprised of a hydrogen-air flame, an optical window for viewing emissions generated in the flame, an optical filter for spectrally selecting the wavelengths of light detected, a photomultiplier tube for measuring the intensity of light emitted and an electrometer for measuring the current output of the photomultiplier.

4.2 The intensity and wavelength of light emitted from the FPD flame depends on the geometric configuration of the flame burner and on the absolute and relative flow rates of gases supplied to the detector. By judicious selection of burner geometry and gas flow rates the FPD flame is usually designed to selectively enhance optical emissions from certain types of

[1] This practice is under the jurisdiction of ASTM Committee E-19 on Chromatography and is the direct responsibility of Subcommittee E19.03 on Methods and Specifications. Gas Chromatography.

Current edition approved Oct. 30, 1981. Published May 1982.

[2] *Annual Book of ASTM Standards*, Vol 14.01.

molecules while suppressing emissions from other molecules.

4.3 Typical FPD flames are normally not hot enough to promote abundant optical emissions from atomic species in the flame. Instead, the optical emissions from an FPD flame usually are due to molecular band emissions or continuum emissions resulting from recombination of atomic or molecular species in the flame. For sulfur detection, light emanating from the $S_2$ molecule is generally detected. For phosphorus detection, light emanating from the HPO molecule is generally detected. Interfering light emissions from general hydrocarbon compounds are mainly comprised of CH and $C_2$ molecular band emissions, and $CO + O \rightarrow CO_2 + h\gamma$ continuum radiation.

4.4 Hydrogen - air or hydrogen - oxygen diffusion flames are normally employed for the FPD. In such diffusion flames, the hydrogen and oxygen do not mix instantaneously, so that these flames are characterized by significant spatial variations in both temperature and chemical species. The important chemical species in a hydrogen - air flame are the H, O, and OH flame radicals. These highly reactive species play a major role in decomposing incoming samples and in the subsequent production of the desired optical emissions. Optical emissions from the HPO and $S_2$ molecular systems are highly favored in those regions of an FPD flame which are locally rich in H-atoms, while CH and $C_2$ light emissions from hydrocarbons originate mainly from those flame regions which are locally rich in O-atoms. The highest sensitivity and specificity for sulfur and phosphorus detection are achieved only when the FPD flame is operated with hydrogen in excess of that stoichiometric amount required for complete combustion of the oxygen supplied to the flame. This assures a large flame volume that is locally abundant in H-atoms, and a minimal flame volume that is locally abundant in O-atoms. The sensitivity and specificity of the FPD are strongly dependent on the absolute and relative flow rates of hydrogen and air. The optimum hydrogen and air flow rates depend on the detailed configuration of the flame burner. For some FPD designs, the flows which are optimum for phosphorus detection are not the same as the flows which are optimum for sulfur detection. Also, the flows which are optimum for one sample compound may not necessarily be optimum for another sample compound.

4.5 Although the detailed chemistry occurring in the FPD flame has not been firmly established, it is known that the intense emissions from the HPO and $S_2$ molecules are the result of chemiluminescent reactions in the flame rather than thermal excitation of these molecules (**1**).[3] The intensity of light radiated from the HPO molecule generally varies as a linear function of P-atom flow into the flame. In the case of the $S_2$ emission, the light intensity is generally a nonlinear function of S-atom flow into the flame, and most often is found to vary as the approximate square of the S-atom flow. Since the FPD response depends on the P-atom or S-atom mass flow per unit time into the detector, the FPD is a mass flow rate type of detector. The upper limit to the intensity of light emitted from both the HPO and $S_2$ molecules is generally determined by the onset of self-absorption effects in the emitting flame. At high concentrations of S and P atoms in the flame, the concentrations of ground state $S_2$ and HPO molecules becomes sufficient to reabsorb light emitted from the radiating states of HPO and $S_2$.

4.6 In the presence of a hydrocarbon background in the FPD flame, the light emissions from the phosphorus and sulfur compounds can be severely quenched (**2, 3**). Such quenching can occur in the gas chromatographic analysis of samples so complex that the GC column does not completely separate the phosphorus or sulfur compounds from overlapping hydrocarbon compounds. Quenching can also occur as the result of an underlying tail of a hydrocarbon solvent peak preceding phosphorus or sulfur compounds in a chromatographic separation. The fact that the phosphorus or sulfur response is reduced by quenching is not always apparent from a chromatogram since the FPD generally gives little response to the hydrocarbon. The existence of quenching can often be revealed by a systematic investigation of the variation of the FPD response as a function of variations in sample volume while the analyte is held at a constant amount.

4.7 The chromatographic detection of trace level phosphorus or sulfur compounds can be

[3] The boldface numerals in parentheses refer to the list of references at the end of this practice.

### 2.1.4. Society for Analytical Chemistry

The Society is now a part of the Royal Society of Chemistry. In 1974 they last published a compilation of standard methods.

> *Official, standardised and recommended methods of analysis* 2nd ed. S. C. Jolly, ed. London: Society for Analytical Chemistry, 1974. 897pp.

A third edition has been in progress for some time and is now scheduled for publication in 1986; but my description and questions must be based on this slightly elderly second edition. This volume includes all those methods still considered valid and still recommended by the Society's Analytical Methods Committee. Most of these methods will have been first published in *The Analyst* (the Society's journal: more about this in Section 2.4), or in separate books. Part 1 contains standardised methods of analysis and Part 2 a bibliography of official, standardised, and recommended methods of analysis. We are concerned only with Part 1 here; it is arranged by chapters with the following headings: Metallic impurities in organic matter, Vitamins, Essential oils, Milk products, Determination of the halogens in organic substances or in solution, Meat products and meat extracts, Oils and fats, Soaps, Pesticide residues, Additives in animal feeding stuffs, Trace elements in fertilisers and feeding stuffs, Evaluation of drugs; Analysis of industrial effluents. The new edition will, presumably, follow a similar pattern.

**SAQ 2.1c**

You have to determine the percentage of free fatty acid in a sample of soap. What reagent do you use in the principal recommended method and what weight of sample is recommended? Are there any alternative methods you could use?

If you cannot get hold of a copy of the book, use the copies provided, although these will make an easy question extremely simple. If you do use the book, remember that I am setting this question some time before the new edition is published. You might find yourself using this new edition, so watch out; some at least, if not all, the details in my response are likely to differ from those in the book you may be using.

# OFFICIAL, STANDARDISED AND RECOMMENDED METHODS OF ANALYSIS

COMPILED AND EDITED FOR THE ANALYTICAL METHODS COMMITTEE OF THE SOCIETY FOR ANALYTICAL CHEMISTRY

BY

N. W. HANSON

B.Sc. Ph.D. F.R.I.C.

THE SOCIETY FOR ANALYTICAL CHEMISTRY

LONDON

1973

# Contents

CONTENTS

CONTENTS

CONTENTS

CONTENTS

CONTENTS

CONTENTS

CONTENTS

# SOAPS

## THE DETERMINATION OF UNSAPONIFIED FAT IN SOAPS*

### DEFINITIONS

*Unsapcnified neutral fat* is the unsaponified neutral glycerides.

*Total unsaponified matter* is the sum of the free fatty acids, unsaponified neutral fat and unsaponifiable matter.

*Total free fat, or total unsaponified saponifiable matter*, is the sum of the free fatty acids and the unsaponified neutral fat.

### PRINCIPLE OF METHOD

In order to obtain the amounts of unsaponified neutral fat, total unsaponified matter and total free fat, the following determinations are made:

(*i*) Free fatty acids by direct titration.

(*ii*) Unsaponified neutral fat, together with unsaponifiable matter, by an extraction method.

(*iii*) Unsaponifiable matter by saponification and extraction of (*ii*).

The unsaponified neutral fat is the difference between (*ii*) and (*iii*), and the total free fat is the difference between (*iii*) and the sum of (*i*) and (*ii*).

### APPLICABILITY

The methods are applicable to most ordinary soaps, but cannot always be applied without some modification to special types, such as filled, medicated, or highly perfumed soaps.

### REAGENTS

All reagents should be of analytical-reagent grade. Water must be distilled or de-ionised.

*Acetone.*

*Ethanol.* 95% *V*/*V*.

*Ether.* Diethyl ether, sp.gr. 0·716.

*Ethanolic potassium hydroxide*, 0·5 N. Reagent solution.

*Ethanolic potassium hydroxide*, 0·1 N. Standard volumetric solution.

*Potassium hydroxide*, 0·5 N. Reagent solution.

*Phenolphthalein indicator solution.* A 0·5% *m*/*V* solution of phenolphthalein in ethanol.

### PROCEDURE

Strict adherence to the prescribed details is essential for accurate results.

#### Determination of Free Fatty Acids

Boil 100 ml of ethanol in a 400-ml conical flask, add 0·5 ml of phenolphthalein indicator solution, allow to cool to 70°C, and neutralise at that temperature with 0·1 N ethanolic potassium hydroxide. Add 5 g of the sample, in thin shavings, and dissolve as rapidly and completely

*Taken from Report No. 2 of the Sub-Committee on the Determination of Unsaponifiable Matter in Oils and Fats and of Unsaponified Fat in Soaps (later the Sub-Committee on Methods of Soap Analysis) to the Analytical Methods Committee (*Analyst*, 1935, **60**, 538). This method was incorporated, with slight modification, in B.S. 1715 : 1963. These modifications are included in this description of the procedure.

as possible by heating (see Note 1). If the solution is not already alkaline, adjust the temperature to 70° C, titrate with 0·1 N ethanolic potassium hydroxide until a faint pink colour persists for 15 seconds, maintaining the temperature at 70° C throughout the titration.

From the amount of 0·1 N ethanolic potassium hydroxide required by the sample, calculate the percentage of free fatty acids in the sample in terms of oleic acid (mol. wt. 282) (see Note 2).

Alternative methods, which may be useful in certain circumstances, are given in the Appendix, p. 180.

**Determination of Unsaponified Neutral Fat and Unsaponifiable Matter**

Dissolve 5 g of the sample, in thin shavings, in about 80 ml of a mixture of 50 ml of ethanol and 100 ml of water without heating more than is necessary, transfer the solution to a 500-ml separating funnel, rinsing in with the remaining 70 ml of the diluted ethanol. While the solution is still warm, add 100 ml of ether, shake, allow to separate, and transfer the lower ethanolic soap layer to a second separating funnel and the ethereal layer to a third separating funnel containing 20 ml of water. Repeat the extraction of the soap solution with 50 ml of ether, returning the lower ethanolic soap solution to the first separating funnel and the ethereal layer to the third separating funnel. Repeat the extraction of the soap solution with a further 50 ml of ether, discard the lower ethanolic soap solution, and again transfer the ethereal layer to the third separating funnel. Swirl the liquid in the third separating funnel gently, allow to separate, and run off and discard the lower aqueous layer. Repeat the washing with successive portions of water until a washing is obtained which is not more than faintly turbid when acidified. Add 20 ml of 0·5 N potassium hydroxide to the ethereal solution, shake vigorously, allow to separate, and run off and discard the lower aqueous layer. Add 20 ml of water, shake vigorously, allow to separate, and run off and discard the lower aqueous layer. Repeat the washing of the ethereal solution with further successive and alternate 20-ml portions each of 0·5 N potassium hydroxide and of water until an alkaline washing is obtained which remains clear when acidified. Finally wash the ethereal solutions with successive 20-ml portions of water until a washing is obtained which does not give a pink colour with phenolphthalein indicator solution. Transfer the ethereal solution to a weighed flask, distil off nearly all the ether, and add 2 to 3 ml of acetone. With the aid of a gentle current of air, remove the solvent completely from the flask, which preferably is almost entirely immersed, held obliquely, and rotated in a boiling water-bath, and weigh the residue of unsaponified neutral fat and unsaponified matter; repeat this operation until the weight is constant.

Calculate the percentage of unsaponified neutral fat and unsaponified matter in the sample.

After the final weighing, dissolve the residue in 10 ml of freshly boiled and neutralised ethanol, and titrate with 0·1 N ethanolic potassium hydroxide, using phenolphthalein indicator solution; not more than 0·1 ml should be required for neutralisation. If more is required, the procedure has not been effectively carried out and must be repeated.

**Determination of Unsaponifiable Matter**

Evaporate the solution resulting from the titration under 'Determination of Unsaponified Neutral Fat and Unsaponifiable Matter' above until most of the ethanol is removed. Add 25 ml of 0·5 N ethanolic potassium hydroxide, and boil under reflux for half an hour. Transfer the ethanolic soap solution to a separating funnel, rinsing in with 50 ml of water, and repeat the procedure of ether extraction, washing, etc., described above under 'Determination of Unsaponified Neutral Fat and Unsaponifiable Matter', commencing at the addition of 100 ml of ether. Weigh the final residue of unsaponifiable matter, and calculate the percentage of unsaponifiable matter in the sample.

CALCULATIONS

Calculate the percentage of unsaponified neutral fat by subtracting the percentage of unsaponifiable matter from the percentage of unsaponified neutral fat and unsaponifiable matter. Calculate the percentage of total free fat by adding together the percentages of unsaponified neutral fat and free fatty acid.

NOTES

1 In the analysis of soaps containing an unusually large proportion of matter insoluble in ethanol, it may be desirable to filter the solution before titration, but such a procedure may cause a slight increase in the apparent free fatty acids present.

2 The results may be expressed in terms of some other acid of different molecular weight, provided that it is made clear that this has been done. In some circumstances it may be desirable to determine the mean molecular weight of the total fatty acids present after separating them from the soap; the mean molecular weight should then be stated.

## Appendix

ALTERNATIVE METHODS FOR THE DETERMINATION OF FREE FATTY ACIDS

BARIUM CHLORIDE METHOD

There is a possibility of the co-existence in soap of free fatty acid and sodium carbonate. When such a soap is dissolved in ethanol some interaction takes place, although if dry sodium carbonate is added to an ethanolic solution of soap containing free fatty acid no reaction takes place.

When it is desired to titrate free fatty acid in the presence of sodium carbonate, the recommended method should be modified by adding 1 ml of a neutral 10% *m/V* solution of barium chloride, $BaCl_2.2H_2O$, in water to neutralised ethanol, dissolving the soap in the boiling mixture, and titrating the free fatty acid without filtration; the effect of sodium carbonate is thus eliminated.

GLYCEROL METHOD

In order that the titration might be carried out at laboratory temperature without gelling of the solution of soap, a mixture of ethanol and glycerol (80:20, by volume) may be used as the solvent. In preparing this mixture, a slight excess of alkali should be added to the ethanol–glycerol mixture, the mixture heated to saponify, made slightly acid, and then adjusted to neutrality to phenolphthalein at 15° C by the cautious addition of 0·1 N ethanolic potassium hydroxide. When warmed with this solvent, 5 g of soap will dissolve in 100 ml. Accurate results are obtained at 15° C.

## THE DETERMINATION OF FREE ALKALI IN SOAPS*

INTRODUCTION

The free alkali in soaps is usually made up of hydroxide and carbonate of sodium or potassium or of both. Alkalinity may also be due to the presence of sodium silicate or other alkaline compounds which may be added to the soap. Special methods applicable to silicated soaps are described on p. 183.

PRINCIPLE OF METHOD

Methods are available for the determination of the 'total free alkali', of 'free caustic (or hydroxide) alkali,' and of 'free carbonate alkali.' When any two of these figures have been determined, the third can be found by difference.

For sodium soaps it is recommended that (*i*) the total free alkali and (*ii*) the total caustic alkali be determined. The free carbonate alkali can then be found from the difference between (*i*) and (*ii*), and may be checked by a determination of ethanol-insoluble alkali or of carbon dioxide. For potassium soaps, the total free alkali and total caustic alkali should be determined, and the free carbonate alkali obtained by difference or from a determination of carbon dioxide.

*Taken from Report No. 3 of the Sub-Committee on Methods of Soap Analysis to the Analytical Methods Committee (*Analyst*, 1937, **62**, 36). This method was incorporated, with slight modification, in B.S. 1715: 1963. These modifications are included in this description of the procedure.

APPLICABILITY

The methods are applicable to most ordinary soaps, but they cannot always be applied, without some modification, to special types, such as silicated or medicated soaps.

### Determination of Total Free Alkali

PRINCIPLE OF METHOD

Slightly more standard mineral acid than is necessary to neutralise the free alkali is added to an ethanolic solution of the soap, and, after heating to remove carbon dioxide, the excess of acid is determined by back-titrating with standard alkali to a phenolphthalein end-point.

REAGENTS

Reagents should be of analytical-reagent grade. Water must be distilled or de-ionised.
*Ethanol.* 95% *V*/*V*.
*Ethanolic potassium hydroxide,* 0·1 N.
*Sodium hydroxide,* N. Standard volumetric solution.
*Sulphuric acid,* N. Standard volumetric solution.
*Phenolphthalein indicator solution.* A 0·5% *m*/*V* solution of phenolphthalein in ethanol.

PROCEDURE

Boil 100 ml of ethanol in a 400-ml conical flask, add 0·5 ml of phenolphthalein indicator solution, allow to cool to 70° C, and neutralise at that temperature with 0·1 N ethanolic potassium hydroxide. Add 10 g of the sample, in thin shavings, and dissolve as rapidly and completely as possible by heating. Immediately add 3·0 ml of N sulphuric acid, and boil under reflux on a water-bath for at least 10 minutes to ensure complete removal of carbon dioxide.

If the solution is colourless, cool to 70° C, and back-titrate with N sodium hydroxide until the pink colour reappears. The excess of sulphuric acid titrated should be not less than 1 ml.

If during the boiling with acid the pink colour reappears, or if there is less than 1 ml of N sulphuric acid in excess, add a further quantity of N sulphuric acid, boil under reflux for at least 10 minutes, and back-titrate with N sodium hydroxide until the pink colour reappears.

From the amount of N sulphuric acid required by the sample, calculate the percentage of total free alkali in terms of $Na_2O$ (or of $K_2O$ in the case of a potassium soap).

### Determination of Free Caustic Alkali

INTRODUCTION

Whereas the solubility of sodium carbonate in ethanol (95% *V*/*V*) is slight, that of potassium carbonate is sufficient to introduce serious errors if the soap is dissolved in ethanol and the free caustic alkali determined by direct titration with acid.

PRINCIPLE OF METHOD

The soap is dissolved in ethanol and the soluble alkalinity is determined by direct titration with standard acid and phenolphthalein. Alternatively, the titration is carried out after the removal of interfering alkalinity due to carbonate by the addition of barium chloride.

APPLICABILITY

The method, in the absence of barium chloride during the titration, is applicable only to sodium soaps and in the presence of not more than about 0·4% of total free alkali, calculated as $Na_2O$. The alternative titration in the presence of barium chloride is applicable to soap containany potash or more than 0·4% of total free alkali.

REAGENTS

Reagents should be of analytical-reagent grade. Water must be distilled or de-ionised.
*Ethanol.* 95% *V*/*V*.
*Sodium hydroxide,* 0·1 N.

F

### 2.1.5. Association of Official Analytical Chemists

The Association of Official Analytical Chemists, based in Arlington, Virginia, USA, comprises federal, state and local government, and industrial analytical scientists who develop, test, and sponsor improved methods of analysing fertilisers, foods, feeds, pesticides, drugs, cosmetics, and other products related to agriculture and public health. It provides regulatory scientists with methods of analysis required for enforcing laws and regulations. Its standards compendium is:

> *Official methods of analysis of the Association of Official Analytical Chemists.* Sidney Williams, ed. 14th ed. Arlington. AOAC, 1984, 1141pp.

This contains over 1,700 methods in a step-by-step format, specifying reagents and apparatus to be used. It is published every five years and is updated by annual supplements. It is arranged by subjects in 52 chapters (see the list of contents in the copies with SAQ 2.1d) and has an index.

**SAQ 2.1d**

This is a nice easy work to use, so this question shouldn't detain us long. Quite simply, find:

(*a*) a method for determining starch in peanut butter;

(*b*) a method for determining salt in nuts.

Try and have a look at the book if you can: this question becomes more-or-less a demonstration if you work from the copies. But the copies are there to be used if you cannot get hold of the book itself. And in either case I shall be able to illustrate the points I want to make in my response.

# OFFICIAL METHODS OF ANALYSIS

OF THE

# ASSOCIATION OF OFFICIAL ANALYTICAL CHEMISTS

EDITED BY SIDNEY WILLIAMS

---

FOURTEENTH EDITION, 1984

---

PUBLISHED BY THE
ASSOCIATION OF OFFICIAL ANALYTICAL CHEMISTS, INC.
1111 NORTH NINETEENTH STREET
SUITE 210
ARLINGTON, VIRGINIA 22209 USA

# Contents

# 27. Nuts and Nut Products

### Volume of Packaged Nuts
### Final Action

**27.001** ***Apparatus***

(**a**) *Graduated cylinders.*—*(1) For less than 500 mL nuts.*—Use 500 mL cylinder with ca 4.7 cm id. *(2) For 500 mL or greater volume of nuts.*—Use 1 L cylinder with ca 5.7 cm id.
(**b**) *Balance.*—Accurate to 0.01 oz or 0.25 g.
(**c**) *Funnel.*—With min. opening of 3.8 cm diam. (suitable funnel may be shaped from smooth cardboard).

**27.002** ***Determination***

Open container and pour nuts loosely into vertical graduate (do not tilt) fitted with funnel. Without shaking, est. location of horizontal plane representing av. ht of product, read vol. nuts, and record max. vol.

Raise graduate 5 cm and drop vertically onto level, firm, resilient surface (do not tamp). Repeat total of 5 times and observe vol. as above. Repeat in successive 5-drop increments until nuts have so settled that vol. decreases ≤2% in last 5-drop increment. Read vol. as above and record as "min. vol." Arithmetical av. of "max." and "min." vol. = mean vol.

Ref.: JAOAC **54,** 584(1971).

### Preservation of Nut Sample
### Procedure

**27.003** ***Storage***

Store sample in air-tight container at 5–10°. Store meats in glass containers only.

### Nuts and Nut Products
### Preparation of Sample
### Procedure

**27.004** ***Preparation***

(**a**) *Nuts in shell.*—Remove meats from shells, and sep. all shell particles from meats. Skin or spermoderm should be included with meat in all nuts, including peanuts and coconuts unless specifically excluded by description. Prep. sepd meats as in (**b**).
(**b**) *Nut meats, shredded coconut, or small pieces.*—Grind ≥250 g twice thru Enterprise No. 5 food chopper, equipped with revolving knife blade and plate with holes ca 3 mm diam. (Other types of food choppers, graters, or comminuting devices that give smooth homogeneous paste without loss of oil may be used.) Mix sample well and store in air-tight glass container.
(**c**) *Nut butters and pastes.*—Transfer sample to container of convenient size and shape, warming semi-solid products, and mix carefully with stiff-blade spatula or knife. (Elec. mixers or stirrers may be used instead if sample is of consistency to give uniform mixt.) Store sample in air-tight glass container.

### Moisture in Nuts and Nut Products
### First Action

**27.005** ***Determination***

(Not applicable to high sugar products or products contg glycerol or propylene glycol)

Dry sample representing ca 2 g dry material to const wt (ca 5 hr) at 95–100° under pressure ≤100 mm Hg (13.3 kPa). Report loss in wt as moisture.

Refs.: JAOAC **8,** 295(1925); **31,** 521(1948); **32,** 527(1949); **33,** 753(1950); **34,** 357(1951); **37,** 845(1954).

### Fat (Crude) in Nuts and Nut Products
### Gravimetric Methods
### First Action

**27.006** ***Determination***

(*Caution: See* **51.011, 51.039,** and **51.054.**)

(**a**) *Direct method.*—If large amts of sol. carbohydrates interfere with complete extn of fat, ext with $H_2O$ before making detn. Ext ca 2 g sample with ether, dried as in **7.061,** 16 hr in Soxhlet-type extractor. Evap. ether, dry residue 30 min at 95–100°, cool in desiccator, and weigh; continue this alternate drying and weighing at 30 min intervals to const wt (1–1.5 hr is usually required).
(**b**) *Indirect method.*—Proceed as in **27.005**; then ext dried substance 16 hr as in (**a**), and dry as in (**a**). Report loss in wt as ether ext.

Refs.: JAOAC **31,** 521(1948); **33,** 753(1950); **34,** 357(1951); **37,** 845(1954).

### Protein (Crude) in Nuts and Nut Products
### Improved Kjeldahl Method
### First Action

**27.007** ***Determination***

Det. N as in **2.057,** and multiply result by 5.46 for peanuts and brazil nuts, 5.18 for almonds, and 5.30 for other tree nuts and coconut. (It may be desirable to defat with pet ether.)

Refs.: JAOAC **33,** 753(1950); **34,** 357(1951); **37,** 845(1954).

### Fiber (Crude) in Nuts and Nut Products
### First Action

**27.008** *See* **7.070.**

### Ash of Nuts and Nut Products
### Gravimetric Method
### First Action

**27.009** ***Determination***

*See* **31.012,** or **31.013** if added chlorides are present.

Refs.: JAOAC **33,** 753(1950); **34,** 357(1951).

### Sugars (Reducing) in Nuts and Nut Products
### Munson-Walker Method
### Final Action

**27.010** ***Determination***

Prep. sample as in 1st and 2nd par. of **7.089**. Proceed as in **31.038,** using 25 mL aliquot (representing 2 g sample). Express results as glucose or invert sugar.

Refs.: JAOAC **33,** 753(1950); **34,** 357(1951).

### Sucrose in Nuts and Nut Products
#### Final Action

**27.011** *See* **7.089.**

### Sodium Chloride in Nuts and Nut Products
#### Titrimetric Method
#### Final Action

**27.012** ***Determination***

To 2 g prepd sample, **27.004**, in Pt dish, add and thoroly incorporate 10 mL *10% Ca(OAc)₂ soln*. For nut butters and pastes, disperse sample in 10 mL acetone before adding $Ca(OAc)_2$, and remove acetone at room temp. with air current. Dry on steam bath, and ash in furnace at lowest visible red heat (550°). (Complete ashing is not necessary.)

Dissolve ash in 25 mL $HNO_3$ (1 + 3), add known vol. 0.1*N* $AgNO_3$, more than enough to ppt all Cl, heat to bp, cool, add 5 mL ferric indicator, **6.019(e)**, and titr. excess Ag with 0.1*N* $NH_4SCN$, **50.030(b)** until soln turns permanent light brown. Calc. Cl as NaCl, after correcting for any Cl in the $Ca(OAc)_2$ soln.

Refs.: JAOAC **33**, 753(1950); **34**, 357(1951).

CAS-7647-14-5 (sodium chloride)

### ★ Inorganic Residue (Water-Insoluble) in Nuts and Nut Products ★
#### First Action

**27.013** *See* **44.032**, 11th ed.

### Mycotoxins in Nuts and Nut Products

**27.014** *See* Chap. **26**.

★Surplus method—*see* inside front cover.

### Examination of Peanut Butter
#### Procedure

**27.015** ***Examination***

Make microscopic examination to detect addn of starch or any off-grade material not identifiable chem.

### Starch in Peanut Butter
#### First Action

**27.016** ***Determination***

Weigh 4–5 g sample by difference into 250 mL centrf. bottle and ext twice with 50 mL portions pet ether, shaking 5 min each time. Wash down sides of bottle with pet ether, centrf., and pour off solv., disregarding opalescence. Warm bottle to drive off remaining solv., transfer residue to mortar, and grind. Return fine powder to bottle with aid of 100 mL 10% NaCl soln. Shake bottle 15 min, wash down sides with NaCl soln, centrf. well, and pour off supernate, disregarding opalescence. Repeat washing twice.

Ext in same manner, once with 70% alcohol and once with $H_2O$, shaking 1–2 min each time. Drain bottle several min., chill, and from pipet add 100 mL HCl (20.5–21.0 g HCl/100 mL) at temp. ≤15°. Shake vigorously 3 min, centrf. well, and pour off soln thru cotton pledget in funnel stem. Cool soln to temp. at which the HCl was added, and pipet off 50 mL into nursing bottle contg 115 mL alcohol. Shake with whirling motion 1 min, let stand 2 min, centrf. 2 min, pour off thru weighed gooch contg thin asbestos pad, and add 50 mL 70% (v/v) alcohol to ppt. Stopper bottle, shake vigorously, wash down sides with the 70% alcohol, centrf. lightly, and pour off thru crucible. Repeat once with 70% alcohol and once with alcohol. Dry crucible and contents 1.5 hr at 130° in air, or 5 hr at 98–100° *in vacuo*. Cover crucible, place in desiccator contg efficient desiccant, and weigh when crucible reaches room temp.

Ref.: JAOAC **37**, 845(1954).

### ★ Glycerol in Shredded Coconut ★

**27.017** *See* **27.017**, 13th ed.

### SELECTED REFERENCE

JAOAC **18**, 418(1935).

### 2.1.6. British Pharmacopoeia

The final major work we shall look at in this Section covers drugs and pharmaceuticals. Its details are:

*British Pharmacopoeia.* London: HMSO, 1980. 2 vols.

It is updated by annual supplements and also has two useful supplementary works:

*Infra-red reference spectra.* 1980. Also updated by annual supplements.

*British Pharmacopoeia approved names.* 1981.

The main *Pharmacopoeia* contains about 1,000 'monographs' (self-contained entries dealing with a single substance or subject) on drugs, preparations, etc. It provides standards for the identity, purity, and potency of substances used in medicine and pharmacy, and data on action, use, solubility, storage, and labelling. Volume 2 contains the formulary (data on the preparation, identification, assay, labelling, etc, of creams, injections, syrups, tablets, etc), blood products, immunological products, radiopharmaceutical preparations, and surgical materials. There is a comprehensive index at the end of Volume 2.

For our final detailed look at a major collection of analytical standards now turn to:

**SAQ 2.1e**

Here you are working in a medical laboratory (I told you we would get around in this Unit didn't I). You are going to analyse some Isoniazid, using a sample taken from a batch prepared for injection. Where do you find a method for the analysis? Is there any difference in acidity between the compound itself and its injection solution? You expect you will have to carry out an infra-red analysis of Isoniazid before long. Is there a published spectrum and what method would you use to prepare the sample?

Once again, this question is going to be pretty easy if you have to rely on the copies, so do try and use the actual books if you can.

# British Pharmacopœia 1980

## Volume I

Published on the recommendation of the
Medicines Commission
pursuant to the Medicines Act 1968

Effective date: 1 December 1980

London Her Majesty's Stationery Office 1980

# Contents

## Contents of Volume I



## Contents of Volume II

## Prepared Ipecacuanha ☆

Prepared Ipecacuanha is Ipecacuanha reduced to a powder not more coarse than a *fine powder*, adjusted if necessary by the addition of powdered Lactose or an ipecacuanha powder of a lower alkaloidal content, to contain 1.90 to 2.10 per cent of total alkaloids, calculated as emetine and with reference to the material dried at 100° to 105°.

Prepared Ipecacuanha complies with the requirements for Identification, Acid-insoluble ash and Sulphated ash stated under Ipecacuanha, and with the following requirements.

**Description** A light grey to yellowish-brown powder; odour, slight.

*Microscopical* Exhibits the characters described under Powdered Ipecacuanha and, when mounted in *glycerol*, crystals of lactose may be seen.

**Assay** Carry out the Assay described under Ipecacuanha. Each ml of 0.1M *hydrochloric acid VS* is equivalent to 0.02403 g of total alkaloids, calculated as emetine.

**Storage** Prepared Ipecacuanha should be kept in a well-closed container, protected from light.

**Action and Use** Expectorant and emetic.

**Usual Dose Range** 25 to 100 mg.

Prepared Ipecacuanha contains in 100 mg 2 mg of the total alkaloids of ipecacuanha, calculated as emetine.

## Isoniazid ☆

(pyridine ring with N; CO·NH·NH$_2$ at position 4)

$C_6H_7N_3O$ 137.1 *54-85-3*

Isoniazid is isonicotinohydrazide. It contains not less than 99.0 per cent of $C_6H_7N_3O$, calculated with reference to the dried substance.

**Description** Colourless crystals or a white, crystalline powder; odourless.

**Solubility** Soluble in 8 parts of *water*, in 45 parts of *ethanol (96 per cent)* and in 1000 parts of *chloroform*; very slightly soluble in *ether*.

**Identification** A. Dissolve 50 mg in 1 ml of *water* with the aid of heat, cool and add 4 ml of *copper reagent*; a greenish precipitate is produced. On warming, gas is evolved and the precipitate becomes reddish-brown.

B. Heat 50 mg with 1 g of *anhydrous sodium carbonate*; pyridine is evolved.

C. *Melting point*, 170° to 174°, Appendix V A, Method I.

D. Dissolve 0.1 g in 2 ml of *water*, add a warm solution of 0.1 g of *vanillin* in 10 ml of *water*, allow to stand, and scratch the inside of the test tube with a glass rod; a yellow precipitate is obtained. *Melting point* of the precipitate after recrystallisation from 5 ml of *ethanol (70 per cent)*, about 227°, Appendix V A, Method I.

**Acidity or alkalinity** pH of a 5 per cent w/v solution, 6.0 to 8.0, Appendix V L.

**Clarity and colour of solution** A 5.0 per cent w/v solution is *clear*, Appendix IV A, Method II A, and *colourless*, Appendix IV B, Method I.

**Heavy metals** A 5.0 per cent w/v solution complies with *limit test A for heavy metals*, Appendix VII (20 ppm). Use 2 ml of *water* in place of the 2 ml of *acetate buffer pH 3.5* and prepare the standard using 20 ml of *lead standard solution (1 ppm Pb)*.

**Hydrazine** Dissolve 2.0 mg in 2 ml of *water*, add 4 ml of *water* and 4 ml of a freshly prepared 0.2 per cent w/v solution of *4-dimethylaminobenzaldehyde* in a mixture of 6 volumes of *hydrochloric acid* and 4 volumes of *water* and allow to stand for three minutes. Measure the *absorbance* of the resulting solution at 450 nm, Appendix II B, using as the blank a mixture of 6 volumes of *water* and 4 volumes of the dimethylaminobenzaldehyde solution; the *absorbance* is not greater than 0.05.

**Loss on drying** When dried to constant weight at 100° to 105°, loses not more than 0.5 per cent of its weight; use 0.5 g.

**Sulphated ash** Not more than 0.1 per cent, Appendix IX A, Method II; use 1 g.

**Assay** Dissolve 0.25 g in sufficient *water* to produce 100 ml. To 20 ml add 100 ml of *water*, 20 ml of *hydrochloric acid*, 0.2 g of *potassium bromide* and titrate slowly with 0.0167M *potassium bromate VS* using 0.1 ml of *ethoxychrysoidine hydrochloride solution* as indicator. Each ml of 0.0167M *potassium bromate VS* is equivalent to 0.003429 g of $C_6H_7N_3O$.

**Storage** Isoniazid should be kept in a well-closed container, protected from light.

**Preparations** Isoniazid Injection
Isoniazid Tablets

**Action and Use** Used in the treatment of tuberculosis.

**Usual Dose Range** 300 to 600 mg daily, in divided doses.
By intramuscular injection, 100 to 300 mg daily, in divided doses; in cases of tubercular meningitis, up to 1 g twice weekly.

## Isoprenaline Hydrochloride

(HO, HO-substituted benzene ring)–CH(OH)·CH$_2$·$\overset{+}{N}$H$_2$Pr$^i$ Cl$^-$

$C_{11}H_{17}NO_3,HCl$ 247.7 *6700-39-6*

Isoprenaline Hydrochloride is $\beta$,3,4-trihydroxy-*N*-isopropylphenethylammonium chloride. It contains not less than 98.0 per cent and not more than the equivalent of 101.5 per cent of $C_{11}H_{17}NO_3,HCl$, calculated with reference to the dried substance.

**Description** A white or almost white, crystalline powder; odourless or almost odourless. Gradually darkens on exposure to air and light; aqueous solutions become pink to brownish-pink on standing exposed to air and almost immediately when made alkaline.

**Solubility** Soluble in less than 1 part of *water* and in 55 parts of *ethanol (96 per cent)*; insoluble in *chloroform* and in *ether*.

**Identification** A. The *infra-red absorption spectrum*, Ap-internal diameter packed with fine particles of silica 5 µm in diameter, the surface of which has been modified by chemically bonded octadecasilyl groups (Ultrasphere ODS is suitable) and maintained at 45° (b) *acetonitrile-phosphate solution* as eluant with a flow rate of 1 ml per minute and (c) a detection wavelength of 210 nm.

In the chromatogram obtained with solution (2), the two largest peaks are due to beef insulin (eluting first) and pork insulin respectively. For beef insulin, in the chromatogram obtained with solution (1), the area of any peak corresponding in position to the peak due to pork insulin in the chromatogram obtained with solution (2) is not greater than 1 per cent of the area of the principal peak. For pork insulin, in the chromatogram obtained with solution (1), the area of any peak corresponding in position to the peak due to beef insulin in the chromatogram obtained with solution (2) is not greater than 1 per cent of the area of the principal peak. The result obtained is valid only if the chromatogram obtained with solution (2) shows a third peak between the two largest peaks.

**Related proteins and peptides** Carry out Method II for *polyacrylamide gel electrophoresis*, Appendix III F. Prepare the following solutions: Solutions (1) to (5) are solutions of *insulin EPBRP* in 100 µl of *sample application buffer* containing 0.5, 1, 3, 5 and 100 µg respectively. Solutions (6) and (7) are solutions of the substance being examined in 100 µl of *sample application buffer* containing 100 and 500 µg respectively. Apply each solution to a tube of the gel. After electrophoresis and staining, evaluate the gels on a cold light illuminator. In the gel prepared with solution (6) the intensity of any band corresponding in position to that of the faster of the two bands behind the principal band in the gel prepared with solution (5) does not exceed that of the principal band in the gel prepared with solution (3). In the gel prepared with solution (7), the intensity of any band corresponding in position to the slower of the two bands behind the principal band in the gel prepared with solution (5) does not exceed that of the principal band in the gel prepared with solution (1). The test is invalid if no band can be detected in the gel prepared with solution (1) or if no gradation is observed in the intensity of staining of the gels prepared with solutions (1) to (4).

**Zinc** 0.3 to 0.6 per cent of Zn, calculated with reference to the dried substance, when determined by the following method. Prepare a solution in 0.01M *hydrochloric acid* to contain 1 mg per ml and determine by *atomic absorption spectrophotometry*, Appendix II D, measuring at 214 nm and using *zinc solution ASp* diluted if necessary with *water* for the standard solutions.

**Nitrogen** 15.0 to 16.5 per cent, calculated with reference to the dried substance, when determined by Method V, Appendix VIII H; use 1 to 3 mg.

**Loss on drying** When dried to constant weight at 105°, loses not more than 10.0 per cent of its weight; use 0.2 g.

**Sulphated ash** Not more than 2.0 per cent, Appendix IX A.

**Assay** Carry out the *biological assay of insulin*, Appendix XIV C 10. The precision of the assay is such that the fiducial limits of error are not less than 90 per cent and not more than 111 per cent of the estimated potency. The upper fiducial limit of error is not less than 26 Units per mg, calculated with reference to the dried substance.

**Storage** Insulin should be kept in a well-closed container and stored at a temperature below 8°.

**Labelling** The label on the container states (1) the number of Units contained in it; (2) the number of Units per mg; (3) the animal source of the insulin; (4) the conditions under which it should be stored; (5) the date after which it is not intended to be used.

**Usual Dose** The dose is determined by the physician in accordance with the needs and response of the patient.

*6/84*

It is intended that Human Insulin, which is clearly excluded from the present definition, will be the subject of a separate monograph in a future publication.

## Isoniazid☆

Change the monograph to:

## Isoniazid☆

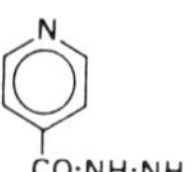


$C_6H_7N_3O$ 137.1 *54-85-3*

Isoniazid is isonicotinohydrazide. It contains not less than 99.0 per cent and not more than the equivalent of 101.0 per cent of $C_6H_7N_3O$, calculated with reference to the dried substance.

**Description** Colourless crystals or a white, crystalline powder; odourless.

**Solubility** Soluble in 8 parts of *water*; sparingly soluble in *ethanol (96 per cent)*; slightly soluble in *chloroform*; very slightly soluble in *ether*.

**Identification** *Test A may be omitted if tests B and C are carried out. Test C may be omitted if tests A and B are carried out.*

A. The *infra-red absorption spectrum*, Appendix II A, exhibits maxima which correspond in position and relative intensity to those in the spectrum obtained with *isoniazid EPCRS*.

B. *Melting point*, 170° to 174°, Appendix V A, Method I.

C. Dissolve 0.1 g in 2 ml of *water*, add a warm solution of 0.1 g of *vanillin* in 10 ml of *water*, allow to stand, and scratch the inside of the test-tube with a glass rod; a yellow precipitate is obtained. *Melting point* of the precipitate, after recrystallisation from 5 ml of *ethanol (70 per cent)* and drying at 100° to 105°, 226° to 231°, Appendix V A, Method I.

**Acidity or alkalinity** pH of a 5.0 per cent w/v solution, 6.0 to 8.0, Appendix V L.

**Clarity and colour of solution** A 5.0 per cent w/v solution is *clear*, Appendix IV A, and not more intensely coloured than *reference solution BY₇*, Appendix IV B, Method II.

**Heavy metals** 2.0 g complies with *limit test C for heavy metals*, Appendix VII (10 ppm). Use 2 ml of *lead standard solution (10 ppm Pb)* for the preparation of the standard.

**Hydrazine and related substances** Carry out the method for *thin-layer chromatography*, Appendix III A, using *silica gel GF254* as the coating substance and a mixture of 50 volumes of *ethyl acetate*, 20 volumes of *acetone*, 20 volumes of *methanol* and 10 volumes of *water* as the mobile phase. Apply separately to the chromatoplate 5 µl of each of two solutions prepared in the following manner. For solution (1) dissolve 1.0 g of the substance being examined in a mixture of equal volumes of *acetone* and *water* and dilute to 10 ml

# British Pharmacopœia 1980

**Volume II**

Published on the recommendation of the
Medicines Commission
pursuant to the Medicines Act 1968

Effective date: 1 December 1980

London Her Majesty's Stationery Office 1980

# Formulary

**Viscosity** At 20°, 7.0 to 11.0 $mm^2 s^{-1}$ (7.0 to 11.0 centistokes), Appendix V H, Method I, using a size C U-tube viscometer.

**Weight per ml** 1.17 to 1.19 g, Appendix V G.

**Arsenic; Copper; Lead; Zinc** Complies with the tests described under Iron Dextran Injection.

**Iron absorption** Complies with the *test for iron absorption*, Appendix XIV J.

**Undue toxicity** Inject 0.02 ml mixed with 0.2 ml of *sodium chloride injection* into a tail vein of each of ten mice; not more than three mice die within five days of injection. If more than three mice die within five days, repeat the test on another group of twenty mice; not more than ten of the thirty mice used in the combined test die within five days of injection.

**Pyrogens** Complies with the *test for pyrogens*, Appendix XIV K, using not less than 0.1 ml per kg of the rabbit's weight.

**Assay** Carry out the Assay described under Iron Dextran Injection.

**Storage** Iron Sorbitol Injection should be stored at a temperature between 15° and 30°; it should not be stored at a low temperature or be allowed to freeze.

**Labelling** The strength is stated as the equivalent amount of iron, Fe, in a suitable dose-volume.

**Action and Use** Used in the treatment of iron-deficiency anæmias.

**Usual Dose Range** By deep intramuscular injection, 1 to 2 ml daily.

## Isoniazid Injection

Isoniazid Inj.

Isoniazid Injection is a sterile solution of Isoniazid in Water for Injections. The acidity of the solution is adjusted to pH 5.6 to 6.0 by the addition of 0.1M hydrochloric acid. The solution is sterilised by *Heating in an Autoclave*.

**Content of isoniazid, $C_6H_7N_3O$** 95.0 to 105.0 per cent of the prescribed or stated amount.

**Identification** A. Evaporate to dryness using a rotary evaporator, a volume of the injection equivalent to 50 mg of Isoniazid, extract the residue with two quantities, each of 10 ml, of *ethanol (96 per cent)* and evaporate the combined ethanol extracts to dryness. The *infra-red absorption spectrum* of the residue, Appendix II A, is concordant with the *reference spectrum* of isoniazid.

B. To a volume equivalent to 25 mg of Isoniazid add 5 ml of *ethanol (96 per cent)*, 0.1 g of *sodium tetraborate* and 5 ml of a 5.0 per cent w/v solution of *1-chloro-2,4-dinitrobenzene* in *ethanol (96 per cent)*, evaporate to dryness on a water-bath, heat for a further ten minutes and dissolve the residue in 10 ml of *methanol*; a reddish-purple colour is produced.

**Acidity** pH, 5.6 to 6.0, Appendix V L.

**Assay** Dilute a volume equivalent to 0.4 g of Isoniazid to 250 ml with *water*; to 25 ml of the solution in a glass-stoppered flask, add 25 ml of 0.05M *bromine VS* and 5 ml of *hydrochloric acid*, shake for one minute, allow to stand for fifteen minutes, add 1 g of *potassium iodide* and titrate with 0.1M *sodium thiosulphate VS* using *starch mucilage* as indicator. Repeat the operation without the injection being examined; the difference between the titrations represents the amount of bromine required. Each ml of 0.05M *bromine VS* is equivalent to 0.003429 g of $C_6H_7N_3O$.

**Usual Dose Range** Isoniazid, by intramuscular injection, 100 to 300 mg daily, in divided doses; in cases of tuberculous meningitis, up to 1 g twice weekly.

**Strength available** An Injection containing 50 mg in 2 ml is usually available.

## Isoprenaline Hydrochloride Injection

Isoprenaline Hydrochlor. Inj.

Isoprenaline Hydrochloride Injection is a sterile solution of Isoprenaline Hydrochloride in Water for Injections containing suitable stabilisers. The solution is sterilised by *Filtration* and distributed aseptically in sterile ampoules, the air in which is replaced by nitrogen or other suitable gas.

**Content of isoprenaline hydrochloride, $C_{11}H_{17}NO_3$, HCl** 95.0 to 105.0 per cent of the prescribed or stated amount.

**Description** A colourless or very pale yellow solution.

**Identification** A. Carry out the method for *thin-layer chromatography*, Appendix III A, using *silica gel G* as the coating substance and a mixture of 4 volumes of 13.5M *ammonia*, 16 volumes of *water*, 30 volumes of *propan-2-ol* and 50 volumes of *ethyl acetate* as the mobile phase. Apply separately to the chromatoplate 2 μl of each of the following solutions. (1) The injection being examined diluted if necessary with sufficient 80 per cent aqueous *methanol* to produce a solution containing 0.1 per cent w/v of Isoprenaline Hydrochloride; (2) a 0.1 per cent w/v solution of *isoprenaline hydrochloride BPCRS* in 80 per cent aqueous *methanol*. After removal of the chromatoplate allow it to dry in air until the odour of the solvent is no longer detectable, place it for a few minutes in an atmosphere saturated with *diethylamine* and spray with *diazotised nitroaniline solution*. The chromatogram obtained with solution (1) exhibits an elongated zone corresponding to that in the chromatogram obtained with solution (2).

B. To 2 ml add 0.1 ml of *iron(III) chloride test-solution*; an emerald green colour develops which, on gradual addition of *sodium hydrogen carbonate solution* changes first to blue and then to red.

**Acidity** pH, 2.5 to 3.0, Appendix V L.

**Assay** To a volume equivalent to 5 mg of Isoprenaline Hydrochloride add sufficient *water* to produce 50 ml. To 20 ml add 0.5 ml of *iron(II) sulphate-citrate solution* and 2 ml of *glycine buffer*, mix and allow to stand for twenty minutes. Add sufficient *water* to produce 25 ml, mix, and measure the *absorbance* of the resulting solution at 540 nm, Appendix II B. Calculate the content of $C_{11}H_{17}NO_3$,HCl from the *absorbance* obtained by repeating the determination using 5 ml of a 0.1 per cent w/v solution of *isoprenaline hydrochloride BPCRS* instead of the injection.

**Storage** Isoprenaline Hydrochloride Injection should be stored in a cool place, protected from light.

**Usual Dose** By intravenous injection or infusion, the dose and rate being determined by the response.

**Strengths available** Injections containing 0.2 mg in 1 ml, 1 mg in 5 ml, and 2 mg in 2 ml are usually available.

# Index

## J



## K



## L

# British Pharmacopœia 1980 Addendum 1983

Published on the recommendation of the
Medicines Commission
pursuant to the Medicines Act 1968

Effective date: 1 June 1984
*see Notice, page xxviii*

London Her Majesty's Stationery Office

## K



## L

# British Pharmacopœia 1980

# Infra-red Reference Spectra

Published on the recommendation of the
Medicines Commission
pursuant to the Medicines Act 1968

Effective date: 1 December 1980

London Her Majesty's Stationery Office 1980

# Contents

# Reference Spectra

**Isoniazid** ***S128***

Phase: Potassium bromide disc

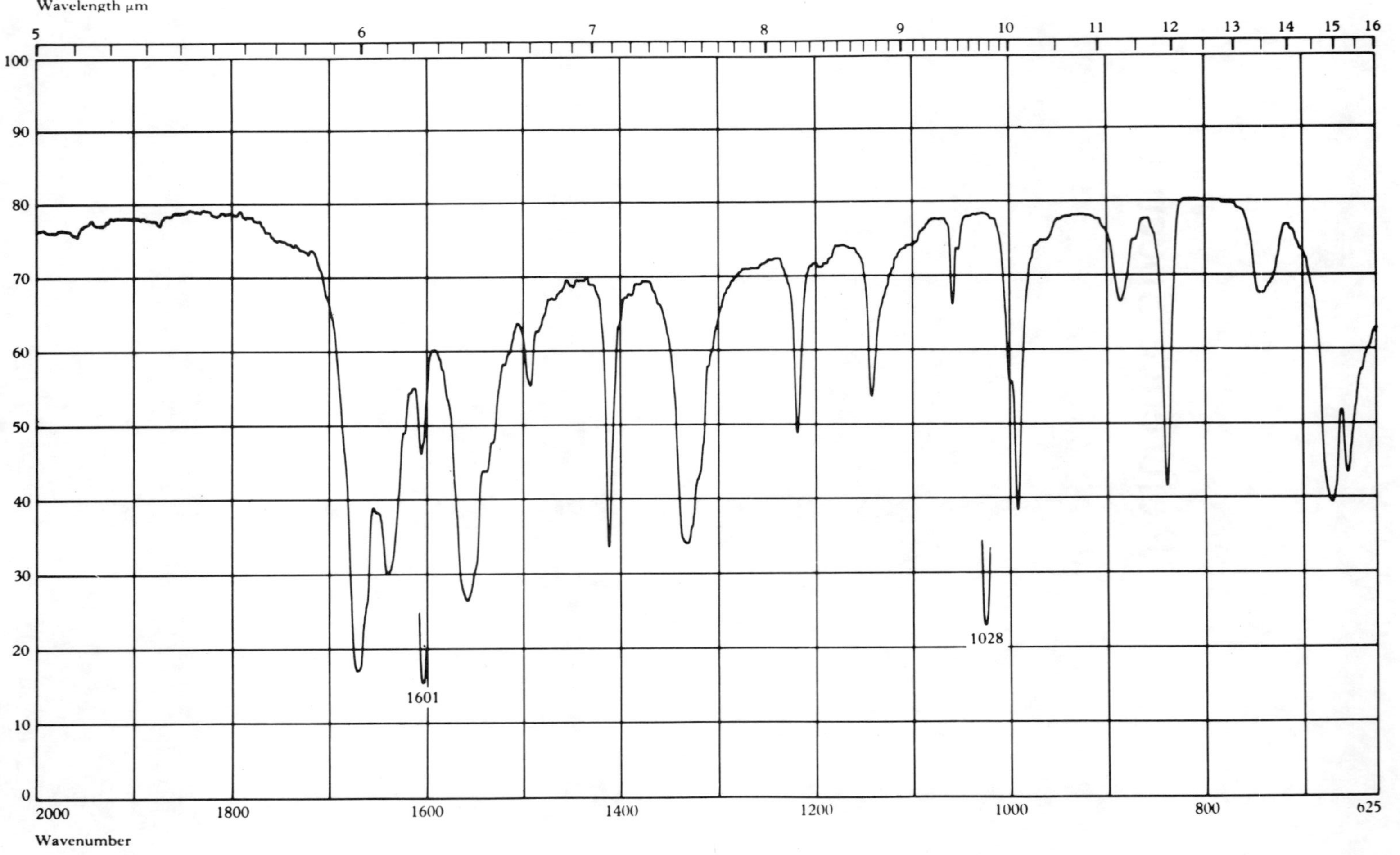

# Methods of Sample Preparation

precipitate with *water* and dry at 105°. Prepare a dispersion of the residue in *potassium bromide IR*.
[*Spectrum No. S127, Iothalamic Acid*]

### M185 Isoniazid Injection

Using a rotary evaporator, evaporate to dryness a volume of the injection equivalent to 50 mg of Isoniazid, extract the residue with two quantities, each of 10 ml, of *ethanol (96 per cent)*, filter the combined extracts and evaporate to dryness. Prepare a dispersion of the residue in *potassium bromide IR*.
[*Spectrum No. S128, Isoniazid*]

### M186 Isoniazid Tablets

Extract a quantity of the powdered tablets equivalent to 0.1 g of Isoniazid with 10 ml of *ethanol (96 per cent)* for fifteen minutes, centrifuge and decant the supernatant liquid. Repeat the procedure with two further quantities, each of 10 ml, of *ethanol (96 per cent)* and evaporate the combined extracts to dryness. Prepare a dispersion of the residue in *potassium bromide IR*.
[*Spectrum No. S128, Isoniazid*]

### M187 Isoprenaline Hydrochloride

Prepare a dispersion in *potassium chloride IR*.
[*Spectrum No. S129, Isoprenaline Hydrochloride*]

### M188 Isoxsuprine Hydrochloride

Prepare a dispersion in *potassium chloride IR*.
[*Spectrum No. S130, Isoxsuprine Hydrochloride*]

If necessary, dissolve 50 mg in 2 ml of *methanol*, add 15 ml of *dichloromethane* and evaporate to dryness. Prepare a dispersion of the residue in *potassium chloride IR*.

### M189 Isoxsuprine Hydrochloride Injection

To a volume of the injection equivalent to 50 mg of Isoxsuprine Hydrochloride add 20 ml of *water* and 10 ml of *ammonia buffer pH 10.0* and extract with three successive quantities, each of 15 ml, of *dichloromethane*. Shake the combined extracts with 5 g of *anhydrous sodium sulphate*, filter and evaporate to dryness. Dissolve the residue in 5 ml of 0.1M *methanolic hydrochloric acid* and evaporate to dryness. Dissolve the residue in 5 ml of *methanol*, evaporate to dryness, redissolve the residue in 5 ml of *methanol*, add 15 ml of *dichloromethane*, evaporate to dryness and dry the residue *in vacuo* at 60° for one hour. Prepare a dispersion of the residue in *potassium chloride IR*.
[*Spectrum No. S130, Isoxsuprine Hydrochloride*]

### M190 Isoxsuprine Hydrochloride Tablets

To a quantity of the powdered tablets equivalent to 50 mg of Isoxsuprine Hydrochloride add 50 ml of 0.1M *hydrochloric acid* and heat on a water-bath for thirty minutes. Cool, filter, and to the filtrate add 10 ml of *ammonia buffer pH 10.0*. Extract with three quantities, each of 15 ml, of *dichloromethane*, and complete the method described under Isoxsuprine Hydrochloride Injection beginning at the words 'Shake the combined extracts . . .'.
[*Spectrum No. S130, Isoxsuprine Hydrochloride*]

### M191 Kanamycin Acid Sulphate

Prepare a dispersion in *potassium bromide IR*.
[*Spectrum No. S131, Kanamycin Acid Sulphate*]

### M192 Kanamycin Sulphate

Prepare a dispersion in *potassium bromide IR*.
[*Spectrum No. S132, Kanamycin Sulphate*]

### M193 Levamisole Hydrochloride

Dissolve 0.5 g in 20 ml of *water*, add 6 ml of M *sodium hydroxide* and extract with 20 ml of *dichloromethane*. Wash the extract with 10 ml of *water*, shake with *anhydrous sodium sulphate*, filter, evaporate to dryness at room temperature, and dry the residue under reduced pressure at a temperature not exceeding 40°. Dissolve the residue in 10 ml of *carbon tetrachloride IR*.
[*Spectrum No. S133, Levamisole*]

### M194 Levodopa

Prepare a dispersion in *potassium bromide IR*.
[*Spectrum No. S134, Levodopa*]

### M195 Levodopa Capsules

Dissolve as completely as possible in 25 ml of M *hydrochloric acid* a quantity of the contents of the capsules equivalent to 0.5 g of Levodopa. Filter and adjust the acidity of the filtrate to pH 3 with 5M *ammonia*, added dropwise with stirring. Allow to stand protected from light for several hours, filter, wash the precipitate with *water* and dry at 105°. Prepare a dispersion of the residue in *potassium bromide IR*.
[*Spectrum No. S134, Levodopa*]

## H



## I



## K



## L



## M



## N

**SAQ 2.1e**

### 2.1.7. Specialised Standards

I now complete this Section with a quick look at a few specialised standards. I have already introduced you to the major standards you are likely to meet, but before we leave the subject you should be made aware that standards are available also on specific subjects from various specialised bodies. Of course you have already met some specialised standards—the *British Pharmacopoeia* and *Official methods of analysis* certainly fall within that definition. But there are other bodies issuing standards too. Here are four examples, again with brief descriptions.

> *Specifications and criteria for biochemical compounds*. 3rd edn. Washington: National Academy of Sciences, 1972. 216pp.

This work establishes criteria and specifications characterising the purity and quality of chemicals available for biochemical research. The criteria state what a compound or material is, or what it does, or both; they may describe any or all of: chemical properties, physical properties, or some kind of activity. Specifications are based on certain criteria and give acceptable ranges for the selected criteria, and procedures for determining that these requirements have been met. The book is arranged in chapters on: Aminoacids and related compounds, Carbohydrates and related compounds, Carotenoids and related compounds; Coenzymes and related compounds, Enzymes; Lipids and related compounds, Nucleotides and related compounds, Porphyrins and related compounds, Radioactive compounds. There is an index of compounds.

A very important compendium of standards for any analytical work in the petroleum industry is:

> *IP Standards for petroleum and its products, Part 1: methods for analysis and testing*. Chichester: Wiley for the Institute of Petroleum, annual. 2 vols.

This contains over 250 methods for analysing and testing petroleum products. It includes chemical and physical methods. Many of the methods are also published as British standards and some are also identical to ASTM methods. Similar to the compendia of analytical standards we have already looked at, this is an invaluable collection of all the methods relevant to a particular industry, in a single work.

Something of a curiosity, but showing that useful information can be found in apparently unlikely places is a bilingual English–Japanese publication, one of a series of volumes on steel standards:

> *Handbook of comparative world steel standards, vol. 6, steel: world standards mutual speedy finder*. Rev. ed. Tokyo: International Technical Information Institute. 1982. 499pp.

Here you will find comparisons of steel standards from the USA, UK, West Germany, France, USSR, and Japan; these comparisons are in a tabular layout showing chemical composition and physical properties of the different steels. This is another work arranged by chapters, covering: structural steel plate, steel plates for boilers and pressure vessels, steel pipes and tubes, and steels for special purposes. There are indexes of standard specification codes to each section.

Finally a most important work which is rather difficult to classify—is it a standard or a reference book? Either way, it is a work you should know about, and because of its importance and authority I include it here as a specialised standard.

> *Hazards in the chemical laboratory*. 3rd ed. L. Bretherick, ed. London: Royal Society of Chemistry, 1981. 567pp.

This indicates and discusses the dangers likely to arise in the chemical laboratory and gives practical advice on their avoidance. It has chapters covering: the Health and Safety at Work Act, 1974, safety planning and management, fire protection, reactive chemical hazards, chemical hazards and toxicology, health care and first aid, hazardous chemicals, and precautions against radiations. The core of the book is printed on yellow paper and comprises chapter 8, hazardous chemicals: this lists chemicals in alphabetical order, indicating their hazards, toxic effects, hazardous reactions, first-aid procedures, fire hazards and spillage-disposal methods.

There, that completes this Section on standards. Perhaps you are beginning to feel rather bemused by all the books and titles to which I am introducing you; and there are a lot more to come! But *don't panic*. You are *not* expected to remember details: you will eventually get to know certain books very well as you actually use them over the years. All I expect you to do here is to practise using different kinds of books and pick up practical hints, experience and, above all, *confidence*, as we go along. Then you will be much better placed to start finding answers in books to real-life questions as they arise.

**Objectives**

When you have worked through this part of the Unit you should be able to:

- understand the function and importance of standards;
- appreciate the value of published standards as sources of technical information;
- recognise the main published standards likely to be of importance to you;
- use these published standards and find the information you might need in them.

## 2.2. MAJOR REFERENCE WORKS

### 2.2.1. Introduction

I hope that if you did not already realise it, you are by now aware that books (and other publications) can be of great assistance to you, and in many different ways. Now we take a step further and look at some major reference works; but first a few general points.

Reference works, even within the field of chemistry, come in all manner of layouts and sizes and with different purposes. We shall now examine five reference works in some detail to demonstrate this point.

The purposes of reference books range from those of single-volume collections of numerical or formula data for quick answers to factual questions, to those of large multi-volume comprehensive and exhaustive (not to mention exhausting!) treatises and handbooks. The layout and arrangement of reference works will reflect their purposes: they might comprise collections of tables and formulae in some kind of logical order, or they might use A-to-Z encyclopaedia or dictionary layouts, with a fully detailed text, or short factual entries. You will meet examples of all these later in this Section.

If a work is arranged by its subjects then relevant information on a particular analytical technique or individual substance, for example, might be found in several, even many, different sections or volumes. To overcome this problem a good comprehensive index is essential; the same can apply to multi-volume (or even single-volume) A-to-Z encyclopaedias. Only with a dictionary-type layout with cross-references in the text can we afford to dispense with an index. At this stage I must make a most important point, exemplified in most of the SAQs in this section: *when looking for information in any reference book always consult the index* (where there is one!). By always consulting the index, you will get quickly to *all* the information you might need and will lessen the risk of missing something important: you will also save yourself a lot of time you might otherwise spend leafing hopefully through a book. Indexes may vary in quality as much as the books themselves: their detail and accuracy can be assessed only by trial and error—by using them.

### 2.2.2. Assessing A Reference Book

When you first come across a new book, or want to compare two or more apparently similar books, there are a few criteria you can use to help to assess them. Of course, the advice I give in this Section may also apply to books other than reference books.

Π You are looking at a reference book for the first time in a library or bookshop. Having been attracted by the title on the spine, you lift the book off the shelf for the first time. What do you look at first? Stop and think for a moment before reading my response.

The best place to start is the title page. There might be a sub-title telling you more exactly what the book is about. The author's name might be familiar to you or you might be told where he is employed (indicating whether he is a teacher or practising chemist). The publisher might be well-known to you as a reputable publisher of scientific reference works. There will be a date either on the title page or its reverse telling you how old the book is (but age need not necessarily be a decisive factor in determining the value of a book, within certain limits, which should normally be obvious in relation to the subject covered). And if a book has been published in several editions this suggests that it has established itself as a useful work in its earlier editions; although it may also mean that certain of the techniques described have become obsolete so necessitating the newer editions.

So you should already have learned something about the book before you ever read a word of its text. Indeed, the text is probably the last part of the book you turn to in your assessment; there are still some more useful sources of information about the book.

Π How do you find out what is in the book you are considering?

No, don't start to read it yet! Look at the list of contents: a well-arranged reference book will have a detailed list of contents; this can guide you to the sections which interest you most, or about which you know most, for a detailed assessment; it can also tell you what is in the book as a whole (and don't forget that index either). Also have a look at the introduction or preface: this should tell you the audience for which the author intends his book, and what his aims are in producing it. By now you should have a good idea of what the book is about and some idea of its likely usefulness to you. Now you can look at the layout and can quickly read (skim-read if you have that technique) a selected section or sections to give yourself an even clearer picture of what the book is like and whether it will really suit your needs. And remember what I said in the Introduction to this Unit about defining the information you think you need and what you need it for, to help you to select the most effective source of information.

### 2.2.3. Some Practical Examples

The remainder of this Section comprises a practical examination of a number of reference works. I have chosen five titles to examine in some detail, either because they are important works in themselves or because they are useful examples of types of reference book, or often for both reasons. In para 2.2.9 I introduce you to a few more general chemical reference works which should prove of use to you. My main problem, in fact, has been to reject a number of reference books, even though I am sure you would find them useful, in order to confine our examination to a few titles. I don't want to overwhelm you with titles or convert what is supposed to be a learning unit into a glorified reading list; but do always bear in mind that there are many more reference books relevant to analytical chemistry than I mention in this Section (and the next).

Where do you find these books? Use any library to which you have access (especially either a good works library, college, or university library, or large public reference library) and see what they have. There is a section of this Unit coming later to tell you how to find books in libraries and how best to use libraries. Use the criteria I mentioned above to assess any titles you find. If you have no, or only very rare, opportunities to visit suitable libraries you can at least use the photocopies accompanying this Section. I have tried to make the selections useful not just to answer the question, but also to exemplify the points I am making about reference books in general and to give you some idea of what the works I am telling you about are like.

If you can get regular access to a good chemical or technical library, so much the better. And if you do want a better idea of what reference books have been published, ask whether your library has a copy of the following book which will serve you as a good starting point:

> Walford, A. J. ed. *Walford's guide to reference material. Vol. 1, Science and technology*. 4th ed. London: Library Association, 1980. 697pp.

For the rest of this Section I am going to give you fairly brief introductions to the various reference books I have selected. *The important thing is for you to use them.* The SAQs and copies will start you off, but if you can get access to copies of any or all of the works themselves, don't limit yourself to my questions. Use the criteria and advice I give you here and in my responses to the SAQs and look up subjects which are of particular interest or concern to you. Now, as to specific titles. To make sure you can identify exactly what works I am talking about, I am going to give you a precise bibliographic reference to each title at the start of theparagraph where I introduce it; the only difference will be that in these references I quote the title first and the editors' (or authors') names next, since most reference books become known by their titles.

### 2.2.4. Comprehensive Analytical Chemistry (Wilson and Wilson)

We start at the top end of the scale with one of the major handbooks to analytical chemistry, a work still in progress 25 years after its inception. Its bibliographic details are as follows:

> *Comprehensive analytical chemistry*. C. L. Wilson and D.W. Wilson, eds; from Vol. 8 ed. by G. Svehla. Amsterdam: Elsevier, 1959 to date. 20v [in 29 vols. to date].

The editors aim to provide a work which should be a self-sufficient reference source; but where they feel this cannot be achieved they intend the work to be a starting point for any analytical investigation. They try to include the widest selection of analytical topics they can and to present their material in sufficient detail to allow it to be used directly. The users of the work are expected to include not only professional analytical chemists but also others whose use of analysis is incidental to their work. I have listed the main contents of the volumes issued up to the end of 1985 showing how the work is divided into a series of thematic volumes, including a few miscellaneous topics such as the history of analytical chemistry (Vol. X) or the application of computers in analytical chemistry (Vol. XVIII). It is intended to include contributions specialising in the analytical methods of certain key industries, and the first such contribution was Vol. XIII, *Analysis of complex hydrocarbon mixtures*. As you can see from the list of volumes, all types of analytical determinations—classical, instrumental, and physical—are covered, both in their theoretical and practical aspects. General theoretical aspects of analytical chemistry are also covered.

COMPREHENSIVE ANALYTICAL CHEMISTRY;
LIST OF VOLUMES
(Includes volumes scheduled for publication during 1985).

Vol. I Classical analysis.

Vol. II Electrical methods; physical separation methods.

Vol. III Electrical analysis with minute samples; standards and standardisation; separations with liquid amalgams; vacuum fusion analysis of gases in metals; electroanalysis in molten salts.

Vol. IV Instrumentation for spectroscopy; analytical atomic absorption and fluorescence spectroscopy; diffuse reflectance spectroscopy.

Vol. V Emission spectroscopy; analytical microwave spectroscopy; analytical applications of electron microscopy.

Vol. VI Analytical infrared spectroscopy.

Vol. VII Thermal methods in analytical chemistry; sub-stoichiometric analytical methods.

Vol. VIII Enzyme electrodes in analytical chemistry; molecular fluorescence spectroscopy; photometric titrations; analytical applications of interferometry.

Vol. IX Ultraviolet photo-electron and photo-ion spectroscopy; Auger electron-spectroscopy; plasma excitation in spectrochemical analysis.

Vol. X Organic spot test analysis; the history of analytical chemistry.

Vol. XI The application of mathematical statistics in analytical chemistry; mass spectrometry; ion-selective electrodes.

Vol. XII Thermal analysis.

Vol. XIII Analysis of complex hydrocarbon mixtures.

Vol. XIV Ion exchangers in analytical chemistry.

Vol. XV Methods of organic analysis.

Vol. XVI Chemical microscopy; thermomicroscopy of organic compounds.

Vol. XVII Gas and liquid analysers.

Vol. XVIII Kinetic methods in chemical analysis; application of computers in analytical chemistry.

Vol. XIX Analytical visible and ultraviolet spectrometry.

Vol. XX Photometric methods in inorganic trace analysis.

Now, try to use the work.

**SAQ 2.2a**

You are about to be interviewed for a job which will entail the analysis of natural gas. It will be prudent for you to know about the methods involved and so you go to a library and find a copy of Wilson and Wilson's *Comprehensive Analytical Chemistry*. (Don't look at the copies yet!)

First you must select the relevant volume. Note the volume number you would select as the most likely source of information on methods of analysis of natural gas, by using the list of volumes above. This is not so easy a task as it would be if you had access to a set of the volumes, but try it anyway—if you get really stuck the copies give the game away.

Next turn to the copies (or better still to the books themselves if you have access to a set) and find the relevant section. Note where you look and the terms under which you search.

Finally, note what is practically the only reliable method for the detailed analysis of natural gas.

Wilson and Wilson's

COMPREHENSIVE ANALYTICAL CHEMISTRY

Edited by

G. SVEHLA, PH.D., D.SC., F.R.S.C.

*Reader in analytical Chemistry*
*The Queen's University of Belfast*

VOLUME XIII

# ANALYSIS OF COMPLEX HYDROCARBON MIXTURES

Part A

SEPARATION METHODS

by

SLAVOJ HÁLA

MEČISLAV KURAŠ

MILAN POPL

*Institute of Chemical Technology, Prague, Czechoslovakia*

ELSEVIER SCIENTIFIC PUBLISHING COMPANY
AMSTERDAM OXFORD NEW YORK
1981

*VOLUME XIII*

ANALYSIS OF COMPLEX HYDROCARBON MIXTURES

*by*

Slavoj Hála, Mečislav Kuraš *and* Milan Popl
Institute of Chemical Technology
Prague, Czechoslovakia

*CONTENTS PART A*



*CONTENTS PART B*

# Part A
# Separation Methods



## Contents

# *Index*

A more detailed index for both parts is given at the end of part B

Chromathermography, 223

Wilson and Wilson's
COMPREHENSIVE ANALYTICAL CHEMISTRY

Edited by
G. SVEHLA, PH.D, D.SC., F.R.S.C.
*Reader in Analytical Chemistry*
*The Queen's University of Belfast*

VOLUME XIII

# ANALYSIS OF COMPLEX HYDROCARBON MIXTURES

Part B
GROUP ANALYSIS AND DETAILED ANALYSIS

by
SLAVOJ HÁLA
MEČISLAV KURAŠ
MILAN POPL

*Institute of Chemical Technology, Prague, Czechoslovakia*

ELSEVIER SCIENTIFIC PUBLISHING COMPANY
AMSTERDAM OXFORD NEW YORK
1981

# Part B
# Group analysis and detailed analysis

## Contents

*Chapter 5*

# Survey of analysis of industrial hydrocarbon mixtures

In this chapter a short survey of the methods of routine analysis of gaseous, liquid and solid hydrocarbon mixtures is given – the products of the processing of mineral oil and other fossile fuels. Mixtures of gaseous hydrocarbons consist only of a small number of individual substances which can be relatively easily identified and determined quantitatively by gas chromatography. Liquid hydrocarbon fractions are much more complex; their analysis is more demanding and becomes more complicated with increasing boiling points of the mixtures analysed. Similarly to gaseous fractions, in light liquid fractions (gasolines) too the majority of individual substances can be determined by capillary gas chromatography without previous separation of individual groups. In higher boiling fractions the analysis usually consists in the determination of groups (hydrocarbon and non-hydrocarbon) and often requires the use of a combination of several separation and identification methods. Examples of combined procedures using the integration and the coordination of various techniques are given in Section 2.10. Solid hydrocarbon mixtures usually consist of high-molecular substances and they often contain considerable amounts of non-hydrocarbon material. Their analysis is still more complex; in the case of asphalts and similar mixtures it is usually restricted to the determination of structural elements of hydrocarbon skeletons.

**5.1 Gaseous hydrocarbon mixtures**

Gaseous mixtures containing hydrocarbons can be roughly classified into two groups, according to their composition: mixtures containing hydrocarbons exclusively or predominantly (natural gas, technical and refinery hydrocarbon gases) and mixtures with a high content of non-hydrocarbon components (exhaust gases, city gas). In this section the standard methods are reviewed which are used for the analysis of three types of gaseous mixtures containing a high percentage of hydrocarbons: natural gas, technical and refinery gases and liquefied hydrocarbon gases.

5.1.1 METHODS OF GAS ANALYSIS

According to the character of the mixture under analysis the following main methods are used for the analysis of gases:

**Orsat's method,** based on the adsorption of the components in various solvents, is one of the first methods for the analysis of gases; it is still often used for the determination of the components of combustion products, simple mixtures or city gas. For the analysis of more complex gaseous hydrocarbon mixtures, for example natural gas, it is no longer sufficient owing to low accuracy and information content of the results. In Orsat's apparatus saturated hydrocarbons in mixtures with $C_4$ and lower hydrocarbons can be determined according to the ASTM Method D 1268[1] (silver mercuric nitrate method). After dilution with nitrogen it is also utilizable for mixtures containing hydrocarbons $C_5$.

**Gas chromatography** is the most perfect method for the analysis of gases and it is suitable for all types of gaseous mixtures. For a detailed analysis of natural gas it is practically the only reliable method (ASTM Method D 1945)[2]. The analysis of natural gas containing liquid hydrocarbons is described in the ASTM Method D 2597[3]. These mixtures cannot be easily injected into the chromatograph owing to the presence of highly volatile components and heavy components as well. The method is suitable for the determination of ethane, propane, *n*-butane, isopentane, *n*-pentane and higher hydrocarbons.

For the analysis of various gaseous hydrocarbons obtained from the refining processes or from natural sources containing only small amounts of $C_5$ olefins and $C_6$ and higher hydrocarbons the UOP Method 539[4] is the method of choice. Detection limit for individual hydrocarbons is 0.1 mol. %. 1-Butene is not separated from isobutanes, argon is determined together with oxygen. The UOP Method 709[5] serves similar purposes. Another standard method (UOP 373)[6] serves for quantitative determination of the composition of gaseous hydrocarbon mixtures containing hydrocarbons $C_2$—$C_5$. With the UOP Method 607[7] the content of aromatic hydrocarbons (benzene, toluene, $C_8$-alkylbenzenes) can be determined in hydrocarbon mixtures or in hydrogen. Similarly the ASTM Method D 2600[8] can be used for the determination of $C_6$—$C_8$ aromates in light saturated hydrocarbons in the concentration range 0.005 to 1.00 vol. % of each aromate. Beckham and Libers[9] analysed mixtures of $C_1$—$C_5$ hydrocarbons containing olefins, dienes, acetylenes and alkanes.

**Infrared analyzers** are also very widely used instruments for the analysis of gaseous mixtures, especially those containing a high percentage of non-hydrocarbon components. Non-dispersion IR analyzers operate without light distribution so that they utilize absorption in a broad region of the spectrum. Several types of commercial instruments for quantitative determination of CO, $CO_2$, $SO_2$, $C_2H_4$, $C_2H_6$ in various mixtures of gases, methane in natural gas, etc. operate on this principle. Dispersion analyzers use prisms, gratings or interference filters for the distribution of light. They are useful mainly in instances where the composition of the analysed gases changes considerably within broad limits, or where several components ahve to be determined. Some types of instruments based on this principle also permit the determination of $H_2S$, $NH_3$ and $Cl_2$ and chlorinated gases in addition to the above mentioned gases.

**Mass spectrometry** was already used for gas analysis in the forties[10]. In spite of certain advantages it is not such a common and universal analytical method as gas chromatography, especially owing to the high price of the instrument and demanding operation. The mass spectrometric method described in the ASTM Method D 2650[11] enables a quantitative analysis of gaseous mixtures containing combinations of the following components: hydrogen, hydrocarbons up to $C_6$, CO, $CO_2$, mercaptans $C_1$ and $C_2$ and air (nitrogen, oxygen, argon). For quantitative determination of individual components a system of linear equations is used, similarly as in liquid mixtures (see Section 3.1). The GC–MS combination is especially suitable from the analytical point of view.

Further two general methods of analysis of gaseous hydrocarbon

# Index

Pp. 1—382 are found in part A and are in italics, pp. 383—822 are found in part B of this book

### 2.2.5. Encyclopedia of Industrial Chemical Analysis

For our second work we stay at the top end of the scale, again beginning with the bibliographic details:

> *Encyclopedia of industrial chemical analysis*. Foster Dee Snell and Clifford L. Hilton, eds. New York: Wiley–Interscience, 1966–1975. 20v.

The aim of this encyclopaedia is to give a comprehensive coverage of the methods and techniques used in industrial laboratories throughout the world, for the analysis and evaluation of chemical products. It discusses all the methods of importance in development, control, and testing laboratories; thus it should help the analytical chemist to select from the wealth of available analytical and testing methods, the most suitable procedure for the solution of a problem. The first three volumes are devoted to methodologies and techniques common to the analysis of many industrial products; they are arranged alphabetically and have their own index at the end of Volume 3. The remaining volumes are devoted mainly to the analysis of specific materials. The index to this part is in Volume 20. The articles in this main section are again arranged alphabetically, but can be divided into five major groups:

Individual compounds selected for their commercial significance or for the complexity of the analytical chemistry involved (eg Acetone (Propanone) Elements and their compounds; Compounds of similar chemical structure eg Carbides and nitrides, phenols); Compounds with the same end-use (eg Adhesives, Plasticisers); Groups of compounds by industry (eg Food and food products, petroleum products).

Once again, look at the work yourself.

**SAQ 2.2b**

> You are working for a food company when, as a result of reorganisation following a merger, you are told that you are being transferred from solid-food analysis to work on the analysis of fizzy drinks. This is a subject you have never worked on before, so you need to find out about the techniques likely to be involved. You find a set of the *Encyclopedia of Industrial Chemical Analysis* in your works library and start looking for information.
>
> Use the copies I have provided, or better still use a set of the encyclopaedias if you can, and look up a useful article. Note where you look, what terms you would look up, and then look at the relevant article. ⟶

**SAQ 2.2b (cont.)**

After reading this article you may decide you want to refresh your memory about techniques of precipitation in gravimetric methods. Again, note where you look and what terms you look up.

# ENCYCLOPEDIA OF INDUSTRIAL CHEMICAL ANALYSIS

*Edited by* Foster Dee Snell *and* Clifford L. Hilton

**VOLUME 1** **General Techniques A–E**

***Interscience Publishers***
***a division of John Wiley & Sons, Inc.***
***New York • London • Sydney***

43. E. Kovats, *Helv. Chim. Acta.* **41,** 1915 (1958).
44. A. Wehrli and E. Kovats, *Helv. Chim. Acta* **42,** 2709 (1959).
45. R. Ryhage, *Anal. Chem.* **36,** 759 (1964).
46. J. T. Walsh and C. Merritt, Jr., *Anal. Chem.* **32,** 1378 (1960).
47. C. Merritt, Jr., and J. T. Walsh, *Anal. Chem.* **34,** 903 (1962).
48. *Ibid.*, 908 (1962). C. Merritt, Jr., J. T. Walsh, and D. H. Robertson, *J. Gas Chromatog.* **2,** 125 (1964).

ACKNOWLEDGMENT: The author is indebted to Dr. John T. Walsh for many helpful suggestions and criticisms offered in review of this manuscript.

C. MERRITT, JR.
*U.S. Army Natick Laboratories*
*Pioneering Research Division*

## GRAVIMETRIC METHODS OF ANALYSIS

Perhaps the most important of the classical methods of analysis are those which are based upon the *weight of a precipitate.* Such methods have come to be known as *gravimetric methods of analysis.* In general, a gravimetric method of analysis consists of the following steps: (a) the sample weight is determined, (b) a sequence of chemical reactions is carried out, (c) the weight of the product of the reactions is determined, and (d) the weight together with the knowledge of the identity of the product are used to calculate the quantitative composition. The key chemical reaction involved is almost always a precipitation, and the product weighed is almost always a precipitate, because a precipitation reaction can be a highly selective means for separating the component of interest from the matrix, and because a precipitate can be weighed readily.

For gravimetric analysis, the precipitation reaction and techniques are chosen to precipitate only the compound of interest, to precipitate it quantitatively, to precipitate particles large enough to be washed, filtered or centrifuged, and dried with little difficulty, and to precipitate a material which is, or which can be converted to, a compound of known composition. The purity of the precipitate depends upon the rates of nucleation and crystal growth, as well as on the equilibria involved. The completeness of the precipitation depends on the solubility equilibrium and also on a number of competing equilibria which, in general, act to increase the solubility of the precipitate. The particle size of the precipitate is governed by the nucleation process which occurs.

In addition to its application in gravimetric analysis, the potential selectivity of precipitation makes it a useful tool for quick qualitative analyses, for separations prior to analysis by physical methods, for masking in complexometric titrations, and for purification of reagents by recrystallization. See also COMPLEXATION IN AQUEOUS MEDIA; PRECIPITATION AND CRYSTALLIZATION; QUALITATIVE ANALYSIS, INORGANIC; QUALITATIVE ANALYSIS, ORGANIC.

For more details than are given here or in the articles mentioned above, the reader is referred to a number of texts and other references (1–10).

### SOLUBILITY

#### Solubility Equilibria

It is desirable that the solubility of a precipitate in the wash solution and mother liquor be negligibly small; thus, for operations on the usual laboratory scale, with balances sensitive to 0.1 mg, the amount of precipitate which dissolves in the mother liquor and wash solution should be less than 0.1 mg.

The solubility of a precipitate is governed by the usual equilibrium considerations. For a precipitate of formula $A_aB_b$, the solubility reaction is written as follows:

$$A_aB_b = aA^+ + bB^-$$

The equilibrium expression for this reaction is

$$K_{sp} = [A^+]^a\,[B^-]^b,$$

where $K_{sp}$, the equilibrium constant, is called the solubility product constant. Many of the precipitates commonly employed in analyses are sufficiently soluble in water that it is necessary to take precautions to avoid solubility losses. Consider the precipitation of 0.100 g of silver chloride from 100 ml of solution, followed by washing with 100 ml of wash solution. For silver chloride, $K_{sp} = 1.8 \times 10^{-10} = [Ag^+][Cl^-]$. Thus, in a saturated solution in water, $[Ag^+] = 1.4 \times 10^{-5}$ *M;* this corresponds to a solubility loss of 0.2 mg in the mother liquor and 0.2 mg in the wash solution, and represents a constant systematic error of four parts per thousand. The solubility loss on washing can be minimized by using a small volume of the wash solution; however, it is generally desirable to precipitate from a dilute solution so that the solubility loss in the mother liquor must be controlled by other methods.

The equilibrium expression indicates that silver chloride is less soluble in a solution containing excess chloride. If, in the above example, a slight excess of chloride is added, so that the chloride concentration at equilibrium is $1.4 \times 10^{-4}$, the solubility of silver chloride drops to $1.4 \times 10^{-6}$ *M*, and the solubility loss in the mother liquor is an acceptable 0.02 mg. This *common ion effect* is generally employed in precipitation methods, but must be used with caution. Because of the existence of competing equilibria, many precipitates become soluble in an excess of precipitant. Most organic precipitants are only moderately soluble in water, and the addition of a large excess of precipitant may yield a precipitate heavily contaminated with the precipitating reagent.

### Competing Equilibria

In most systems, several reactions may compete for the substance of interest. These competing reactions involve either hydrolysis or complexation and result in an increase in solubility. A quite complete compilation of equilibrium constants of these reactions, as well as of precipitation reactions, is available (1); Laitinen (6) and Sillen (9) present methods for calculating from equilibrium considerations the optimum pH and excess of common ion to minimize the solubility of the precipitate.

Competing equilibria may also be employed to minimize interferences, by complexing ions which might otherwise tend to precipitate with the substance of interest. Thus, calcium in a limestone may be precipitated as oxalate without interference from appreciable amounts of iron by adding sulfosalicylic acid to complex the iron (11).

Competing equilibria of the following types are commonly encountered:

1. Complexation with excess precipitant. This is well known in the case of amphoteric oxide precipitates, such as alumina. The phenomenon is not confined to such compounds, because almost everything tends to form soluble complexes with almost everything else (1). Thus, silver chloride, AgCl, is about one-third more soluble

in 0.5 *M* sodium chloride than in pure water, due to the formation of the species $AgCl_2^-$.

2. Hydrolysis of the cation. Most metals, including the alkaline earths, tend to form soluble hydroxy complexes; this results in increased solubility. Formation of these complexes can be minimized by precipitating at low pH; however, this effect is frequently unimportant because of the instability of many of these complexes.

3. Hydrolysis of the anion. Many common precipitants, such as $[OH]^-$, $S^{2-}$, $[C_2O_4]^{2-}$, and $[Cr_2O_4]^{2-}$, are the anions of weak acids. The solubilities of salts containing these anions are strongly pH dependent, increasing as the pH is lowered. This ability to change solubility by changing the pH permits such separations as magnesium from calcium by oxalate precipitation.

4. Complexation with foreign ligands. It is sometimes possible to prevent interferences by complexing the interfering ion with an auxiliary complexing agent. Thus, coprecipitation of lead with barium sulfate is markedly reduced by complexing the lead with diglycolate.

### Factors Affecting $K_{sp}$

The equilibrium constants are influenced by the size of the crystalline particles, crystal structure, temperature, nature of the solvent, and the presence of inert salts.

Because of the relatively greater effect of surface forces, small particles are more soluble than large ones. For example, freshly precipitated lead chromate is approximately 70% more soluble than an aged lead chromate precipitate consisting of much larger particles (12). This decrease in solubility is one of the factors necessitating aging of the precipitate in contact with the mother liquor.

Published values of $K_{sp}$ usually refer to the most stable form at the temperature of measurement. Frequently, precipitates first come down in a more soluble, metastable form and revert to the less soluble form on standing. Freshly precipitated nickel sulfide is readily soluble in acid, but on standing it reverts to an acid-insoluble form. This phenomenon of polymorphism is rather widespread, and is another factor which necessitates aging of the precipitate.

Solubility in general increases with increasing temperature. For most inorganic precipitates the effect is small, and precipitation and filtration can take place from hot solutions. For more soluble compounds the change in solubility with the temperature is greater; this makes possible the purification of compounds by *recrystallization.* A hot saturated solution of the compound is prepared and filtered to remove insoluble impurities. The solution is gradually cooled to such a temperature that the compound is no longer soluble; some precipitation then takes place. This first crop of crystals, which is enriched in the impurities which are less soluble than the compound, is discarded. By further cooling, a second, relatively pure, crop of crystals is obtained. No attempt is made to precipitate the reagent completely, since any additional crops of crystals would be enriched in impurities less soluble than the compound.

The solubility of substances depends, of course, on the nature of the solvent. Organic solvents are sometimes useful in carrying out separations which would not be feasible in aqueous systems. The use of ethyl acetate or *n*-butyl alcohol as a solvent permits the separation of sodium from potassium, as perchlorates; potassium perchlorate is very insoluble in these solvents. Organic solvents are also quite useful for separations involving the use of organic precipitants. In a few cases, mixed solvents are used to decrease the solubility of a precipitate. Lead and calcium sulfates

# ENCYCLOPEDIA OF INDUSTRIAL CHEMICAL ANALYSIS

*Edited by* Foster Dee Snell *and* Clifford L. Hilton

VOLUME 3

**General Techniques**
**P–Z**

**Index to Volumes 1–3**

***Interscience Publishers***
***a division of John Wiley & Sons, Inc.***
***New York · London · Sydney***

# INDEX TO VOLUMES 1–3

Entries are indexed by volume and page numbers; boldface numbers indicate volumes, and numbers that follow colons indicate pages.

Titles of articles are indicated by italic type; *Adsorption chromatography* **1**:78–98. Step-by-step procedures are indicated by an asterisk following the entry.

## A

## Q

5. C. Pierce, *J. Phys. Chem.* **63**, 1076 (1959).
6. *Military Spec. Mil C-17605 B (Ships)*, U. S. Govt. Printing Office, Washington, D.C., Sept. 10, 1964, 4.6.2.
7. *ASTM D 2355–65T, Evaluation of Activated Carbon*, American Society for Testing and Materials, Philadelphia, Pa., 1968.
8. *U.S. Pharmacopeia*, 17th rev. ed., U.S. Pharmacopeial Convention, Mack Printing Co., Easton, Pa., 1961, p. 109.
9. *Bulletin AWWA B600-53*, American Water Works Association, New York; made standard May 15, 1953.
10. J. W. Hassler, *Active Carbon—The Modern Purifier*, West Virginia Pulp & Paper Co., New York, 1941, pp. 90–91.
11. *Standard Methods for the Examination of Water and Waste Water*, 11th ed., American Public Health Association, New York, 1961, pp. 253–260.
12. U.S. Pat. 2,933,454 (April 19, 1960), A. J. Repik and C. T. Sloan, Jr. (to Pittsburgh Coke and Chemical Co.).
13. Ref. 11, p. 93.
14. J. P. Black and R. L. Hlozek, *Proc. Sugar Ind. Tech.* **24** (1), 17 (1965).
15. *Chemical Warfare Service Pamphlet No. 2, Sect. L*, U. S. Govt. Printing Office, Washington, D.C.
16. *Bone Char Project, Quarterly Report No. 39*, National Bureau of Standards, Washington, D.C., July–Sept. 1955, pp. 4–39.

G. H. Sheffler
*Atlas Chemical Industries, Inc.*

## CARBONATED BEVERAGES

Carbonated beverages, commonly known as soft drinks, consist mainly of water impregnated with carbon dioxide gas and may contain added color, flavor, sweetener, and acid. Carbonated beverages are a development of studies on the waters from the naturally occurring effervescent springs of Europe. Early studies of carbon dioxide impregnated waters are attributed to Paracelsus (1493–1541) who was concerned with the therapeutic value of the effervescent waters. The alchemist Thurneysser (1530–1596) is reported to have made the first attempt to synthesize the naturally carbonated waters. Many of the notable 16th to 18th century philosopher-scientists contributed to the knowledge of gas-saturated liquids. Among the most prominent were Libavius, Boyle, Black, Lavoisier, Bergman, Scheele, Cavendish, and Priestly (1). To Thomas Henry, an apothecary at Manchester, England, belongs the distinction of being the first entrepreneur offering artificial mineral waters for sale. The introduction of carbonated waters in the United States appears to have been accomplished almost simultaneously by Benjamin Stillman in New Haven and Joseph Hawkins in Philadelphia around 1807. The early use of carbonated water was restricted to medicinal purposes, but with the availability of liquid carbon dioxide and flavoring extracts in the mid 1800s the beverage grew in popularity as a refreshing beverage.

In 1965 the industry employed an estimated 40,000 production workers and the wholesale volume of beverages produced was $3,458,632,000. The estimated 1967 volume of soft drink production and consumption was 2,470,452,000 cases or a per capita consumption of 298.1 containers. These figures are based on a case of twenty-four 8-oz containers and a 1967 population of 198,751,000.

For additional information on the chemistry and technology of carbonated beverages, References 2 and 3 should be consulted.

## Properties and Composition

For purposes of analysis, carbonated beverages can be classified as sugar-sweetened, flavored beverages; synthetically sweetened, flavored beverages; and carbonated waters of the club soda, seltzer, and sparkling water variety.

Sugar-sweetened carbonated beverages contain 83–93% water, 7–17% sugar as sucrose, 0–200 grains of acid, reported as citric acid, per gal, and 0.1–5 gas vol of carbon dioxide. The pH may vary from 7.0 to 2.5. Birch beer, cola, ginger ale, and root beer

**Table 1.** Ingredients in Various Carbonated Beverages

| Beverage | Flavors | Color | Sugar, % | Acid | Acid, oz/gal of syrup | $CO_2$, gas vol[a] |
|---|---|---|---|---|---|---|
| sparkling water | sodium bicarbonate or sodium sulfate, 300–400 ppm[b] | none | none | none | none | 4–5 |
| cream soda | vanillin or ethyl vanillin | caramel or amaranth | 11–13 | citric | ¼ | 2½ |
| cola | extract of kola nut, lime oil, and spice oils, caffeine | caramel | 11–13 | phosphoric | 0.6 | 3½ |
| ginger ale | ginger root, oil of ginger, and lime oil | caramel | 7–11 | citric | 1 | 4–4½ |
| root beer | oil of wintergreen, vanilla, nutmeg, cloves, or anise | caramel | 11–13 | citric | ¼ | 3 |
| orange | oil of orange and orange juice | Sunset Yellow FCF and some Ponceau SX or tartrazine | 12–14 | citric | 1 | 1½–2½ |
| strawberry | "aldehyde $C_{16}$" (ethyl $\beta$-methyl-$\beta$-phenylglycidate, ethyl $\alpha,\beta$-epoxy-$\beta$-methylhydrocinnamate) | amaranth | 11–13 | tartaric or citric | 1 | 2½ |
| grape | methyl anthranilate and oil of cognac, grape juice sometimes added | amaranth and Brilliant Blue FCF | 11–13 | tartaric | ¾ | 1–2½ |
| peach | "aldehyde $C_{14}$" ($\gamma$-heptylbutyrolactone) | tartrazine | 11–13 | citric | 1 | 2½ |
| tonic | lemon oil, lime oil, quinine hydrochloride | none | 9–11 | citric | 4 | 4½ |
| cherry | benzaldehyde or oil of bitter almond | amaranth | 11–13 | tartaric or citric | 1 | 2½ |
| lemon | oil of lemon | none | 11 | citric | 1 | 3½ |
| lemon | lemon juice and oil of lemon | tartrazine (optional) | 11–13 | citric[c] | 1 | 1½ |
| tom collins | lemon juice | none | 7–9 | citric[c] | 2½ | 4–4½ |
| lime rickey | lime juice | none | 8–12 | citric[c] | 1 | 4–4½ |

[a] 1 gas vol is the amount of carbon dioxide that water will absorb at 60° F and 1 atm; this corresponds to an equal volume of carbon dioxide.

[b] Either sodium bicarbonate or sodium sulfate is added in an effort to simulate the natural mineral waters.

[c] This acidity is not due to the addition of acid to the other ingredients, but to the natural acid in the lemon or lime juice that is used as the flavor.

are some common flavors. In addition, there are a large number of fruit flavors such as lemon, orange, raspberry, and grape. The quantity of flavoring extract present is highly variable depending upon the particular flavor. Some citrus-type beverages contain up to 5–6% reconstituted juice and essential oils. A few highly concentrated synthetic flavors may be present at levels of only a few parts per thousand in the finished beverage. Colorants may be present in highly variable concentrations. The food preservative, sodium benzoate, may or may not be present; when used the concentration is usually 0.1 or 0.05%.

The synthetically sweetened beverages in general have sugar replaced with saccharin, which has roughly 300 times the sweetening power of sucrose, and/or one of the cyclamates, which have a sweetening power about thirty times that of sucrose. They may also occasionally be distinguished from their sugar-sweetened counterparts by modification of flavor and acid level as well as by the addition of gums and stabilizers.

Carbonated waters to which inorganic and/or organic salts are sometimes added constitute the class of beverage commonly known as sparkling water or club soda. When salts are added, the concentration rarely exceeds a few thousand ppm. The salts are principally carbonates, bicarbonates, and sulfates of calcium, magnesium, and potassium. Sodium citrate is the most frequently encountered organic salt.

Table 1 lists the usual ingredients in various carbonated beverages (3).

## Methods of Manufacture

The principal steps in the production of carbonated beverages are shown in Figure 1. In addition to the preparation of the product, extensive operations are required in the packaging phase of soft drink production. For example, bottles are automatically cleaned and sanitized in 2–5% solutions of sodium hydroxide at a minimum temperature of 120°F. After filling, the bottled beverages are sealed with cork or plastic-lined closures.

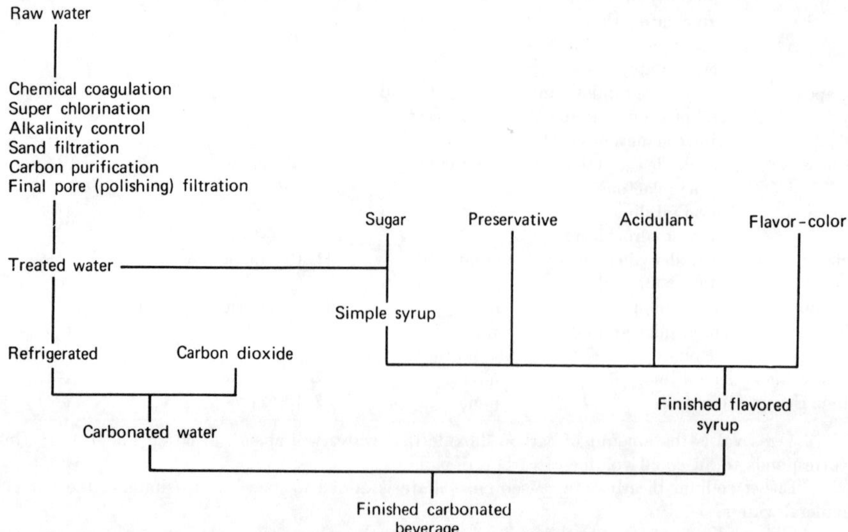


**Fig. 1.** Flow diagram for carbonated beverage preparation.

Water treatment is accomplished in standardized systems designed for this purpose such as the Infilco Accelerator and Permutit Spaulding Precipitator. Treatment consists of coagulation with ferrous sulfate or aluminum sulfate, superchlorination with calcium hypochlorite, and alkalinity reduction through precipitation with lime or the alkaline salts of calcium and magnesium. Finally, the water is subjected to filtration through pressure sand filters, activated carbon purifiers, and fine-pore (polishing) filters. This treatment is designed to produce a commercially sterile water, free of turbidity and suspended solids, and with a total alkalinity of less than 50 ppm expressed as calcium carbonate. The total dissolved solids of the raw water should not exceed 500–750 ppm. In other respects the water should conform to APHA standards for potable water (4).

Carbonation is provided by use of commercial grades of carbon dioxide. The gas is occasionally subjected to treatment with silica gel or activated carbon filters to remove trace impurities, but this practice is not general. The quantity of gas present in the beverage is measured in gas vol; 1 gas vol is the amount of carbon dioxide that water will absorb at atmospheric pressure and 60°F. At this pressure and temperature water will absorb an equal volume of carbon dioxide.

Acidity of the carbonated beverage is most often provided by citric, tartaric, or phosphoric acid. Small quantities of lactic and acetic acid are sometimes used. Phosphoric acid is used predominately in the cola-type beverages and citric acid is used mainly for the fruit flavors. Tartaric acid is used particularly in grape beverages.

The nutritive sweeteners are restricted almost entirely to sucrose or its derivatives and dextrose, with the former far more prevalent. The quantity of dextrose used rarely exceeds 20% of the total sugar solids. Partially inverted liquid sucrose solutions are frequently used. With the preparation of acidified syrups and the finished beverage, total inversion of sucrose is accomplished in a short period.

Sugar of poor quality has a very detrimental effect on the taste, odor, and stability of a carbonated beverage. Therefore, extensive work has been done by the American Bottlers of Carbonated Beverages in establishing standards for sugar (5). Important excerpts from these standards are listed below (3). These standards apply to dry sugar as produced, immediately prior to packaging.

*Excerpts from the Tentative Standards for Dry Granulated Sugar*

1. *Ash:* The ash content of "Bottlers" sugar shall not be more than 0.015%.
2. *Color:* The solution color of "Bottlers" sugar shall not be more than 35 reference basis units. (Reference basis units have been developed by joint research of carbonated beverage and sugar industries, making possible expression of test results on a common basis.)
3. *Sediment:* The sediment content of "Bottlers" sugar shall not be more than shown on a prepared sediment disc available from American Bottlers of Carbonated Beverages, 1128 Sixteenth Street, N.W., Washington, D.C. 20036 (2 ppm).
4. *Taste and odor:* "Bottlers" sugar shall have no obviously objectionable taste or odor in either dry form or in a 10% sugar solution prepared with tasteless, odorless water.
5. *Bacteriology:* "Bottlers" sugar shall not contain more than the following for each 10 g of sugar: 200 mesophilic bacteria, 10 yeast organisms, and 10 mold organisms.
6. *Floc:* "Bottlers" beet sugar shall be floc-tested by the manufacturer whose responsibility it shall be to determine the suitability of such sugar for use in carbonated beverages. A "negative" determination is one basis for claiming such sugar suitable for use in carbonated beverages.
7. *Turbidity:* A universally accepted method of testing for turbidity has not yet been developed. Meanwhile, it is imperative that "Bottlers" sugar be as nearly free of turbidity as possible.

*Excerpts from the Tentative Standards for Liquid Sugar*

1. *Shipping unit:* The product shall be shipped in clean sanitary shipping units. The shipping unit shall be so constructed as to eliminate any possibility of adding any impurity to the liquid sugar.
2. *Identification:* The product shipped shall be identified on the bill of lading as "Bottlers Liquid Sugar," and shall specify whether it is sucrose type or 50% invert type.
3. *Color:* The color of Bottlers Liquid Sugar shall not be more than 35 reference basis units at the manufacturer's storage tank at the point of manufacture.
4. *Sediment:* The sediment content of Bottlers Liquid Sugar shall not be more than shown on a prepared sediment disc (approximately 2 ppm) available from American Bottlers of Carbonated Beverages, upon request.
5. *Taste and odor:* Bottlers Liquid Sugar shall have no objectionable taste or odor in an undiluted form or in a 10% solution acidified to pH 2.5 with USP phosphoric acid.
6. *Microbiological procedures and standards:* All microbiological testing to be done according to the procedures of the Sugar-Bottling Industries Committee. Standards for yeast, mold, and mesophilic bacteria are available.
7. *Floc evaluation of liquid beet sugar:* Same as for granulated sugar.

Nonnutritive or synthetic sweeteners are restricted to soluble saccharin, calcium cyclamate, sodium cyclamate, and more recently cyclamic acid. A combination of one of the cyclamates with saccharin is frequently used to minimize the undesirable flavor features of both compounds. The ratio of their combination is usually around 10 parts cyclamate to 1 part saccharin thus providing an equal sweetness contribution by each compound.

Color of the carbonated beverage is achieved exclusively through the use of caramel or FD&C certified colors. Most dyes used are water soluble types. The range of colors prepared is large and considerable variation of color shade and intensity is encountered within any class of beverages. The caramel used in carbonated beverages is ordinarily an acid-proof type designed to withstand prolonged exposure to beverage acidulants.

Whereas the carbonated water, sweeteners, and acidulants provide basic organoleptic sensations to the carbonated beverage, the characteristic tastes of the flavored carbonated beverages are derived from the flavoring materials added. The principal types used by the bottler are alcoholic extracts, emulsions, aqueous solutions, concentrated juices, and solutions of flavors in glycerol and propylene glycol. The flavoring compounds may be natural or synthetic or a combination of both. The natural flavors are generally classified as essential oils, oleoresins, and true fruit flavors. A comprehensive discussion of carbonated beverage flavors has been prepared by Jacobs (2).

## Methods of Analysis

The common methods for analyzing the manufactured beverage are given in this section. Package tests and methods for the analysis of intermediates are also given. Additional methods are available from various sources (2,6,7).

### Analysis of Carbonated Beverages

#### *Acidity*

Acidity values of carbonated beverages are determined titrimetrically after boiling the sample to remove carbon dioxide. The results are reported as citric acid in either grains/gal or g/100 ml.

***Procedure***

Boil a known volume of the beverage to remove carbon dioxide. Cool the beverage and restore to the initial volume by adding water. Withdraw a 10-ml aliquot and add 1 drop of 0.5% phenolphthalein indicator solution. Titrate with 0.05 *N* sodium hydroxide solution. If a colored beverage is being examined, use a pH meter and titrate to a pH of 8.3.

$$\text{Citric acid, g/100 ml of sample} = \frac{(\text{ml of NaOH})(N)(6.404)}{\text{vol of sample, in ml}}$$

To convert from g/100 ml to grains/gal, multiply by 584.27. The factor for converting grains/gal to g/100 ml is 0.00171.

### *Air Content*

Air content of carbonated beverages is determined by a selective absorption of carbon dioxide from headspace gases purged from bottle or can into a gas buret. The remaining gases are reported as ml of total air in the sample. A Zahm air tester is used in the method given; see also Vol. 7, p. 629.

***Procedure***

Close the needle valve of the air tester, open the buret cock, and place approximately 250 ml of a 20% sodium hydroxide solution in the reservoir. Elevate the reservoir to permit the solution to flow into the graduated column. Close the buret cock, lower the reservoir, and after submerging the piercing needle in a small beaker of water, open the needle valve until all air in the piercing mechanism is displaced with water drawn from the beaker; then close the needle valve. Vent the gas buret and fill it completely with the sodium hydroxide solution.

Place the bottle or can on the base pad and puncture the crown or top with the piercing device. Open the needle valve and allow gas to flow into the buret, reducing the gage pressure to about 5 lb; then close the needle valve. Shake the apparatus a few times to allow absorption of the carbon dioxide by the sodium hydroxide solution. Repeat release of the gas into the buret until a constant reading is obtained. Report as total air content in ml.

### *Ascorbic Acid*

Ascorbic acid, or vitamin C, is added to some carbonated beverages as an antioxidant. In the method given, ascorbic acid is titrated with iodine solution, using starch indicator solution. Alternatively, the end point may be determined potentiometrically, using platinum and saturated calomel electrodes.

***Procedure***

Prepare a standard solution of ascorbic acid by dissolving 100 mg of the reagent in 100 ml of 3% metaphosphoric acid using a volumetric flask. Prepare approximately 0.01 *N* iodine solution by diluting 1 ml of a *N* iodine solution to 100 ml. Then standardize the prepared iodine solution by titrating 5-ml aliquots of the standard ascorbic acid solution, using starch indicator solution.

Degas approximately 50 ml of the beverage sample using a vacuum apparatus. Using volumetric pipets, mix the degassed beverage with 3% metaphosphoric acid in the ratio of 2 parts beverage to 3 parts acid. Then titrate 5-ml aliquots of the mixture with the iodine solution to the starch end point.

Calculate the ascorbic acid by the following equation:

$$\text{Ascorbic acid, mg/ml} = \frac{(\text{ml of iodine solution})\ (N)\ (88.1)}{\text{ml of sample in aliquot}}$$

### *Biological Examination*

The presence of microorganisms in carbonated beverages is considered of no significance from a public health standpoint since no pathogenic species are known to be capable of survival in the finished beverage. Some yeasts and acid-tolerant bacteria are, however, regarded as spoilage agents in certain beverages. The microbiological counts are also treated as an index of the sanitary maintenance program in the bottling plant.

#### *Procedure*

Aseptically open the bottled beverage and lightly flame the top. To eliminate the gas from the beverage, aseptically pour a small portion into a sterile, wide-mouth 2-oz jar. Cover the jar, shake briefly; then using sterile pipets, transfer 4-ml portions to sterile petri plates. Should very high cell counts be anticipated, prepare serial dilutions of the product with sterile dilution bottles or tubes.

Then prepare standard pour plates by adding melted Mycophil agar or Difco wort at no more than 40°C (pH adjusted to 4.0 with sterile 10% lactic acid solution). Incubate the plates at 30°C for 96 hr. Then make counts and report as yeast, mold, and acidophilic bacteria per ml of beverage.

### *Caffeine*

The usual procedure for the determination of the caffeine content of carbonated beverages involves a chloroform extraction of a sample made strongly alkaline with sodium hydroxide. With subsequent evaporation of the chloroform extract, a simple gravimetric determination is made of the residue and the total extractables are expressed as caffeine. Should significant quantities of chloroform extractables other than caffeine be present, the residue is redissolved in water, acidified, and a colorimetric determination performed using a phosphomolybdic acid solution. The procedures given are modified AOAC methods (8).

#### *Procedures*

GRAVIMETRIC METHOD. Pipet 25 ml of degassed sample into a 50-ml beaker and adjust the pH to 10.0 with 10% sodium hydroxide. Transfer this solution to a separatory funnel and extract with two 40-ml and two 30-ml portions of chloroform. Draw the successive chloroform extracts off into a previously dried and weighed evaporating dish. Evaporate the extract to dryness. Calculate the gain in weight, corresponding to total extractables, as caffeine, in mg/100 ml. If significant chloroform extractables other than caffeine are present, proceed with the colorimetric method which follows.

COLORIMETRIC METHOD. Redissolve the residue in water with two successive 5-ml washings; transfer the solution to a 50-ml beaker. Add 1 ml of 1:1 hydrochloric acid. Cover the beaker with a watch glass and heat on a steam bath. Add 2 ml of 20% phosphomolybdic acid solution dropwise with stirring; replace the watch glass and continue heating for 20 min. Filter the solution through a fine, fritted glass, Büchner funnel with suction and wash the precipitate with three 5-ml portions of 1:9 hydrochloric acid. Transfer the Büchner funnel to a side-arm test tube. Dissolve the precipitate in three 5-ml portions of acetone; transfer to a 25-ml volumetric flask and dilute to volume with acetone. Determine the absorbance at 440 m$\mu$ against acetone and determine caffeine from a standard curve. Calculate as mg/100 ml of beverage.

# ENCYCLOPEDIA OF INDUSTRIAL CHEMICAL ANALYSIS

*Edited by* Foster Dee Snell *and* Leslie S. Ettre

**VOLUME 20**

**Index to Volumes 4–19**

***Interscience Publishers***
***division of John Wiley & Sons***
***New York • London • Sydney • Toronto***

### 2.2.6. **Kirk-Othmer** *Encyclopedia of Chemical Technology*

Our third major encyclopaedia is also the most recently published:

> *Encyclopedia of chemical technology*. [Kirk-Othmer] 3rd ed. New York: Wiley–Interscience. 1978–1984. 24v.

This is a traditional-style A-to-Z encyclopaedia containing long and detailed articles on many and various topics concerned with industrial chemistry and chemical technology. About half the articles deal with chemical substances; other articles cover industrial processes and uses, foods and other human uses, pharmaceuticals, dyes, fibres, unit operations and processes, fundamentals, analytical methods, scientific topics, and miscellaneous subjects, eg computers, patents; there is also a most useful article in Vol. 13 (pp 278–336) on information retrieval, which I recommend you to read when you have completed this Unit on Using Literature.

I hope you are getting used to these encyclopaedias by now (and if you were an experienced encyclopaedia user before starting this Unit I hope you are picking up a few more specialised and practical tips); anyway, you should by now appreciate their potential (and perhaps actual) usefulness to you, so on again with a practical example.

**SAQ 2.2c**

You are working in a forensic laboratory when your director suddenly tells you that in a day or two he wants you to help in an analysis of a suspect painting. You have never worked with paintings before and point this out to your director; he just tells you that it will be an ir and pigment-analysis task.

You decide that you must know more about the work involved before you start and have access to a copy of the Kirk-Othmer *Encyclopedia of Chemical Technology*.

Use the copies provided (or better still, if you have access to a complete set of the 3rd edition, use it and ignore the copies) and find the information you would need. Just note how you first approached the work and the terms you looked up and where.

Of course it won't be difficult to find if you use the photocopies, but the list of contents gives you an indication of what articles are, and are not, included.

KIRK-OTHMER

# ENCYCLOPEDIA OF CHEMICAL TECHNOLOGY

THIRD EDITION

## VOLUME 1

A TO ALKANOLAMINES

A WILEY-INTERSCIENCE PUBLICATION
John Wiley & Sons
NEW YORK • CHICHESTER • BRISBANE • TORONTO • SINGAPORE

# CONTENTS

KIRK-OTHMER

# ENCYCLOPEDIA OF CHEMICAL TECHNOLOGY

THIRD EDITION

VOLUME 1

A TO ALKANOLAMINES

A WILEY-INTERSCIENCE PUBLICATION
John Wiley & Sons
NEW YORK • CHICHESTER • BRISBANE • TORONTO

# CONTENTS OF THE ENCYCLOPEDIA

## VOLUME 1

## VOLUME 2



## VOLUME 3

**CONTENTS OF THE ENCYCLOPEDIA** **vii**



## VOLUME 4

## VOLUME 5



## VOLUME 6

## VOLUME 7

## VOLUME 8



## VOLUME 9

## VOLUME 10



## VOLUME 11

## VOLUME 12

**CONTENTS OF THE ENCYCLOPEDIA** **xiii**



## VOLUME 13



## VOLUME 14

## VOLUME 15



## VOLUME 16

## VOLUME 17

## VOLUME 18



## VOLUME 19

## VOLUME 20



## VOLUME 21

## VOLUME 23



## VOLUME 22

## VOLUME 24



## Supplement Volume

100. Jpn. Pat. 56 015,416 (July 12, 1979), (to Unitika Kabushiki Kaisha).
101. Ger. Pat. 2,713,435 (Apr. 7, 1976), (to Kureha Kagaku Kogyo).

K. PORTER
ICI Fibres

# FINE ART EXAMINATION AND CONSERVATION

## SCIENTIFIC EXAMINATION

In few interdisciplinary fields have two so apparently different disciplines been brought together as in the application of physical sciences to the study of art objects (1–3). The scientist, accustomed to think in terms of abstractions, is trained to approach a stated problem by analyzing it for the identification of measurable variables, and devising means to obtain numerical values for the latter; on the other hand, the art historian relies on the trained eye, which enables the visual recognition of stylistic characteristics and the mental comparison of these with observations about numerous other art works, an intellectual process of which the procedural mechanism does not allow for an easy explanation other than "you have to see it."

Practitioners of the two disciplines often have had difficulties in understanding each other's problems and solutions of these, and the lack of communication has in the past sometimes led to wasted effort and mistrust on both sides. Remedying this problem has required mutual efforts, and the fact that this field may now truly be regarded as having come of age must be credited to those art historians who have made themselves familiar with the principles of scientific methodology, as well as to the scientists who have taken an actual, professional interest in the humanities, realizing that this is only possible by learning the language and systematics of their counterparts.

All this does not imply that scientific examination of art objects is a completely new field of study. It is, however, in the twentieth century that the number of technological studies in art have grown virtually exponentially.

One can surmise several reasons for this phenomenon. First, the great museums had to be established, bringing together the large collections for the academic purposes of study and education rather than for the personal aesthetic gratification of a private owner. Once this had happened, it did not take long for the first museum-based laboratories to be established; the first such laboratory was founded in 1888 at the Royal Museum of Berlin, followed by the British Museum during World War I, and the Boston Museum of Fine Arts in 1929.

Second, the explosive development in analytical techniques, especially during the second half of this century, has given birth to many instrumental methods which, through high sensitivities and small sample requirements, are extremely appropriate for the study of art objects. For example, metal-alloy analysis has a long history of application in numismatics, where the evaluation of relative values of contemporary coinages depends on the knowledge of their intrinsic values. With the traditional wet chemical analysis, this often meant the loss of half a coin. Although the analytical results are certainly very accurate and precise, and hence allow meaningful and valuable conclusions, the reluctance of a curator to sacrifice an irreplaceable part of humanity's cultural heritage for this purpose is not only understandable, but justified. In the museum, responsibility for the objects in the collection belongs to the curator, who must constantly be aware of the goals of a museum: to further scholarly study of the material in the collection; to use it for educational purposes through the organization of exhibitions; and, possibly the most important mandate, to preserve it for future generations so that they, too, can see and study the cultural records from the past. Because of the extreme importance of the proper preservation and prevention of damage, a consequence of the irreplaceability of the objects, it is absolutely necessary that any study involving the removal of even the tiniest sample be discussed by the curator and scientist in order to determine the goals of the proposed project; the value of the results, especially in furthering the understanding of the object and its historical context; and the extent of the damage which will be inflicted. The latter has to be balanced carefully against the benefits to be derived from the study. There are several possible reasons why a scientific study of an art work may be desirable. An obvious one is in cases where the authenticity of an object is doubted on stylistic grounds but no unanimous opinion exists. The scientist will identify the materials, analyze their chemical composition, and investigate whether this corresponds to what has been found in comparable objects of unquestioned provenance. If the sources for the materials can be characterized, eg, through trace element composition or structure, it may be possible to determine whether the sources involved in the procurement of the materials for comparable objects with known provenance are the same. Comparative examination of the technological processes involved in the manufacture allows for conclusions whether the object was made with techniques actually used by the people who supposedly created it. Finally, dating techniques may allow the establishment of the date of manufacture.

Several interesting cases of well-known forgeries are discussed in ref. 4. Probably the most notorious case is that of the forger Han van Meegeren, who faked works by Vermeer and de Hoogh. Several of his works were acquired by reputable Dutch museums, and one was sold to Hermann Goering during the German occupation of the Netherlands. When, after the war, van Meegeren was arrested and charged with collaboration, he claimed his innocence, protesting that this painting, and many others, were not originals, but rather forgeries by his hand. These claims were hotly contested, and he had to prove, in captivity, that he could indeed fake Vermeer's style so well as to deceive the stylistic experts. The technical examination which proved these works to be forgeries was the work of Paul Coremans (5). Later, $^{210}$Pb dating confirmed the recent date of manufacture (6).

A famous case of fake ceramic objects is that of the large Etruscan Warriors in the Metropolitan Museum of Art. The proof of nonauthenticity was provided by a spectrochemical analysis of the glaze. Although Greek and Etruscan artists applied the decoration using a slip of essentially the same clay as the one from which the object was crafted, and obtained the red and black colors through a special cycle of oxidizing- and reducing-firing conditions, the black color on the Warriors was found to be the result of the use of a manganese black glaze (7).

The Drake Plate is a brass plate, supposedly left behind in 1579 by Sir Francis Drake in California, claiming the land for the British crown. Discovered in 1937, it has since been in the custody of the University of California at Berkeley. Recently, chemical analysis has found that both the principal and trace-element compositions of this plate are incompatible with the attribution (8–9).

The Vinland Map, supposed evidence of the Viking discovery of America long before Columbus, was proven a forgery through the detection of the modern pigment titanium white (10).

The famous bronze Greek horse in the Metropolitan Museum of Art was declared a forgery in 1967, after an examination which led to the conclusion that the object was cast by means of a sand-casting technique, unknown to the ancient Greeks (11). Subsequent technical studies by a team of experts have, however, not only negated all arguments brought against the sculpture, but even authenticated the bronze by means of thermoluminescence dating (12).

Though the cases in which scientific analysis is called upon to assist in the authentication process tend to draw the most attention and have, occasionally, spectacular results, they are by no means the most important applications of the scientific method in the study of art works. Most fruitful and satisfying are those truly interdisciplinary projects in which an object or group of objects, of unquestionable provenance, is studied jointly by the historian and the scientist, in order to better evaluate its historical context. Classifications according to material and technical properties may refine stylistic differentiations, and geographic attributions may lead to the evaluation of past cultural cross-fertilizations (13–15). Without the study of objects of unquestionable attribution, a judgment which generally only the stylistic expert can make, the criteria used in authenticity studies could not be established.

Probably of more direct interest to the scientist than the art historian, but nevertheless of value in the larger framework of the humanities, is the study of the history of technology (16). Generally, this part of the historical record has been overlooked in the past, but affords profound insights into the history of the development of civilization. The earliest existing written records, treatises of craftspeople and artists on the techniques and materials with which they worked, date back to medieval times (17–20). For the millenia of human activities before that time, the record is in the objects which remain, and only through the study of these can knowledge in this regard be furthered.

Last, there is a very practical reason for the need of scientific support in the preservation of art objects (21). In order for the preservation specialist (conservator) to be able to arrest decay processes at work in an object, an understanding of these processes must be reached through study of the material properties and identification of the factors that influence the deterioration process. Subsequently, an evaluation of possible treatments with regard to their potential effectiveness, both immediate and long term, as well as other possible consequences, is made. For this, the active involvement of scientists is necessary and has become, to some degree, standard practice.

## Methodologies

**Optical Techniques.** While stylistic experts perform their observations with the naked eye and under normal illumination, the eye is often aided by optical devices and special illuminations in the laboratory. The most important tool in a museum laboratory is the low power stereomicroscope. This instrument, usually used at magnifications of 3–50 ×, has enough depth of field to be useful for the study of surface phenomena on many types of objects without the need for sample preparation (see Fig. 1).

The information thus obtained can relate to toolmarks and manufacturing

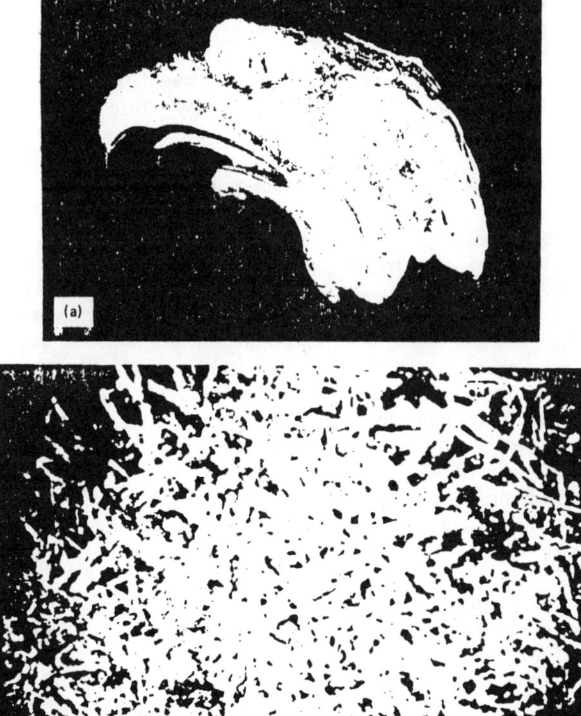

**Figure 1.** **(a)** A bronze hawk's head, forgery of a Roman finial. **(b)** The patina, studied through stereomicroscope, shows a structure typical for an artificially induced corrosion, probably effected l pickling. Courtesy of the Research Laboratory, Museum of Fine Arts.

example, the technique is frequently used to date the wood used in paintings on panels (39–40). Measurement of the relative thickness variations in a number of consecutive rings which can be observed on the edge of the panel allows the placing of this pattern in the chronological sequence developed for the area of origin.

Thermoluminescence dating (33–34,41) has, in the short time since its development, had a tremendous impact on the technical study of ceramic materials (see Analytical methods).

Absorption of radiation energy, such as from naturally occurring radioisotopes, in a nonconductor results, via a sequence of primary ionizations and secondary excitations, in the production of electrons and electron-hole pairs with energy levels in the conduction band. With such energies, the electrons and electron-hole pairs are able to move freely through the crystal, which allows them to enter crystal imperfections. At such imperfections, metastable energy levels exist between the valency and conduction bands. One such type of metastable level is the so-called trap: an electron or electron hole which de-excites partially at such an imperfection until it reaches the trap level cannot subsequently de-excite completely because transitions between the trap level and valency band are forbidden. The name of the trap level derives from the consequent restriction in freedom of both movement and energy of the trapped electron or hole. Only transfer of extra energy to the system, as through heating of the crystal, can free them again via re-excitation to conduction band levels. Although most de-excitations to the valency band, followed by recombination, will be nonradiative in nature, some take place in the luminescence centers, which are crystal imperfections of another type.

In a clay, the radiation dose rate is determined, to a large extent, by the concentration of radioisotopes, eg, $^{40}K$, and the members of the decay chains of uranium and thorium contained within the constituent minerals. This results in fairly constant rates of ionizations, and hence of trap population. Because the process has been taking place over a geological time period, the trap-population density in a clay is very high. However, firing, the last step in the manufacture of a ceramic object, results in a depopulation, which resets the clock for thermoluminescence dating. Immediately after cooling, repopulation commences, again at a constant rate determined by the dose rate. Thus, at any time after firing, the trap-population density will be a function of the time elapsed since the firing.

In thermoluminescence dating, a sample of the material is heated, and the light that is emitted by the sample as a result of de-excitations of the electrons or holes, freed from the trap, at luminescence centers is measured to provide a measure of the trap-population density. The signal is compared with that obtained from the same sample after a laboratory irradiation of known dose. The annual dose rate for the clay is calculated from determined concentrations of radioisotopes in the material and assumed or measured environmental radiation intensities.

The application range of the technique extends well beyond the first human ceramic production; in fact, it has been used in geological studies such as the dating of lava flows. The accuracy obtained in the dating of ceramics depends largely on knowledge of the environmental history, eg, burial conditions, and physical behavior of the material; the information on burial conditions is needed for the calculation of certain correction factors on the annual dose rate which otherwise have to be estimated. At best, the accuracy of thermoluminescence dating is ca 5–10%. Though for many historical contexts a far more precise dating is possible on stylistical arguments, thermoluminescence dating is very useful in prehistoric archaeology. Accuracy is of less importance in authenticity questions, and it is here that the technique finds extremely widespread use.

A related technique depends on the same principles, only the trap population is measured by means of esr spectrometry (42).

In amino acid racemization dating, the degree to which racemization has progressed since the death of an organism serves as a measure of the time elapsed after that event (43). Because the reaction rates are dependent upon the environmental conditions of the material, eg, average temperature, humidity, and pH of the burial soil, the working range of this technique varies with these locality-bound data, whereas the accuracy depends on their variation in time. The technique is applied, in particular, to the dating of shell, bone, or ivory objects.

Another relative dating technique for fossil materials is known as nitrogen–fluorine dating (44). Bone or ivory will, under burial conditions, lose nitrogen because of groundwater leaching of the amino acids resulting from hydrolysis of proteins, especially collagen. At the same time, fluorine will be absorbed from the groundwater by the hydroxyapatite in the bone, forming fluoroapatite. Thus, the ratio of nitrogen and fluorine concentrations in the bone decreases with time. Another elemental concentration which changes is that of uranium, which can be absorbed by the hydroxyapatite. The reaction rates are dependent on local burial conditions, and the technique only has value as a relative dating for materials excavated or otherwise recovered from the same site.

A few techniques exist that do not provide for direct dating but rather give information as to whether the object is of recent manufacture. One of these is $^{210}Pb$ dating (45). In the decay series of uranium, the first long-lived member after $^{226}Ra$, with its 1622-yr half-life, is $^{210}Pb$ with a 22-yr half-life. In a lead ore, $^{210}Pb$ is in radioactive equilibrium with $^{226}Ra$, but refining of lead results in a chemical separation in which the greatest fraction of the radium remains in the slag. Thus, the equilibrium is disturbed, with the lead having a much higher $^{210}Pb$ activity than $^{226}Ra$ activity. This disequilibrium continues until the much faster decay of the excess $^{210}Pb$ results in a reduction of this activity to a level where it is in equilibrium again with the $^{226}Ra$ activity. The time needed to reach this state is of the order of 4–5 half-lives, ie, ca 100 yr. If an object contains refined lead, eg, in the pigment lead-white in a painting, measurement of both $^{226}Ra$ and $^{210}Pb$ activities will indicate whether radioactive equilibrium exists between these two radioisotopes and hence whether the lead was refined within the last 100 yr or earlier.

Indirect dating, or dating by inference, is not really dating but rather the reaching of certain conclusions regarding the date of manufacture of an object from other information obtained in its study. The presence on a painting of certain pigments of which the date of invention is known provides a *terminus post quem*, ie, the painting could have been executed at any time after the invention (46). On the other hand, the presence of a material of which the manufacture was discontinued at a certain time can provide a *terminus ante quem*.

### Types of Objects—Methods Used

**Paintings.** Before an expostulation of the procedures used in the examination of paintings, a synopsis of materials and techniques used in their creation is in order (17,47–52).

A painting is a composite of many materials of various types. As a first approach, one can distinguish three groups of components: supports, paint media, and pigments.

The support is the substrate upon which the paint layers are laid down. This can be a specially prepared area on a wall for a wall painting, a wooden panel as in a panel painting, or a fabric in canvas paintings. Paper is a prevalent support in Oriental painting. Other supports are encountered less frequently, eg, metal panels (especially copper sheet).

In the primitive cave paintings, which are the earliest known paintings, the paint was applied directly onto the wall, with little or no preparation. In ancient Egypt, however, as early as the Old Kingdom, wall surfaces were specially prepared with a coating of plaster. In time, the refinement and complexity of the preparation layers increased; in the Renaissance, several layers of different composition and fineness were superimposed. Other preparations used, especially in the Far East, consisted of a clay layer.

The use of wooden panels as support for painting goes back again to Egypt. Although the painting served originally as decoration of a wooden object such as a sarcophagus, later the wooden panel became only a support for the painting, as in the mummy portraits from the Fayum, executed during the Roman period. In Europe's Classical period, wooden panels must have been the supports for portable paintings, and remarks by Pliny indicate so; however, none of these works of art has survived. The tradition was carried through into the Christian era, and wooden panels were virtually the only portable supports used in the Middle Ages and early Renaissance.

The types of wood used in European panel paintings varied with the geographic area. In Italy and southern France, the most popular wood was poplar. In the north, oak was the first choice, followed by pine and linden. Another difference between north and south can be found in the preparation of the panel: whereas the ground layer in the southern paintings consists of real gesso, ie, finely ground burnt gypsum in a glue medium, the northern artists prepared their ground layer with chalk.

The earliest remaining example of painting on a fabric support is from the twelfth dynasty in Egypt. In Europa, although the technique was known, these supports were not frequently used until the Renaissance, when the increasing size of paintings resulted in a tremendous rise in the popularity of canvas supports, a popularity which has lasted until the present. The fabric used almost exclusively by western painters was linen; in the Orient, silk became a frequently used support. The nature and composition of the ground or priming layer on European canvases varied significantly.

The medium is the binder which provides for the adhesion of the pigments. The most important types of medium are the temper media (glue, egg, and gum), the oils, and wax. In addition, there is the true fresco technique, where the pigments are laid down in a fresh, wet plaster preparation layer. Several other media have been used, but much less frequently, eg, casein temper. In modern paints, a number of synthetic resins are used for this purpose.

The use of the various tempera and of wax has been identified on objects dating back to ancient Egypt. The Fayum mummy portraits are beautiful examples of encaustic painting, ie, with molten wax as medium. A rather special variation was the technique used by the Romans for their wall paintings. In these, the medium, referred to by Pliny as "Punic wax," probably consisted of partially saponified wax. In Europe, wax ceased to be used by the ninth century.

The polysaccharides in gum arabic formed a medium used for illuminated manuscripts and inks as well as for painting. Gum is also the binder in watercolors.

The proteinaceous gelatins in the various animal glues were also widely used as paint media, as well as in illuminations. Glues are the traditional media in Oriental painting. They remained the prevalent binders for ground layers in European painting long after oils had become virtually the only medium for the color layers.

Of the various tempera, egg temper was the most important in European painting, both in wall and panel painting. It was little used outside Europe. The main period of its use was in the Middle Ages and early Renaissance. After the sixteenth century, however, it was rarely used, as oils had become the preeminent media.

The earliest written references to the use of oils as paint media date from the twelfth century. The van Eycks, who traditionally have been credited as the inventors of oil painting, did indeed improve the technique to such a degree that oil quickly replaced egg temper as the prevalent medium.

Pigments can be categorized in two different ways: inorganic or organic and natural or synthetic. Organic pigments generally consist of a dyestuff, precipitated on an inorganic substrate. The earlier pigments tend to be inorganic and natural, although important exceptions in both regards exist. Thus, a pigment prepared by the precipitation of madder onto a gypsum substrate was an important color in Egypt during the Graeco-Roman period, and in the Old Kingdom a synthetic pigment, Egyptian blue (copper calcium silicate), was made in large quantities.

For many pigments (qv), a period of time in which they had their widest use can be indicated (47,53). Dates of introduction are known either from documentary sources or from identification on paintings of known dates (46). For some pigments, an approximate date for the discontinuation of their use can be assigned; in some cases, knowledge of the preparation process or even their very existence was lost over an appreciable time span.

A varnish is often applied on top of the paint layers. A varnish serves two purposes: as a protective coating and also for an optical effect that enriches the colors of the painting. A traditional varnish consists of a natural plant resin, dissolved or fused in a liquid for application to the surface. There are two types of varnish resins: the hard ones (the most important of which is copal), and the soft ones (notably dammar and mastic). The hard resins are fossil; to convert them to a fluid state, they are fused in oil at high temperature. The soft resins are dissolved in a suitable organic solvent, eg, turpentine. A typical technical examination of a painting (54), intended to ascertain its period, condition, and the degree of previous restoration would likely entail most of the following steps.

***Visual Examination.*** A close inspection under normal illumination reveals many indications of the condition of the painting and previous repairs. Also, because oil paints become more transparent with age, pentimenti, which originally would have been invisible after the overpainting, can be observed. Raking light illumination is extremely useful at this stage to determine the extent of cracking, distortions of the support, delaminations of the paint layers, etc. This stage of the examination is often done in close cooperation with the stylistic experts; thus, obvious problematic areas can be identified before the other tests are started.

***Examination with the Stereomicroscope at Low Magnification.*** The obtained information relates to the characteristics of the craquelure, the pigments used (grain size, morphology), buildup of paint layers (visible in damages and along cracks), technique of the artist (brush work, use of pure and mixed colors, use of glazes, etc), and condition (eg, damages and losses, amount of inpainting or overpainting).

***Examination under Uv Illumination.*** This, as explained above, principally provides information regarding more recent restorations or changes.

***Infrared Reflectography.*** The ir photograph, or the image from an ir-sensitive vidicon system, gives more evidence of restorations as well as pentimenti. The most important use of this technique is in the revealing of underdrawing: the pigments typically used for underdrawing, such as charcoal or certain earth pigments, do not reflect ir radiation very efficiently. A variation of this technique is transmitted ir photography, where the light source and the camera are placed on the opposite sides of the painting.

***X-Radiography.*** The radiograph reveals evidence of damages and losses in the paint layers and the support. Dimensional changes may also be detected, eg, a cut-down painting on canvas will, on one or more sides, lack the telltale deformations of the canvas caused by the tacking to the stretcher. The radiograph illustrates the artist's compositional approach (eg, blocking out of compositional elements, amount of changes in composition during execution of the painting) as well as the painting technique (eg, preparation of support, brush work, thickness of paint layers, variations in thickness between different areas in the painting) (see Fig. 4). Through the study of x-radiographs, characteristics of a given artist's works can often be established that can serve as benchmarks in the examination of paintings of uncertain attribution.

***Pigment Analysis.*** Identification of pigments present on paintings of unquestioned attribution provides reference information regarding the use of certain pigments in given periods or schools. In the study of paintings with uncertain provenance, the pigments identified in it may either negate the possibility of the proposed attribution or lend more credibility to it. Most pigment analyses are done in one of three ways: microscopy and microchemical tests, x-ray diffraction-powder analysis, or energy-dispersive x-ray fluorescence spectrometry. Although the latter technique does not directly identify the pigment but rather the chemical elements present, this information often allows deduction of the identity of the pigment. The technique can be used nondestructively, without any need for sample removal (see Nondestructive testing). Thus, many areas on a painting can be analyzed, giving better overall information.

***Microscopic Examination of Cross Sections through the Paint Layers.*** This gives definite information regarding the paint-layer sequence in the area from which the sample was taken (22,55). This information illustrates the artist's use of underlayers and glazes, superposition of compositional elements, and changes in composition.

***Medium Analysis.*** The techniques frequently used for identification of the binding media include relatively simple solubility tests, ir absorption spectrophotometry, specific staining techniques (56), tlc (57), gc (58), and ms. Significant work, eg, entailing the identification of the particular oils used in a painting, has recently been made possible through the use of combination gc–ms (59).

**Figure 4.** **(a)** "Mrs. Freake and Baby Mary," Boston, 1671–1674. Worcester Art Museum, Gift of Mr. and Mrs. Albert W. Rice. **(b)** The x-radiograph of this painting shows considerable changes in the composition: in an earlier version, the robe of the sitter has a different collar while her right hand does not hold the child but rather lies in her lap, holding a fan. **(c)** Line drawings of the two versions. The baby was not included in the first version of the portrait; when she was added, the mother's dress was also updated (drawing by Carol Greger, copyright 1983). Courtesy of the Worcester Art Museum and the Museum of Fine Arts.

KIRK-OTHMER

# ENCYCLOPEDIA OF CHEMICAL TECHNOLOGY

THIRD EDITION

# INDEX

TO VOLUMES 1–24
AND
SUPPLEMENT

A WILEY-INTERSCIENCE PUBLICATION
John Wiley & Sons
NEW YORK • CHICHESTER • BRISBANE • TORONTO • SINGAPORE

## 98 Arsenic

## 626 Indoxyl

**636 Ionization chambers**

**842 P.A.C**

**890 Phytopathogens**

## 892 Pigment Metal 1

**SAQ 2.2c**

### 2.2.7. Meites *Handbook of Analytical Chemistry*

Now we move away from the major encyclopaedias and turn to a rather dated but still very useful single-volume work:

> *Handbook of analytical chemistry*. Louis Meites, ed. New York: McGraw-Hill, 1963. Various pagings.

Probably one of the first things you have noticed about this book is its age: obviously many changes take place all the time in chemistry, as in all sciences, and usually you would expect to be better off with a recently published work. But it does not always follow that a new book is automatically better than an old one. Many basic chemical and physical data and many techniques may remain unchanged for considerable periods and, over twenty years old though it may be, this *Handbook* is still a mine of useful information. It aims to provide analytical chemists and workers in related sciences with concise and convenient summaries of the most important and useful fundamental data and practical procedures. The sections on quantitative analysis (Methods for the determination of specific substances, and methods for the analysis of technical materials) are written from the point of view of the sample: they summarise the most widely used and the most meritorious of the procedures for the determination and analysis of various specific substances and are primarily intended to aid you in selecting an analytical procedure to meet the requirements of a specific problem. Most of the remainder of the volume, written from the point of view of the technique employed, presents the fundamental data that characterise the behaviour of different substances towards the techniques of separation and measurement most widely used in chemical analysis, and also outlines the most important and most reliable of the analytical methods and procedures based on them.

The arrangement of the *Handbook* is in a series of 15 sections bringing related tables of data together; you will find a list of contents among the accompanying photocopies. This book has been out of print for some years and newer analytical techniques have been covered in specialised handbooks by the same editor and publisher; but this is still a book you will find in use in many works, and other, libraries. So, without further delay let us look at the book itself.

**SAQ 2.2d**

You want a fairly quick review of the methods of analysis and determination of glucose: you would also like a note of any analysis in which glucose is used as a reagent.

Look in the Meites *Handbook of Analytical Chemistry* if you can, but if you do not have easy access to this then use the copies provided instead.

Note the page numbers where you find relevant information and look the references up in the text. How many methods are given for analysis of aldoses (eg glucose) and which one requires the longest heating? What methods can you use to determine glucose? What substance can be determined by using glucose as a reagent?

# HANDBOOK OF ANALYTICAL CHEMISTRY

Edited by

**LOUIS MEITES**

*Professor of Analytical Chemistry*
*Polytechnic Institute of Brooklyn*

FIRST EDITION

McGRAW-HILL BOOK COMPANY

New York San Francisco Toronto London Sydney

## CONTENTS

### References for Table 5-6

1. Kolthoff and VanBeck, *ZaC*, **70**, 369 (1927).
2. Lange and Berger, *Z. Elektrochem.*, **36**, 980 (1930).
3. Clark, *JCS*, 768 (1926).
4. Martin, *AC*, **30**, 233 (1958).
5. Muller, "Elektrometrische Massanalyse," Steinkopf, Dresden, Germany, 1942.
6. Kolthoff and Furman, "Potentiometric Titrations," Wiley, New York, 1931.
7. Karchmer, *AC*, **29**, 425 (1957).
8. Schnell, *Österr. Chemiker-Ztg.*, **54**, 52 (1953).
9. Crane, *AC*, **28**, 1794 (1956).
10. Kirsten, Berggren, and Nilsson, *AC*, **30**, 237 (1958).
11. Doležal, Hencl, and Simon, *CL*, **46**, 267 (1952).
12. Bush, Zuehlke, and Ballard, *AC*, **31**, 1369 (1959).
13. Kolthoff and Verzijl, *Rec. trav. chim.*, **42**, 1056 (1923).
14. DeFord and Horn, *AC*, **28**, 797 (1956).
15. Przybylowicz and Rogers, *AC*, **30**, 1064 (1958).
16. Mayr and Burger, *Monatsh.*, **56**, 113 (1930).
17. Atanasiu, *J. chim. phys.*, **23**, 501 (1926).
18. Lingane and Hartley, *ACA*, **11**, 475 (1954).
19. Davis and McLendon, *T*, **2**, 124 (1959).
20. Velisek, *CL*, **24**, 443 (1930).
21. Fischer and Babcock, *AC*, **30**, 1732 (1958).

## Table 5-7. Potentiometric Redox Titrations

Potentiometric oxidation-reduction titrations are carried out using some "noble" metal electrode, the potential of which is measured *versus* a suitable reference electrode. The noble metal indicator theoretically serves simply to transfer electrons from dissolved species in solution to the external circuit. Usually platinum or gold is used as an indicator electrode. It should always be remembered that both platinum and gold can become oxidized by strong oxidants. This oxidation sometimes obscures end points or at least causes steady potentials to be obtained rather slowly. Certain difficulties are also observed with platinum electrodes in strongly reducing solutions such as chromous solutions. The reduction of hydrogen ion is catalyzed at platinum surfaces, causing the observed potentials to be less reducing than they should be. Mercury electrodes do not have the latter disadvantage.

The general technique of redox titrations is essentially the same as that for other titrations. Usually the end point is located as the point of maximum potential change for a specific volume increment. Sometimes potential changes are so great that accurate results may be obtained by approaching the end point dropwise and establishing the drop which causes a large potential change. The end point may also be established by other methods such as titrating to a calculated or, better, a predetermined potential. The end-point potentials given in the fifth column of this table are calculated values (see p. 3-58 *ff*.).

This table gives information about a few of the most useful or interesting potentiometric redox titrations.

More complete but general information may be found in Lingane, "Electroanalytical Chemistry," 2d ed., Interscience, New York, 1958, or in Kolthoff and Furman, "Potentiometric Titrations," Wiley, New York, 1931. The reviews published in *Analytical Chemistry* are also an excellent source of information concerning recent developments and specific procedures.

Table 5-7. Potentiometric Redox Titrations (*Continued*)

| Method number | Substance determined | Reagent | Supporting electrolyte, procedure, etc. | End-point potential, volts *vs.* S.C.E. | Remarks | Reference (p. 5-24) |
|---|---|---|---|---|---|---|
| 1 | As(III) | Ce(IV) | 4 $M$ HCl; ICl cat. | $\Delta E$ max. | Many other redg. agts. can be titrd. | 12 |
| 2 | Au(III) | Ascorbic a. | pH 1.6–3; 50°C. | $\Delta E$ max. | Cu(II), Fe(III), Hg(II), and 0.1 $M$ $Cl^-$ do not interfere. | 30 |
| 3 | $Br^-$ | $MnO_4^-$ | 10 ml. samp. + 5 ml. 10% KCN + 10 ml. concd. $H_2SO_4$ + 100 ml. $H_2O$ | $\Delta E$ max. | $2MnO_4^- + 11H^+ + 5Br^- + 5HCN \rightarrow 2Mn^{++} + 8H_2O + 5BrCN$ | 2 |
| 4 | Carbohydrates | Fe(II) | Add 25.00 ml. 0.3 $M$ $Cr_2O_7^=$ to samp., then 10 ml. concd. $H_2SO_4$, stir, let stand 10 mins., add 150 ml. $H_2O$, and back-titr. excess Cr(VI) with std. Fe(II). | $\Delta E$ max. | Many other org. substs. react. | 11 |
| ... | $C_2O_4^=$ | ......... | See meth. 15. | ................ | ............................ | ...... |
| 5 | Ce(III) | Feic or $MnO_4^-$ | Add samp. to enough 4 $M$ $K_2CO_3$ to ensure 1.5 $M$ $K_2CO_3$ at end pt. Titr. in absence of air. | +0.04 v. | Glucose may be detd. indirectly. | 15 |
| ... |  | ......... | See also meth. 19. | ................ | ............................ | ...... |
| 6 | Co(II) | Fe(II) | Ox. samp. with $NaBO_3$ in alk. soln., b., deaerate, add excess std. Fe(II) + excess 3 $M$ $H_2SO_4$ + 10 ml. 25% $H_3PO_4$ + excess std. $Cr_2O_7^=$, and back-titr. with std. Fe(II). | $\Delta E$ max. | W(VI) does not interfere. | 19 |
| ... | Cr(III) | ......... | See meth. 19. | ................ | ............................ | ...... |
| 7 | $Cr_2O_7^=$ | As(III) | 20% $H_2SO_4$ | $\Delta E$ max. | Cr may be detd. in steels in presence of V [titrd. subsequently with Fe(II)] and Mn (correct for induced rxn.) | 17 |

## Table 6-12. Colorimetric Methods of Organic Analysis

**Scope.** This table contains a mixture of reliable older methods and newer methods untested by time, in the 400- to 800-mμ region. Many of the older methods may be found in vols. III and IV of Snell and Snell, "Colorimetric Methods of Analysis," 3d ed., Van Nostrand, Princeton, N.J., 1953–1954. Selected newer methods appearing in the literature up to and including 1960 are also included.

The table is not always exhaustive; it includes mainly general methods and methods which determine an unusual functionality. The simple amino acid is included, for example, but the more complex protein is not. No attempt has been made to include all the methods that can be used to determine a particular function, but only to present methods which are representative, reliable, extremely convenient, or which are more recent. Functions such as esters and alkenes are covered more exhaustively because of the paucity of other methods available for their determination. Some less reliable or qualitative methods are included, such as those for aromatic hydrocarbons, where no other colorimetric methods exist.

**Classification.** The various types of compounds are arranged in the following order:

| *Compounds* | *Code designation* |
|---|---|
| 1. Acids (carboxylic) | A |
| 2. Acid derivatives | AD |
| 3. Alkenes and alkynes | Al |
| 4. Amines | Am |
| 5. Amino acids | AA |
| 6. Aromatic hydrocarbons | Ar |
| 7. Carbonyl groups | C |
| 8. Carbonyl derivatives | CD |
| 9. Esters and lactones | Es |
| 10. Ethers and epoxides | E |
| 11. Halogen compounds | Ha |
| 12. Heterocyclic compounds | He |
| 13. Hydrazine and derivatives | Hy |
| 14. Hydroxyl groups | H |
| 15. Nitro groups and other organo-nitrogen compounds | N |
| 16. Peroxides | Pe |
| 17. Phosphorus compounds | P |
| 18. Quinones | Q |
| 19. Silicon compounds | Si |
| 20. Sugars | Su |
| 21. Sulfur compounds | S |
| 22. Miscellaneous compounds | M |

The code designations are employed in numbering the individual methods for the sake of cross-references within the table. For example, ascorbic acid may be sought under "Acids," but as the methods for its colorimetric determination are all based on its behavior as a lactone, they are listed under "Esters and lactones" rather than under "Acids"; they are therefore designated as "Es 4.1," "Es 4.2," etc., to indicate this placement, and these code designations are given in the appropriate place under "Acids" to direct the user to the entries he seeks.

Within each group of compounds, general methods are usually given before methods for the determination of individual compounds. Typical compounds to which general methods are applicable are named in parentheses in the first column. Individual compounds are listed alphabetically where this arrangement is justified by their number. As in the case of ascorbic acid, which was mentioned above, compounds are generally arranged on the basis of the functional group which is most significant to the reaction in question. In many cases the placement has had to be somewhat arbitrary: for example, although pyridine is an amine and is also aromatic, it is classed here as a heterocyclic compound to permit grouping it with substituted pyridines which may not react as either amines or aromatic compounds. Cross-references are always provided for such compounds in the other sections of the table in which they may be sought.

Esters are classified by the acid moiety: thus, a method for methyl esters is a general

***Spectrophotometric Determination of Organic Substances*** **Table 6-12**

method for any acid moiety, but a method for the acetyl group is a specific method for acetate esters.

**Method or Reagent; Conditions.** In the third column of the table, the name of the method or of the important reagent is given first. Then, after a semicolon, the reagents added (in the order of their use) and the important experimental conditions are listed to give some idea of the complexity of the method. Unless the sample is mentioned, it is assumed to be contained in an aqueous solution to which the reagents are added. A term such as "samp./alc." means that an ethanolic solution of the sample is added at the point indicated. For full experimental details, the reader should consult the compilation of Snell and Snell or the original reference.

**Wavelength and Absorptivity.** The wavelength, in millimicrons, recommended for absorbance measurements is given in the fourth column of the table; with rare exceptions, this is the wavelength of maximum absorption in the 400- to 800-m$\mu$ range. In many cases it was possible to give only the color of the final solution as a guide to the selection of filters or wavelength settings for the absorbance measurements.

Numbers appearing in parentheses in this column are molar absorptivities, expressed in square centimeters per millimole. They correspond to the absorbances that would be measured in 1-cm. cells for hypothetical final solutions containing 1 mole/liter of the substance determined if Beer's law were obeyed up to that concentration.

A value appearing in this column enclosed in brackets (*e.g.*, [460]) denotes an alternate wavelength at which measurements may be made. This is almost always the wavelength of maximum absorption on a subsidiary band, or it may occasionally represent a shoulder whose use may be convenient in the analysis of solutions so concentrated that their absorbances would be inconveniently high if measured at an absorption maximum. A bracketed value in this column is always to be used in conjunction with the bracketed values in the "Range" column which follows.

Where a range of values (*e.g.*, 430–580) is quoted in this column, it is the range over which the position of the absorption maximum may vary for different members of the class of compounds in question.

**Range.** The range of concentrations recommended for use is given in parts per million (mg./l. or $\gamma$/ml.) in the fifth column of the table. These concentrations refer to the solution that is finally prepared and subjected to measurement. As these limits sometimes represent no more than the extreme concentrations that the authors happened to investigate, it should not be assumed that they represent the limits of obedience to Beer's law, or even that Beer's law is obeyed within these limits. Moreover, the ranges sometimes had to be estimated in cases where insufficient data for a complete calibration curve were available. For general methods, the molarities sometimes given in the literature were converted to parts per million for the sake of uniformity by employing typical rounded molecular weights, such as 100, 200, etc.

**Interferences.** The sixth column lists important substances or classes of substances that interfere in the methods in question. Important non-interfering substances which may not be obvious are often then given after a period and "*Not:*".

**References.** References in English have been given wherever possible. *Chemical Abstracts* or *Analytical Abstracts* references are given in many cases in which the original journals may be relatively difficult to obtain. The abbreviation "EOC" denotes "Elsevier's Encyclopedia of Organic Chemistry."

**Special Abbreviations.** The following special abbreviations are employed:

| | |
|---|---|
| Chr. a. | (Chromotropic acid) 1,8-dihydroxynaphthalene-3,6-disulfonic acid |
| DMF | *N*,*N*-dimethylformamide |
| DMG | dimethylglyoxime |
| 2,4-DNP | 2,4-dinitrophenylhydrazine |
| PE | petroleum ether |

In addition, times and temperatures are abbreviated as indicated by the following typical examples:

| | |
|---|---|
| " . . . , 10 mins., . . . " | . . . , let stand (or let the reaction proceed) for 10 minutes, . . . |
| " . . . , 30°-10 mins., . . . " | . . . , let stand at 30°C, for 10 minutes, . . . |

In the second example, if the temperature given is 90°, the sample is to be allowed to stand on a water or steam bath for the designated length of time. If a time appears at the end of an entry in the third column (headed "Method or reagent; conditions"), the final solution is to be allowed to stand for that length of time before its absorbance is measured.

## Table 6-12. Colorimetric Methods of Organic Analysis (*Continued*)

| Functional group or compound determined | Code number | Method or reagent; conditions | Wavelength and absorptivity | Range, p.p.m. in final soln. | Interferences | Reference |
|---|---|---|---|---|---|---|
| **Acids, Carboxylic** | | | | | | |
| *General Methods* | | | | | | |
| Fatty acids | A 1.1 | Esterification–ferric hydroxamate; evap. to dryness with $SOCl_2$ or $CH_2N_2$, heat with MeOH, evap., $Et_2O$, 2.5% NaOH/alc., 2.5% $NH_2OH{\cdot}HCl$/alc., 65°–evap., dissolve in 0.04% $Fe(ClO_4)_3$/alc.–0.35% $HClO_4$. | 520 | 5–100 | $H_2O$, ROH, see Es 1.1. | *AC*, **18**, 317 (1946) |
| $C_7$–$C_{18}$ acids | A 1.2 | Cu or Co soap extn.; samp./KOH-MeOH, pH 9, MeOH, 2% $Cu(NO_3)_2$, ext. with $CHCl_3$, filt. | 525 (Co), 675 (Cu) | 4,000–20,000 (Co), 800–4,000 (Cu) | Much NaCl or KCl | *ACA*, **15**, 77 (1956) |
| $C_{17}$–$C_{18}$ acids (stearic a.) | A 1.3 | Rosaniline (I); $H_2SO_4$, methylal/MeOH, $H_2O$, ext. with PE, evap. PE, 2-PrOH, solid I, 46°–30 mins., $C_6H_6$. | 520 | 30–110 | ........ | *CA*, **53**, 6332g (1959) |
| Acids in di-Me terephthalate | A 1.4 | Bromothymol blue (I); 0.8% neutd. I/DMF, 25% toluene/DMF, Ascarite, Drierite | 610 | 0–4 | $CO_2$, $H_2O$ | *AC*, **32**, 542 (1960) |
| Binary mixtures of acids (HOAc + benzoic or chloroacetic a.) | A 1.5 | Diphenyldiazomethane (I) rxn. rate; I, alc., $C_6H_6$, meas. rate of change of *A* at const. temp. (30°). | 525 | 0.05–0.1 *M* total a. | A third acid | *AC*, **24**, 360 (1952) |
| *Aliphatic Monobasic Acids* | | | | | | |
| Acetic | ....... | See A 1.1. | | | | |
| Acetoacetic | A 2.1 | 4-Nitroaniline (I) coupling; Ox-$PO_4^{-3}$-$CO_3^{-}$ buffer, 0.1% I/0.05 *M* $H_2SO_4$, 4% $NaNO_2$, 0°–10 mins., 5 *M* NaOH, 0°–2 mins., BuOH-$C_6H_6$ extn., 0.05 *M* NaOH/alc. | 650 | 50–800 | —$CCOCH_2CO$— group | *JBC*, **179**, 1235 (1949) |
| Ascorbic | ....... | See Es 4.1–4.6. | | | | |
| Bromoacetic | A 2.2 | Conversion to glycine; $Et_2O$ extn., $NH_3$, evap. $Et_2O$, ninhydrin, $Cd(OAc)_2$. | ........ | ........ | ........ | *AA*, **7**, 181 (1960) |
| Crotonic | A 2.3 | Fast Green FCF (I); 0.002 *M* $KMnO_4$/2 *M* $H_2SO_4$, room temp.–20 mins., 0.1% I/1 *M* $H_2SO_4$, 2 mins., 1 *M* $Na_2SO_4$, $H_2O$ | 682 | 0–1.5 | $H_3Cit$, α-hydroxy acids | *AC*, **23**, 1853 (1951) |

## Table 6-12. Colorimetric Methods of Organic Analysis (*Continued*)

| Functional group or compound determined | Code number | Method or reagent; conditions | Wavelength and absorptivity | Range, p.p.m. in final soln. | Interferences | Reference |
|---|---|---|---|---|---|---|
| | | $(NH_4)_2Ox$, 5 % $Fe(NH_4)_2(SO_4)_2$/2 % $H_2SO_4$. | | | | |
| **Phenoxy groups in organosilicon compounds** | Si 1.3 | Oxdn.; samp./EtOH, $NH_3$, reflux, filt., wash, aq. $Cl_2$ or aq. $Br_2$. | Blue-green | ............... | .................... | *CA*, **52,** 8852d (1958) |
| **Silicon and silicates from organosilicones** | Si 1.4 | Mo blue; samp./10 % KOH, 5 % $(NH_4)_2MoO_4$/10 % HOAc, $Na_2SO_3$ + $Na_2SO_4$, heat 5 mins., 2 % $(NH_4)_2Ox$, 2 % $Na_2CO_3$, 15 % glycerol, 1 hr. | 650–700 | ............... | .................... | *CA*, **52,** 14430g (1958) |
| Trimethylchlorosilane, free and combined in **siloxanes** | Si 1.5 | Mo blue; samp./$C_6H_6$, 0.1 g. $P_2O_5$, reflux 10 mins., $C_6H_6$, $(NH_4)_2MoO_4$, $H_2SO_4$, 90°–5 mins., 2.5 *M* $H_2SO_4$, 5 % $SnCl_2$, 1 *M* $H_2SO_4$, 30 mins. | 570 | ............... | *Not:* $Me_2SiCl_2$, <400 % $MeSiCl_3$ | *CA*, **53,** 6914f (1959) |
| **Sugars** | | | | | | |
| ***General Methods:*** | | | | | | |
| ***Pentoses**** | | | | | | |
| **Pentoses** (arabinose, xylose) | Su 1.1 | Formn. of furfural (I)—condensation with aniline (II); 13 % HCl, redistd. xylene, reflux 2.5 hrs., sep. I by distn. or extn., make alk., buffer, 10 % II/HOAc, 20°–1 hr. | 510 | 0.2–3 | Hexoses, light. *Not:* Me pentoses | *IEC,AE*, **15,** 162 (1943) |
| | Su 1.2 | Picric a. (I); satd. I/20 % aq. $Na_2CO_3$, 90°–30 mins. | ............. | *ca.* 40 | Hexoses | *JBC*, **38,** 33 (1919) |
| | Su 1.3 | $H_2SO_4$–cysteine (I); cool samp., $H_2SO_4$, cool to room temp., 3 % I hydrochloride. | Meas. $A_{390} - A_{425}$ | 2–100 | *Not:* hexoses | *JBC*, **181,** 379 (1949) |
| | Su 1.4 | Anthrone (I); 4-hr.-old 0.05 % I/$H_2SO_4$, mix 10 secs., heat to 95°, chill to room temp., meas. within 45 mins. | 620 | 0.003–0.03 | Carbohydrates | *AC*, **24,** 2004 (1952) |
| | Su 1.5 | Orcinol (I)–$Fe^{+++}$; 1 % I + 0.1 % $FeCl_3$/HCl, 90°–20 mins. | 610 | 10–20 | Carbohydrates | *JBC*, **167,** 369 (1947) |
| **Aldopentoses, aldohexoses** | Su 1.6 | *o*-Aminodiphenyl (I); 0.4 % I/HOAc, 90°–30 to 120 mins., cool. | 380 | 0–300 | Sucrose, Me aldopentoses | *AC*, **28,** 1916 (1956) |
| ***Hexoses: Reducing Sugars, Aldoses†*** | | | | | | |
| **Hexoses** (glucose)..... | Su 2.1 | Diphenylamine (I); samp./0.3 *M* HCl, | ............. | ............... | *Not:* pentoses | *CA*, **37,** $3188^{6}$ (1943) |

**Spectrophotometric Determination of Organic Substances** **Table 6-12**

| | | | | | | |
|---|---|---|---|---|---|---|
| | | 0.01 % I/40 % EtOH, 1.8 *M* HCl, 90°–5 mins. | | | | |
| Aldoses (glucose) | Su 2.2 | Alk. Feic; excess 0.12 % $K_3$Feic, 3 % $Na_2CO_3$, 90°–5 mins. | 420 | 0–25 | *Not:* ketoses | *IEC,AE*, **10**, 411 (1938) |
| | Su 2.3 | Oxdn. of $Cu_2O$—Mo blue formn.; 0.6 % $CuSO_4$/Fehling's soln., 90°–6 mins., cool, 15 % $H_2MoO_4$/26 % $H_3PO_4$ contg. 7.5 % $Na_2CO_3$, 1 min., $H_2O$, meas. within 10 mins. | 660 | ............ | ............ | *JBC*, **76**, 457 (1928) |
| | Su 2.4 | 3,5-Dinitrosalicylic a. (I); 1 % I, 4.5 % NaOH, KNaTart, 10 % phenol, 90°–5 mins., cool, 3 mins. | ............ | ............ | ............ | *JBC*, **65**, 393 (1925) |
| | Su 2.5 | 4-Aminosalicylic a. (I); mix 1 part aq. samp., 1 part Na salt of I, and 10 parts HOAc, 90°–20 mins. | 520 | ............ | MeCHO. *Not:* $Me_2CO$, ketoses | *AA*, **5**, 3379 (1958) |
| Reducing sugars | Su 2.6 | Ferric hydroxamate; 0.8 *M* KCN, 0.42 *M* HOAc, overnight, evap., 1.4 % $NH_2OH \cdot HCl$/MeOH, 1.1 % KOH/MeOH, evap. 20 mins. at 50°, 1.5 % $FeCl_3 \cdot 6H_2O$, 10 mins. | 500 | 10–100 | Fructose; see also Es 1.1. | *AC*, **30**, 1538 (1958) |
| Aldoses | Su 2.7 | Hypoiodite oxdn.; 0.75 *M* $Na_2CO_3$, 0.01 *M* $KI_3$, 25°–1 hr., 0.4 *M* $H_3PO_4$. | 484 [404] [352] | 0–200 [0–20] [—] | Redg. agts. | *AC*, **31**, 1790 (1959) |
| *Hexoses: Ketoses* | | | | | | |
| Ketoses (fructose) | Su 3.1 | Resorcinol (I)–$Fe^{+++}$; 0.1 % I/EtOH, 0.00075 % $FeCl_3$/HCl, 80°–8 mins. | 480 | 5–20 | Furfural, >5 mg./ml. galactose, >3 mg./ml. glucose | *JBC*, **107**, 15 (1934) |
| | Su 3.2 | Diphenylamine (I); 1 % I/60 % EtOH–40 % HCl, 90°–15 mins. | 610 | ............ | Glucose, $CCl_3COOH$ | *JBC*, **127**, 606 (1939) |
| | Su 3.3 | Orcein (I); 0.2 % I, $H_3PO_4$, 90°–10 mins., 20 % NaOH. | Yellow | ............ | Sucrose. *Not:* glucose, lactose, maltose | *CA*, **37**, 5336[8] (1943) |
| Fructose in glucose | Su 3.4 | Carbazole meth.: $A_{fructose}/A_{glucose}$ = 280. | ............ | ............ | ............ | *AA*, **6**, 2208 (1959) |
| Fructose | Su 3.5 | *p*-Anisidine (I); 0.5 % I/$H_3PO_4$, heat to boiling, cool 30 mins. | 450 | 0–50 | *Not:* glucose, other aldoses | *AC*, **31**, 1234 (1959) |
| | ....... | See also Su 2.6. | | | | |
| *Disaccharides* | | | | | | |
| Lactose | Su 4.1 | $MeNH_2$ (I); 1.6 % I, 21.6 % NaOH, let stand under $N_2$ 15 mins., 55°–30 mins., meas. within 2–15 mins. | 540 | 50–600 | Pentoses, hexoses | *A*, **67**, 130 (1942) |

* Most general methods apply to pentoses.
† Most hexose methods apply to reducing sugars and aldoses. See also Su 1.1 to 1.6.

## Table 6-35. Fluorometric Methods for the Determination of Organic and General Biochemical Substances (*Continued*)

| Substance determined | Reagent | Conditions | Excitation, mμ | Emission, mμ | Sensitivity | Interferences | References |
|---|---|---|---|---|---|---|---|
| 1-Naphthol | ... | 0.1 *M* NaOH in 20 % EtOH | 365 | 480 | 0.7 γ/ml. | Some naphthalene compds. | *AC*, **30**, 96 (1958) |
| 2-Naphthol | ... | 0.1 *M* NaOH in 20 % EtOH | 365 | 426 | 0.7 γ/ml. | Some naphthalene compds. | *Ibid.* |
| | Malic a. | 92 % $H_2SO_4$, 10 mins. at 40° | 365 | Green | 1 γ/ml. | 1-Naphthol | *AC*, **21**, 1375 (1949) |
| Nerve gases (sarin, soman, or tabun) | Indole + $NaBO_3$ | Let stand 1 min. | 365 | 480 | 0.005 γ/ml. | ... | *AC*, **29**, 276 (1957) |
| 2-Nitronaphthalene | 60 % oleum, then dil. and add Zn dust. | pH 4.6 | 365 | 450 | 0.1 γ/ml. | ... | *AC*, **23**, 717 (1951) |
| *o*-Nitrophenol | Zn + HCl; then benzoic a. at 160° | $C_6H_6$ soln. | 365 | Green-yellow | ... | ... | *IEC,AE*, **12**, 403 (1940) |
| Phenylalanine | ... | $H_2O$ soln. | 215, 260 | 282 | 0.1 γ | ... | *Biochem. J.*, **65**, 476 (1957) |
| Piperonyl butoxide | ... | MeOH soln. | 292 | 318 | 0.01 γ/ml. | ... | *JAFC*, **6**, 32 (1958) |
| Proteins | ... | $H_2O$ soln. | 280, 240 | 313, 350 | ... | ... | *Doklady Akad. Nauk USSR*, **116**, 594 (1957); *Biochem. J.*, **76**, 381 (1960) |
| Pyruvaldehyde | Chromotropic a. | $H_2SO_4$ | 436 | 540 | 0.1 γ/ml. | Formaldehyde, redg. sugars | *AC*, **22**, 899 (1950) |
| Pyruvic a. | DPNH + enzymic coupling | ... | 365 | 460 | ... | Other metabolites | *Biochem. J.*, **64**, 56P (1956) |
| Salicylic a. (*o*-hydroxybenzoic a.) | ... | pH 11 | 310 | 435 | 0.01 γ/ml. | ... | *J. Pharm. Exptl. Therap.*, **120**, 26 (1957) |
| | ... | pH 5.5 | 314 | 410 | 1 γ/ml. | *m*-Hydroxybenzoic a. | *AC*, **30**, 1361 (1958) |
| Salicyloyl hydrazide | ... | pH 10 | 350 | 425 | 1 γ/ml. | ... | *AC*, **31**, 296 (1959) |
| Serotonin (5-hydroxytryptamine) | ... | 3 *M* HCl | 295 | 550 | 0.2 γ | Other 5-hydroxyindoles | *J. Pharm. Exptl. Therap.*, **117**, 82 (1956); *Arch. Biochem. Biophys.*, **68**, 1 (1957); *JBC*, **230**, 865 (1958) |
| | ... | $H_2O$ soln. | 295 | 340 | 0.005 γ/ml. | ... | *JBC*, **215**, 336 (1955); *Arch. Biochem. Biophys.*, **68**, 1 (1957); *Science*, **125**, 442 (1957) |
| Skatole | ... | $H_2O$ soln. | 290 | 370 | 0.001 γ/ml. | ... | *Science*, **125**, 442 (1957) |
| Succinic a. | Resorcinol | Concd. $H_2SO_4$, 1 hr. at 130° | 470 | Green | 0.01 γ/ml. | ... | *Plant Physiol.*, **23**, 443 (1948) |
| Thioglycolic a. | Na 1,2-naphthoquinone-4-sulfonate | pH 7.5 | 365 | Blue-white | 0.3 γ | Some redg. agts. | *ZaC*, **139**, 263 (1953) |

***Fluorescence Analysis*** **Table 6-35**

| | | | | | | | |
|---|---|---|---|---|---|---|---|
| Thymidine........ | $Br_2$ water, then *o*-aminobenzaldehyde in 0.15 *M* NaOH | $PO_4^{-3}$ buffer | 365 | 420 | 0.02 γ | Acetol, some aldehydes | *JBC*, **233**, 483 (1958) |
| Thymidylic a. ..... | $Br_2$ water, then *o*-aminobenzaldehyde in 0.15 *M* NaOH | $PO_4^{-3}$ buffer | 365 | 420 | 0.02 γ | Acetol, some aldehydes | *Ibid.* |
| Thymine.......... | $Br_2$ water, then *o*-aminobenzaldehyde in 0.15 *M* NaOH | $PO_4^{-3}$ buffer | 365 | 420 | 0.02 γ | Acetol, some aldehydes | *Ibid.* |
| Thymol.......... | .................... | pH 7 | 265 | 300 | 0.1 γ/ml. | ................ | *J. Pharm. Exptl. Therap.*, **120**, 26 (1957) |
| Tryptamine....... | .................... | pH 7 | 290 | 360 | 0.002 γ/ml. | ................ | *Arch. Biochem. Biophys.* **68**, 1 (1957); *Science*, **125**, 442 (1957) |
| Tryptophan....... | .................... | pH 11 | 285 | 365 | 0.003 γ/ml. | Redg. and oxdg. agts. | *Arch. Biochem. Biophys.*, **68**, 1 (1957); *JBC*, **223**, 313 (1956); *Biochem. J.*, **65**, 476 (1957) |
| | Glucose at pH 1.38, 4 hrs. at 118° | pH 1.80 | 365 | 440 | 4 γ | Other amino acids | *AC*, **28**, 884 (1956) |
| Tryptophan and derivatives | 70 % $HClO_4$, 1 hr. at 40° | ................ | 400 | Yellow-green | 100 γ | Indoleacetic a. | *JACS*, **70**, 2615 (1948); *JBC*, **180**, 1065 (1949) |
| Tryptophan derivs. (proteins)........ | $CF_3COOH$ | $O_2$ absent | 420 | 530 | ........... | ................ | *Science*, **129**, 641 (1959) |
| Tyramine......... | .................... | pH 1 | 275 | 310 | 0.02 γ/ml. | ................ | *Arch. Biochem. Biophys.*, **68**, 1 (1957) |
| Tyrosine.......... | .................... | pH 7 | 275 | 310 | 0.005 γ/ml. | ................ | *Ibid.*; *Biochem. J.*, **65**, 476 (1957); *JBC*, **223**, 313 (1956) |
| | 1-Nitroso-2-naphthol | 30 mins. at 55° | 460 | 570 | 1 γ | Tyramine | *J. Lab. Clin. Med.*, **50**, 733 (1957) |
| Umbelliferone...... | | See 7-Hydroxycoumarin. | | | | | |
| Uric a. ........... | .................... | pH 1 | 325 | 370 | 0.7 γ/ml. | ................ | *Arch. Biochem. Biophys.*, **68**, 1 (1957) |
| Urobilin........... | $Zn(OAc)_2$ in EtOH | Ext. with $CHCl_3$. | 365 | Yellow | ........... | ................ | *Lancet*, **264**, 71 (1953) |
| Xanthine.......... | .................... | pH 1 | 315 | 435 | 0.08 γ/ml. | ................ | *Arch. Biochem. Biophys.*, **68**, 1 (1957) |
| Xanthopterin...... | .................... | 20 % $H_2SO_4$ | 365 | 460, 520 | ........... | ................ | *A*, **72**, 382 (1947); *Acta Med. Jugoslav.*, **2**, 87 (1948) |
| Xanthurenic a. ..... | .................... | pH 11 | 350 | 460 | 0.005 γ/ml. | ................ | *Arch. Biochem. Biophys.*, **68**, 1 (1957) |
| | .................... | 50 % satd. NaOH; let stand 1 hr. | 370 | 530 | 1 γ | ................ | *JBC*, **230**, 781 (1958) |
| Warfarin [3-(α-acetonylbenzyl)-4-hydroxycoumarin]........ | .................... | MeOH soln. | 320 | 385 | 0.04 γ/ml. | ................ | *JAFC*, **6**, 32 (1958) |

## Table 13-39. Liver Function Tests

| Test | Technique | Procedure | Sample | Interferences | Reference |
|---|---|---|---|---|---|
| Bilirubin......... | Photometric | Meas. color formed by diazo rxn. in aq. soln. (direct) or after addn. of MeOH (indirect). | S 1 | Hemolysis | 3 |
| Bromsulfophthalein | Photometric | Meas. difference between $A_{580}$ at pH 10.4 and at pH 7.4. | S 0.5 | Severe lipemia or hemolysis | 7 |
| Cephalin flocculation | Visual | Estimate flocculation produced with cephalin-cholesterol emulsion. | S 0.2 | None | 4 |
| Galactose......... | Photometric | Remove glucose by yeast fermentation; det. galactose by Cu redn. meth. | B 1 | None | 6 |
| Hippuric a. ...... | Volumetric | Ppt. hippuric a. by addn. of salt and a., and titr. ppt. with std. alk. | U 10 | Benzoic a. | 9 |
| Icterus index...... | Photometric | Meas. yellow color of dild. serum at 460 mμ. | S 0.5 | Hemolysis, lipemia, lipochromes | 2 |
| Thymol turbidity . | Photometric | Meas. turbidity formed with thymol-barbital buffer rgt. | S 0.1 | Lipemia | 8 |
| Urobilinogen...... | Photometric | Red. with $Fe(OH)_2$, ext. with petroleum ether, meas. red color formed with *p*-dimethylaminobenzaldehyde. | F 5, U 50 | None | 1 |
| $ZnSO_4$ turbidity... | Photometric | Meas. turbidity formed with $ZnSO_4$-barbital buffer rgt. | S 0.1 | Lipemia | 5 |

### References for Table 13-39


1. Balikov, in Seligson (ed.), "Standard Methods of Clinical Chemistry," vol. II, Academic Press, New York, 1958, p. 192; *cf.* also Henry *et al.*, *Clin. Chem.*, **7**, 231 (1961).
2. Henry *et al.*, *Am. J. Clin. Pathol.*, **23**, 841 (1953).
3. Kingsley *et al.*, in Reiner (ed.), "Standard Methods of Clinical Chemistry," vol. I, Academic Press, New York, 1953, p. 11.
4. Knowlton, in Seligson (ed.), "Standard Methods of Clinical Chemistry," vol. II, Academic Press, New York, 1958, p. 12.
5. Kunkel, *Proc. Soc. Exptl. Biol. Med.*, **66**, 217 (1947).
6. Meranze *et al.*, *Am. J. Clin. Pathol.*, **12**, 261 (1942).
7. Seligson and Marino, in Seligson (ed.), "Standard Methods of Clinical Chemistry," vol. II, Academic Press, New York, 1958, p. 186.
8. Shank and Hoagland, *JBC*, **162**, 133 (1946).
9. Weichselbaum and Probstein, *J. Lab. Clin. Med.*, **24**, 636 (1939).

## Table 13-40. Methods for the Determination of Carbohydrates and Intermediates

| Test | Technique | Procedure | Sample | Interferences | Reference |
|---|---|---|---|---|---|
| Acetone and acetoacetic a. | Photometric | Isothermally dist. a mixt. of samp. and aniline into salicylaldehyde rgt.; add alk. | B 0.2 | None | 3 |
| Citrate........... | Photometric | Convert to pentabromoacetone and add NaI to form colored complex. | S 1.5, U 0.5 | β-Hydroxybutyric a. | 4 |
| Glucose.......... | Photometric | To $Zn(OH)_2$ filtrate add $Cu^{++}$ ($\rightarrow Cu_2O$), then treat $Cu_2O$ with phosphomolybdic a. and det. Mo blue produced. | B 1 | Saccharoids (very sl.) | 5 |
| Lactic a. ........ | Photometric | Remove glucose from protein-free filtrate by $CuSO_4$ and $Ca(OH)_2$. Convert lactic a. to MeCHO by heat and $H_2SO_4$, and det. MeCHO by rxn. with $Cu^{++}$ + p-hydroxydiphenyl. | B 0.5 | None | 1 |
| Pyruvic a. ....... | Photometric | Form the dinitrophenylhydrazone in a $CCl_3COOH$ filtrate, ext. into xylene, and add NaOH. | B 2 | High concns. of other keto acids | 2 |

### References for Table 13-40


1. Barker and Summerson, *JBC*, **138,** 535 (1941).
2. Friedemann and Haugen, *JBC*, **147,** 415 (1943).
3. Pawan, *Biochem. J.*, **68,** 33P (1958).
4. Taussky and Shorr, *JBC*, **169,** 103 (1947).
5. Young, in Reiner (ed.), "Standard Methods of Clinical Chemistry," vol. I, Academic Press, New York, 1953, p. 60.


## Table 13-41. Methods for the Determination of Vitamins

See also Table 9-11.

| Test | Technique | Procedure | Sample | Interferences | Reference |
|---|---|---|---|---|---|
| Ascorbic a. ...... | Photometric | Meas. decrease of color of 2,6-dichlorophenol indophenol added to $HPO_3$ filtrate. | S or U, 2 | S, none; U, sulfhydryl compds. | 1 |
| Vitamin A and carotene | Photometric | Meas. $A_{550}$ and $A_{600}$ after rxn. with "activated glycerol dichlorohydrin" in a petroleum ether ext. ($\rightarrow$ pink color). | S 1 | Other carotenoids | 2 |

### References for Table 13-41


1. Owen and Iggo, *Biochem. J.*, **62,** 675 (1956).
2. Sobel and Snow, *JBC*, **171,** 617 (1947).

# INDEX

## Introduction

The function of an index is to guide the user of a book to the information that it contains and to ensure that nothing pertinent to his purpose can escape his notice. The 1700-odd pages of this book contain a large number of pieces of information, and these may be used for a variety of different purposes. These facts are responsible for the extreme length of this index, which contains some tens of thousands of entries, and this in turn has prompted the adoption of a number of measures aimed at keeping the number of its pages within reasonable bounds without impairing its utility.

This is a subject index; authors' names do not appear in it except in a few cases in which they are widely used to designate substances or methods (*e.g.*, Bindschedler's Green, Romijn method). It is an essentially complete index; with a few trivial exceptions, every substance mentioned, no matter how incidentally, in the body of the volume is cited in the index (and liberal cross-referencing is given). This includes every substance for which a numerical value of any property is given, every substance for which a qualitative test, a quantitative method, or a method of analysis is given—including those that "can be similarly determined"—every substance mentioned as a reagent or as an indicator, and so on. The exceptions are as follows:

1. Groups of closely related organic compounds, such as the steroids in Table 6-47 and the imidazoles in Table 10-43, whose individual members are not mentioned elsewhere are indexed as groups rather than as individual compounds. However, compounds that appear elsewhere as well, and also compounds (like abietic acid and *n*-decylamine) that appear only once but that could not conveniently be located by means of a group entry, are indexed individually.

2. Inorganic compounds that appear only once, such as silver arsenate in Table 1-9, are not individually indexed: in this case there are index entries under "silver salts," "arsenate," and "solubility products" that will suffice for its location. But compounds that are mentioned more than once in different tables, such as cesium chloride in Tables 1-7, 1-8, 6-53, and 6-57, are individually indexed. So are a few compounds, such as erbium hydroxide in Table 1-9, where an index entry under a group classification such as "erbium compounds" would have sacrificed preciseness without saving space.

3. Not every appearance of a widely used procedure, reagent, or indicator is indexed: it would serve no useful purpose to cite every example of the use of thiosulfate for a back-titration of iodine. In such cases, only those citations are given which seem to the editor to be the most significant by virtue of giving the most fundamental or general information regarding the procedure or substance, indicating its utility in unusual circumstances, or pointing to interferences with or modifications of a familiar procedure in dealing with special kinds of samples.

Nomenclature is an exacerbating problem in the construction of an index. There are good reasons for preferring "hexacyanoferrate(III)" to "ferricyanide," but it is the latter that is used here because of its greater familiarity. With organic com-

***Index***

pounds such simple choices are rarely available: what one chemist thinks of as "butadiene dimer" is more familiar to another as "4-vinylcyclohexene." In every case, an attempt has been made to provide an entry under the name that is likely to be most useful. Several criteria are involved in this choice, including familiarity, accuracy and unambiguousness, and—peculiarly important to the analytical chemist—juxtaposition to other closely related compounds for which the entries may provide information that is directly or indirectly relevant to the user's purpose. In addition, cross-references are given liberally where it has seemed desirable. Although this is not a glossary of organic and technical nomenclature, a number of alternate names are given in parentheses for two reasons. One is because the name appearing in the body of the volume may differ from that used here: for example, the compound indexed as 2,7-dihydroxy-4-methylquinoline is named as a carbostyril derivative in the text, and so the latter name is given parenthetically to ensure its recognition. Another is to ensure the proper interpretation of the positions of substituent groups in compounds named in sub-entries. Finally, it may be mentioned that (except for the use of *o*-, *m*-, and *p*- in the names of disubstituted benzene derivatives and the use of Greek letters to denote side-chain substitution as in "$\alpha,\alpha$-dichlorotoluene") the positions of substituent groups are denoted by numerical prefixes: *e.g.*, "3-aminobenzoic acid" and "3-aminobutyric acid" instead of "*m*-aminobenzoic acid" and "$\beta$-aminobutyric acid."

The names and arrangement resulting from the application of these rules and criteria differ in some respects from those employed by *Chemical Abstracts* or by The Chemical Society (*cf. Handbook for Chemical Society Authors*, The Chemical Society, London, 1960), but they are believed to be the most practical for the present purpose. Special thanks are due to Professor Ernest I. Becker for his invaluable help with problems of organic nomenclature and indexing.

To have given in each index entry a full description of the information to be found in each place cited would have produced an index consuming nearly as much space as the material indexed. Consequently, the entries are given in a special form, of which the following is a typical example:

Azide, ***Detn*** **3**-5, -68; **12**-133. ***Pr*** **1**-14*ff*.; **5**-10, -31, -102. ***Rgt*** **2**-20; **3**-75; **5**-217; **6**-54; **12**-154, -171

Definitions of the abbreviations printed in boldface italic type (***Detn, Pr, Rgt,*** and half a dozen others) appear at the foot of each left-hand page of the index. The numbers in boldface type, which are always followed by hyphens (**3**-, **12**-, etc.), are section numbers; the condensed titles of the 15 sections of the volume are given at the foot of each right-hand page of the index. The numbers in lightface type, which are always preceded by hyphens (-5, -68, -133, etc.), are page numbers; all references within any pair of semicolons are to the same section. With the aid of these lists of abbreviations and section titles, the above entry for azide might be mentally translated into something like

| Azide | |
|---|---|
| determination of | |
| gravimetric or volumetric | 3-5, 3-68 |
| in functional-group analysis* | 12-133 |
| properties of | |
| fundamental† | 1-14*ff*. |
| electrochemical‡ | 5-10, 5-31, 5-102 |

* A translation requiring a little familiarity with the arrangement of Section 12.
† Solubility products of some metal azides.
‡ Standard potential, equivalent conductance, and polarographic characteristics, respectively.

use as reagent
  in qualitative analysis 2-20
  in inorganic gravimetric or volumetric analysis 3-73
  in electroanalysis 5-217
  in an optical method of analysis 6-54
  in functional-group analysis* 12-154, 12-171

Much condensation is effected in this way without, in the editor's opinion, serious prejudice to the ready accessibility of the information sought.

The following comments will be found to be useful aids in the deciphering of these entries:

***An*** (analysis of) A procedure in which, for example, a sample of steel is subjected to one or more operations for the purpose of finding its lead content is an *analysis of* steel and a *determination of* lead, and would be so indexed here. Both of these terms are reserved for relatively detailed practical methods (see "***Detn***" and "***Pr***" below). Methods for the analysis of metal alloys are indexed under the names of the principal constituents of the alloys: for example, all methods for the analysis of brass, bronze, and other copper-base alloys are indexed together under "Copper, ***An***."

***Detn*** (determination of) Under this heading there are collected all references to practical methods for the (quantitative) determination of the substance named. No distinction is drawn between methods that have been used for the analysis of complex materials and those that have been used only for the analysis of pure solutions. The methods appearing in Section 13 naturally fall into the former class. Methods for the practical analysis of complex materials cited in other sections can most conveniently be located by searching for entries listed under both the analysis of the sample and the determination of the constituent sought: thus, for example, a search under "Uranium, ***Detn***" and "Minerals, ***An***" would quickly yield just two citations (pages 5-133 and 13-124) that might contain descriptions of methods for the determination of uranium in a mineralogical sample.

The methods indexed under this heading often include information about the relevant properties of the substance in question, but those that do are not also indexed under "***Pr***" (properties of). In this way the latter heading is reserved for references giving information but not actual analytical methods. For example, it is clear that the polarographic data cited on page 5-102 could be used as the basis for determining azide in some samples, but it is equally clear that for practical analytical purposes such information is quite different from a description of a method that has been tested and found to be reliable.

***DI*** (detection and identification of) This category includes references to descriptions of qualitative tests and methods for the differentiation of the substance named from others that resemble or accompany it. In accord with what was said above under "***Detn,***" simple data on the properties of a substance, though they may often be used for these purposes, are never indexed here.

***Ind*** (as indicator) This heading has generally been reserved for entries giving information on the formula, preparation, and behavior as an indicator of the substance named. Systematic names of indicators are generally not indexed, the familiar trivial ones being far more likely to be sought; but an exception is made for a number of metal indicators that are designated by their systematic names in Tables 3-32 through 3-38. This is to provide the user with a convenient way of locating further information in Tables 3-32 through 3-34 about an indicator that has been selected on the basis of an entry in one of the tables that follow. It may be mentioned that all of the trivial names of the indicators appearing in these tables have been indexed here, in the belief that this may be of service to the chemist wishing to perform a titration for which the indicator was designated in the original literature by a name with which he and his chemical supplier are unfamiliar.

## Index



1, fundamental data; 2, qualitative; 3, inorganic gravimetric and volumetric; 4, gas; 5, electrometric; 6, optical; 7, nuclear and magnetic; 8, thermal; 9, biological; 10, separations; 11, pH; 12, specific substances; 13, technical materials; 14, statistics; 15, definitions

## *Index*



*An*, analysis of; *Detn*, determination of; *DI*, detection or identification of; *Ind*, as indicator; *M*, masking of; *Meas*, measurement of; *Pr*, properties of; *Rgt*, as reagent; *Sp*, separation of

***Index***



**1**, fundamental data; **2**, qualitative; **3**, inorganic gravimetric and volumetric; **4**, gas; **5**, electrometric; **6**, optical; **7**, nuclear and magnetic; **8**, thermal; **9**, biological; **10**, separations; **11**, pH; **12**, specific substances; **13**, technical materials; **14**, statistics; **15**, definitions

### 2.2.8. Merck Index

The last reference book we consider in some detail is a standard single-volume encyclopaedia which is one of the best-known and most-consulted chemical reference books:

> *The Merck index: an encyclopaedia of chemicals and drugs*. Martin Windholz, ed. 10th ed. Rahway, New Jersey: Merck, 1983, 1937pp.

This is a comprehensive interdisciplinary encyclopaedia of drugs, chemicals, and biological substances. It gives concise descriptions of the preparation and general properties of compounds and correlates their trivial, generic, and chemical names with structures, trademarks, and company affiliations. It also provides information on the use, principal pharmacological action, and toxicity of these substances. It uses an A-to-Z arrangement with a comprehensive name index; also it has indexes of *Chemical Abstracts* registry numbers and formulae.

So, to the final detailed look at a reference book in this Section.

**SAQ 2.2e**

The *Merck Index* is primarily intended for quick reference to data on a comprehensive range of subjects: so here are just a few sample quick reference questions to answer from it and get an idea of its contents and layout. Again, if you can use the book itself, do so; if not, use the copies provided. Jot down answers to *all* the questions before reading my response.

You are given a tablet marked 'Valadol'. What is it?

What are the chemical constituents of raspberry?

You are given a sample labelled as a stable aqueous solution of curare:

What concentration would you expect it to have? We all know about curare as a poison used on arrows and darts, but what therapeutic use does it have?

What is the formula of, and what are the alternative names for Valium?

What substance, which occurs in nutmeg butter, is represented by the formula $C_{14}H_{28}O_2$?

What is the Ruzicka large-ring synthesis?

# THE MERCK INDEX

AN ENCYCLOPEDIA OF
CHEMICALS, DRUGS, AND BIOLOGICALS

TENTH EDITION

Martha Windholz, *Editor*
Susan Budavari, *Co-Editor*
Rosemary F. Blumetti, *Associate Editor*
Elizabeth S. Otterbein, *Assistant Editor*

*Published by*
MERCK & CO., INC.
RAHWAY, N.J., U.S.A.

1983

# TABLE OF CONTENTS

Crystals from acetone, mp 104-105.5° (Fr. pat. **M2332**), mp 112° (Offe, *loc. cit.*). $LD_{50}$ i.p. in mice: 7 g/kg.
Sodium salt, $C_8H_{14}NNaO_3$, *Plastenan.*
THERAP CAT: Anti-inflammatory.

**38. 4-Acetamido-2-methyl-1-naphthol.** ***4-Hydroxy-3-methyl-1-naphthaleneacetamide;*** 4-acetylamino-2-methyl-1-naphthol; 1-hydroxy-2-methyl-4-acetylaminonaphthalene; *N*-acetyl vitamin $K_5$; K-Vitrat. $C_{13}H_{13}NO_2$; mol wt 215.24. C 72.54%, H 6.09%, N 6.51%, O 14.87%. Prepd from 4-amino-2-methyl-1-naphthol hydrochloride by treatment with acetic anhydride and anhydr sodium acetate in glacial acetic acid: Veldstra, Wiardi, *Rec. Trav. Chim.* **62,** 75 (1943).

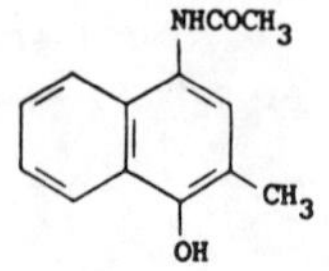

Needles, 208-210°.
THERAP CAT: *See* Vitamin $K_5$.

**39. Acetaminophen.** ***N-(4-Hydroxyphenyl)acetamide; 4'-hydroxyacetanilide;*** *p*-hydroxyacetanilide; *p*-acetamidophenol; *p*-acetaminophenol; *p*-acetylaminophenol; *N*-acetyl-*p*-aminophenol; paracetamol; Acamol; Alpiny; Anhiba; Ben-u-ron; Bickie-mol; Dial-a-gesic; Doliprane; Temlo; Valadol; Cetadol; Acetalgin; Enelfa; Exdol; Finimal; Datril; Dirox; Nobedon; APAP; Tabalgin; Tempra; Calpol; Dymadon; Momentum; Naprinol; Panets; Homoolan; Febrilix; Abensanil; Anaflon; Amadil; Eneril; Tralgon; Hedex; Pacemo; Panex; Panofen; Parmol; Tylenol; Panadol; Lyteca; Apamide; Tapar; Gelocatil; Korum; Paraspen. $C_8H_9NO_2$; mol wt 151.16. C 63.56%, H 6.00%, N 9.27%, O 21.17%. Prepn from *p*-nitrophenol: Morse, *Ber.* **11,** 232 (1878); Tingle, Williams, *Am. Chem. J.* **37,** 63 (1907); from *p*-aminophenol: Lumière *et al., Bull. Soc. Chim. France* [3] **33,** 785 (1905); Fierz-David, Kuster, *Helv. Chim. Acta* **22,** 94 (1939); Wilbert, De Angelis, U.S. pat. **2,998,450** (1961 to Warner-Lambert); Bergmann, **Ger.** pat. **453,577;** *Chem. Zentr.* **1928, I,** 2663; *Frdl.* **16,** 238; from *p*-hydroxyacetophenone hydrazone: Pearson *et al., J. Am. Chem. Soc.* **75,** 5907 (1953). Comprehensive description: J. E. Fairbrother in *Analytical Profiles of Drug Substances* **vol. 3,** K. Florey, Ed. (Academic Press, New York, 1974) pp 1-109. Review of pharmacology: B. Ameer, D. J. Greenblatt, *Ann. Int. Med.* **87,** 202-209 (1977). Review of acetaminophen-induced hepatotoxicity: J. A. Hinson, *Rev. Biochem. Toxicol.* **2,** 103-129 (1980); *idem, Life Sci.* **29,** 107-116 (1981). For proposed toxic metabolite *see* acetimidoquinone.

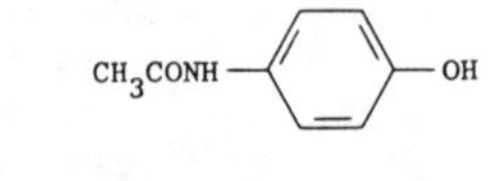

Large monoclinic prisms from water, mp 169-170.5°. $d_4^{21}$ 1.293. uv max (ethanol): 250 nm (ε 13,800). Very slightly sol in cold water, considerably more sol in hot water. Sol in methanol, ethanol, dimethylformamide, ethylene dichloride, acetone, ethyl acetate. Slightly sol in ether. Insol in petr ether, pentane, benzene. $LD_{50}$ in mice (mg/kg): 338 orally, G. A. Starmer *et al., Toxicol. Appl. Pharmacol.* **19,** 20 (1971); 500 i.p., D. C. Dahlin, S. D. Nelson, *J. Med. Chem.* **25,** 885 (1982).
USE: Manuf azo dyes, photographic chemicals.
THERAP CAT: Analgesic, antipyretic.

**40. Acetaminosalol.** *p*-Acetamidophenyl salicylate; acetylaminophenyl salicylate; acetyl-*p*-aminosalol; *p*-acetylaminophenol salicylic acid ester; phenetsal; Salophen; Phenosal. $C_{15}H_{13}NO_4$; mol wt 271.26. C 66.41%, H 4.83%, N 5.16%, O 23.59%. Prepn: Brewster, *J. Am. Chem. Soc.* **40,** 1136 (1918).

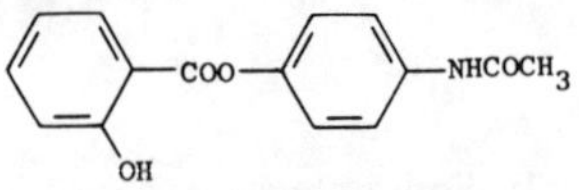

Crystals from hot ethanol, mp 187°. Practically insol in petr ether, cold water, more sol in warm water. Sol in alcohol, ether, benzene. Incompatible with alkalies and alkaline solns which dissolve it with decompn. The alkaline soln gradually becomes blue when boiled, the blue color being discharged upon continued boiling and again produced upon cooling and exposure to air.
THERAP CAT: Analgesic; antipyretic; antirheumatic.
THERAP CAT (VET): Analgesic, antipyretic.

**41. Acetanilide.** ***N-Phenylacetamide;*** antifebrin; acetylaniline; acetylaminobenzene. $C_8H_9NO$; mol wt 135.16. C 71.09%, H 6.71%, N 10.36%, O 11.84%. Usually prepd from aniline and acetic acids: A. I. Vogel, *Practical Organic Chemistry* (London, 3rd ed., 1959) p 577. From aniline and acetyl chloride: Gattermann-Wieland, *Praxis des Organischen Chemikers* (Berlin, 40th ed., 1961) p 114. From aniline, acetone and ketene: Hurd, *Org. Syn.,* **coll. vol. I** (New York, 2nd ed., 1941) p 332. Monograph: Gross, *Acetanilid* (Hillhouse Press, New Haven, 1946).

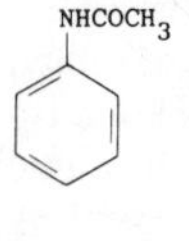

Orthorhombic plates, scales from water. mp 113-115°; bp 304-305°. Slightly burning taste. Appreciably volatile at 95°. $d_4^{15}$ 1.219. Kb at 28°: $1 \times 10^{-13}$. One gram dissolves in 185 ml water, 20 ml boiling water, 3.4 ml alcohol, 3 ml methanol, 0.6 ml boiling alcohol, 3.7 ml chloroform, 4 ml acetone, 5 ml glycerol, 8 ml dioxane, 18 ml ether, 47 ml benzene. Very sparingly sol in petr ether. Chloral hydrate increases the soly of acetanilide in water. $LD_{50}$ orally in rats: 800 mg/kg.
*Pharmaceut. Incompat.* Ethyl nitrate, amyl nitrite, acid solns of nitriles, alkalies (aniline liberated), alkali bromide and iodides in aq soln (insol compounds formed); chloral hydrate, phenol, resorcinol, thymol (these produce a liquid or a soft mass on trituration); ferric salts (red color or precipitate).
USE: Manufacture of other medicinals and of dyes; stabilizer for $H_2O_2$ soln; as addition to cellulose ester varnishes.
THERAP CAT: Antipyretic, analgesic.
THERAP CAT (VET): Antipyretic, analgesic.

**42. *p*-Acetanisidine.** ***N-(4-Methoxyphenyl)acetamide; p-acetanisidide;*** methacetin; *p*-methoxyacetanilide. $C_9H_{11}NO_2$; mol wt 165.19. C 65.44%, H 6.71%, N 8.48%, O 19.37%. $CH_3OC_6H_4NHCOCH_3$. Prepn by acetylation of *p*-anisidine: Reverdin, Bucky, *Ber.* **39,** 2679 (1906); from *p*-acetylanisole and hydroxylamine-*O*-sulfonic acid: Sanford *et al., J. Am. Chem. Soc.* **67,** 1941 (1945).
Cryst powder, feeble, bitter taste, mp 130-132°. Slightly sol in water; sol in alcohol, acetone, chloroform, dil acids, alkalies.

**43. Acetarsone.** ***[3-(Acetylamino)-4-hydroxyphenyl]-arsonic acid; N-acetyl-4-hydroxy-m-arsanilic acid;*** 3-acetamido-4-hydroxyphenylarsonic acid; 3-acetamido-4-hydroxybenzenearsonic acid; acetarsol; acetphenarsine; Ehrlich 594; Fourneau 190; 190 F; F 190; Stovarsol; Amarsan; Arsaphen; Dynarsan; Goyl; Kharophen; Limarsol; Malagride; Gynoplix; Oralcid; Devegan; Orarsan; Osarsal; Osarsol; Osvarsan; Paroxyl; Sanogyl; Spirocid; S.V.C.; Monargan; Ginarsol; Stovarsolan. $C_8H_{10}AsNO_5$; mol wt 275.08. C 34.93%, H 3.66%, As 27.23%, N 5.09%, O 29.08%. The commercial product contains from 27.1 to 27.4% As. Various synthetic routes: Raiziss, Fisher, *J. Am. Chem. Soc.* **48,** 1323 (1926); Hewitt, King, *J. Chem. Soc.* **1926,** 823. Several expired patents.

water; sol in $NH_4OH$; in HCl forming CuCl which dissolves in excess HCl. With dil $H_2SO_4$ or dil $HNO_3$, the cupric salt is formed and half the copper is pptd as the metal. $LD_{50}$ orally in rats: 0.47 g/kg, H. F. Smyth *et al., Am. Ind. Hygiene Assoc. J.* **30,** 470 (1969).

USE: Fungicide; antiseptic for fishnets; in antifouling paints for marine use; in photoelectric cells; as red pigment for glass, ceramic glazes; in brazing pastes; in rectifiers; as catalyst.

**2658. Cuprous Potassium Cyanide.** ***Potassium bis(cyano-C)cuprate(1 −); potassium dicyanocuprate(I);*** potassiocuprous cyanide; potassium cyanocuprate(I). $C_2CuKN_2$; mol wt 154.67. C 15.53%, Cu 41.08%, K 25.28%, N 18.11%. $KCu(CN)_2$. Prepd by evaporation of an aq soln of CuCN and KCN: Staritzky, Walker, *Anal. Chem.* **28,** 419 (1956).

Monoclinic, prismatic crystals. d 2.38. Practically insol in water; dec to CuCN by cold dil $H_2SO_4$ or heating in water; sol in DMSO.

USE: In electroplating copper and brass.

**2659. Cuprous Selenide.** Berzeline; selenkupfer. $Cu_2Se$; mol wt 206.04. Cu 61.68%, Se 38.32%. Occurs in nature as the mineral ***berzelianite*** which in addition to copper and selenium contains 4.73-8.50% silver, 0.35-0.54% iron, up to 0.38% thallium and traces of lead and magnesium. Prepd by reduction of an ammoniacal soln of a cupric salt and $H_2SeO_3$ with hydrazine: Benzing *et al., J. Am. Chem. Soc.* **80,** 2657 (1958); Kulifay, *J. Inorg. Nucl. Chem.* **25,** 75 (1963); alternate methods: Glemser, Sauer in *Handbook of Preparative Inorganic Chemistry,* **vol. 2,** G. Brauer, Ed. (Academic Press, New York, 2nd ed., 1965) p 1019.

Blue-black to black, tetragonal or cubic crystals with a metallic luster. mp 1113°. $d_4^{21}$ 6.84. Sol in HCl with evolution of $H_2Se$, in $H_2SO_4$ with evolution of $SO_2$, in KCN soln; oxidized by $HNO_3$ to $CuSeO_3$.

USE: In semiconductors.

**2660. Cuprous Sulfide.** $Cu_2S$; mol wt 159.15. Cu 79.85%, S 20.15%. Occurs in nature as the mineral ***chalcocite*** also called ***copper glance*** (grey, black, green, blue, or violet rhombic crystals). Prepn: Glemser, Sauer in *Handbook of Preparative Inorganic Chemistry,* **vol. 2,** G. Brauer, Ed. (Academic Press, New York, 2nd ed., 1965) p 1016.

Blue to grayish-black lustrous powder, granules, or orthorhombic crystals. mp about 1100°. $d_4^{20}$ 5.6. On heating in the absence of air it forms Cu and CuS; in the presence of air it forms CuO, $CuSO_4$ and $SO_2$. Practically insol in water, acetic acid; very sparingly sol in HCl; dec by $HNO_3$, concd $H_2SO_4$. Partially sol in $NH_4OH$, readily in cyanide solns due to complex ion formation.

USE: In luminous paints; in electrodes for thermoelements; in preparation of $CuSO_4$; in solid-lubricant compositions; as catalyst.

**2661. Cuprous Sulfite.** $Cu_2O_3S$; mol wt 207.14. Cu 61.35%, O 23.17%, S 15.48%. $Cu_2SO_3$. Prepn and review: Dasent, Morrison, *J. Inorg. Nucl. Chem.* **26,** 1122 (1964).

Hemihydrate, ***Etard's salt.*** White to pale-yellow hexagonal crystals. Slightly sol in water; sol in HCl, $NH_4OH$, alkali solns; practically insol in ether, alcohol.

***Cupro-cupric sulfate.*** $Cu_3O_6S_2$, ***Chevreul's salt.*** Dihydrate, red microcryst powder or prismatic crystals. Practically insol in water, alcohol; sol in $NH_4OH$, HCl.

***Rogojski's salt.*** Formerly considered to be $Cu_2SO_3.H_2O$; recently shown to be an equimolar mixture of Chevreul's salt and metallic copper (Dasent, Morrison, *loc. cit.).* Brick-red solid.

USE: As fungicide for grape vines; in dyeing polyacrylic fibers; polymerization catalyst. Chevreul's salt is a selective molluscicide.

**2662. Cuprous Thiocyanate.** Cuprous sulfocyanate; cuprous sulfocyanide. CCuNS; mol wt 121.62. C 9.87%, Cu 52.24%, N 11.52%, S 26.36%. CuSCN. Prepn: Demmerle *et al., Ind. Eng. Chem.* **42,** 2 (1950); Newman, *Analyst* **88,** 500 (1963).

White to yellow amorphous powder. d 2.85. Practically insol in water, dil acids, alc, acetone; sol in $NH_4OH$, ether, solns of alkali thiocyanates; dec by concd mineral acids.

USE: In marine antifouling paints; in primer compositions for explosives industry.

**2663. Cuproxoline.** ***Tetrahydrogen bis[8-hydroxy-5,7-quinolinedisulfonato(3 −)]cuprate(4 −) compd with N-ethylethanamine (1:4); 8-hydroxy-5,7-quinolinedisulfonic acid copper derivative compound with diethylamine;*** cupric bis-[8-hydroxyquinoline di(diethylammonium sulfonate)]; Dicuprene; Cuprimyl. $C_{34}H_{56}CuN_6O_{14}S_4$; mol wt 964.67. C 42.33%, H 5.85%, Cu 6.59%, N 8.71%, O 23.22%, S 13.30%. Commercial development: U.C.B., Brussels; E. Fougera & Co., New York.

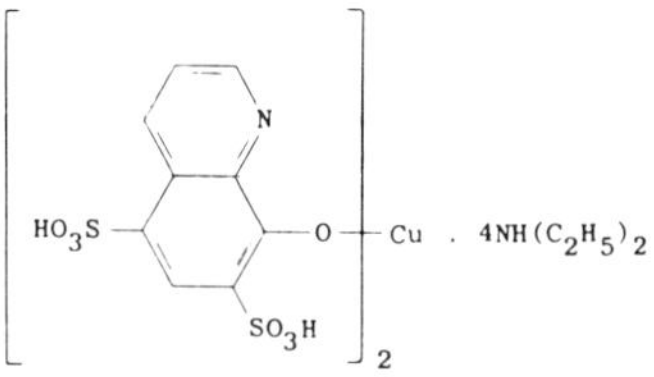

$HO_3S$ … N … O — Cu . $4NH(C_2H_5)_2$ … $SO_3H$ … 2

Green platelets. Sol in water, giving a dark-green soln. A 10% aq soln is almost neutral and can be sterilized by autoclaving.

THERAP CAT: Antiarthritic.

**2664. Curare.** Ourari; urari; woorari; woorali; wourara. Rendering of an Indian name given to the unstandardized extracts derived mainly from the bark of various spp. of Strychnos and Chondodendron, prepared for use as arrow poisons by Indians in the Amazon and Orinoco valleys, and in the Guianas; the physiologically active principle of which is (+)-tubocurarine chloride (*q.v.*). Three kinds of curare have appeared in commerce, distinguished by the kind of containers in which they were packed: ***Tube curare*** or ***bamboo curare, pot curare,*** and ***gourd curare*** or ***calabash curare.*** Listing of members of the family *Menispermaceae*—botanical components of tube or bamboo curare and of pot curare: Krukoff, Moldenke, *Studies of American Menispermaceae, with special reference to species used in the preparation of arrow poisons* in *Brittonia* **3,** 1-74 (1938), also suppl. no. 1-5. (Note that the curare available for medical use under the name Intocostrin is a physiologically standardized extract from *Chondodendron tomentosum* R. & P., *Menispermaceae*). Listing of *Strychnos* spp. *Loganiaceae*—botanical components of calabash curare (and of pot curare): Krukoff, Monachino, *The American Species of Strychnos* in *Brittonia* **4,** 248-322 (1942), also suppl. no. 1-6. Alkaloids from calabash curare and bark of *Strychnos* spp.: P. Karrer, *J. Pharm. Pharmacol.* **8,** 161-164 (1956); Schmid, Karrer, *Helv. Chim. Acta* **29,** 1853 (1946); **30,** 1162 (1947); Marino-Bettolo, *Festschrift Arthur Stoll* (Birkhäuser-Verlag, Basel 1957) pp 257-280. Review of history, chemistry, and use of arrow-poison curare: McIntyre, *Curare* (Chicago, 1947); Bovet *et al., Curare and Curare-like Agents* (Van Nostrand, Princeton, 1959).

Curare is sol in water and in dil alcohol. Stable aq solns (in ampuls) are standardized to contain 20 units per ml.

THERAP CAT: Skeletal muscle relaxant.

THERAP CAT (VET): Muscle relaxant.

**2665. *C*-Curarine I.** $[C_{40}H_{44}N_4O]^{2+}$; mol wt 596.78. Isoln from calabash-curare: Wieland, Pistor, *Ann.* **536,** 68 (1938); Karrer, Schmid, *Helv. Chim. Acta* **29,** 1853 (1946); Zürcher *et al., J. Am. Chem. Soc.* **80,** 1500 (1958). Structure: Nagyvàry *et al., Tetrahedron* **14,** 138 (1961); Grdinic *et al., J. Am. Chem. Soc.* **86,** 3357 (1964); Jones, Nowacki, *Chem. Commun.* **1972,** 805. Prepn of C-dihydrotoxiferine chloride: Bernauer *et al., Helv. Chim. Acta* **40,** 1999 (1957).

12.21%. Prepn: Hoefle, **Ger.** pat. **1,158,927** (1963 to Parke, Davis), *C.A.* **60,** 10608a (1964). Pharmacology, toxicology, preliminary human tolerance studies: L. T. Blouin *et al., J. New Drugs* **3,** 302 (1963). Evaluation in congestive heart failure: E. L. Coodley *et al., J. Clin. Pharmacol.* **19,** 127 (1979).

Crystals from aq acetic acid, mp 165-166°. $LD_{50}$ in mice, rats (mg/kg): 2580, 1400 orally; i.p. in mice: 520 mg/kg, L. T. Blouin *et al., loc. cit.*
THERAP CAT: Diuretic; antihypertensive.

**2960. Diaphen.** *α-Chloro-α-phenylbenzeneacetic acid 2-(diethylamino)ethyl ester;* 2-diethylaminoethyl α-chlorodiphenylacetate; α-chlorodiphenylacetic acid 2-diethylaminoethyl ester; diethylaminoethyl diphenylchloroacetate. $C_{20}H_{24}ClNO_2$; mol wt 345.88. C 69.45%, H 6.99%, Cl 10.25%, N 4.05%, O 9.25%. Prepn: Burtner, Cusic, *J. Am. Chem. Soc.* **65,** 262 (1943).

Crystals, mp 149-151°. $LD_{50}$ i.p. in mice: 80 mg/kg.
THERAP CAT: Hydrochloride as anticholinergic; antihistaminic.

**2961. Diastase of Malt.** Maltin. A mixture contg amylolytic enzymes obtained from malt. Converts at least 50 times its wt of potato starch into sugars (dextrin and maltose) in 30 minutes.
Yellowish-white, amorphous powder or translucent scales. Loses amylolytic power on keeping, on heating its soln above 85°, or on adding much acid. Sol in water with some turbidity; almost insol in alc. *Keep well closed.*
USE: Manuf of sol starch; converting starch into sugar; removing starch from fabrics.

**2962. Diathymosulfone.** *[Sulfonylbis(p-phenyleneazo)]-dithymol;* bis[*p*-(4-hydroxy-2-methyl-5-isopropylphenylazo)phenyl] sulfone; 4,4'-bis(*p*-isopropyl-*m*-cresylazo)diphenyl sulfone; thymol sulfone; Diatox. $C_{32}H_{34}N_4O_4S$; mol wt 570.72. C 67.34%, H 6.00%, N 9.82%, O 11.21%, S 5.62%. Prepn: **Brit.** pat. **758,744** (1956 to Lab. Laborec); Chiarlo *et al., Arch. Ital. Sci. Farmacol.* **7,** 167 (1957); *C.A.* **51,** 16949c (1957).

Red-brown microcrystalline powder from ethanol + water, mp 222-224°. Absorption max (alcohol): 400 nm. Practically insol in water; sol in alkali solns with red color, in dioxane, acetone; less sol in alcohol, ether.
Di-silver salt, $C_{32}H_{32}Ag_2N_4O_4S$, *thymolated silver sulfone.*
THERAP CAT: Antimycobacterial.

**2963. Diatretyne I.** *(E)-8-Amino-8-oxo-2-octene-4,6-diynoic acid; trans-7-carbamoyl-2-heptene-4,6-diynoic acid;* 7-carbamidohepta-2-ene-4,6-diynoic acid; diatretyne amide. $C_8H_5NO_3$; mol wt 163.13. C 58.90%, H 3.09%, N 8.59%, O 29.42%. HOOC—CH=CHC≡CC≡CCONH$_2$. Antibiotic substance produced by the fungus *Clitocybe diatreta:* Anchel, *J. Am. Chem. Soc.* **75,** 4621 (1953); *Arch. Biochem. Biophys.* **78,** 100 (1958). Structure: Bu'Lock *et al., Chem. & Ind. (London)* **1954,** 990. Synthesis: Ashworth *et al., J. Chem. Soc.* **1958,** 950; Prevost et al., *Bull. Soc. Chim. France* **1961,** 2171.
Crystals from methanol, explodes 198°. uv max (95% ethanol): 224, 274, 291, 309.5 nm (ε × $10^{-3}$ 37.7, 16.5, 23, 19).

**2964. Diatretyne II.** *7-Cyano-2-heptene-4,6-diynoic acid;* diatretyne nitrile; nudic acid B. $C_8H_3NO_2$; mol wt 145.12. C 66.21%, H 2.08%, N 9.65%, O 22.05%. HOOC-CH=CHC≡CC≡CCN. Antibiotic substance produced by the basidiomycete *Tricholoma nudum* and by the fungus *Clitocybe diatreta.* Isoln: Anchel, *J. Am. Chem. Soc.* **74,** 1588 (1952). Structure: Anchel, *Science* **121,** 607 (1955). Identity of nudic acid B and diatretyne II: Heatly, Stephenson, *Nature* **179,** 1078 (1957). Synthesis: Ashworth *et al., J. Chem. Soc.* **1958,** 950.
Short needles from ether + light petr, dec 179-180°. Darkens and turns black within 2 hrs on exposure to sunlight. Sol in the usual organic solvents except hexane; dissolves slowly in sodium carbonate soln; practically insol in water and dil acids.

**2965. Diatrizoate Sodium.** *3,5-Bis(acetylamino)-2,4,6-triiodobenzoic acid sodium salt; 3,5-diacetamido-2,4,6-triiodobenzoic acid sodium salt;* sodium 3,5-diacetamido-2,4,6-triiodobenzoate; urografic acid sodium salt; sodium amidotrizoate; sodium diatrizoate; Hypaque Sodium. $C_{11}H_8I_3N_2NaO_4$; mol wt 635.92. C 20.78%, H 1.27%, I 59.87%, N 4.41%, Na 3.62%, O 10.06%. Prepn of the acid: Larsen *et al., J. Am. Chem. Soc.* **78,** 3210 (1956). Description: Langecker *et al., Arch. Exp. Pathol. Pharmakol.* **222,** 584 (1954). Comprehensive description: H. H. Lerner in *Analytical Profiles of Drug Substances* **vol. 4,** K. Florey, Ed. (Academic Press, New York, 1975) pp 137-167.

Rhombic needles, slightly salty taste, mp 261-262° (dec). Soly in water at 20°: 60 g/100 ml. pH of a 50% aq soln 7.0 to 7.5. On chilling solns may become cloudy or form a precipitate which disappears at room temp. Solns may be sterilized by autoclaving. Should be protected from light. $LD_{50}$ i.v. in rats: 14.7 g/kg, Langecker *et al., loc. cit.*
Free acid, crystals from dil DMF, mp > 300°.
THERAP CAT: Diagnostic aid (radiopaque medium).

**2966. Diaveridine.** *5-[(3,4-Dimethoxyphenyl)methyl]-2,4-pyrimidinediamine; 2,4-diamino-5-veratrylpyrimidine;* 2,4-diamino-5-(3',4'-dimethoxybenzyl)pyrimidine. $C_{13}H_{16}N_4O_2$; mol wt 260.29. C 59.98%, H 6.20%, N 21.53%, O 12.29%. Prepn: Falco *et al., J. Am. Chem. Soc.* **73,** 3758 (1951). Hitchings, Falco, **U.S.** pat. **2,624,732** (1953 to Burroughs Wellcome); Stenbuck *et al., J. Org. Chem.* **28,** 1983 (1963); Hoffer, **U.S.** pat. **3,341,541** (1967 to Hoffmann-La Roche); M. Hoffer *et al., J. Med. Chem.* **14,** 462 (1971). Crystal structure: T. F. Koetzle, G. Williams, *Acta Crystallogr.* **B34,** 323 (1978).

Crystals, mp 233°.
Ingredient of ***Darvisul*** which also contains sulfaquinoxaline.
THERAP CAT (VET): Antiprotozoan (in combination with sulfaquinoxaline).

**2967. Diazepam.** *7-Chloro-1,3-dihydro-1-methyl-5-phenyl-2H-1,4-benzodiazepin-2-one;* 7-chloro-1-methyl-5-phenyl-3*H*-1,4-benzodiazepin-2(1*H*)-one; methyl diazepinone; diacepin; La III; Ro 5-2807; Wy 3467; NSC-77518; Alupram; Ansiolin; Apaurin; Apozepam; Atensine; Atilen; Bialzepam; Calmpose; Ceregulart; Diazemuls; Dipam; Eridan; Evacalm; Faustan; Lembrol; Levium; Morosan; Neurolytril; Noan; Pacitran; Paxate; Paxel; Relanium; Sedapam; Seduxen; Setonil; Stesolid; Stesolin; Tranimul; Tranquase; Tranquo-Tablinen; Valium; Valrelease; Vival; Vivol. $C_{16}H_{13}ClN_2O$; mol wt 284.76. C 67.49%, H 4.60%, Cl 12.45%, N 9.84%, O 5.62%. Prepn: Sternbach, Reeder, *J. Org. Chem.* **26**, 4936 (1961); Reeder, Sternbach, **U.S.** pat. **3,371,-085** (1968 to Hoffmann-La Roche); prepd but not claimed: *eidem*, **U.S.** pats. **3,109,843** and **3,136,815** (1963, 1964 to Hoffmann-La Roche); M. Gates, *J. Org. Chem.* **45**, 1675 (1980); M. Ishikura *et al., ibid.* **47**, 2456 (1982). Purification: Chase, U.S. pat. **3,102,116** (1963 to Hoffmann-La Roche). Pharmacology: Hudson, Wolpert, *Arch. Int. Pharmacodyn. Ther.* **186**, 388 (1970). Metabolism: Randall *et al., Curr. Ther. Res. Clin. Exp.* **7**, 590 (1965); Van der Kleijn, *Ann. N.Y. Acad. Sci.* **179**, 115 (1971). Comprehensive description: A. MacDonald *et al.*, in *Analytical Profiles of Drug Substances* **vol. 1**, K. Florey, Ed. (Academic Press, New York, 1972) pp 79-99. Clinical pharmacokinetics: M. Mandelli *et al., Clin. Pharmacokinet.* **3**, 72 (1978).

Plates from acetone + petr ether, mp 125-126°. $pK_a$ 3.4. Sol in chloroform, DMF, benzene, acetone, alc; slightly sol in water. $LD_{50}$ orally in rats: 710 mg/kg, E. I. Goldenthal, *Toxicol. Appl. Pharmacol.* **18**, 185 (1971).

*Caution:* Abuse may lead to habituation or addiction.

THERAP CAT: Sedative; tranquilizer (minor).

**2968. Diazinon®.** *Phosphorothioic acid O,O-diethyl O-[6-methyl-2-(1-methylethyl)-4-pyrimidinyl] ester;* thiophosphoric acid 2-isopropyl-4-methyl-6-pyrimidyl diethyl ester; *O,O*-diethyl *O*-2-isopropyl-4-methyl-6-pyrimidyl thiophosphate; diethyl 2-isopropyl-4-methyl-6-pyrimidyl thionophosphate; dimpylate; diazinon (outside U.S.); G-24480; Basudin; Diazol; Garden Tox; Sarolex; Spectracide. $C_{12}H_{21}N_2O_3PS$; mol wt 304.36. C 47.36%, H 6.95%, N 9.20%, O 15.77%, P 10.18%, S 10.54%. Prepn: Gysin, Margot, U.S. pat. **2,754,243** (1956 to Geigy). Insecticidal properties: Gasser, *Z. Naturforsch.* **8b**, 225 (1953).

Liquid. Faint ester-like odor. $d_4^{20}$ 1.116-1.118. $bp_{0.002}$ 83-84°. $n_D^{20}$ 1.4978-1.4981. Vapor pressure at 20° = 1.4 × $10^{-4}$, at 40° = 1.1 × $10^{-3}$ (about 5 times vapor pressure of parathion). Volatility at 20° = 2.4 mg/cubic meter; at 40° = 17.6 mg/cubic meter. Decomposes above 120°. Soly in water at 20° = 0.004%. Miscible with alcohol, ether, petr ether, cyclohexane, benzene, and similar hydrocarbons. More stable in alkaline formulations, then when at neutral or acid pH. For $LD_{50}$ values see Gaines, *Toxicol. Appl. Pharmacol.* **14**, 515 (1969).

USE: Insecticide. *Caution:* Cholinesterase inhibitor.

**2969. Diazoacetic Ester.** Ethyl diazoacetate; ethyl diazoethanoate. $C_4H_6N_2O_2$; mol wt 114.10. C 42.10%, H 5.30%, N 24.55%, O 28.04%. $^-N{=}N^+{=}CHCOOC_2H_5$. Prepd by the action of sodium nitrite on glycine ethyl ester hydrochloride: Curtius *J. Prakt. Chem.* [2] **38**, 396 (1888); Silberrad, *J. Chem. Soc.* **81**, 600 (1902); Womack, Nelson, *Org. Syn.* **24**, 56 (1944).

Yellow oil. Pungent odor. Very volatile. Explosive. *Distillation, even under reduced pressure, is dangerous.* mp −22°. $d_4^{17.6}$ 1.0852. $bp_5$ 42°; $bp_{12}$ 45°; $bp_{88}$ 85-86°; $bp_{720}$ 140-141°. $n_D^{17.6}$ 1.4588. Slightly sol in water. Neutral reaction. Volatile with steam, ether, and benzene vapors. Miscible with alcohol, benzene, ligroin, ether. Explodes on contact with concd $H_2SO_4$.

**2970. Diazoaminobenzene.** *1,3-Diphenyl-1-triazene;* anilinoazobenzene; benzeneazoaniline. $C_{12}H_{11}N_3$; mol wt 197.11. C 73.07%, H 5.62%, N 21.31%. Made by diazotizing aniline dissolved in HCl with $NaNO_2$ and then adding a concd soln of sodium acetate: Hartman, Dickey, *Org. Syn.* **coll. vol. II**, 163 (1943).

Golden-yellow, small crystals. mp 98°; explodes when heated to 150°. Insol in water; freely sol in benzene, ether, hot alc.

**2971. *p*-Diazobenzenesulfonic Acid.** *p-Sulfobenzenediazonium hydroxide inner salt;* sulfanilic acid diazide. $C_6H_4N_2O_3S$; mol wt 184.17. C 39.13%, H 2.19%, N 15.21%, O 26.06%, S 17.41%. Prepn by diazotization of sulfanilic acid and structure: Whetsel *et al., J. Am. Chem. Soc.* **78**, 3360 (1956); Bugg *et al., Acta Cryst.* **17**, 767 (1964).

White or slightly red crystals; when dry, easily explodes by heat, a blow, or on rubbing; hence marketed as a paste. Slightly sol in cold water, alcohol; freely sol in hot water, dil alkalies, HCl.

USE: As reagent for phenol, fecal matter in water, glucose, aldehydes, and albumin; also in manuf of azo and other dyes.

**2972. Diazomethane.** Azimethylene. $CH_2N_2$; mol wt 42.04. C 28.57%, H 4.80%, N 66.64%. $CH_2{=}N^+{=}N^-$. Prepd from chloroform and hydrazine by reaction with potassium hydroxide: Staudinger, Kupfer, *Ber.* **45**, 501 (1912); McManus *et al., J. Org. Chem.* **33**, 4272 (1968); from KOH and nitrosomethylurea: Dessaux, Durand, *Bull. Soc. Chim. France* **1963**, 41. These methods yield gaseous diazomethane. The following procedures yield ether solns of diazomethane. From *N*-nitroso-*β*-methylaminoisobutyl methyl ketone in ether and isopropanol by reaction with sodium isopropoxide or from the same ketone in ether by reaction with sodium cyclohexoxide: Redemann *et al., Org. Syn.* **25**, 28 (1945). By KOH saponification of nitrosomethylurea in ether: Arndt, *Org. Syn.* **coll. vol. II**, 165 (1943), or of nitrosomethylurethan in ether: von Pechmann, *Ber.* **27**, 1888 (1894); **28**, 855 (1895); Meerwein, Burneleit, *Ber.* **61**, 1845 (1928). In the laboratory diazomethane may be prepd most simply by the action of alkali on the commercially available *N*-methyl-*N*-nitroso-*N'*-nitroguanidine, *cf.* McKay, *J. Am. Chem. Soc.* **70**, 1974 (1948); McKay *et al., Can. J. Res.* **28**, 683 (1950); Fieser and Fieser, *Organic Chemistry* (Reinhold, New York, 3rd ed., 1956) p 176. Alternate intermediate: *p*-tolylsulfonylmethylnitrosamide ("Diazald" Aldrich Chemical Co.), *see* De Boer, Backer, *Org. Syn.* **coll. vol. IV**, 250 (1963).

Very toxic yellow gas. mp −145°; bp −23°. Undil liq and concd solns may explode violently, especially if impurities are present. Gaseous diazomethane may explode on heating to 100° or on rough glass surfaces. Ground glass apparatus and glass stirrers with glass sleeve bearings where grinding may occur, should not be used. Alkali metals also produce explosions with diazomethane. Sol in ether, dioxane. Such solns dec only slowly at low temps. Decompn is more rapid



*Consult the cross index before using this section.*

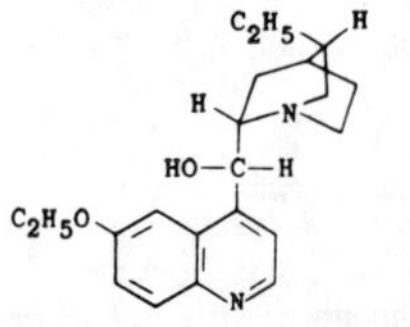

3755 **3,3′-Ethylidenebis(4-hydroxycoumarin)**

White bitter cryst powder. mp 123-128° when solvent-free. $[\alpha]_D^{25}$ −136.2° (alcohol). Practically insol in water; sol in alcohol, benzene, chloroform, ether, dil acids, oils, fats. LD s.c. in mice: 5.0 g/kg.

Hydrochloride, $C_{21}H_{29}ClN_2O_2$, *Neumolisina.* Rhombic crystals from acetone + ether, mp 252-254°. $[\alpha]_D^{21}$ −123.6° (water). One gram dissolves in 2 ml water, 5 ml alcohol, 2.5 ml chloroform; sparingly sol in dry acetone. Practically insol in ether, petr ether. Aq solns are sterilizable without decompn. *Protect from light.* LD s.c. in mice: 5.0 g/kg.
THERAP CAT: Antiseptic.

**3755. 3,3′-Ethylidenebis(4-hydroxycoumarin).** *3,3′-Ethylidenebis[4-hydroxy-2H-1-benzopyran-2-one];* 3,3-ethylidenebis(4-oxycoumarin); Pertrombon. $C_{20}H_{14}O_6$; mol wt 350.31. C 68.57%, H 4.03%, O 27.40%. Prepn: Jansen, Jensen, *Z. Phys. Chem.* **277,** 66 (1943); Sullivan *et al., J. Am. Chem. Soc.* **66,** 2288 (1943); Stahmann *et al., ibid.* **67,** 900 (1944); **Austrian** pat. **196,874** (1958 to Spofa), *C.A.* **52,** 10212d (1958); Fucik, **Czech.** pat. **85,918** (1957), *C.A.* **52,** 2930d (1958).

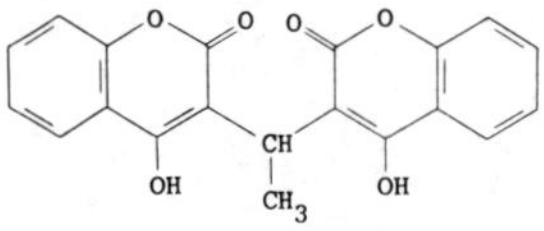

Crystals from ethanol + dioxane, mp 178°.
Dibenzoate, $C_{34}H_{22}O_8$, mp 209-210°.
THERAP CAT: Anticoagulant.

**3756. Ethylidene Chloride.** *1,1-Dichloroethane.* $C_2H_4Cl_2$; mol wt 98.97. C 24.27%, H 4.07%, Cl 71.65%. $CH_3CHCl_2$. Prepd by the action of $PCl_5$ on acetaldehyde.

Oily liquid; odor and taste as of chloroform. $d_4^{20}$ 1.1757; $d_4^{25}$ 1.1680. bp 57.3°: Mumford, Phillips, *J. Chem. Soc.* **1950,** 75. Solidif about −98°. $n_D^{20}$ 1.4167. Sol in about 200 parts water; miscible with alcohol.
*Caution:* Narcotic in high concns.

**3757. Ethylidene Diacetate.** *1,1-Ethanediol diacetate;* 1,1-diacetoxyethane. $C_6H_{10}O_4$; mol wt 146.14. C 49.31%, H 6.90%, O 43.79%. $(CH_3COO)_2CHCH_3$. Made by heating acetaldehyde and acetic anhydride. Control of fungal and bacterial growth in crops and animal feedstuffs: D. L. Kensler *et al.,* **U.S.** pats. **3,931,412** and **4,012,526** (1976 and 1977 to Chevron).

Liq; sharp, fruity odor. $d_4^{12}$ 1.061. bp 167-169°. Slightly sol in water; miscible with alcohol. With NaOH it yields acetaldehyde.
USE: Agricultural fungicide.

**3758. Ethyl Iodide.** *Iodoethane.* $C_2H_5I$; mol wt 155.98. C 15.40%. H 3.23%, I 81.37%. $CH_3CH_2I$. Made by the action of iodine on alcohol in the presence of red phosphorus. Ethyl iodide contains about 98% $CH_3CH_2I$ and about 2% alcohol.

Heavy, clear, very refractive liquid; ethereal odor. When freshly prepared it is colorless, but soon becomes red (on exposure to light and air) due to liberation of iodine. Silver leaf prevents or greatly retards decompn. $d_{20}^{20}$ 1.950. bp 72°. mp −108°. $n_D^{15}$ 1.5168. Sol in 250 parts water with gradual decompn; miscible with alcohol and most organic solvents. *Keep dry and cool, tightly closed, protected from light.*
THERAP CAT: Diagnostic aid (determination of cardiac output).

**3759. Ethyl Isobutyrate.** *2-Methylpropanoic acid ethyl ester.* $C_6H_{12}O_2$; mol wt 116.16. C 62.04%, H 10.41%, O 27.55%. $(CH_3)_2CHCOOC_2H_5$.

Liquid; aromatic, fruity odor. $d_{20}^{20}$ 0.870. bp 110-111°. mp −88°. $n_D^{20}$ 1.3903. Slightly sol in water; miscible with alcohol, ether.
USE: Manuf flavoring compds and essences.

**3760. Ethyl Isothiocyanate.** *Isothiocyanatoethane;* ethyl mustard oil. $C_3H_5NS$; mol wt 87.15. C 41.34%, H 5.78%, N 16.08%, S 36.80%. $C_2H_5NCS$. Obtained by the action of mercuric chloride on the product of the reaction of carbon disulfide with ethylamine.

Colorless to yellowish liq; pungent odor. $d_4^{18}$ 1.003. bp 130-132°. mp −6°. $n_D^{18}$ 1.5142. Insol in water; miscible with alcohol, ether.
USE: Military poison gas. *Caution:* Vapors irritate the eyes; blister the skin.

**3761. Ethyl Isovalerate.** *3-Methylbutanoic acid ethyl ester.* $C_7H_{14}O_2$; mol wt 130.18. C 64.58%, H 10.84%, O 24.58%. $(CH_3)_2CHCH_2COOC_2H_5$.

Colorless, oily liq; apple odor. $d_{20}^{20}$ 0.868. bp 135°. mp −99°. $n_D^{20}$ 1.4009. Sol in about 350 parts water; miscible with alcohol, benzene, ether.
USE: In alc soln for flavoring confectionery and beverages.

**3762. Ethyl Lactate.** *2-Hydroxypropanoic acid ethyl ester;* ethyl α-hydroxypropionate. $C_5H_{10}O_3$; mol wt 118.13. C 50.83%, H 8.53%, O 40.63%. $CH_3CHOHCOOC_2H_5$.

Colorless liquid; characteristic odor. $d_4^{14}$ 1.042. bp 154°. $[\alpha]_D^{14}$ −10°. Flash pt, closed cup: 117°F (47°C). Miscible with water (with partial decompn), alcohol, ether. *Keep well closed.*
USE: As solvent for nitrocellulose and cellulose acetate.

**3763. Ethyl Laurate.** *Dodecanoic acid ethyl ester.* $C_{14}H_{28}O_2$; mol wt 228.36. C 73.63%, H 12.36%, O 14.01%. $CH_3(CH_2)_{10}COOC_2H_5$.

Oil. $d_{19}^{19}$ 0.867. mp −10°. bp 269°; $bp_{25}$ 163°. $n_D$ 1.4321. Insol in water; very sol in alcohol, ether.

**3764. Ethyl Levulinate.** *4-Oxopentanoic acid ethyl ester.* $C_7H_{12}O_3$; mol wt 144.17. C 58.31%, H 8.39%, O 33.29%. $CH_3COCH_2CH_2COOC_2H_5$.

Liquid. $d_{20}^{20}$ 1.012. bp 205-206°. $n_D^{20}$ 1.4229. Freely sol in water; miscible with alcohol.

**3765. Ethyl Linoleate.** *9,12-Octadecadienoic acid ethyl ester;* mandenol. $C_{20}H_{36}O_2$; mol wt 308.49%. C 77.86%, H 11.76%, O 10.37%. Prepn from sunflower seed oil: McCutcheon, *Org. Syn.* **coll. vol. III,** 526 (1955); *see also* Parker *et al., Biochem. Prepn.* **4,** 88 (1955).

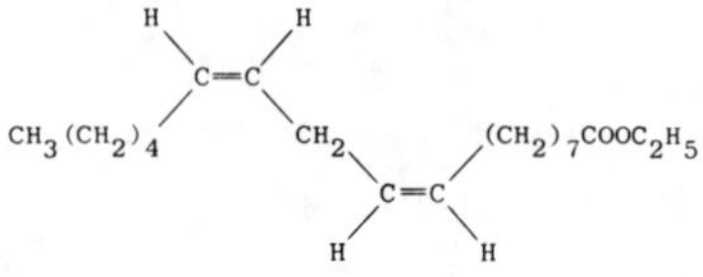

Colorless oil. More stable to air oxidation than linoleic acid. $d_4^{15.5}$ 0.8846; $d^{20}$ 0.8919. $bp_6$ 193°; $bp_{2.5}$ 175°; $bp_{0.001}$ 133°. $n_D^{20}$ 1.46753; $n_D^{48}$ 1.4489. Iodine value 162.5. Miscible with dimethylformamide, fat solvents, oils.
USE: In the vitamin industry.

**3766. Ethyl Loflazepate.** *7-Chloro-5-(2-fluorophenyl)-2,3-dihydro-2-oxo-1H-1,4-benzodiazepine-3-carboxylic acid ethyl ester;* ethyl 7-chloro-5-(*o*-fluorophenyl)-2,3-dihydro-2-oxo-1*H*-1,4-benzodiazepine-3-carboxylate; CM-6912; Victan. $C_{18}H_{14}ClFN_2O_3$; mol wt 360.77. C 59.93%, H 3.91%, Cl 9.83%, F 5.27%, N 7.76%, O 13.30%. Benzodiazepine with anxiolytic activity. Prepn: J. Hellerbach *et al.,* **Ger.** pat. **2,001,276** corresp to **U.S.** pat. **3,657,223** (1970, 1972 both to Hoffmann-La Roche). Analysis in plasma and urine: W. Cantreels, J. R. Jeanniot, *Biomed. Mass Spectrom.* **7,** 565 (1980). GC analysis of metabolites: J. P. Cano, *J. Chromatog.* **226,** 413 (1981). Toxicologic evaluation: G. Mozue *et al., Int. J. Clin. Pharmacol. Ther. Toxicol.* **19,** 453 (1981).

Muscle" in *The Proteins,* **vol. 4,** H. Neurath, R. L. Hill, Eds. (Academic Press, New York, 1979) p 245-409. Polymorphism of mammalian myosins and distribution of isoforms in different muscle fibers reviewed by A. G. Weeds in *Plasticity of Muscle,* D. Pette, Ed. (DeGruyter, Berlin, 1980) pp 55-68. Role of myosin light chains in calcium regulation: Kendrick-Jones, *Nature* **249,** 631 (1974). *See also* Tropomyosin.

**6180. β-Myrcene.** *7-Methyl-3-methylene-1,6-octadiene;* 2-methyl-6-methylene-2,7-octadiene. $C_{10}H_{16}$; mol wt 136.23. C 88.16%, H 11.84%. Found in oil of bay, verbena, hop, and others. Isoln: Power, Kleber, *Pharm. Rundschau* **13,** 60 (1895); *see also* Ruzicka, Stoll, *Helv. Chim. Acta* **7,** 272 (1924); Goulding, Roberts, *J. Chem. Soc.* **105,** 2614 (1914); Booker *et al., ibid.* **1940,** 1453; Kugler, Kováts, *Helv. Chim. Acta* **46,** 1480 (1963). Obtained by pyrolysis of β-pinene: Goldblatt, Palkin, *J. Am. Chem. Soc.* **63,** 3517 (1941); Rummelsburg, **U.S.** pat. **2,444,790** (1948 to Hercules Powder); Houlihan *et al., J. Am. Chem. Soc.* **81,** 4692 (1959). Separation of isomers: Ohloff *et al., Ann.* **675,** 83 (1964). Synthesis: O. P. Vig *et al., Indian J. Chem.* **7,** 450 (1969); **11,** 104 (1973); **13,** 1244 (1975); T. Mandai *et al., Tetrahedron Letters* **22,** 763 (1981). Formation from isoprene: K. Takabe *et al., Synthesis* **1977,** 307.

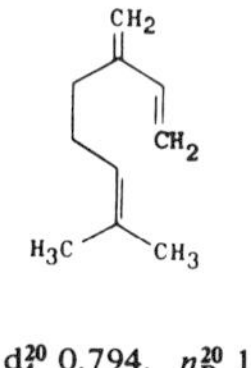

Oil. Pleasant odor. $d_4^{20}$ 0.794. $n_D^{20}$ 1.4709. uv max (ethanol): 226 nm (ε 16,100). Practically insol in water; sol in alcohol, chloroform, ether, glacial acetic acid.

*α-Myrcene. 2-Methyl-6-methylene-1,7-octadiene.* $bp_{10}$ 44°. $n_D^{25}$ 1.4661. $d_{25}^{25}$ 0.7959. uv max (isooctane): 224.5 nm (ε 18,600). Not found in nature. Prepn: Mitzner *et al., J. Org. Chem.* **30,** 646 (1965); Vig *et al., J. Indian Chem. Soc.* **50,** 329 (1973).

USE: β-Myrcene as an intermediate in the manuf of perfume chemicals.

**6181. Myricetin.** *3,5,7-Trihydroxy-2-(3,4,5-trihydroxyphenyl)-4H-1-benzopyran-4-one; 3,3′,4′,5,5′,7-hexahydroxyflavone;* cannabiscetin; delphidenolon 1575. $C_{15}H_{10}O_8$; mol wt 318.23. C 56.61%, H 3.17%, O 40.22%. From the bark of *Myrica nagi* Thunb., *Myriaceae:* Perkin, Hummel, *J. Chem. Soc.* **69,** 1287 (1896). Structure: Perkin, *ibid.* **81,** 203 (1902). Identity with cannabiscetin: Seshadri, Venkateswarlu, *Proc. Indian Acad. Sci.* **23A,** 296 (1946); *C.A.* **40,** 6447 (1946). Synthesis: Kalff, Robinson, *J. Chem. Soc.* **127,** 181 (1925); Rao *et al., J. Sci. Ind. Res.* **8B,** No. 6, 113 (1949). Occurrence in *Hamamelidaceae* and *Anacardiaceae:* Reznek, Egger, *Z. Naturforsch.* **15b,** 247 (1960). Metabolism: Smith, Griffiths, *Biochem. J.* **118,** 53p (1970).

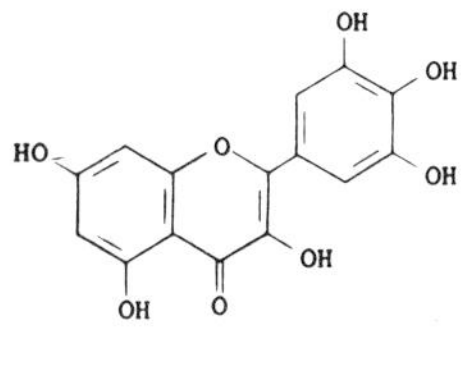

Yellow needles from dil alc, mp 357°. uv max (ethanol): 375, 255 nm. Sparingly sol in boiling water; sol in alcohol; practically insol in chloroform, acetic acid.

Hexaacetate, $C_{27}H_{22}O_{14}$, crystals, mp 213°.

Hexaethyl ether, $C_{27}H_{34}O_8$, needles from alcohol, mp 149-151°.

3-Rhamnoside, $C_{21}H_{20}O_{12}$, *myricitrin.* Structure: Hattori, Hayashi, *Acta Phytochim.* **5,** 213 (1931), *C.A.* **26,** 990[8] (1932). Pale yellow leaflets from water, mp 199-200°. uv max (alc): 262, 352 nm. Sparingly sol in water, abs alcohol.

**6182. Myristica.** Nutmeg; nux moschata; nuces (semen) nucistae. Dried ripe seed of *Myristica fragrans* Houtt., *Myristicaceae,* deprived of its seed coat and with or without a thin coating of lime. *Habit.* Southern Asia, Moluccas, cultivated in many tropical countries. *Constit.* 25-35% fixed oil and 5-15% volatile oils. The myristicin fraction of the latter, together with elemicin (> 25% content) is supposedly responsible for the purported hallucinogenic properties of nutmeg seed: Shulgin, *Nature* **197,** 4865 (1963); Weil, *Econ. Bot.* **19,** 194 (1965). Review of botany, history, chemistry and psychopharmacology: *Public Health Service Publ. No. 1645,* D. H. Efron, Ed., 183-229 (1967).

USE: As a spice in food; as source of myristica oil. *Caution:* Ingestion of large quantities causes drowsiness, stupor, death.

THERAP CAT: Carminative.

**6183. Myristic Acid.** *Tetradecanoic acid.* $C_{14}H_{28}O_2$; mol wt 228.36. C 73.63%, H 12.36%, O 14.01%. $CH_3(CH_2)_{12}COOH$. Occurs in nutmeg butter *(Myristica fragrans* Houtt.) to the extent of 70-80%; predominates in the fats of the *Myristicaceae;* in palm seed fats it may comprise 20% of the total fatty acids; in milk fats between 8-12% of the total acids. Occurs in most animal and vegetable fats; has been found in considerable amounts (up to 15%) in sperm whale oil, *see* Markley, *Fatty Acids* (New York, 1947). Prepn: G. D. Beal, *Org. Syn.* **coll. vol. I,** 379 (2nd ed., 1941). Prepn from tall-oil fatty acids: Segessmann, Molnar, **U.S.** pat. **2,481,356** (1949); from 9-ketotetradecanoic acid: Ames *et al., J. Chem. Soc.* **1950,** 174; by electrolysis of methyl hydrogen adipate + decanoic acid: Greaves *et al., ibid.* **1950,** 3326; by Maurer oxidation of myristyl alc: Langenbeck, Richter, *Ber.* **89,** 202 (1956); from cetanol: Selwitz, **U.S.** pat. **2,969,380** (1961 to Gulf). Wide-line NMR spectrum: A. V. Bailey, R. A. Pittman, *J. Am. Oil Chem. Soc.* **48,** 775 (1971).

Crystals from methanol, mp 58.5°. $bp_{100}$ 250.5°; $bp_{16}$ 199°; $bp_4$ 184°. $d_4^{54}$ 0.8622. $d_4^{70}$ 0.8528. $d_4^{90}$ 0.8394. $n_D^{60}$ 1.4305; $n_D^{70}$ 1.4273. Soluble in abs alcohol, methanol, ether, petr ether, benzene, chloroform. Neutralization value: 245.68. Absorption spectrum: Markley, *op. cit.* $LD_{50}$ i.v. in mice: 43±2.6 mg/kg, L. Orö, A. Wretlind, *Acta Pharmacol. Toxicol.* **18,** 141 (1961).

Ethyl ester, $C_{16}H_{32}O_2$, *ethyl myristate.* Liquid, mp 12°; bp 295°; $bp_{30}$ 195°. $d_4^{20}$ 0.856. Insol in water; sol in alcohol; slightly sol in ether.

USE: As ingredient in soaps and shaving creams; in lubricants; in coatings for anodized aluminum.

**6184. Myristyl Alcohol.** *1-Tetradecanol;* tetradecyl alcohol. $C_{14}H_{30}O$; mol wt 214.38. C 78.43%, H 14.11%, O 7.46%. $CH_3(CH_2)_{12}CH_2OH$. Prepn by sodium reduction of fatty acid esters: Hansley, *Ind. Eng. Chem.* **39,** 55 (1947); by $LiAlH_4$ reduction of fatty acids: Watanabe, *Bull. Chem. Soc. Japan* **34,** 398 (1961); from acetaldehyde + dimethylamine: Langenbeck *et al., J. Prakt. Chem.* **8,** 112 (1959).

White crystals, mp 38°. d 0.824. $bp_{15}$ 167°. Practically insol in water; sol in ether, slightly in alcohol.

USE: As emollient for cold creams, etc., also for making the sulfated alcohol whose sodium salt is applicable as a "wetter" in textiles.

**6185. Myristyltrimethylammonium Bromide.** *N,N,N-Trimethyl-1-tetradecanaminium bromide; trimethyltetradecylammonium bromide;* tetradonium bromide; Morpan T; Mytab. $C_{17}H_{38}BrN$; mol wt 336.41. C 60.70%, H 11.39%, Br 23.75%, N 4.16%. $[CH_3(CH_2)_{13}N(CH_3)_3]^+Br^-$. Cationic germicidal detergent. Prepn and antibacterial activity: R. S. Shelton *et al., J. Am. Chem. Soc.* **68,** 753 (1946). Toxicity and pharmacology: B. Isomaa, K. Bjondahl, *Acta Pharmacol. Toxicol.* **47,** 17 (1980).

White powder, mp 245-250°. Sol in 5 parts water. *Caution: Corrosive!* $LD_{50}$ i.v. in mice, rats: 12.0, 15.0 mg/kg, B. Issomaa, K. Bjondahl, *loc. cit.*

USE: Disinfectant; deodorant; laboratory reagent.

**6186. Myrophine.** *7,8-Didehydro-4,5-epoxy-17-methyl-3-(phenylmethoxy)morphinan-6-ol tetradecanoate(ester);* 3-benzylmorphine 6-myristate; benzylmorphine myristic acid ester; benzylmorphinyl myristate; myristyl benzylmorphine; 3-benzyloxy-6-myristoyloxy-*N*-methyl-4,5-epoxy-7-morphinene; 3-benzyloxy-6-myristoyloxymorphine; myro-

B. J. Price *et al.*, **Fr.** pat. **2,384,765** corresp to **U.S.** pat. **4,128,658** (both 1978 to Allen & Hanburys). HPLC determn in plasma: P. F. Carey, L. E. Martin, *J. Liq. Chromatog.* **1979,** 1291. Pharmacological studies: J. Bradshaw *et al., Brit. J. Pharmacol.* **66,** 464 (1979); M. J. Daly *et al., Gut* **21,** 408 (1980). Efficacy in treatment of duodenal ulcers: A. Berstad *et al., Scand. J. Gastroenterol.* **15,** 637 (1980); R. P. Walt *et al., Gut* **22,** 49 (1981).

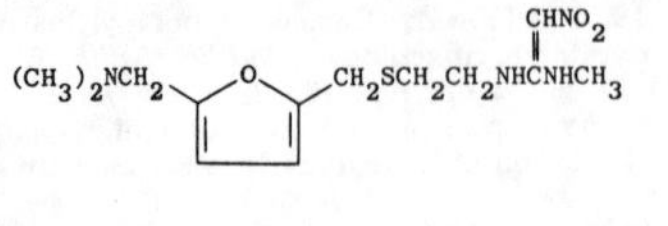

$(CH_3)_2NCH_2$–(furan)–$CH_2SCH_2CH_2NHC(=CHNO_2)NHCH_3$

Solid, mp 69-70°.

Hydrochloride, $C_{13}H_{23}ClN_4O_3S$, ***AH 19065, Ranidil, Zantac.*** Off-white solid, mp 133-134°.

THERAP CAT: Antagonist (to histamine $H_2$ receptors), esp in treatment of duodenal ulcers.

**8020. Rapeseed Oil.** Colza oil. Oil expressed from seeds of *Brassica campestris* L., *Cruciferae.*

Pale yellow, rather viscid liq. d 0.913-0.917; $n_D^{20}$ 1.4720-1.4752. Solidif −2° to −10°. Sapon no. 170-177. Iodine no. 97-105. Sol in chloroform, ether, $CS_2$.

USE: Lubricant; manuf rubber substitutes, margarine, soft soaps, blown oils; oiling woolens.

**8021. Raspberry.** Fresh ripe fruit of varieties of *Rubus idaeus* L., or *R. stigosus* Michx., *Rosaceae. Habit.* Europe, Asia, cultivated in Canada and U.S. *Constit.* Sugar, malic and citric acids.

USE: Flavoring.

THERAP CAT: Pharmaceutic aid (flavored vehicle).

**8022. Raunescine.** ***17α-Hydroxy-18β-[(3,4,5-trimethoxybenzoyl)oxy]-3β,20α-yohimban-16β-carboxylic acid methyl ester;*** 3,4,5-trimethoxybenzoyl methyl raunescate. $C_{31}H_{36}N_2O_8$; mol wt 564.62. C 65.94%, H 6.43%, N 4.96%, O 22.67%. Isoln from *Rauwolfia canescens* L., *R. tetraphylla* and other *Rauwolfia* spp, *Apocynaceae.* Isoln and structure: Hosansky, Smith, *J. Am. Pharm. Assoc. Sci. Ed.* **44,** 639 (1955); Huebner, Schlittler, *J. Am. Chem. Soc.* **79,** 250 (1957); van Tamelen, Taylor, *ibid.* **79,** 5256 (1957). Extraction: **Brit.** pat. **833,149** (1960 to Ciba).

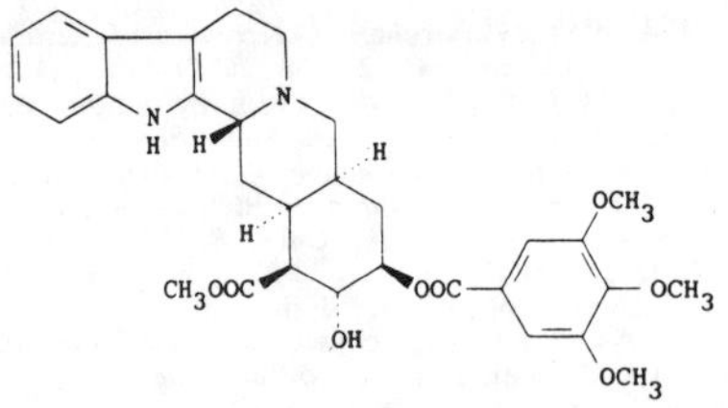

[Structure: N, H, H, H, H, $OCH_3$, $CH_3OOC$, OOC, $OCH_3$, OH, $OCH_3$]

Monohydrate, hexagonal prisms from 90% methanol, mp 160-170°. $[\alpha]_D^{25}$ −74° (chloroform). uv max (U.S.P. alc): 218, 271 nm (log ε 4.77, 4.26).

Nitrate, $C_{31}H_{36}N_2O_8.HNO_3$, mp 223-225°; (mp 208-210°). $[\alpha]_D^{25}$ −80° (5*N* acetic acid).

Monoacetate monohydrate, $C_{33}H_{38}N_2O_9.H_2O$, mp 159-162°. $[\alpha]_D^{25}$ −157° (chloroform).

**8023. Rauwolfia serpentina.** *Rauwolfia serpentina* (L.) Benth. *(Ophioxylon serpentinum* L.) *Apocynaceae.* A small shrub native to the Orient from India to Sumatra. The crude botanical drug is usually the root. Contains many indole alkaloids, e.g., reserpine, reserpinine, yohimbine, ajmaline, serpentine, serpetinine. Review and bibliography with botanical information: Monachino, *Econ. Bot.* **8,** 349-365 (1954). The genus *Rauwolfia* is large and has a wide distribution in the tropics.

Extracts are sold under trade names, such as ***Austrawolf, Egalin, Gendon, Hiwolfia, Koglucoid, Ralfen, Raucolyt, Raudixin, Raulfin, Raurixin, Rauserpol, Rautensine, Rauverid, Rauwiloid, Ra-Valeas, Rauwoldin, Rawotal, Rivadescin, Roxinil, Sarpagan, Serpina, Wolfin.***

THERAP CAT: Antihypertensive.

THERAP CAT (VET): *See* Reserpine.

**8024. Rayon.** Regenerated cellulose. The Code of Federal Regulations defines rayon as a manufactured fiber composed of regenerated cellulose, as well as manufactured fibers composed of regenerated cellulose in which substituents have replaced not more than 15 percent of the hydrogens of the hydroxyl groups. Now produced almost exclusively by the viscose process. *Review:* J. Lundberg, A. Turbak in Kirk-Othmer *Encyclopedia of Chemical Technology* **Vol. 19** (Wiley-Interscience, New York, 3rd ed., 1982) pp 855-880.

*Note:* ***Avisco Rayon*** is ***purified rayon,*** a medicinal grade of rayon.

USE: Purified rayon as surgical aid.

**8025. Rayopake®.** 3,4-Dihydro-3-iodo-6-methyl-2,4-dioxo-1(2*H*)-pyridineacetic acid 2,2′-iminodiethanol salt; 2-methyl-4,6-dioxo-5-iodotetrahydro-1-pyridineacetic acid diethanolamine salt; 2,4-dioxo-3-iodo-6-methyltetrahydropyridine-1-acetic acid diethanolamine salt. $C_{12}H_{19}IN_2O_6$; mol wt 414.21. C 34.80%, H 4.62%, I 30.64%, N 6.76%, O 23.18%. Prepn: **Swiss.** pat. **234,886** (1945 to Hoffmann-La Roche).

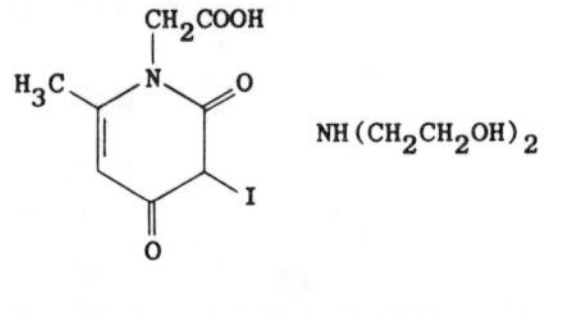

[Structure: $CH_2COOH$, $H_3C$, N, O, I, O; $NH(CH_2CH_2OH)_2$]

Crystals from abs ethanol, mp 194-195°.

THERAP CAT: Diagnostic aid (radiopaque medium).

**8026. Razoxane.** ***4,4′-(1-Methyl-1,2-ethanediyl)bis-2,6-piperazinedione; (±)-4,4′-propylenedi-2,6-piperazinedione;*** (±)-(3,5,3′,5′-tetraoxo)-1,2-dipiperazinopropane; (±)-1,2-bis(3,5-dioxopiperazinyl)propane; ICI 59118; ICRF 159; NSC 129943; Razoxin. $C_{11}H_{16}N_4O_4$; mol wt 268.28. C 49.25%, H 6.01%, N 20.88%, O 23.86%. Prepn: A. M. Creighton *et al., Nature* **222,** 384 (1969); A. M. Creighton, **Brit.** pat. **1,234,935** corresp to **U.S.** pat. **3,941,790** (1971, 1976 both to Natl. Res. Dev. Corp.). Mode of action: H. B. A. Sharpe *et al., Nature* **226,** 524 (1970). Metabolism: R. E. Bellet *et al., Eur. J. Cancer* **13,** 1293 (1977). Pharmacology: A. Atherton, *ibid.* **11,** 383 (1975). Clinical studies: M. T. Bakowski *et al., Int. J. Radiat. Oncol. Biol. Phys.* **4,** 115 (1978); H. W. Bruckner *et al., Cancer Treat. Rep.* **66,** 1713 (1982). Toxicity study: E. Hassenstein, K. Renner, *Strahlentherapie* **154,** 122 (1978). Mitigating effects on daunomycin-induced cardiomyopathy in mice: V. W. Fischer *et al., Drug Chem. Toxicol.* **5,** 155 (1982). *Review:* M. T. Bakowski, *Cancer Treat. Rev.* **3,** 95-107 (1976).

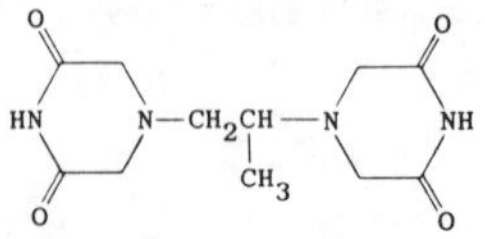

[Structure: HN, N—$CH_2CH$—N, NH, $CH_3$, O, O, O, O]

Pale cream microcrystalline solid, mp 237-239°. 320 mg/kg/day administered orally to beagle dogs was lethal, E. J. Gralla *et al., Cancer Chemother. Rep.* **5,** 1 (1974).

THERAP CAT: Antineoplastic.

**8027. Reddingite.** $3(Mn,Fe)O.P_2O_5.3H_2O$—manganese iron phosphate trihydrate.

**8028. Reductic Acid.** ***2,3-Dihydroxy-2-cyclopenten-1-one;*** 2-cyclopenten-2,3-diol-1-one. $C_5H_6O_3$; mol wt 114.10. C 52.63%, H 5.30%, O 42.07%. Antioxidant obtained from pectic substances: Reichstein, Oppenauer, *Helv. Chim. Acta* **16,** 988 (1933); **17,** 390 (1934); Goldstein, **U.S.** pat. **2,854,484** (1958). Alternate syntheses: Hesse *et al., Ann.* **563,** 31 (1949); **592,** 137 (1955); **736,** 134 (1970). Crystal

## Organic Name Reactions

(Academic Press, New York, 1967) pp 398–404; A. Rachlin *et al., Org. Syn.* **51**, 8 (1971); J. A. Peters, H. Van Bekkum, *Rec. Trav. Chim.* **100**, 21 (1981).

**Rosenmund-von Braun Synthesis.**
K. W. Rosenmund and E. Struck, *Ber.* **52**, 1749 (1916); J. von Braun and G. Manz, *Ann.* **488**, 111 (1931).
Formation of aromatic nitriles from aryl halides and cuprous cyanide:

$$ArX + CuCN \xrightarrow{250°} ArCN + CuX$$

*Reviews:* D. T. Moury, *Chem. Revs.* **42**, 207 (1948); J. E. Callen *et al., Org. Syn.* **III**, 212 (1955); M. S. Newman, *ibid.* 631; K. Takagi *et al., Bull. Chem. Soc. Japan* **48**, 3298 (1975).
*Cf.* Ullmann Reaction.

**Roth, H.** *see* Kuhn-Roth.

**Rothemund Reactions.**
P. Rothemund, *J. Am. Chem. Soc.* **57**, 2010 (1935); **61**, 2912 (1939).
Preparation of *meso*-tetrasubstituted porphyrins by condensation of pyrrole with an aldehyde.

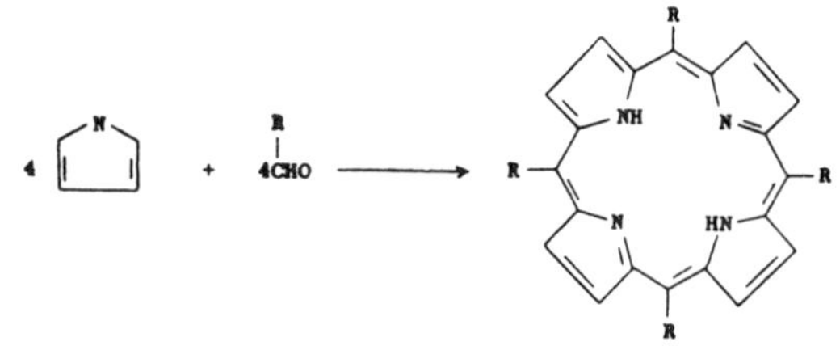

This reaction has found wide application with aldehydes in which R is an aromatic nucleus.
Ball *et al., J. Am. Chem. Soc.* **68**, 2278 (1946); Thomas, Martell *ibid.* **78**, 1335 (1956). Mechanism: Badger *et al., Austral. J. Chem.* **17**, 1028 (1964); R. G. Little, *J. Heterocycl. Chem.* **18**, 833 (1981).

**Rowe Rearrangement.**
F. M. Rowe, E. Levin, A. C. Burns, J. S. H. Davies and W. Tepper, *J. Chem. Soc.* **1926**, 690.
Intramolecular rearrangement of pseudo-phthalazones into phthalazones by heating at 180° in the presence of dilute aqueous acids:

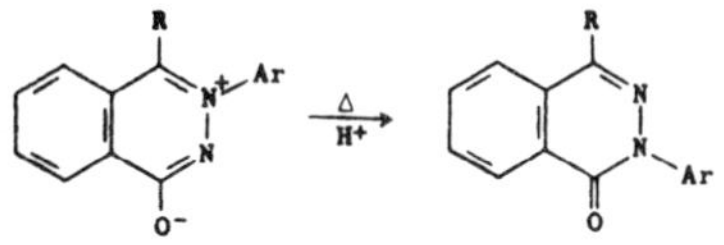

F. M. Rowe, D. A. W. Adams, A. T. Peters and A. E. Gillam, *J. Chem. Soc.* **1937**, 90; W. R. Vaughan, *Chem. Revs.* **43**, 487 (1948); W. R. Vaughan, D. I. McCane and G. J. Sloan, *J. Am. Chem. Soc.* **73**, 2298 (1951); R. C. Elderfield and S. L. Wythe in Elderfield, *Heterocyclic Compounds* **6**, 206 (New York, 1957).

**Ruff-Fenton Degradation.**
O. Ruff, *Ber.* **31**, 1573 (1898); **32**, 550 (1899); H. J. H. Fenton, *Proc. Chem. Soc.* **9**, 113 (1893).
Shortening of the carbon chain of sugars by the oxidation of aldonic acids (as calcium salts) with hydrogen peroxide and ferric salts:

$$RCHOHCOOH + H_2O_2 \xrightarrow{Fe^{3+}} RCHO + CO_2 + 2H_2O$$

W. Pigman, *The Carbohydrates* (Academic Press, New York, 1957), p 118; H. S. Isbell, M. A. Salam, *Carbohyd. Res.* **90**, 123 (1981).
*Cf.* Fenton Reaction.

**Rügheimer, L.** *see* Staedel-Rügheimer.

**Rupe Rearrangement (Reaction)** *see* Meyer-Schuster Rearrangement (Reaction).

**Ruzicka Large Ring Synthesis.**
L. Ruzicka, M. Stoll, and H. Schinz, *Helv. Chim. Acta* **9**, 249, 339, 389, 499 (1926).
Formation of large ring alicyclic ketones from dicarboxylic acids by thermal decomposition of salts with metals of the second and fourth groups of the periodic system (Ca, Th, Ce).

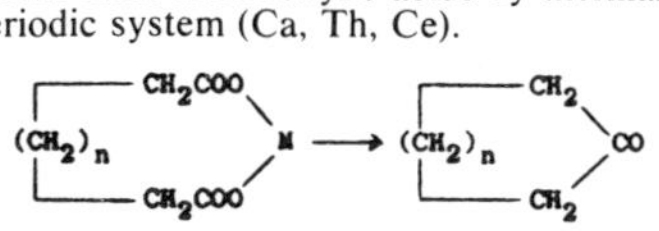

**Organic Name Reactions**

L. Ruzicka, *Chemistry & Industry* (*London*) **54**, 2 (1935); H. Gilman, *Organic Chemistry* **1**, p 78 (New York, 1943); K. Ziegler, *Houben-Weyl-Müller* **IV** (**II**), 754 (1955).
*Cf.* Blanc Reaction.

**Sabatier-Senderens Reduction.**
P. Sabatier, J. B. Senderens, *Compt. Rend.* **128**, 1173 (1899).
Catalytic hydrogenation of organic compounds in the vapor phase by passage over hot, finely divided nickel (the oldest of all hydrogenation methods).
E. B. Maxted in G. M. Schwab, *Handbuch der Katalyse* **7**, 624 (Vienna, 1943); H. Roth *et al.*, in Houben-Weyl, *Methoden der Organischen Chemie* **II**, 288 (1953); G. Schiller, *ibid.* **IV/II**, 284 (1955); H. O. House, *Modern Synthetic Reactions* (W. A. Benjamin, Calif., 2nd ed., 1972) Chapter 1.

**Sachs, F.** *see* Ehrlich-Sachs.

**Sand, J.** *see* Hofmann-Sand.

**Sandmeyer Diphenylurea Isatin Synthesis.**
T. Sandmeyer, *Z. Farb. Textile Chem.* **2**, 129 (1903).
Formation of a cyanoformamidine by treatment of a symmetrical diphenylthiourea with potassium cyanide in alcohol containing lead carbonate, reduction with ammonium sulfide and ring-closure with concentrated sulfuric acid to isatin-2-anil; also formed smoothly by ring closure of the cyanoformamidine with aluminum chloride in benzene or carbon disulfide:

$$(C_6H_5NH)_2CS \xrightarrow[PbCO_3]{KCN} C_6H_5NHC({=}NPh)CN \xrightarrow{(NH_4)_2S} C_6H_5NHC({=}NPh)CSNH_2$$

AlCl$_3$ / H$_2$SO$_4$ → isatin-2-anil (NC$_6$H$_5$, N–H) $\xrightarrow{H^{\oplus}}$ isatin

J. R. Geigy & Co., Ger. pat. **115,169**, **116,563** (1900); *Friedländer* **6**, 574, 575 (1900–1902); A. Reissert, *Ber.* **37**, 3708 (1904); G. Schultz, G. Rhode and G. Hertzog, *J. Prakt. Chem.* [2] **74**, 74, 76 (1906); C. Hollins, *The Synthesis of Nitrogen Ring Compounds*, p 102 (London, 1924); C. S. Marvel and G. S. Hiers, *Organic Syntheses* **coll. vol. I**, 2nd ed., p 327 (1943); P. L. Julian *et al.*, in Elderfield, *Heterocyclic Compounds* **3**, 207 (New York, 1952); O. Bayer, W. Eckert in Houben-Weyl, *Methoden der Organischen Chemie* **vol. 7/4** (Georg Thieme, Stuttgart, 1968) p 11.

**Sandmeyer Isonitrosoacetanilide Isatin Synthesis.**
T. Sandmeyer, *Helv. Chim. Acta* **2**, 234 (1919).
Formation of isonitrosoacetodiphenylamidine by condensation of chloral hydrate, hydroxylamine and aniline, cyclization with concentrated sulfuric acid, and quantitative hydrolysis to isatin on dilution:

C$_6$H$_5$–NH–C(=NC$_6$H$_5$)–HC=NOH $\xrightarrow{H_2SO_4}$ 3-imino-isatin-2-anil (NH, NC$_6$H$_5$) $\xrightarrow{H_2O}$ isatin-2-anil (O, NC$_6$H$_5$) $\xrightarrow{H_2O}$ isatin

J. Martinet, P. Cousset, *Compt. Rend.* **172**, 1234 (1921); C. Hollins, *The Synthesis of Nitrogen Ring Compounds* (London, 1924) p 103; C. S. Marvel, G. S. Hiers, *Organic Syntheses* **coll. vol. I**, 2nd ed. (1943) p 327; P. L. Julian *et al.*, in Elderfield, *Heterocyclic Compounds* **3** (New York, 1952) p 208; F. E. Sheibley, J. S. McNulty, *J. Org. Chem.* **21**, 171 (1956); O. Bayer, W. Eckert in Houben-Weyl, *Methoden der Organischen Chemie* **vol. 7/4** (Georg Thieme, Stuttgart, 1968) p 14.

**Sandmeyer Reaction** (Gattermann Reaction, Körner-Contardi Reaction).
T. Sandmeyer, *Ber.* **17**, 1633, 2650 (1884); L. Gattermann, *Ber.* **23**, 1218 (1890); G. Körner, A. Contardi, *Atti Accad. nazl. Lincei* **23 II**, 464 (1914); *C.A.* **9**, 1478 (1915).
Replacement of diazonium groups in aromatic compounds by halo or cyano groups in the presence of cuprous salts (Sandmeyer), copper powder (Gattermann) or cupric salts (Körner-Contardi):

$$C_6H_5N_2^+X^- + KCN \xrightarrow{Cu} C_6H_5CN + N_2 + KCl$$

*Reviews:* D. T. Moury, *Chem. Revs.* **42**, 213 (1948); E. Pfeil, *Angew. Chem.* **65**, 255 (1953); H. H. Hodgson, *Chem. Revs.* **40**, 251 (1947); W. A. Coudrey, D. S. Davies, *Quart. Rev.* **6**, 358 (1952); A. Roedig in Houben-Weyl, *Methoden der Organischen Chemie* **vol. 5/4** (Georg Thieme, Stuttgart, 1960) p 438; R. Stroh, *ibid.* **5/3**, 846 (1962); J. K. Kochi, *J. Am. Chem. Soc.* **79**, 2942 (1957).
For the direct conversion: arylamines → aryl halides, *see* M. P. Doyle, *J. Org. Chem.* **42**, 2426 (1977).

**Sarett Oxidation.**
G. I. Poos, G. E. Arth, R. E. Beyler and L. H. Sarett, *J. Am. Chem. Soc.* **75**, 422 (1953).
Oxidation of primary and secondary alcohols to aldehydes and ketones by means of $CrO_3$-pyridine complex:

# FORMULA INDEX

## A



*** Denotes derivative of title compound**

** Denotes derivative of title compound*

# CROSS INDEX OF NAMES

The Cross Index of Names refers the reader to monograph numbers with two types of citations:

**Ethacrynic Acid,** 3643 (format for titles)
**Ethacrynate Sodium** *see* 3643 (format for all synonyms)

Company name included in citation denotes trademark ownership:

**Edecrin** [MSD] *see* 3643

An asterisk (*) preceding an entry signifies that the name does not appear in the monograph.

**SAQ 2.2e**

### 2.2.9. Other General Reference Works

We have now examined five major analytical reference works in some detail; I hope you have picked up some general tips and experience for using reference works as well as more specific familiarisation with these works. But before leaving this Section I must introduce, briefly at least, four more works. My initial aim here is to show you how reference works which are not specifically about analytical chemistry can still be of considerable use to you: also, the four titles help to illustrate some of the general points I made earlier in 2.2.1 to 2.2.3; and, very importantly, they help me to re-emphasise my earlier point that I have been very selective with titles and that there are many more very useful reference books to be found than just those I name in this section. If you can get access to copies of these works, do examine them.

We will start at the top again with one of the best general scientific and technical encyclopaedias currently available:

> *McGraw-Hill Encyclopedia of science and technology*. 5th edn. New York: McGraw-Hill, 1982. 15v.

This is a traditional A-to-Z encyclopaedia with its final volume comprising a comprehensive index to the whole work. It features detailed articles on all aspects of science and technology; the article on analytical chemistry shown here extends to five columns of text. It includes cross-references in the text to related articles and there are also references to further reading at the end of each major article. The aim of the encyclopaedia is to provide the widest possible range of articles, understandable and useful to any person of modest technical training, wanting to obtain information outside his particular field. You should find it a most valuable introduction to any scientific or technical topic you meet for the first time.

We stay with encyclopaedias for out next title, this time a single-volume work covering chemistry only:

> *Van Nostrand Reinhold Encyclopedia of Chemistry*. Douglas M. Considine and Glenn D. Considine, eds. 4th edn. New York: Van Nostrand Reinhold, 1984. 1982pp.

The articles in this book are described by the editors as 'concise', although the one on analysis extends to six columns of text and two columns of tabular data; as in the *McGraw-Hill Encyclopedia*, references for further reading are given for the major articles. Two useful points about this work are that it is up-to-date and makes considerable use of graphic and tabular material. It also has a general index. In addition to those topics expected in an encyclopaedia of chemistry, particular emphasis is placed on the following subjects: advanced processes, strategic raw materials, chemistry of metals, energy sources and conversion, wastes and pollution, analytical instrumentation, growing use of food chemicals, structure of matter, new and improved materials, plant chemistry, biochemistry and biotechnology.

Moving away from encyclopaedias now, here is another single-volume compendium of data, in some respects similar to the Meites *Handbook*, but covering a much wider field:

> *CRC Handbook of chemistry and physics: a ready-reference book of chemical and physical data*. Robert C. Weast, ed. 63rd edn. Cleveland: CRC Press, 1982. Various pagings.

This is a well-established handbook with new editions more-or-less annually, although by no means all the data are revised for every edition. Reference books are expensive and libraries don't always even try to keep up-to-date with every new edition of every reference book as they appear: something else you have to remember to check when using a library. But remember also that much (even most) of the data in the 58th edition is as valid as that in the 63rd. There is much very useful tabular and formula data in this handbook, but it is not the easiest reference work in which to find your way around. There is a general index but it is not very detailed and the list of contents is, to say the least, brief. You have to think before you look anything up in this work (although I hope you would stop and think before looking anything up anywhere).

Finally, a dictionary:

> *The Condensed chemical dictionary*. Revised by Gessner G. Hawley. 10th ed. New York: Van Nostrand Reinhold, 1982. 1135pp.

This is an A-to-Z compendium of technical data and descriptive information covering chemicals and chemical phenomena. Three kinds of information are included: descriptions of chemicals, raw materials, processes, and equipment; expanded definitions of chemical entities, phenomena, and terminology; and descriptions or identification of trademarked products used in chemical industries. There is no index and all references are in the text, a layout designed to give quick access to brief entries. So here is another quick-reference work covering a wide field, and therefore giving only brief entries; but if you are looking only for a very quick answer to a specific question, then this is the kind of work to turn to first.

So by now you have worked with various reference books and looked at some more. I have made all the important points either in the text or in responses to the SAQs, but let me round this section off with a final comment. You might find yourself in one of two extreme situations (or perhaps somewhere between the two, but I think the moral will remain obvious): served by an excellent library containing all the works I have discussed, and more besides; or with no adequate library conveniently available and only a minimal number of books to refer to—perhaps one or two of those I have discussed, or perhaps a couple of completely different titles. In the first instance especially, you have first to decide exactly what information you need and how much you need, before deciding what reference work is likely to be best suited to your purposes; and even then you might need to proceed by trial-and-error until you are thoroughly familiar with the various reference books available to you. In the second case you cannot of course be choosy, but you still need to decide what information you need, how much and for what purpose, and it is surprising how much information you can find in a small choice of books once you become thoroughly conversant with their contents and their idiosyncracies.

In any event, practise using and get to know whatever reference books are available to you and remember the criteria and practical hints I have given you in this section: then you should be able to find information whenever you need it with the minimum of delay. If the message I want to hammer home throughout this Unit is still not obvious, let me repeat it here—*use these books as often as you can*; this Unit is aimed at introducing you to types of materials and giving you the ability and confidence to find them and use them.

Of course, if you can use general reference books to find specialised information, then you should also be able to use specialised reference books (if they are the only ones available to you) to find more general information. And that leads us into Section 2.3, a quick look at some specialised reference works.

But before we proceed, I think it desirable to recall some of my general points, before you find yourself immersed in detail. So let's have a final SAQ for this Section.

**SAQ 2.2f**

Right, let's see how much of my advice you remember.

You want to buy a reference book (the subject is immaterial to the principle of this question). You are in a well-stocked bookshop and have narrowed your choice down to three titles, A, B and C.

A is the 5th edition, 1984, of an American book published by Wiley–Interscience and originally issued in 1960. It has 600 pages and costs thirty pounds (or fifty dollars).

B is the first edition of a work published in 1978 by the then Chemical Society. It has 300 pages and costs twelve pounds (or twenty dollars).

C is the 2nd English edition, dated 1980, of a work first published in German in 1970. The book is published in Amsterdam by Elsevier, has 400 pages and costs thirty pounds (or fifty dollars).

First of all, note in *your own* order of preference, *four* criteria you would use (costs, length, etc.) to judge these works. Then make a final decision and note which book—A, B or C—you would buy.

**Objectives**

When you have worked through this part of the Unit you should be able to:

- understand the importance and usefulness of reference works;
- appreciate the wide range of reference works available
- recognise certain criteria to help you to select titles likely to give you the information you need to tackle most problems you may meet in analytical chemistry;
- make profitable use of any reference work you select by realising the importance of certain features, especially layout and indexes;
- find information by using the major reference books described.

## 2.3. SPECIALIST REFERENCE WORKS

### 2.3.1. Introduction

By now you should have a good idea of the kinds of reference books available to you and of how to use them. To complete this topic I shall now introduce you very briefly to a few more titles. Then at the end of this Section we will make sure you have grasped the major points arising from Sections 2.1 to 2.3.

I said earlier that I did not want to turn this Unit into a glorified book list, but I am afraid this Section is going to look rather like that! I shall list various titles, partly because they are useful in themselves and partly to show you the range of specialised works you can find. I am setting SAQs on only two of them. Again, if you can get access to any of the books, do have a look at them; a few minutes browsing through a copy of a book can tell you far more than can my brief descriptions.

You will probably find for yourself a number of reference books I have not mentioned in this Unit: use the advice and general comments I give here to evaluate any such reference books for yourself. Because I don't mention a particular title in this Unit, it does not mean that I do not recommend that book, simply that space has been limited, to avoid swamping you with titles.

### 2.3.2. A Selection of Specialist Reference Books

In the following list I once again refer to works by their titles, as I have done so far throughout Section 2. Again, I am starting with some major handbooks and encyclopaedias.

> *Materials and technology: a systematic encyclopaedia of the technology of materials used in industry and commerce, including foodstuffs and fuels*. London: Longmans, 1968–1975. 8v.

Although becoming a little dated, this is still a major encyclopaedia providing for both the layman and the specialist, information on many materials. It describes the sources, manufacture, processing, and uses of both natural and synthetic materials. Its subject coverage is: air and water, inorganic chemicals, nucleonics (Vol 1); non-metallic minerals and rocks; (Vol 2); metals and ores (Vol 3); petroleum and organic chemicals (Vol 4); natural organic materials and related synthetic products (Vol 5); wood, papers, textiles, plastics, and photographic materials (Vol 6); vegetable food products and luxuries (Vol 7); edible oils and fats, animal food products, and material resources, (Vol 8). It is arranged in a systematic rather than an alphabetical order, but it has indexes to each volume and a general index in volume 8.

Do you recall the *CRC Handbook of chemistry and physics* in Section 2.2.9? Well, its publishers produce a large number of general and specialised handbooks. They have a detailed series on chromatography, which began with the following book:

> *CRC Handbook of chromatography*. Gunter Zweig and Joseph Sherma, eds. Cleveland: CRC Press, 1972. 2v.

Volume 1 contains some 550 tables of chromatographic data expressed in the uniform terms of $R_f$ values, retention times, and retention volumes; over 12,000 compounds are cross-indexed. Volume 2 is intended to give the researcher or the novice a working knowledge of the theory and practice of the various fields of chromatography. It also has a section on detection reagents for paper and thin-layer chromatography, and another with descriptions of selected methods of sample preparation. Sub-sections on practical applications supply information on commercial sources of all types of chromatographic supplies. Since this introductory work the *Chromatography Series* has been continued with specialised volumes. So far the following have been published:

> *Terpenoids*. Carmine J. Cosica, ed, 1984, 200pp;
>
> *Polymers*. Charles G. Smith, *et al*, editors, 1982, 200pp;
>
> *Phenols and organic acids*. Toshihiko Hanai, ed, 1982, 304pp;
>
> *Carbohydrates*. Shirley C. Churms, ed, 1982, 288pp;

*Drugs*. Ram N. Gupta, ed, 1981, 2v;

*Amino acids and amines*. Stanley Blackburn, ed, 1983, 312pp;

*Pesticides and related organic chemicals*. Joanne M Follweiler and Joseph Sherma, eds, vol 1, 1984, 376pp;

*Lipids*. Helmut K. Mangold, ed, 1984, 2v.

A third specialised encyclopaedia is:

*Encyclopaedia of polymer science and technology: plastics, resins, rubbers, fibers*. New York: Wiley–Interscience, 1964–1972. 16v.

Again, this is getting rather dated, but it remains a valuable source of information. There are about 450 articles which give a comprehensive account of all facets of polymer science and technology. These articles may be divided by types into five major categories: chemical substances, polymer properties, methods and processes (eg extrusion, chromatography); uses (eg aerospace applications, rubber goods); and general background (eg classification of polymerisations, nomenclature of polymers). The arrangement of the work is as a traditional A-to-Z encyclopaedia.

Necessarily much more up-to-date for such a rapidly developing discipline is:

*Biochemical engineering and biotechnology handbook*. Bernard Atkinson and Ferda Mavituna. London: Macmillan, 1983. 119pp.

This work aims to assist the development of both large and small biotechnological processes by establishing for the unit processes and unit operations involved the following criteria: (*a*) their scientific and engineering bases, (*b*) their performance and operating characteristics, (*c*) the factors which influence their performance, (*d*) their integration into complete processes. Among other aims, this information should allow users of the *Handbook* to establish the appropriate methodology for a particular process development and identify equipment needs. The book is arranged in 14 chapters, with a brief glossary at the beginning of the chapter where appropriate. To give you an idea of the coverage, the titles of the fourteen chapters are: Properties of industrially important micro organisms, Thermodynamic aspects of microbial metabolism, Stoichiometric aspects of microbial metabolism, Microbial activity, Product information, Enzyme activity, Reactors, Flow-behaviour of fermentation fluids, Gas–liquid mass-transfer; Solid-and liquid-phase mass-transfer, Heat transfer; Downstream process engineering, Product recovery processes and unit operations, and Processes. The information is presented predominantly as tables and figures, and literature references are given. There is an index of topics.

There is also a shorter reference book available for biochemists:

*Concise encyclopedia of biochemistry*. Thomas Scott and Mary Brewer, eds, Berlin: Walter de Gruyter, 1983. 516pp.

This contains descriptions of processes and properties, and the structures etc, of substances and compounds. It has an encyclopaedic dictionary arrangement with cross-references in the text but no index.

Another concise specialist reference work is:

> *A–Z of clinical chemistry: a guide for the trainee*. W. Hood. Lancaster: MTP Press, 1980. 386pp.

This gives a dictionary-style presentation of basic information for trainees on all aspects of clinical chemistry, including new techniques, tests, terminology, methods of diagnosing diseases, and other topics. It also gives suggestions for further reading, including general lists of analytical and clinical textbooks.

You will have had, or can expect to have, experience of working with various chemical techniques and substances. Perhaps you have already worked with various common and uncommon substances in your employment: how safe are they and under what conditions do substances you have worked with become dangerous? If you come across a substance new to you, how will you know whether it is likely to present any hazards to you? Well, here is a reference book to help you answer exactly those questions:

> *Handbook of reactive chemical hazards: an indexed guide to published data*. L. Bretherick. 2nd edn. London: Butterworth, 1979. 1281pp.

This is intended to give the research student, the practising chemist, and the safety officer access to a wide selection of documented information, to allow ready assessment of the likely reaction-hazard potential associated with an existing or proposed chemical compound or reaction system. It has a general A-to-Z section and a main formula-based section on individual compounds. There are indexes to the different sections.

**SAQ 2.3a**

An easy one to start with, but there's a useful point to be made in the responses. If you don't already know, what will happen if you mix hydrogen peroxide with acetone?

You are going to work with ketone peroxides and, in particular, with a chemical compound $C_9H_{18}O_6$ whose IUPAC name you have checked and discovered to be

3,3,6,6,9,9-Hexamethyl-1,2,4,5,7,8-Hexaoxacyclononane.

What hazards must you beware of? Before you try to look up this compound I must warn you that we have unwittingly

⟶

**SAQ 2.3a (cont.)**

stumbled into a doubtful area of chemical terminology. To avoid over-complications at this stage I will just advise you that in the *Handbook of Reactive Chemical Hazards* you will find this compound named

3,3,6,6,9,9-Hexamethyl-1,2,4,5,7,8-Hexaoxaonane.

This is just another example of the kind of complication that can beset you when you come to look for published information.

Remember to use the book if you can, and use the copies only if you have to.

AN INDEXED GUIDE TO PUBLISHED DATA

# Handbook of Reactive Chemical Hazards

Second Edition

L BRETHERICK, BSC, CChem, FRSC

Senior Project Leader,
New Technology Division,
Sunbury Research Centre,
British Petroleum Company, Ltd.

Butterworths
London Boston Sydney Wellington Durban Toronto

# Contents

The case history involved dry resin which expanded and split a glass column when wetted with a salt solution.

*See* NITRIC ACID, $HNO_3$ : Ion exchange resins

## ISOCYANIDES $RN{=}C:$

Acids

Sidgwick, 1950, 673

Acid-catalysed hydrolysis of isocyanides ('carbylamines'), to primary amine and formic acid is very rapid, sometimes explosively so.

## ISOPROPYLIDENE COMPOUNDS $Me_2C{=}C$

Ozone

*See* OZONE, $O_3$ : Isopropylidene compounds

## KETONE PEROXIDES

Davies, 1961, 72

Kirk-Othmer, 1967, Vol. 14, 777

The variety of peroxides (monomeric, dimeric and polymeric) which can be produced from interaction of a given ketone with hydrogen peroxide is very wide (see group-types below). The proportions of the products in the reaction mixture depend on the reaction conditions used, as well as the structure of the ketone. Many of the products appear to coexist in equilibrium, and several types of structure are explosive and sensitive in varying degrees to heat or shock. Extreme caution is therefore required in handling ketone peroxides in high concentrations, particularly those derived from ketones of low molecular weight. Acetone is thus entirely unsuitable as a reaction or cleaning solvent whenever hydrogen peroxide is used.

*See* CYCLIC PEROXIDES
1-OXYPEROXY COMPOUNDS
POLYPEROXIDES
3,3,6,6-TETRAMETHYL-1,2,4,5-TETRAOXANE, $C_6H_{12}O_4$
3,6-DIETHYL-3,6-DIMETHYL-1,2,4,5-TETRAOXANE, $C_8H_{16}O_4$
3,3,6,6,9,9-HEXAMETHYL-1,2,4,5,7,8-HEXAOXAONANE, $C_9H_{18}O_6$
3,6-DI(SPIROCYCLOHEXANE) TETRAOXANE, $C_{12}H_{20}O_4$
3,6,9-TRIETHYL-3,6,9-TRIMETHYL-1,2,4,5,7,8-HEXAOXAONANE, $C_{12}H_{24}O_6$
TRI(SPIROCYCLOPENTANE)-1,1,4,4,7,7-HEXAOXAONANE, $C_{15}H_{24}O_6$

## KETOXIMINOSILANES $RSi(ON{=}CMeEt)_3$

Tyler, L. J., *Chem. Eng. News*, 1974, **52**(35), 3

Vacuum-distilled material kept dark for 6 months was found to have largely polymerised to a yellow solid, and it exploded 15 days later.
*See other* CYANO COMPOUNDS

**2- or 5-ETHYLTETRAZOLE** **$C_3H_6N_4$**

**CH=NNEtN=N or HNN=NN=CEt**

Aluminium hydride
*See* ALUMINIUM HYDRIDE, $AlH_3$: Tetrazole derivatives

**AZIDOACETONE OXIME** **$C_3H_6N_4O$**

**$N_3CH_2C(Me)$=NOH**

Forster, M. O. *et al., J. Chem. Soc.*, 1908, **93**, 83
During distillation at 84°C/2.6 mbar, the large residue darkened and finally exploded violently.
*See also* BROMOACETONE OXIME, $C_3H_6BrNO$
*See other* 2-AZIDOCARBONYL COMPOUNDS

**1,3,5-TRINITROSOHEXAHYDRO-1,3,5-TRIAZINE** **$C_3H_6N_6O_3$**

**$CH_2N(NO)CH_2N(NO)CH_2NNO$**

Sulphuric acid
Urbanski, 1967, Vol. 3, 122
Concentrated sulphuric acid causes explosive decomposition.
*See other* NITROSO COMPOUNDS

**† ACETONE** **$CH_3COCH_3$** **$C_3H_6O$**
*MCA SD-87*, 1962

Bromoform
*See* BROMOFORM, $CHBr_3$: Acetone, etc.

Carbon,
Air
Boiston, D. A., *Brit. Chem. Eng.*, 1968, **13**, 85
Fires in plant to recover acetone from air with active carbon are due to the bulk surface effect of oxidative heating when air flow is too low to cool effectively.

Chloroform
*See* CHLOROFORM, $CHCl_3$: Acetone, etc.

2-Methyl-1,3-butadiene

*See* 2-METHYL-1,3-BUTADIENE (ISOPRENE), $C_5H_8$: Acetone

Nitric acid,
Sulphuric acid,

Fawcett, H.H., *Ind. Eng. Chem.*, 1959, **51**(4), 89A
Acetone will decompose with explosive violence if brought into contact with mixed nitric and sulphuric acids, particularly under confinement.

*See* Oxidants, below

Oxidants

*See* BROMINE TRIFLUORIDE, $BrF_3$: Solvents
BROMINE, $Br_2$: Acetone
NITROSYL CHLORIDE, ClNO: Acetone, etc.
NITROSYL PERCHLORATE, $ClNO_5$: Organic materials
NITRYL PERCHLORATE, $ClNO_6$: Organic solvents
CHROMYL CHLORIDE, $Cl_2CrO_2$: Organic solvents
CHROMIUM TRIOXIDE, $CrO_3$: Acetone
DIOXYGEN DIFLUORIDE, $F_2O_2$
NITRIC ACID, $HNO_3$: Acetone, etc.
HYDROGEN PEROXIDE, $H_2O_2$: Acetone, etc.
: Ketones
: Oxygenated compounds
PEROXOMONOSULPHURIC ACID, $H_2O_5S$: Acetone
OXYGEN (Gas), $O_2$: Acetone, Acetylene

*See* Nitric acid, Sulphuric acid, above

Potassium *tert*-butoxide

*See* POTASSIUM *tert*-BUTOXIDE, $C_4H_9KO$: Acids, etc.

Sulphur dichloride

*See* SULPHUR DICHLORIDE, $Cl_2S$: Acetone

Thiotrithiazyl perchlorate

*See* THIOTRITHIAZYL PERCHLORATE, $ClN_3O_4S_4$: Organic solvents

## † 2-PROPEN-1-OL (ALLYL ALCOHOL) $CH_2{=}CHCH_2OH$ $C_3H_6O$

Alkali,
2,4,6-Trichloro-1,3,5-triazine (Cyanuric chloride)

*See* 2,4,6-TRICHLORO-1,3,5-TRIAZINE (CYANURIC CHLORIDE), $C_3Cl_3N_3$: Allyl alcohol, etc.

Sulphuric acid

Senderens, J. B., *Compt. Rend.*, 1925, **181**, 698–700
During preparation of diallyl ether by dehydration of the alcohol with sulphuric acid, a violent explosion may occur (possibly due to polymerisation).

† 1,3,5-TRIMETHYLCYCLOHEXANE $C_9H_{18}$

$CH_2CHMeCH_2CHMeCH_2CHMe$

DI-*tert*-BUTYL DIPEROXYCARBONATE $C_9H_{18}O_5$

$(Me_3COO)_2CO$

A partially decomposed sample exploded violently at 135° C.

*See* PEROXYCARBONATE ESTERS

3,3,6,6,9,9-HEXAMETHYL-1,2,4,5,7,8-HEXAOXAONANE $C_9H_{18}O_6$

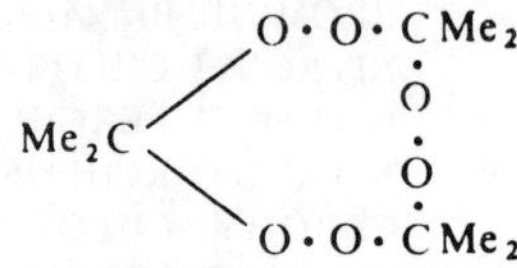


Dilthey, W. *et al.*, *J. Prakt. Chem.*, 1940, **154**, 219

Rohrlich, M. *et al.*, *Z. Ges. Schiess u. Sprengstoffw.*, 1943, **38**, 97–99

This trimeric acetone peroxide is powerfully explosive and will perforate a steel plate when heated on it.

*See also* 6-AMINOPENICILLANIC ACID *S*-OXIDE, $C_8H_{12}N_2O_4S$
OZONE, $O_3$: Citronellic acid

*See other* CYCLIC PEROXIDES

3,6,9-TRIETHYL-1,2,4,5,7,8-HEXAOXAONANE $C_9H_{18}O_6$

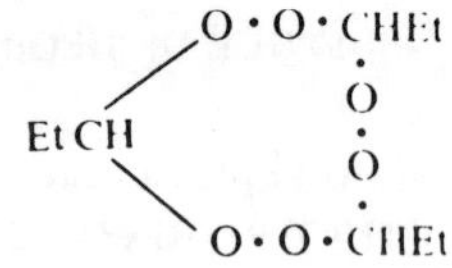


Rieche, A. *et al.*, *Ber.*, 1939, **72**, 1938

This trimeric 'propylidene peroxide' is an extremely explosive and friction-sensitive oil.

*See other* CYCLIC PEROXIDES

2,2,4,4,6,6-HEXAMETHYLTRITHIANE $C_9H_{18}S_3$

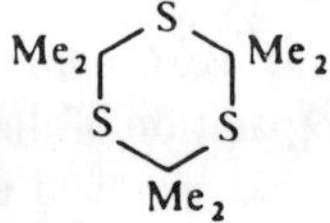


Nitric acid

*See* NITRIC ACID, $HNO_3$: Hexamethyltrithiane

**SODIUM PHOSPHINATE ('HYPOPHOSPHITE')** **$H_2NaO_2P$**

**NaOP(H)OH**

Mellor, 1940, Vol. 8, 881

Evaporation of aqueous solutions by heating may cause an explosion, phosphine being evolved.

Oxidants

1. Costa, R. L., *Chem. Eng. News*, 1947, **25**, 3177
2. Mellor, 1940, Vol. 8, 881
3. Mellor, 1971, Vol. 8, Suppl. 3, 624

Evaporation of a moist mixture of sodium phosphinate and a trace of sodium chlorate by slow heating caused a violent explosion. It was concluded that, once started, the decomposition of the phosphinate proceeds spontaneously [1]. Similar reactions have been reported with nitrates instead of chlorates. Such mixtures had previously been proposed as explosives [2].

Interaction of iodine with the anhydrous salt is violently exothermic, causing ignition [3].

*See* PERCHLORIC ACID, $ClHO_4$: Sodium phosphinate

*See other* REDUCANTS

**SODIUM DIHYDROGENPHOSPHIDE** **$NaPH_2$** **$H_2NaP$**

Albers, H. *et al.*, *Ber.*, 1943, **76**, 23

It ignites in air.

*See related* NON-METAL HYDRIDES

**OXOSILANE** **$H_2Si{=}O$** **$H_2OSi$**

Kautsky, K., *Z. Anorg. Chem.*, 1921, **117**, 209

It ignites in air.

*See related* NON-METAL HYDRIDES

**HYDROGEN PEROXIDE** **HOOH** **$H_2O_2$**

1. Shanley, E. S. *et al.*, *Ind. Eng. Chem.*, 1947, **39**, 1536
2. Naistat, S. S. *et al.*, *Chem. Eng. Prog.*, 1961, **57**(8), 76
3. Kirk-Othmer, 1966, Vol. 11, 407; *MCA SD-53*, 1969
4. Anon., *Fire Prot. Ass. J.*, 1954, 215
5. Smith, I. C. P., private comm., 1973
6. *MCA Case History No. 1121*
7. Campbell, G. A. *et al.*, *Proc. 4th I.Ch.E. Symp. on Chem. Proc. Hazards*, 1971, 37–43
8. Clark, M. C. *et al.*, *Chem. & Ind.*, 1974, 113
9. Cookson, P. G. *et al.*, *J. Organomet. Chem.*, 1975, **99**, C31–32

The hazards attendant upon use of concentrated hydrogen peroxide solutions have been reviewed [1–3]. Salient points include:

Release of enough energy during catalytic decomposition of 65% peroxide to evaporate all water present and formed, and subsequent liability of ignition of combustible materials.

Most cellulosic materials contain enough catalyst to cause spontaneous ignition with 90% peroxide.

Contamination of concentrated peroxide causes possibility of explosion. Readily oxidisable materials, or alkaline substances containing heavy metals, may react violently.

Soluble fuels (acetone, ethanol, glycerol) will detonate on admixture with peroxide of over 30% concentration, the violence increasing with concentration.

Handling systems must exclude fittings of iron, brass, copper, Monel, and screwed joints caulked with red lead.

Concentrated peroxide may decompose violently in contact with iron, copper, chromium, and most other metals or their salts, and dust (which frequently contains rust). Absolute cleanliness, suitable equipment (PVC, butyl or Neoprene rubber) and personal protection are essential for safe handling [4]. During concentration under vacuum of aqueous [5] or of aqueous–alcoholic [6] solutions of hydrogen peroxide, violent explosions occurred when the concentration was sufficiently high (probably above 90%) [3]. Detonation of hydrogen peroxide vapour has been studied experimentally [7].

Explosion of a screw-capped winchester of 35% peroxide solution after 2 years owing to internal pressure of liberated oxygen emphasises the need to date-label materials of limited stability, and to vent the container automatically by fitting a Bunsen valve or similar device [8]. Possible hazards in use and handling of concentrated hydrogen peroxide can be avoided by using the solid 2:1 complex of hydrogen peroxide with 1,4-diazabicyclo[2.2.2]octane. This is hygroscopic but stable for at least several months in storage, although it has been reported to decompose above 60°C [9].

Acetal,
Acetic acid

Ashley, J. N. *et al.*, *Chem. & Ind.*, 1957, 702

An organic sulphide containing an acetal group in the molecule had been oxidised to the sulphone with 30% hydrogen peroxide in acetic acid. After the residue had been concentrated by vacuum distillation at 50–60°C, the residue exploded during handling. This was attributed to formation of the peroxide of the acetal (formally a *gem*-diether) or of the aldehyde formed by hydrolysis, but peracetic acid may also have been involved.

**Acetaldehyde,**
**Desiccants**

**Karnojitzky, V. J., *Chim. Ind. (Paris)*, 1962, 88, 235**

**Interaction gives the extremely explosive poly(ethylidene) peroxide, also formed on warming peroxidised diethyl ether.** **$H_2O_2$**

Acetic acid

Grundmann, C. *et al.*, *Ber.*, 1939, **69**, 1755

During preparation of peracetic acid, the temperature should not be too low to prevent reaction as the reagents are mixed, because reaction may begin later with explosive violence.

*See* Oxygenated compounds, etc., below
PEROXYACETIC ACID, $C_2H_4O_3$

Acetic acid,
*N*-Heterocycles

1. Wommack, J. B., *Chem. Eng. News*, 1977, **55**(50), 5
2. Dholakia, S. *et al.*, *Chem. & Ind.*, 1977, 963

During isolation of the di-*N*-oxide of 2,5-dimethylpyrazine [1] and of the mono-*N*-oxide of 2,2′-bipyridyl [2], prepared from action of hydrogen peroxide in acetic acid on the heterocycles, violent explosions occurred during vacuum evaporation of the excess peroxyacetic acid. For a method not involving this considerable explosion risk,

*See* *N*-OXIDES

Acetic acid,
3-Hydroxythietane

1. Dittmer, D. C. *et al.*, *J. Org. Chem.*, 1971, **36**, 1324
2. Lamm, B. *et al.*, *Acta Chem. Scand. Ser. B*, 1974, **28**, 701–703

During preparation of the 1,1-dioxide by oxidation of the precursory 3-hydroxythietane with hydrogen peroxide in acetic acid, evaporation led to a peracetic acid explosion [1]. An alternative oxidation with 30% peroxide solution catalysed by sodium tungstate avoids this hazard [2].

Acetic anhydride

1. Prett, K., *Textilveredlung*, 1966, **1**, 288–290
2. Swern, D., *Chem. Rev.*, 1945, **45**, 5

During preparation of peracetic acid solutions for textile bleaching operations, the reaction mixture must be kept acid. Under alkaline conditions, highly explosive diacetyl peroxide separates from solution [1]. An excess of the anhydride has the same effect [2].

*See* DIACETYL PEROXIDE, $C_4H_6O_4$

Acetone,
Other reagents

1. Anon., *Angew. Chem. (Nachr.)*, 1970, **18**, 3
2. *MCA Case History No. 233*
3. *MCA Case History No. 223*

4. Stirling, C. J. M., *Chem. Brit.*, 1969, **5**, 36
5. Seidl, H., *Angew. Chem.* (*Intern. Ed.*), 1964, **3**, 640
6. Treibs, W., ibid., 802
7. Brewer, A. D., *Chem. Brit.*, 1975, **11**, 335

Acetone and hydrogen peroxide readily form explosive dimeric and trimeric peroxides, particularly during evaporation of the mixture. Many explosions have occurred during work-up of peroxide reactions run in acetone, including partial hydrolysis of a nitrile [1] and oxidation of 2,2′-thiodiethanol [2] and of an unspecified material [3]. The reaction mixture from oxidation of a sulphide with hydrogen peroxide in acetone exploded violently during vacuum evaporation at 90°C. On another occasion oxidation of a sulphide in acetone in presence of molybdate catalyst proceeded with explosive violence. A general warning about use of acetone as solvent for peroxide oxidations is given [4]. During the isolation of 1-tetralone, produced by oxidation of tetralin with hydrogen peroxide in acetone, a violent explosion occurred which was attributed to acetone peroxide [5]. The originator of the method later gave detailed instructions for a safe procedure, which include the exclusion of mineral acids, even in traces [6].

During oxidation of a sulphur heterocycle (unspecified) in acetone with excess 35% hydrogen peroxide, a white solid separated during 3 days standing in a cool place. The solid (20 g) appears to have been acetone peroxide, because it exploded with great violence during drying in a vacuum oven. A previous warning [4] on the incompatibility of acetone and hydrogen peroxide is repeated [7].

*See* Oxygenated compounds, etc., below
KETONE PEROXIDES
*See also* 3,3,6,6,9,9-HEXAMETHYL-1,2,4,5,7,8-HEXAOXAONANE, $C_9H_{18}O_6$

Alcohols

1. *MCA SD-53*, 1969
2. Spengler, G. *et al.*, *Brennst. Chem.*, 1965, **46**, 117

Homogeneous mixtures of concentrated peroxide and alcohols or other peroxide-miscible organic liquids are capable of detonation by shock or heat [1]. Furfuryl alcohol ignites in contact with 85% peroxide within 1 s [2].

*See* Oxygenated compounds, etc., below

Alcohols,
Sulphuric acid

Hedaya, E. *et al.*, *Chem. Eng. News*, 1967, **45**(43), 73

During conversion of alcohols to hydroperoxides, the order of mixing the reagents is important. Addition of concentrated acid to mixtures of an alcohol and concentrated peroxide almost inevitably leads to

## Appendix 4

# Index of Class, Group and Topic Titles used in Section I

# Appendix 5

# Index of Chemical Names and Synonyms used in Section 2

The majority of the names for chemicals in this alphabetically arranged index conform to one of the systematic series permitted under the various sections of the IUPAC Definitive Rules for Nomenclature. Where there is a marked difference between these names and the alternative names recommended in the IUPAC-based BS 2474:1965 or ASE lists, or long-established traditional names, these are given as cross-references back to the IUPAC names used as bold titles in the text.

It should be noted that italicised hyphenated prefixes which indicate structure, such as *cis*-, *o*-, *m*-, *tert*-, *mixo*-, *N*-, etc., have been ignored during alphabetical arrangement of successive names in this index, while the roman structural prefixes iso and neo, and roman multiplying prefixes, such as di, tris, tetra and hexakis, have been included for indexing purposes.

There are several general cross-entries in the index for which no page numbers are given, to facilitate finding the systematic names for some of the less common traditional names. There are, however, a few general entries with page numbers which do appear as items in the text.

Some particular industries and products also have their own reference books. One example is:

*Food industries manual*. M.D Ranken, ed. 21st edn. Glasgow: L. Hill, 1984. 530pp.

This is a reference manual on technical aspects of every branch of the food industry, stressing principles rather than details so as to give a clear idea of the different processes and operations involved. It is presented in chapters, with each chapter's contents arranged alphabetically under sub-headings for easy reference (except the chapter on nutrition which is divided according to a standard classification of nutritional concepts). The titles of these chapters are: Meat and meat products, Fish and fish products, Dairy products, Flour and baked goods, Fats and fatty foods, Coffee, tea, and cocoa, Fruit juices and soft drinks, Preserves, Pickles and sauces, Confectionery products, Snack foods, Nutrition, Freezing and refrigeration; Dehydration and dried products, Canning, Handling and storage. There is a comprehensive index.

If you are looking for information on adhesives, whether on materials or specific products, you can turn to:

*Adhesives handbook*. J. Shields. 3rd edn. London: Butterworth, 1984. 360pp.

This is primarily intended to give basic guidance for the designer concerned with adhesive bonding as an assembly process, but it also of value to the chemist and anyone else seeking general information on adhesives. There is no index, but the work has fairly detailed and informative contents pages. It is arranged in chapters (see the contents pages on the copies accompanying SAQ 2.3b).

**SAQ 2.3b**

You have some urea-formaldehyde resin adhesive named 'Beetle', and that is all you know about it.

What would it be used for and who manufactured it?

Is it marketed in any form other than a resin?

How long could you store the resin?

To what British Standard would it need to conform?

You need to know if there is a firm in Cambridge which manufactures urea-formaldehyde adhesives: can you find the name of one?

Use the copies if you must, but try and use the book if you can.

**J. Shields,** B.Sc.

# Adhesives handbook

Third edition

**Butterworths**
London Boston Durban Singapore Sydney Toronto Wellington

# Contents

Component of a blood albumen-soyabean adhesive for interior plywoods based on softwood.

*Remarks*
Where maximum bond strength is unwanted, the glue may be extended with fillers such as clays and wood flows, to produce glues for low strength applications, e.g. paper and softboard laminating, cardboard box fabrication and as particle binders.

**4.3.55 Starch**

*Physical form*
Powder (blended flour) which is mixed with water or chemicals such as sodium hydroxide.

*Shelf life, 20°C*
Dry powder, up to 1 yr.

*Working life, 20°C*
Several days, to indefinite if protected from biodeterioration.

*Closed assembly time, 20°C*
10 min.

*Setting action*
Loss of water.

*Processing conditions*
For general purposes may be set at room temperature; develops tack rapidly. Plywood manufacture, 1–2 d at room temperature and 500–700 kPa bonding pressure.

*Coverage*
0·15–0·25 kg/m$^2$.

*Permanence*
Poor resistance to weathering, moisture and temperature changes. Under damp conditions, biodeterioration is a hazard and the adhesive is limited to use under dry interior conditions only.

*Applications*
Paper carton and bottle labelling; stationery; interior grade plywood fabrication.

*Remarks*
Addition of sodium hydroxide to starch adhesives increases the risk of staining timber substrates. Joint strengths are low compared with other vegetable adhesive types but are usually adequate for the applications mentioned above. Poor resistance to water and biodeterioration is improved by the addition of preservatives.

**4.3.56 Thermoplastic resins (miscellaneous)**

In addition to the thermoplastic adhesives discussed already, many other synthetic and natural materials of this type find application for particular purposes. Some of the forms, main uses and properties are conveniently summarised here.

COUMARONE-INDENE

*Physical form*
Aromatic solvent solution (benzene toluene, xylene), or mastic.

*Properties*
Forms rigid, tough films which soften at high temperatures. Resistant to water and alkalies but not oils or greases. Bond strengths are low.

*Applications*
Bonding wood, fabrics, felt. Component of hot-melt adhesive formulations. Flooring applications.

*Remarks*
Predominantly an adhesive constituent rather than an adhesive.

SHELLAC

*Physical form*
Alcoholic solvent solution or as hot-melt mastic.

*Properties*
Good electrical insulating properties. Lacks flexibility unless compounded with other materials. Resistant to water, grease and oils. Softens on heating and is soluble in many solvents. Bond strengths are moderate.

*Applications*
Porous materials, metals, ceramics, cork and mica. Gasket cements. Primer for metals and mica intended for adhesive bonding. Insulating sealing waxes. Component of hot-melt adhesives. Electrical industry applications.

*Remarks*
Declined in use because of high cost. Basis of de Khotinsky cement. Available in various degrees of purity and colour.

ROSIN (COLOPHONY)

*Physical form*
Solvent solution or hot-melt mastic.

*Properties*
Poor resistance to solvents and oils; good resistance to water. Subject to oxidation and has poor ageing properties. Brittle and is usually modified with plasticisers. Bond strengths are moderate and develop fairly rapidly.

*Applications*
Paper (temporary adhesive and label varnish). Com-

ponent of pressure sensitive adhesives based on styrene-butadiene co-polymers. Component of hot-melt adhesives.

*Remarks*
Synthetic materials have replaced many of the rosin adhesives where the base material is scarce. Esters and other rosin derivatives find application as adhesive components.

OLEO-RESINS (VEGETABLE OILS + ROSIN, PHENOLIC OR ALKYD RESINS)

*Physical form*
Viscous liquids with solvent or as mastics (with filler).

*Properties*
Excellent tack and rapid development of low bond strength. Soft plastic consistency results in low bond strength. Good ageing properties and retain flexibility for long periods (years). Fair resistance to water and oils but poor resistance to solvents and alkalies. Pliable at low temperatures (−46°C) but embrittle at 40–70°C over long exposure period.

*Applications*
Linoleum, felt, cork and similar materials to each other and metals; plastic tiles; acoustical wall coverings. Caulking and sealing.

*Remarks*
Usually require a solvent to reduce viscosity of oil/resin formulation. Good wetting properties for most surfaces.

BITUMEN (INCLUDING ASPHALT)

*Physical form*
Water emulsions, hot-melts or solvent solutions.

*Properties*
Dark coloured materials having poor strength. Depending on formulation, may be soft, sticky, or hard and brittle. Resistant to water, acids and alkalies but not oils and many solvents. Generally poor resistance to high (above 45°C) and low temperatures.

*Applications*
Concrete, glass, metals, felt, paper; flooring applications; tiles to walls; laminating paper and metal foil; corkboard insulation; waterproofing walls; road binding materials. Hot-melt component.

*Remarks*
Suitable for low stress applications where durability is important. Tendency to cold flow even at room temperature. Some emulsions resist flow up to 105°C. May be compounded with epoxy resins to improve strength properties.

#### *4.3.57 Urea formaldehyde*

*Type*
Synthetic thermosetting resin.

*Physical form*
Two part products consisting of the resin and hardening agent (liquid or powder). Also available as spray-dried powder, with incorporated hardener, which is activated by mixing with water.

*Shelf life, 20°C*
Resins, 3–6 mth; powders, at least 1 yr; catalysts, indefinite.

*Working life, 20°C*
From 0–48 h according to composition and setting temperature. Room temperature setting types range from 3–10 h.

*Closed assembly time, 20°C*
From 0–24 h depending on composition. From 0–30 min where resin and catalyst are applied separately to each adherend.

*Setting process*
Condensation polymerisation with elimination of water.

*Processing conditions*
Suitable for use with radio-frequency equipment. For general purposes, 2–4 h at 20°C and 350–700 kPa bonding pressure. For plywood manufacture, 5–10 min at 120°C and up to 1600 kPa. Timber: (Hardwood) 15–24 h at 20°C and 1400 kPa. (Softwood) 15–24 h at 20°C and 700 kPa. Bonding pressures depend on type of wood, shape of parts and similar factors.

*Coverage*
0·1 kg/m$^2$ general purpose bonding. 0·075 kg/m$^2$ plywood fabrication. 0·125–0·2 kg/m$^2$ timber construction work.

*Permanence*
Urea formaldehyde adhesives lack the durability of combinations with melamine, phenolic, or resorcinol resins, and are unsuitable for service conditions which are extreme, e.g. high humidity, boiling water and temperatures above 60°C. Resistant to cold water and biodeterioration and may be compounded with additives to improve durability. Hardening agents used are important components and manufacturers advice should be followed. Resins cured with ammonium thiocyanate have superior hot water resistance to those hardened with diammo-

nium phosphate. Use of starch extender reduces water resistance, increases resin viscosity to overcome adhesive absorption into porous materials (wood) and lowers cost of adhesive. Use of phenolic or resorcinol additives, promotes resistance to water, weather and temperature changes; durability improved but increases cost of adhesive. Use of melamine fortifier, increases resistance to water and temperature and improves durability. Melamine or resorcinol fortifiers do not enhance durability beyond that of phenolic resins.

*Applications*
Wood and related materials with moisture content of 7–15%. Manufacture of light woods, thin veneers and plywood furniture assembly and joinery. Timber construction subject to low stresses.

*Remarks*
Development of full bond strength is gradual over a few days but assemblies may be handled after 6–12 h. Bond strengths usually exceed the strength of the wood when glue-line thickness ranges from 0·05–0·16 mm. Gap-filling properties are poor where glue-lines exceed 0·37 mm. Thick glue films will craze and weaken bond strength unless special modifiers, such as furfural alcohol resins are used. Gap-filling properties are important where the adhesive is extended with wood or walnut shell flow. Adhesives are non-staining and light coloured; unsuitable for exterior conditions or extreme temperatures.

*Specification*
BS 1204:1979 (Adhesives for constructional work in wood). BS 1203:1979 (Adhesives for plywood).

#### 4.3.58 Water and solvent based adhesives

At some stage in their application, all adhesives must be fluid enough to wet the substrate prior to bond formation. Most adhesives are fluid when applied and these are represented by the water based, solvent based, prepolymer fluids (100% solids) and hot-melts. For water or solvent based adhesives, assembly of parts can be carried out without removing the liquid vehicle provided one of the substrates is porous; for non-porous substrates (e.g. metals) the removal of the vehicle is necessary before bonding the parts. Here, bond formation depends on the reactivation of the dried adhesive film under heat and pressure.

Water based adhesives are made from materials that can be dissolved or dispersed in water alone. Some of these materials are the basis for solvent based adhesives and the principal ones which are used for liquid adhesive formulations are given below.

*Water based*
Casein
Dextrin and Starch
Glue (animal)
Gums
Lignin
Polyvinyl alcohol
Sodium carboxymethyl cellulose
Sodium silicate

*Organic solvent based*
Cellulose acetate-butyrate
Cellulose nitrate
Cyclised rubber
Polyisobutylene

*Water or organic solvent*
Asphalt* and coal tar resins*
Amino resins* (melamine and urea formaldehyde)
Phenolic resins* (phenol and resorcinol formaldehyde)
Polyamides
Polyacrylate† (and polymethacrylate†)
Rubbers (natural and reclaimed,* synthetic elastomers based on polychloroprene,† acrylonitrile-butadiene†, styrene-butadiene† and butyl*)
Vinyl polymers and copolymers (polyvinyl acetate,† polyvinyl chloride,† polyvinyl ethers, polyvinylidene chloride†)

The reader is referred to the relevant pages for more detailed description of the properties and applications of specific adhesives. The advantages and disadvantages of liquid-based adhesives are conveniently summarised in Table 4.8.

#### 4.3.59 Waxes

Waxes are thermoplastic materials of natural or synthetic origin which are usually applied as hot-melt adhesives above 54°C; setting action results from the cooling of the molten material. Numerous types are available which are derived from mineral, vegetable, petroleum sources or based on synthetic low-molecular weight polymers, and they may be compounded with elastomers to improve their flexibility.

Resistance to water and biodeterioration is usually good but solvent resistance is poor and the heat resistance very limited.

Adhesives of this type are useful for temporary bonds with non-porous materials such as metals and glass, and for packaging applications involving paper, cellophane, and foils and films based on metals and plastics. Sirawax softens at +30°C; is a

*Available as water dispersions (lattices)—made by emulsion polymerisation.
†Available as water dispersions—made by emulsifying or dispersing solid polymers.

**Table 4.8 Advantages and disadvantages of liquid-based adhesives**

| *Factor* | *Water-based* | *Solvent-based* |
|---|---|---|
| Cost of production | Inexpensive, easily manufactured | Medium cost adhesives. |
| Inflammability | Non-inflammable | Often a fire and explosive hazard. |
| Toxicity | Non-toxic solvent | Usually toxic and constitute a health hazard. Venting equipment required. |
| Storage properties | Prone to metal contamination from container vessels | Shelf lives up to 1 yr where solvent is retained. |
| Solids content | Wide range | Wide range. |
| Viscosity | Wide range | Moderately wide range. |
| Drying rates and open-times | Slow-drying and long setting times | Quick drying; variable over a wide range |
| Maintenance of applicator equipment | Easy clean-up of equipment | May present difficulties where adhesive is quick drying. |
| Development of bond strength | Slow. Generally poorer tack properties than solvent-types | Rapid. Tack properties are usually good. |
| Substrate compatibility | Unsuitable for hydrophobic surfaces (many plastics) where wetting is often inadequate. Causes shrinkage of some substrates (textiles, paper and cellulosics). Corrosive action towards some metals. Suitable for many substrates (including foamed plastics) | Suitable for many hydrophobic surfaces; good wetting properties. Often unsuitable for plastic foams where solvent attack occurs. Compatible with metallic substrates; basis for the contact adhesives which display generally good adhesion to a wide variety of materials. |
| Water resistance | Poor | Good. |
| Cold resistance | Subject to freezing with embrittlement of adhesive. | Generally good retention of resilience when frozen. |
| Electrical properties | Poor | Good. |

typical and generally useful laboratory wax adhesive for temporary fixation of glass, ceramics, metals, etc., (e.g. as components of small assemblies and instruments).

Other uses of waxes concern their use as temporary encapsulants and formulation in hot-melt compositions (see HOT MELT ADHESIVES).

## 4.4 References

### *4.4.1 General*

SKEIST, I. and MIRON, J., 'Introduction to adhesives', p. 3–17, *in Handbook of Adhesives* (2nd edn), Reinhold, New York, (1977).

WARSON, H., 'Adhesives', *App. Polym. Emulsions*, **6**(5), 43–52, (1976). 85 refs.

ALLEN, K. W., 'Nature of adhesives', *Br. Polym. J.*, **11**, 50–53, (1979). 8 refs.

WAKE, W. C., 'Recent developments in adhesives', *Plast. Rubber Inst.*, **5**, 157–161, (1980).

FENN, D. J., 'Structural adhesives', *Engng. Des.*, **6**, 18–21, (1980).

JACKSON, B. S., (Ed.), *Industrial Adhesives and Sealants*, Hutchinson Benham, London, (1976). A general account of polymerised industrial polymer adhesives and sealants with particular reference to paper, packaging, and converting industries, shoe manufacturing, and domestic applications.

### *4.4.2 Natural products*

BOLKER, H., *Natural and Synthetic Products*, Dekker, New York, (1974).

CORNWALL, E. D., 'Animal glues and their industrial application', *Adhesion and Adhesives*, (Houwink and Salomon, Eds), Elsevier, (1967).

174 ADHESIVES HANDBOOK

**Table 8.24 Volume resistivity**

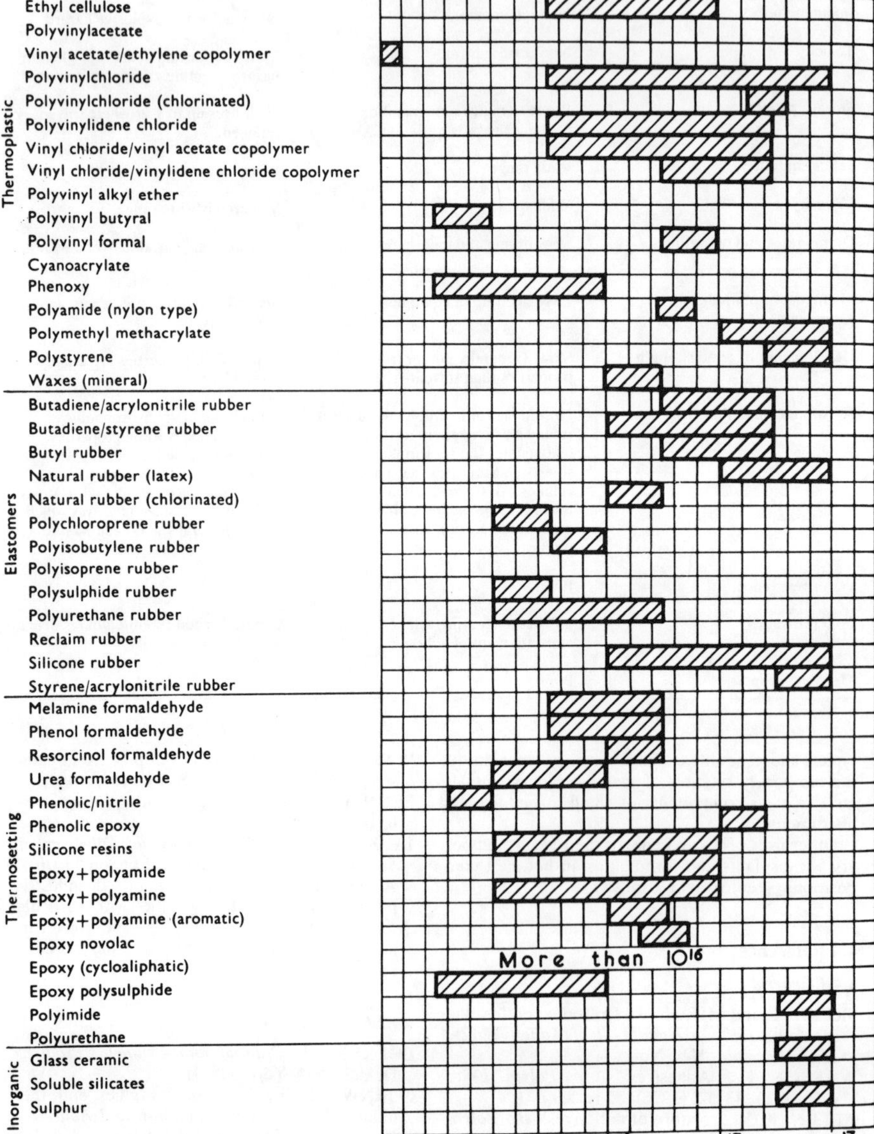

# 9

# Adhesive products directory

## 9.1 Introduction

The data presented in this chapter have been selected from manufacturers' trade literature and supplemented with other published information which is believed to be accurate. In addition, some performance data derived from evaluation work at Sira have been included.

Throughout this handbook an attempt has been made to carry out tests and to quote literature or other sources of information correctly. It is appreciated that there is always the possibility of error and the author apologises to anyone harmed by any such errors which were not apparent at the time of publishing. Similarly an apology is made to those whose products have not been mentioned. It is, of course, impossible to deal with all products and selection of some must of necessity mean omission of many.

It should be noted that adhesives' manufacturers generally disclaim responsibility for the results obtained by the user, who should carry out his own evaluation to determine suitability of the product. Adhesives' properties vary between wide limits and their performance depends on the conditions under which they are prepared and used; it is preferble that the manufacturer should always be approached for detailed information about his product. The data given here should thus be regarded as an indication of the probable values obtainable under certain conditions rather than as absolute values obtainable under all conditions. Hence, it should be recogised from the outset that differences in results from those described here are often possible.

Only a small percentage of the total number of adhesives products available has been described and space limitations have restricted the amount of detail it has been possible to include. Some manufacturers supply such a large number of types that several pages would be required to describe them. As far as possible an attempt has been made to include a description of a representative sample of each type of adhesive. Whenever practicable, commercial products of a particular adhesive type have been grouped together in order to demonstrate the potential versatility of performance. Many manufacturers, although mainly concerned with a few basic adhesive types, will formulate specialised adhesives to meet their customers' requirements. The size of any entry in this section is not indicative of the manufacturers' importance in the adhesives' field since it is impossible to know whether the adhesives are produced as main or secondary products.

## 9.2 SI and related units of measurement: ready reference conversion factors

Table 9.1 is selective and concerned with those units in use throughout the handbook. It is intended to facilitate rapid reference to conversions frequently needed in everyday work.

A number of publications have been issued by the British Standards Institution and the National Physical Laboratory. The following list will assist those requiring further information on metric units in general.

(1) *The Metric System in the United Kingdom: The Use of SI Units*. British Standards Institution. PD 5686. January, (1978).
(2) *Changing to the Metric System. Conversion Factors, Symbols and Definitions*, P. Anderton and P. H. Bigg. National Physical Laboratory HMSO (1980).
(3) The International System (SI) Units BS 3763: 1976.
(4) Conversion Factors and Tables.

BS 350: Part 1: 1974, Bases of Tables, Conversion factors
BS 350: Part 2: 1962, Detailed conversion tables
BS 350: Supplement No. 1: 1967, Additional tables for SI conversions. Amendments AMD 342: 1969 and AMD 1255: 1973.

**Table 9.1**

| *Property* | *Non SI unit* | *Factor for conversion to SI unit* |
|---|---|---|
| *MECHANICAL* | | |
| Modulus of elasticity | E( × $10^6$ lb/in$^2$) | E( × 74 Pa) |
| Peel strength | 1 lb/in | 0·175127 N/mm |
| | 1 kg/cm | 0·980665 N/mm |
| Shear strength | 1 lb/in$^2$ | 6·89476 kPa |
| | 1 kgf/cm$^2$ | 98·0665 kPa |
| Compressive | 1 lb/in$^2$ | 6·89476 kPa |
| Tensile strength | 1 lb/in$^2$ | 6·89476 kPa |
| Flexural strength | 1 lb/in$^2$ | 6·89476 kPa |
| Impact strength | ft. lb/inch notch | 1·355 J for 25·4 mm notch |
| Tear strength | 1 lb/in | 0·175127 N/mm |
| Fatigue strength after load cycling | 1 lb/in$^2$ at $X$ Hz | 6·89476 kPa at $X$ Hz |
| | 1 kgf/cm$^2$ at $X$ Hz | 0·098065 MPa at $X$ Hz |
| *ELECTRICAL* | | |
| Dielectric strength | 1 volt/mil | 0·0383701 kv/mm |
| Dielectric constant | 1 at $10^n$ c.p.s. | 1 at $10^n$ Hz |
| Dissipation factor | 1 at $10^n$ c.p.s. | 1 at $10^n$ *Hz* |
| Volume resistivity | 1 Ω/cm | $10^{-1}$ Ω/mm or, $10^2$ Ω/m |
| *HEAT* | | |
| Thermal conductivity | 1 Btu in ft$^{-2}$h$^{-1}$ °F$^{-1}$ | 0·144228 W/m°C |
| | 1 Cal cm$^{-1}$ sec$^{-1}$ 1°C$^{-1}$ | 418·68 W/m°C |
| Thermal expansion | 1°F$^{-1}$ | 1·8°C$^{-1}$ |
| *OTHER PROPERTIES* | | |
| Viscosity (dynamic) | 1 poise | 0·1 Pas |
| Vicosity (kinematic) | 1 Stokes | $10^{-4}$ m$^2$/s |
| Capacity | 1 lb UK gal$^{-1}$ | 0·09978 kg/litre |
| | 1 lb US gal$^{-1}$ | 0·0831 kg/litre |
| Bulk density | lb/ft$^3$ | 16·0185 kg/m$^3$ |
| Length | 1 in | 25·4 mm |
| | 1 ft | 0·3048 m |
| Area | 1 in$^2$ | 6·4516 × $10^{-4}$m$^2$ |
| | 1 ft$^2$ | 0·0929 m$^2$ |
| Weight | 1 oz | 28·35 g |
| | 1 lb | 453·6 g |
| | 1 g | $10^{-3}$ kg |
| Volume | 1 pt | 0·569 litre |
| | 1 gal | 4·546 litre |

*Note:* Impact strengths.
Impact strengths should be given as energy/area = force/length. Impact strengths with notch are also to be given in energy/area if related to the cross-section but carry the unit energy/length if they refer to the width of the notch.
1 kg-cm/cm = 9·80665 N/cm = 980665 kg s$^{-2}$[a]
1 ft-lb/in. notch = 53·378659 N = 53·378659 J/m[a]
[a] Under substitution of masses of old units by forces of new units

(5) *How to Write Metric—a Style Guide for Writing and Using SI Units*. Metrication Board, HMSO, (1977).
(6) *SI—The International System of Units*, (Goldman and Bell, Eds), National Physical Laboratory, (1982).
(7) *Metrication—Managing the Industrial Transition*, (Liptai and Pearson, Eds), ASTM Special Technical Publication STP 574, (1974).
(8) *Standard for Metric Practice*, ASTM Specification E380, (1982).
(9) Walshaw, A. C., *SI Calculations in Engineering Science*, Butterworths, London, (1977).

In the SI system (Système International d'Unités) the kilogram force is replaced by the Newton (N = kg m/s$^2$) to form derived units which include,

energy, (N.m denoted by unit symbol J for joule)
power, (N.m/s or unit symbol W for watt)
force, e.g. pressure or stress, (N/m$^2$ or unit symbol Pa for pascal). In certain cases, pressures and stresses are conveniently expressed in N/mm$^2$ or MN/m$^2$

The unit symbols can be used to provide a range of stress units, namely . . . μ, Pa, mPa, Pa, kPa, MPa . . . appropriate to a given measurement.

## 9.3 Key to adhesive product charts

The following alphabetical symbols have been used in the tabulated description of properties and performance data of adhesive products (p. 178).

| *Symbol* | *Property* |
|---|---|
| a | Peel strength |
| b | Shear strength |
| c | Compressive strength |
| d | Tensile strength |
| e | Torsion shear |
| f | Flexural strength |
| g | Impact strength |
| h | Tear strength |
| j | Modulus of elasticity |
| k | Elongation |
| l | Hardness (Shore D or L) |
| m | Dielectric strength |
| n | Dielectric constant |
| o | Dissipation factor |
| p | Volume resistivity |
| q | Thermal conductivity |
| r | Thermal expansion |
| s | Viscosity (dynamic) |
| t | Viscosity (kinematic) |
| T | Heat Deflection Temperature |
| u | Capacity |
| v | Compression set ASTM D395–55 |
| w | Fatigue strength after load cycling |
| x | Modulus of rupture |
| y | Bulk density |
| z | Refractive Index |

**Table 9.2 Adhesive materials and products index**

| *Code nos.* | *Adhesive type* | *Basic information page nos.* | *Examples of trade products page nos.* |
|---|---|---|---|
| 1. | Animal/Fish glue | 36, 49 | 178 |
| 2. | Casein | 38 | 180 |
| 3. | Blood Albumen | 37 | * |
| 4. | Soy(a) bean glue | 75 | * |
| 5. | Dextrine | 78 | 180 |
| 6. | Starch | 76 | 180 |
| 7. | Gum arabic | 78* | 178 |
| 8. | Canada balsam | 40 | |
| 9. | Cellulose acetate | 39 | 184 |
| 10. | Cellulose acetate-butyrate | 39 | * |
| 11. | Cellulose caprate | 40 | * |
| 12. | Cellulose nitrate | 40 | 184 |
| 13. | Methyl cellulose | 40 | 182 |
| 14. | Hydroxy ethyl cellulose | 40 | 182 |
| 15. | Ethyl cellulose | 40 | 182 |
| 16. | Natural rubber (latex) | 55 | 200 |
| 17. | Rubber hydrochloride | 74 | * |
| 18. | Chlorinated rubber | 73 | 202 |
| 19. | Reclaim rubber | 72 | 200 |
| 20. | Butyl rubber | 38 | 202, 204 |
| 21. | Polyisobutylene rubber | 38 | * |
| 22. | Nitrile rubber | 55 | 210 |
| 23. | Polyisoprene rubber | * | * |
| 24. | Butadiene styrene rubber | 37 | 202 |
| 25. | Polyurethane rubber | 68 | 216 |
| 26. | Polysulphide rubber | 68 | * |
| 27. | Silicone rubber | 74 | 224 |
| 28. | Polychloroprene (neoprene) rubber | 65 | 206 |
| 29. | Polyvinyl acetate | 71 | 190 |
| 30. | Vinyl acetate—acrylic acid | 49 | 220 |
| 31. | Vinyl acetate—ethylene | 49 | 198 |
| 32. | Polyvinyl chloride (PVC) | 72 | 186 |
| 33. | Chlorinated PVC | * | 186 |
| 34. | Vinyl chloride—vinyl acetate | 72 | * |
| 35. | Vinyl chloride—vinylidene | 72 | * |
| 36. | Polyvinylidene chloride | 72, 78 | 188 |
| 37. | Polyvinyl formal | 70 | * |
| 38. | Polyvinyl butyral | 70 | * |
| 39. | Polyvinyl alcohol | 72 | * |
| 40. | Polyvinyl alkyl ether | 71 | * |
| 41. | Polyvinyl pyridine | | * |
| 42. | Polyacrylate | 34, 78 | * |
| 43. | Polyacrylate (carboxylic) | 34 | 226 |
| 44. | Polyacrylic esters | 35 | 230 |
| 45. | Polymethylmethacrylate | 34 | 226 |
| 46. | Cyanoacrylate | 42 | 238 |
| 47. | Acrylamide | 34 | |
| 48. | Ionomer resins | 52 | Trade source 142 |
| 49. | Polyamide (nylon based) | 56 | * |
| 50. | Polyamide (versamid based) | 61 | 194, 198 |
| 51. | Phenolic polyamide | 59 | |
| 52. | Polystyrene | 67 | 185, 186 |
| 53. | Polyhydroxy ether | 71 | 244 |
| 54. | Polyesters (linear) | 36, 66 | 194, 198, 286 |
| 55. | Polyester + isocyanate | 53, 54 | * |
| 56. | Polyester + monomer | 66 | * |
| 57. | Urea formaldehyde | 77 | 244 |
| 58. | Melamine formaldehyde | 54 | 248 |
| 59. | Urea—melamine formaldehydes | 77 | 248 |
| 60. | Phenol formaldehyde | 57, 58 | 248 |
| 61. | Resorcinol formaldehyde | 73 | 248 |
| 62. | Phenolic-resorcinol formaldehydes | 73 | 246 |
| 63. | Phenolic—epoxy | 56 | 254, 256 |
| 64. | Silicone resins | 74 | 224 |
| 65. | Epoxy (+ polyamine) | 44 | 268 |
| 66. | Epoxy (+ polyanhydride) | 44 | |
| 67. | Epoxy (+ polyamide) | 47 | 266 |
| 68. | Epoxy—alkyl ester) | | * |
| 69. | Epoxy (cycloaliphatic) | 44 | * |
| 70. | Epoxy—bitumen | 77 | 258 |
| 71. | Epoxy—nylon | 46 | 294 |
| 72. | Epoxy—polysulphide | 47 | 266, 268 |
| 73. | Epoxy—polyurethane | 48 | 266 |
| 74. | Polyethylene imine | | |
| 75. | Polyisocyanate | 53 | 220 |
| 76. | Furane resins | 49 | * |
| 77. | Phenolic isocyanate | 54 | |
| 78. | Phenolic—nitrile | 59 | 212, 214 |
| 79. | Phenolic—neoprene | 58 | * |
| 80. | Phenolic—polyvinyl butyral | 60 | 244 |
| 81. | Phenolic—polyvinyl formal | 60 | 244 |
| 82. | Polyimide | 63 | 244, 284 |
| 83. | Polybenzothiazole | 64 | |
| 84. | Polybenzimidazole | 64 | * |
| 85. | Bitumen (asphalt) | 77 | 182 |
| 86. | Silicates (soluble) | 51 | 288 |
| 87. | Mineral waxes | 78 | * |
| 88. | Ceramics | 41 | 292 |
| 89. | Inorganic | 51 | 290 |

*Consult manufacturers for information (pp. 311–313)

## 9.2 Adhesive products table

Adhesive products have been listed in an alphabetical order which approximately corresponds to the sequence of adhesive materials described in the previous section. The information appearing in the table following is given under the following headings:

*Trade source*
A list of code numbers representing commercial suppliers of adhesive products will be found on p. 315 (Table 10.4).

*Trade name or designation*
Information on the main chemical ingredient of the adhesive where available. Some products are described in general terms where the manufacturer has not disclosed the composition of the product (e.g. synthetic rubber-resin, wax, etc).

*Number of components*
The number of ingredients which are supplied to the user must be mixed to constitute the adhesive (e.g. components such as resins, hardeners and catalysts).

*Physical form, consistency or viscosity*
Viscosity values given refer to the mixed adhesive where multiple-component formulations are concerned.

*Working life*
The term refers to the pot-life at room temperature (20–25°C) of approximately 100–150 g batches for multiple-component adhesives unless otherwise stated. For one-component pre-mixed adhesives the working life is specified in terms of the time within which the joint must be assembled after adhesive application (e.g. open-time for solvent-based adhesives).

| *Basic type* | *Trade name* Source (*key p. 315*) | *Colour* | *Physical form/viscosity* (m.Pa.s) | *Method of application* | *Curing cycle* | *Bonding pressure* (kPa) | *Working life* | *Service temperature* (°C) |
|---|---|---|---|---|---|---|---|---|
| Animal glue + catalyst | Plastick 5 | Amber | Thick liquid | Brush Machine | Hot setting. 2 min at 54°C or, 1 min at 79°C | Contact–10 | 4 h at 15°C–2 h at 27°C | +70 |
| Animal glue | Croid 409 116 | Brown | High viscosity liquid pH = 7 to 7·5 | Brush Roller | Applied at 20°C or preferably at 49–55°C | Contact–2 | Unlimited pot life | +50 |
| Animal glue | Calbar 71 | Amber | Liquid | Brush Spreader | Sets at 20°C after pressing and storage | 3·4–12 | Assembly of joint in 10–15 min after application | +50 |
| Animal glue | Glumaster 338 | Amber | Range of viscosities available | Brush Roller | Cold or hot setting products | Contact | Indefinite | +50 |
| Fish glue | Samson 309 | Pale buff | High viscosity liquid | Brush Machine | Dried at 20°C for 1 h | Contact | Open time 0·5 h at 20°C | Up to 60 |
| Fish glue | Certofix 96 | Yellow | Thick liquid 50% w/w solids | Dispensed from container | Dried at 20°C for 1 h | Contact | Open time 0·5 h at 20°C | Up to 60 |
| Gum Arabic | Samson No. 1 309 | Pale straw | Low viscosity liquid | Brush Machine | Sets in air after joint assembly | Contact | Indefinite | Up to 50 |

*Method of application*
The means by which an adhesive may be applied are briefly indicated. Mechanical applicators are feasible for many of the adhesives listed as hand-applied materials, but consultation with the manufacturer is recommended.

*Processing*
Curing cycles specify the shortest and longest time/temperature conditions for curing thermosetting adhesives; intermediate cycles are usually available from manufacturers. The column also indicates the drying time or heat-activation conditions for other types of adhesive (e.g. solvent-based resin-rubbers) where appropriate. Bonding pressures and times are stated in an adjacent column.

*Main uses*
Possible uses are restricted to the more important industrial, domestic and special purpose applications.

*Adherends*
Materials which may be bonded are listed in column form as a quick reference for adhesive selection purposes. Joints which are the subject of performance data described in the adjacent columns are indicated under the heading Test specimen/specification.

*Physical data*
Wherever possible bulk physical properties of the adhesive and bond strength data are reported against test specifications, conditions and temperatures described in the adjacent columns. The key to the alphabetical code used to denote the properties will be found on p. 176 together with a standard table for conversions between the SI system and other units of measurement. The data are intended as a guide to general properties and performance of adhesive products. They should not be used to make comparisons between adhesives where the processing conditions and test methods employed are different.

| *Adherend* | *Test specimen/ specification* | *Test temperature (°C)* | *Physical data* | *Key (p. 176)* | *Remarks* |
|---|---|---|---|---|---|
| Wood (oak, pine, beech)<br>Veneer materials | Wood (beech)–wood | 20 | 3·8 MPa | b | Unsuitable for use with iron vessels or platens where oak veneering is undertaken (staining hazard). Employed for wood veneer applications and laminate fixation |
| Wood | Wood–wood | 20 | 4·2 MPa<br>Gelation temperature is 20–22°C<br>Poor resistance to water<br>Slightly alkaline | b | May be diluted with water (25% w/w).<br>Alternative products of varying viscosity are available |
| Wood<br>Veneers<br>Plastics laminates | Wood–wood | 20 | +4 MPa | b | Used for gluing veneers and decorative laminates. Dilution up to 25% w/w with water permissible. Normally 1–2 h pressing time for joint assembly |
| Paper | | | | | One of a range of animal glues for packaging industry. Other viscosities are available |
| Wood<br>Chipboard<br>Paper | Wood–wood | 20 | High tack material with moderate resistance to water<br>8 MPa | b | General purpose glue for porous materials. Rapid setting with good flexibility in cured glueline |
| Paper<br>Leather<br>Textiles<br>Wood | Wood–wood | 20 | 9–7 MPa<br>Poor resistance to water | b | General purpose glue for absorbent materials |
| Paper<br>Cardboard | | | Sets to a medium flexibility material with a moderate resistance to water<br>Slightly acid, tacky gum | | Used for paper sheet and strip gumming for tapes, rolls, etc. Remoistened prior to bonding. Labelling and litho-plate preparation applications. One of a range of gums of various viscosities and tack qualities |

*continued*

| *Basic type* | *Trade name* Source (*key p. 315*) | *Colour* | *Physical form/viscosity* (m.Pa.s) | *Method of application* | *Curing cycle* | *Bonding pressure* (kPa) | *Working life* | *Service temperature* (°C) |
|---|---|---|---|---|---|---|---|---|
| **Polyimide** | FM34 121 | Olive green | Filled resin on glass cloth carrier 14% w/w volatiles | Cut out shape is sandwiched in joint | Heated to 280°C over 30 min and cured at 280°C for 90 min | 2800 | 6 mth at −20°C or 2 wk at 20°C | −40–300 Short periods at 350 |
| **Phenolic-polyvinyl formal** | Redux 775 100 | White | Liquid with powder | Brush or roller to apply liquid. Wet substrate is then dipped into powder | Cured for 30 min at 150±5°C Pressure is removed when joints have cooled to 70°C | 1·7–700 Depends on substrate | 30 min drying period. Prepared substrates can be stored 3 d before assembly | Up to 120 |
| **Phenolic-polyvinyl butyral** | Bakelite J11185 44 | Orange to brown | Liquid | Brush | Dried for 0·5–2 h at 60–80°C, then cured for 18 h at 100°C or 20 min at 148°C under pressure | 4–70 | Indefinite | Up to 100 |
| Urea formaldehyde | Beetle Cement W54 43 | Grey | Low viscosity liquid 68 + 2% solids content | Brush Spreader | Under pressure 15s at 149°C to 90s at 71°C for catalyst used | Up to 700 | ¾ h at 32°C to 48 h at 10°C for catalyst used | |
| Urea formaldehyde | Cascamite One Shot 71 | White | Powder | Brush Roller Spreader | 9 h at 10°C to 1 h at 21°C after mixing powder with water (22%) | 40–400 for 18 h at 10°C to 3 h at 21°C | | |
| Urea formaldehyde | Aerolite H-F + Hardener HF 100 | | Low viscosity liquid | Spreader | 8 min at 71°C to 1 min at 110°C | Up to 700 | 18–20 h at 21°C | |

| *Adherends* | *Test specimen/ specification* | *Test tem- perature (°C)* | *Physical data* | *Key (p. 176)* | *Remarks* |
|---|---|---|---|---|---|
| Metals (high temperature structural applications) | Steel—steel | 25<br>260<br><br>260<br><br>25<br>25<br>300 | 24 MPa<br>14·6 MPa after 10 min at 260°C<br>14 MPa after 3 yrs at 260°C<br>1·4 N/mm<br>80 MPa<br>35 MPa<br>Complies with MM-A-132 Type 4 | b<br>b<br><br>b<br><br>a<br>e<br>e | Primers available to promote strength and heat durability. Correct storage essential for optimum performance. Loses volatiles during cure so that gluelines are subject to porosity. Excellent resistance to common organic solvents. Available in roll form—various widths and lengths |
| Metals<br>Wood<br>Thermosetting plastics<br>Rubbers<br>Asbestos, friction materials | Steel (mild)—steel<br>Magnesium alloy (HK31A–H24)—magnesium alloy<br>Titanium—titanium(K1, 130)<br>(2024-T3) Alclad—Alclad<br><br>Aluminium<br>Creep rupture; 192 h at 11·05 MPa load | 24<br>24<br><br>24<br><br>24<br>82<br><br>24<br>24 | 34·6 MPa<br>22·5 MPa<br><br>32 MPa conforms to DTD 775B<br>34·7 MPa<br>15·9 MPa conforms to MIL-A-5090B<br>5·3 N/mm<br>Less than 0·0254 cm max deformation<br>3300 MPa<br>415 W/(m°C)<br>Conforms to MMM-A-132 Typel, Class 2 | b<br>b<br><br>b<br><br>b<br>b<br><br>a<br><br>j<br>q | Alternative Redux systems are available as liquid/ powder or film systems. These adhesives have been the subject of considerable investigation. Extensive peromance data are available from supplier. Redux Liquid K6 + Powder C is excellent bonding agent for vulcanised rubbers (neoprene, natural, etc.) or thermoset plastics. Redux 64 is suitable for friction brake lining pads. Aircraft and industrial applications form main usage areas. Excellent resistance to fuels, oils, salt spray and relative humidity (100% RH at 50°C). Accelerator available to promote cure |
| Steel<br>Phenolic (laminates)<br>Asbestos<br>Graphite<br>Wood<br>Metal (foils) | Aluminium (alloy)—aluminium<br>Steel—steel | 20<br><br>20<br>20<br><br>20 | 34·7 MPa<br><br>42·7 MPa<br>12.9 MPa cured 18 h at 100°C<br>15·1 MPa cured 0·5 h at 149°C<br>2·1 J for 25·4 mm notch | b<br><br>d<br>b<br><br>b<br><br>g | A phenolic stoving cement for assembly of metals, phenolic laminates and inorganic substrates. Used for metal foil-to-plastic printed circuit board (PCB) fabrication. For copper-to-board bonding in PCB manufacture, pressures up to 100 kPa may be required.<br>Adhesive can be thinned with alcohol in 1:1 ratio |
| Wood<br>Chipboard<br>Hardboard | | | Satisfies BS 1203; 1979 (Type MR) and BS 1204: 1979 Part 1 and Part 2 (Type MR) for moisture resistant joints Melamine hardener is available to satisfy BS 1203: 1979 (Type BR) Boil resistance joints | | Similar in type to products W2 and W37. Is suited to woodworking assembly where long working life with short clamping times is required |
| Wood<br>Plywood<br>Chipboard<br>Phenolic laminate | | | Complies with BS 1204: 1979 (Type MR) | | Used for wood gluing and bonding of plastic laminates to wood. Other uses concern boat building and timber engineering. May be extended with wood flour for veneering assemblies. Coverage, 7 $m^2$/kg |
| Wood laminates | | | | | Rapid curing adhesive suitable for use with radio frequency heating of the glueline. Alternative catalysts, L38, L50, L48 may be used. |

*continued*

# 10

# Adhesives and trade sources

## 10.1 Trade names list

Adhesives are known and sold commercially under a variety of trade names which, to the uninitiated, give little information as to their chemical nature and application. The table below is based on an analysis of current trade literature and other published material and lists the trade names of adhesives which are presently available in the United Kingdom.

The trade names, arranged alphabetically, are followed in turn by descriptions which designate the basic chemical composition and/or give some indication of the form and use of the adhesive. The list includes only those materials which are actual adhesive products and, with few exceptions, no representation has been given to the considerable number of raw materials which are known to form the basis of adhesive systems.

A list of manufacturers against code numbers is given on p. 315 (Table 10.4).

Additional sources of information on adhesive materials will be found in the references at the end of Table 10.1.

**Table 10.1.**

| *Trade names* | *Type, description or uses* | *Code no.* |
|---|---|---|
| **Aatex** | Natural and synthetic latex adhesives for footwear manufacture | 24 |
| **Acalor** | Range of chemically resistant cements for tiles | 4 |
| **Acebrand** | Animal glues for paper and paint manufacture | 5 |
| **Acrifix** | Range of adhesives based on acrylic resins | 111 |
| **Acrofix** | Dental cement | 130 |
| **Acronal** | Acrylic ester polymers and co-polymer emulsions, solids and solutions | 37 |
| **Acrulite** | Cold curing acrylic resins used as adhesives for perspex | 305 |
| **Adamastic** | Polishing wheel liquid cement | 5 |
| **Adam's** | Liquid skin glue | 5 |
| **Adcol** | Range of starch-based adhesives for paper bag manufacture, labelling and bookbinding | 6 |
| **Adcote** | Range of 2-part water-based adhesives for flexible packaging materials | 383 |
| **Admelt** | Hot-melt adhesives | 5 |
| **Adhesive MV** | Paper-to-paper laminating adhesive | 13 |
| **Adhesive SV** | Paper-to-paper or fabric lamination | 13 |
| **Adhesive SF** | Paper laminating adhesive | 13 |
| **Adhesive STK** | General-purpose and wood lamination | 13 |
| **Adhesive SN** | Pressure-sensitive adhesive | 13 |
| **Adsol** | Solvent based rubber adhesives | 9 |
| **Aero** | Animal glue for woodworking trades | 116 |
| **Aerocol** | Polyvinyl acetate adhesives | 100 |
| **Aerodux** | Resorcinol formaldehyde wood glues | 100 |
| **Aerolastic** | Rubber-pitch joint sealant | 155 |
| **Aerolite** | Urea formaldehyde wood glues | 100 |

**Table 10.1** *continued*

| *Trade Names* | *Type, description or uses* | *Code no* |
|---|---|---|
| Aerophen | Phenol formaldehyde resins | 100 |
| Aerosol, 873B | Synthetic rubber/resin adhesive in aerosol form | 122 |
| Aerospace | One part silicone rubber adhesive and sealant | 137 |
| AF | Epoxy resin | 352 |
| Agf | Eurea formaldehyde gap-filling adhesive for porous materials | 363 |
| Agomet | Epoxy resin based adhesives for metals, ceramics, plastics, etc. | 83 |
| Airborne L229 | One-part solvent, and 2-part curing adhesives | 14 |
| Airborne L79 | One-part polyurethane adhesive in non-flammable solvent | 14 |
| Airborne L28 | Two-part polyurethane adhesive; air curing | 14 |
| Airborne L532 | One-part polyurethane adhesive; air curing | 14 |
| Albertol | Phenolic resins with resin modifier | 190 |
| Alcotex | Polyvinyl alcohol products | 296 |
| Alcyn | Cyanoacrylate based adhesives | 154 |
| Alfadite | Range of casein glues | 5 |
| Alfatalat | Alkyd resin products | 296 |
| Alfrax | Aloxite cements | 90 |
| Alganol | Dental cement | 130 |
| Alpha-Ace | Cyanocrylate adhesive | 328 |
| Amco | Adhesive for bonding expanded polystyrene to metal | 71 |
| Amset | Range of fast setting animal glue adhesives for packaging industry | 338 |
| Amtex | Range of adhesives based on latex and polyvinyl acetate | 71 |
| Ancaflex | Curing agents for expoxide resins | 22 |
| Ancaminep | Amine curing agents for epoxide resins | 22 |
| Anderson | Bitumen compounds and hot-melt adhesives | 23 |
| Anderson-Mastic | Adhesive for bituminous substrates | 23 |
| Anglo | Natural and synthetic rubber solvent based adhesives for the footwear industry | 24 |
| AP Adhesive | Nitrocellulose/synthetic rubber adhesive | 177 |
| Aquapaint | Range of epoxy resin adhesives and coatings | 292 |
| Arabol | Trade name for self-adhesive tapes | 71 |
| Arafix | General purpose impact adhesive | 71 |
| Araldite | Range of epoxy resin adhesives and casting resins | 100 |
| Araprene | Range of adhesives based on neoprene rubber | 71 |
| Arasite | Adhesives for bonding insulation materials to metal | 71 |
| Arastik | Adhesive for bonding insulation materials to metals | 71 |
| Arastix | Rubber-based adhesive | 71 |
| Ardux | Urea adhesive for bonding rigid panels to vertical or uneven wall surfaces | 100 |
| Armourglaze | Range of epoxy resin coatings and adhesives | 220 |
| As | Synthetic resin pastes and emulsions | 9 |
| Atlas | Two-part synthetic resin adhesive | 30 |
| Atlas-Ago | Synthetic resin adhesives | 83 |
| Autoplax | Range of adhesives for polythene and polyvinyl chloride films based on epoxy and other synthetic resins | 33 |
| Autostic | Sodium silicate based cement | 92 |
| Avabond | Adhesives for furniture, packing, and woodwork | 34 |
| Bakelite | Range of synthetic resin based adhesives | 44 |
| Bakelite cement | Stoving cement based on phenolic resins | 50 |
| Bal-brik exterior | Grey cement based adhesive for decorative cladding | 80 |
| Bal-brik interior | Polyvinyl acetate adhesive for internal cladding | 80 |
| Bal-cem | Grey cement based floor tile adhesive | 80 |
| Bal-cem white | White cement based wall tile adhesive | 80 |
| Bal-epoxy | Epoxy resin adhesive | 80 |
| Bal-fix | Lightweight ceramic tile adhesive | 80 |
| Bal-flex | Two-part rubber-latex cement based ceramic tile adhesive | 80 |
| Bal-floor | Bitumen based adhesive for ceramic and mosaic floors | 80 |
| Bal-grip | Ready mixed, water resistant adhesive for bathroom applications | 80 |
| Bal-mix | Cement based wall tile fixative for interior or exterior use | 80 |
| Bal-proof | Synthetic rubber/resin based adhesive for ceramic materials and tiles | 80 |
| Balsa | General-purpose adhesive for balsa wood models | 177 |
| Bal-super set | Thin or thick bed cement adhesive for interior or exterior use | 80 |
| Bal-tad | Polyvinyl acetate based adhesive for ceramic tiles | 80 |
| Bal-wall | White, polyvinyl acetate, thin bed adhesive | 80 |

*continued*

**Table 10.1** *continued*

| *Trade Names* | *Type, description or uses* | *Code no* |
|---|---|---|
| Bal CTF2 | White thin bed adhesive | 80 |
| Beckopox | Epoxy resins and hardeners | 190 |
| Beetle | Range of adhesives based on urea or melamine formaldehyde in liquid or powder form for furniture assembly | 43 |
| Belzona | Adhesive/sealant with metallic powder base | |
| Berger | Range of high performance, acrylic copolymer, pressure-sensitive adhesives | 294 |
| Bettabond | Range of adhesives for packaging, furniture, building and general manufacturing | 60 |
| Bevaloid | Vinyl acetate copolymer based adhesives for packaging industry | 189 |
| Bexol | Range of synthetic/rubber adhesives and solutions | 50 |
| Bilaseal | Epoxy resin based putty | 41 |
| Bimul | Range of adhesives based on bitumen emulsions | 71 |
| Bisol | Vinyl acetate | 46 |
| Bodex | Range of dextrine gums for bottle labelling | 71 |
| Bondaglass | Glass fibre and synthetic resin repair kit for metals | 66A |
| Bondapaste | Two-part polyester resin adhesive and filler for metal repairs | 66A |
| Bond-a-snack | Formulated adhesives and polymers for packaging and structural applications | 259 |
| Bond-a-stik | Adhesives and polymers for packaging | 259 |
| Bondcrete | Polyvinyl acetate adhesive sealant for concrete and woodwork | 38 |
| Bondex | Bituminous plastic adhesive sealant | 23 |
| Bond GT | Thermosetting resin adhesive for film lamination | 312 |
| Bonding Agent 007 | Epoxy resin based adhesive | 276 |
| Bondmaster | Range of adhesives based on synthetic resins/rubbers | 310 |
| Boscoprene | Two-part synthetic resin/rubber cements for sealing and coating | 72 |
| Boscotex | Aqueous synthetic polymer dispersions for PVC | 72 |
| Bostik | Range of synthetic resin rubber adhesives, sealing and coating compounds | 72 |
| Boston | Range of adhesives for the footwear industry | 72 |
| Breon | Range of nitrile rubbers and lattices | 46 |
| Brimor | High temperature strain gauge cements based on aluminium phosphate and silica | 301 |
| Britfix | Epoxy resin and polystyrene based adhesives for general purpose use | 195 |
| BSL | Range of film adhesives and surface protection solutions | 100 |
| Bulldog | Synthetic rubber cement for bonding polyurethane foam; gummed tapes for packaging | 72 |
| Butterfly | Range of gummed, self adhesive and heat-seal papers and tapes | 214, 309 |
| Cabufix | Range of adhesives for cellulose acetate butyrate bonding | 244 |
| Calbar | Liquid animal glue | 53 |
| Carbolfrax | Silicon carbide refractory cement | 90 |
| Cariflex | Range of synthetic rubber formulations based on styrene butadiene, polybutadiene, polyisoprene | 318 |
| Carpetex | Rubber latex adhesive for backing carpets and rugs | 224 |
| Casco | Casein glues in powder form for timber and plywood | 71 |
| Cascodex | Vegetable glues for bookbinding and packaging | 71 |
| Cascogel | Animal glues for packaging and bookbinding | 71 |
| Cascomelt | Range of hot-melt adhesives for packaging and bookbinding | 71 |
| Cascomite | Range of urea formaldehyde glues for wood | 71 |
| Casco-ML | Anaerobic polyester adhesive/sealants for locking metal screws | 71 |
| Cascophen | Range of phenol/resorcinol/cresylic resin based adhesives | 71 |
| Casco-resin | Range of urea formaldehyde based adhesives for wood | 71 |
| Casco-rez | Range of polyvinyl acetate adhesives for woodworking | 71 |
| Cascosel | Casein, latex, polyvinyl acetate, rubber/resins, based adhesives for packaging | 71 |
| Cascoset | Emulsion adhesives for floor tiles | 71 |
| Cascotape | Self adhesive tapes | 71 |
| Cascote | Furane resins | 71 |
| Catabond | Phenolic and resorcinol formaldehyde adhesives and binders for wood, metals | 95 |
| Catabrase | Phenolic formaldehyde adhesives for the abrasives industry | 95 |
| Catacol | Phenolic resin based adhesives for bonding wood, metal foils to phenolic plastics | 95 |
| Cebond | Polyvinyl acetate based adhesives, cement and concrete additives | 98 |
| Ceemar | Range of adhesive cements for the building industry | 335 |
| Cellodex | Vegetable adhesives | 53 |
| Cellofas | Carboxy methyl cellulose adhesive for wallpaper applications | 201 |
| Cellu-gum | Adhesive for stationery applications | 259 |
| Cerafix | Polyvinyl acetate paste for ceramic tiles | 363 |

*continued*

**Table 10.1** *continued*

| *Trade Names* | *Type, description or uses* | *Code no* |
|---|---|---|
| Ceramabond 503 | High temperature alumina cement for ceramics and refractories | 245 |
| Certite | Polyester resin based materials | 307 |
| Certofix | Fish glue for general purposes | 96 |
| Chain | Pearl and powder animal glue products | 298 |
| Chelatex | Rubber latex–water based products | 94 |
| Chelsea | Range of adhesives for shoe trade, based on neoprene and natural rubbers | 94 |
| Chelsea melt | Hot-melt adhesives | 94 |
| Chemicgum | Nitrile butadiene rubbers and latices | 171 |
| Chemlok | Polymer/resin in solvent for bonding metal to polyurethane | 144 |
| Chemlok 220 | Polymer/resin in solvent for rubber to metal bonding | 144 |
| Chemlok 233 | Polymer/resin in solvent for rubber to metal bonding | 144 |
| Chemlok 402 | Polymer/resin in solvent for rubber to textile bonding | 144 |
| Chemlok 607 | Polymer/resin in solvent for fluorosilicone rubber to metal bonding | 144 |
| Chemlok 810 | Aqueous polymer/resin adhesive for nitrile rubber-to-metal bonding | 144 |
| Chemtite | Bottle labelling adhesive resistant to water/ice | 259 |
| Chinafix | Neoprene based adhesive for repairing porcelain and pottery | 211 |
| Chromix | High temperature, mineral based cement | 164 |
| Chuk'ka | Cement for repairing china, glass, porcelain, pottery, etc. | 110 |
| Clam 143 | Starch based adhesives for wallpaper | 101 |
| Clam 2 | Latex adhesive for wallcovering furnishings | 101 |
| Clam 7 | Polyvinyl acetate general purpose adhesive | 101 |
| Clam 58 | Vinyl adhesive for acrylic foam binding | 101 |
| Clam 3 | General-purpose contact adhesive | 101 |
| Claysil | Sodium silicate based material with clay filler | 117 |
| Clipfas | Liquid adhesive for bonding nylon | 30 |
| Clipper | Neoprene rubber contact adhesive | 30 |
| Coat-rez | Solvent-resin coating formulations for textiles and food packaging materials | 88 |
| Colegrip | Industrial adhesives | 103 |
| Collapress 95 | Adhesives for woodwork | 328 |
| Collerseal | Waterproof adhesive cloth tapes | 172 |
| Colmafix | Epoxy resin based adhesive | 321 |
| Coma Grout | Filled epoxy resin levelling compound for building applications | 321 |
| Colorbind | Coloured adhesive cloth tapes for bookbinding | 172 |
| Colset | Range of adhesives based on bitumen | 62 |
| Comet | Liquid skin glue | 5 |
| Comet Cold | Cold liquid skin glue | 5 |
| Copybind | Carpet binding paste | 110 |
| Copydex | General purpose latex adhesive for textiles, carpet binding, paper, bookbinding | 110 |
| Corkbond | Synthetic solvent based adhesives | 308 |
| Cow | Range of adhesives based on synthetic rubbers and resins | 113 |
| Creteform | Epoxy based adhesive with filler, for the building trades | 188 |
| Crispin | General purpose adhesives | 115 |
| Crodafix | Polyvinyl acetate adhesives | 116 |
| Crodaglu | Jelly glues | 116 |
| Crodagrip | Water based and solvent based adhesives | 116 |
| Crodalam | Laminating adhesives | 116 |
| Crodamelt | Hot-melt adhesives | 116 |
| Crodaseal | Latex adhesives | 116 |
| Croid | Range of industrial adhesives based on natural and synthetic polymers and rubbers | 116 |
| Crystal Clear 513 | Methyl cellulose based adhesive | 328 |
| CTF | Ceramic tile fix plasticised cement for floor and wall tiles | 80 |
| CT-S | Synthetic rubber blend adhesive | 140, 141 |
| Cubix | Animal glues for woodworking and bookbinding, in solid form | 319 |
| Culminal | Methyl cellulose based adhesive for wallpaper application | 328 |
| CXL | Range of epoxy resin adhesives and coatings | 103 |
| CXL 75 | Adhesive for phenolic laminate bonding | 103 |
| CXL 194 | Epoxy resin adhesives for concrete or metals subject to wet conditions of exposure | 103 |
| CXL 400 | General-purpose contact adhesive | 103 |
| CXL 1350 | Synthetic rubber adhesives | 103 |
| Cyanolit | Ethyl or methyl cyanoacrylate based adhesives | 204 |
| Cybond | Modified polyurethane based adhesive | 121 |

*continued*

312 ADHESIVES HANDBOOK

**Table 10.1** *continued*

| *Trade Names* | *Type, description or uses* | *Code no* |
|---|---|---|
| Unibond | Polyvinyl acetate emulsion | 364 |
| Unicol | Casein emulsion adhesives | 367 |
| Uni-filla | Water-soluble-powder adhesive sealant for vinyl materials | 364 |
| Unifix | Natural product liquid and jelly adhesives | 367 |
| Uni-last | Neoprene adhesive | 364 |
| Unilite Arbrisso | Range of synthetic adhesives for industrial use; wood flooring, packaging, plastics, building, etc. | 367 |
| Unilite | Range of adhesives for joinery, flooring, building and textiles | 367 |
| Unilok | Range of adhesives for wood, rubbers, fabrics, plastics, metals and paper | 368 |
| Unimatic | Bonding compound for engineering applications | 365 |
| Union | Animal and fish glue products | 367 |
| Unipaste | Starch based wallpaper adhesive | 364 |
| Uniprene | Synthetic resin/rubber based adhesive for tile and sheet floorings | 367 |
| Unirod 100 | Thermoplastic cement for the footwear industry | 72 |
| Uniset | One-part epoxide adhesives and insulation compounds | 21 |
| Unistik | General-purpose neoprene adhesive | 364 |
| Unitak | Aqueous, non-flammable contact adhesive | 364 |
| Unitimb | Adhesive for wood | 364 |
| Universal | Liquid animal glue | 116 |
| Univip | Polyvinyl acetate emulsion for expanded polystyrene | 364 |
| U.P.7 | Neoprene based contact adhesive | 364 |
| Uvefix | Adhesive for ethyl cellulose films | 244 |
| | | |
| Vale | Urea formaldehyde resin | 53 |
| Vandike | Polyvinyl acetate or copolymer adhesives | 77 |
| Vantac | Aqueous pressure-sensitive adhesive products | 77 |
| Versalon | Polyamide resin for hot-melt adhesives | 111 |
| Versamid | Polyamide resins for hot-melt and structural adhesives | 111 |
| Vincit | Range of adhesives based on polyvinyl acetate emulsions, acrylic resins, neoprene rubber, epoxy resins and hot-melt materials | 90 |
| Vinnaplas | Polyvinyl acetate resin and copolymers available as solids or dispersions | 83 |
| Vinnol | Vinyl chloride/vinyl acetate copolymers | 83 |
| V-mat | Epoxy resin based adhesives for concrete repair | 285 |
| Vulcabond E | Aqueous solution of phenolic polymer to promote adhesion of rubbers to polyester textiles | 374 |
| Vulcabond TX | Mixed isocyanates in xylene-adhesion promotor for textiles | 374 |
| Vulcabond VP | Isocyanurate adhesion promotor for PVC plastisols to textiles | 374 |
| Vylok | Adhesives for packaging and structural uses | 259 |
| Vytard | Adhesives for packaging industry | 259 |
| | | |
| Wallcol | Synthetic resin emulsion for polyvinyl chloride and textiles | 367 |
| Wallmount | Adhesive film | 72 |
| Waterproof Cerafix | Ceramic wall tile adhesive | 364 |
| WBSP | Magnesite cement | 164 |
| Weatherban | Polysulphide rubber based sealing compounds | 352 |
| Weldtite | Polystyrene cements for model construction | 49 |
| Whitecraft | Polyvinyl acetate based adhesive | 177 |
| Widespread | Thixotropic contact adhesive for the building industry | 140 |
| Wilson | Range of adhesives for the packaging trade | 384 |
| Witcobond | Range of solvent adhesives based on neoprene, nitrile, natural, polyurethane rubbers and various copolymers | 386 |
| Witcodex | Starch and dextrine based adhesives | 386 |
| Witcogrip | Range of adhesives based on polyvinyl acetate, acrylic resins, latex, neoprene, reclaim rubber and other polymers in aqueous solution | 386 |
| Wood-lok | Range of copolymer emulsions for woodworking | 259 |
| Woodworker | Adhesive for woodworking and building trade | 140 |
| Wresilac | Amino based adhesive for woodworking | 294 |
| Wresimul | Emulsion adhesive for refractory materials | 294 |
| Wresinoid | Range of adhesives for foundry applications and abrasive wheels | 294 |
| Wresinol | Abrasive paper adhesive | 294 |
| Wresitex | Adhesive for non-woven fabrics and textiles | 294 |

*continued*

**Table 10.1** *continued*

| *Trade Names* | *Type, description or uses* | *Code no* |
|---|---|---|
| YDC | Range of starch, dextrine, polyurethane, urea formaldehyde and other resin based adhesives for joinery and plastics | 387 |
| Zipgrip 10 | Cyanoacrylate adhesive | 131 |

## 10.2 References


*Adhesives Directory* (1982), Wheatland Journals Ltd, Rickmansworth, Herts. Presents a series of classified lists of basic adhesive materials, trade names and uses. Trade sources for raw materials, plant and processing equipment are indexed, and addresses for associations. Consultants and trade names briefly listed.

*New Trade Names in the Rubber and Plastics Industries,* RAPRA, (1982), (annual publication).

*European Plastics Year Book,* IPC Scientific and Technical Press, Guildford, (1982), (annual publication).

*Surface Coating Resin Index,* The British Resin Manufacturers' Association, London, (1982) (annual publication).


## 10.3 Basic adhesive types and trade sources

Manufacturers of adhesives products based on the raw materials listed below are specified by the code number system given on page 315 (Table 10.4).

**Table 10.2 Basic adhesive types and trade sources**

| *Basic materials* | *Trade source* |
|---|---|
| Acrylic | 5, 7, 10, 18, 23, 24, 38, 64, 71, 72, 77, 101, 102A, 114, 115, 116, 121, 136, 140, 141, 149, 154, 155, 169A, 182, 191, 200, 201, 203, 214, 235, 236, 237, 243, 253, 259, 269, 272, 277, 279, 289, 296, 302, 309, 325, 328, 329, 338, 352, 353, 364, 367, 384, 386, 387 |
| Alkyds | 72, 111, 114, 187, 191, 353, 357 |
| Alkyd/phenolic | 353 |
| Animal glues | 5, 23, 53, 71, 116, 195, 214, 253, 259, 277, 297, 298, 309, 319, 325, 328, 338, 367, 384 |
| Bitumen and asphalts | 30, 51, 72, 80, 101, 102A, 154, 155, 200, 201, 243, 277, 299, 315, 352, 384 |
| Bitumen/latex | 71, 138, 154, 155, 200, 277, 306, 352, 386 |
| Butyl rubber | 7, 72, 113, 116, 140, 141, 155, 157, 203, 253, 259, 277, 352, 353, 386 |
| Casein | 7, 9, 29, 51, 53, 60, 71, 72, 85, 101, 115, 116, 147, 186, 203, 214, 259, 298, 308, 309, 325, 329, 338, 367, 384 |
| Casein/latex | 9, 53, 60, 71, 72, 94, 116, 140, 141, 147, 154, 200, 203, 211, 214, 289, 306, 308, 309, 338, 353, 384 |
| Cellulose acetate | 7, 148, 290, 353, 360 |
| Cellulose acetate–butyrate | 259 |
| Cellulose nitrate | 7, 195, 201, 230, 248 |
| Chlorinated rubber | 7, 51, 72, 94, 113, 115, 140, 141, 155, 157, 169A, 193, 214, 259, 277, 278, 282, 289, 309, 338, 353 |
| Coumarone–indene | 60, 72, 289, 353, 367 |
| Cresylic resins | 71, 201, 285, 367 |
| Cyanoacrylate | 7, 72, 100, 136, 148, 149, 204, 235, 272, 277, 327, 328 |
| Dextrin | 5, 23, 71, 85, 101, 116, 169A, 195, 214, 236, 248, 253, 259, 297, 298, 309, 319, 328, 329, 338, 367, 382, 384, 387 |
| Epoxide | 9, 16, 30, 47, 51, 60, 71, 72, 90, 95, 100, 102A, 114, 121, 131, 132, 134, 137, 140, 141, 144, 149, 154, 155, 156, 162, 169A, 174, 184, 188, 191. 193, 195, 203, 204, 237, 243, 248, 272, 274A, 277, 282, 285, 290, 307, 310, 315, 320, 325, 328, 335, 336, 340, 346, 352, 353, 357, 364, 366, 367, 383 |
| Epoxide/alkyd ester | 140, 141, 353 |
| Epoxide/phenolic | 121, 140, 141, 203, 285, 352, 353 |
| Epoxide/polyamide | 60, 71, 72, 100, 121, 140, 141, 155, 169A, 188, 225, 277, 285, 315, 340, 352, 353, 383 |
| Epoxide/polyester | 140, 141, 203, 285, 353, 383 |

*continued*

**Table 10.2** *continued*

| *Basic material* | *Trade source* |
|---|---|
| Epoxide/polysulphide | 71, 72, 100, 140, 141, 285, 353 |
| Ethyl cellulose | 7 |
| Ethylene vinyl acetate copolymer | 5, 7, 24, 71, 76, 94, 115, 116, 154, 203, 214, 253, 255, 259, 269, 272A, 289, 309, 338, 384, 387 |
| Fish glues | 116, 253, 298, 328, 367, 384 |
| Furane resins | 95, 285, 357 |
| Gums (natural) | 5, 116, 230, 253, 298, 328, 337, 348, 367, 384 |
| Hot-melt | 53, 60, 71, 72, 90, 94, 111, 116, 140, 141, 147, 148, 154, 188, 193, 201, 214, 253, 259, 272A, 274A, 277, 282, 309, 310, 325, 328, 338, 353, 367, 384, 387 |
| Hydroxy propyl methyl cellulose | 132, 278, 328, 384 |
| Inorganic–mineral | 5, 7, 51, 80, 90, 114, 149, 164, 243, 285, 307, 318, 353, 364 |
| Isocyanate | 50, 51, 53, 71, 72, 111, 116, 190, 201, 203, 259, 277, 308, 353, 367, 386 |
| Melamine/formaldehyde | 43, 71, 76, 95, 100, 101, 116, 191, 201, 274A, 325, 328, 387 |
| Methyl cellulose | 53, 60, 71, 147, 193, 201, 243, 282, 289, 328, 338, 353, 384 |
| Neoprene/phenolic | 9, 14, 24, 53, 60, 72, 94, 113, 116, 140, 141, 154, 155, 171, 190, 203, 259, 273, 277, 289, 310, 338 |
| Neoprene rubber (polychloroprene) | 5, 7, 24, 30, 51, 71, 72, 80, 94, 101, 113, 115, 116, 140, 141, 154, 155, 157, 169A, 171, 182, 193, 203, 230, 243, 248, 253, 259, 272A, 277, 289, 299, 305, 328, 338, 352, 353, 364, 367, 382, 386 |
| Nitrile/phenolic | 9, 24, 50, 53, 72, 94, 113, 121, 140, 141, 154, 171, 190, 203, 259, 272A, 285, 310, 338, 352, 253 |
| Nitrile rubber (acrylonitrile butadiene) | 5, 7, 9, 24, 51, 71, 72, 94, 113, 116, 121, 140, 141, 154, 157, 169A, 171, 193, 200, 243, 253, 259, 277, 282, 289, 299, 305, 328, 338, 352, 353, 382, 386 |
| Oleoresins | 30, 71, 155, 352, 357, 367, 382 |
| Phenol formaldehyde | 50, 71, 100, 101, 116, 125, 133, 155, 325, 367, 387 |
| Phenolic/polyvinyl | 50, 95, 100, 277, 310 |
| Phenolic/polyamide | 100 |
| Polyamides | 7, 22, 23, 30, 71, 72, 94, 114, 121, 154, 195, 259, 272A, 277, 278, 320, 338, 353, 357, 367, 387 |
| Polybenzimidazole | 380 |
| Polyester (unsaturated) | 7, 24, 44, 50, 51, 53, 72, 94, 111, 140, 141, 148, 149, 154, 169A, 187, 188, 190, 191, 195, 203, 235, 248, 259, 277, 285, 299, 307, 310, 312A, 313, 318, 320, 328, 353, 367, 383, 386, 387 |
| Polyimide | 121, 142, 201, 244, 255 |
| Polyisobutylene | 7, 24, 72, 94, 115, 140, 141, 154, 155, 169A, 203, 214, 230, 253, 259, 277, 278, 309, 338, 352, 353, 357, 367 |
| Polyphenylene resins | 255 |
| Polystyrene | 7, 10, 24, 30, 72, 101, 116, 140, 141, 195, 203, 230, 255, 259, 277, 289, 338, 367 |
| Polysulphide rubber | 72, 113, 140, 141, 155, 310 |
| Polyurethane | 7, 14, 24, 30, 34, 51, 72, 79, 94, 113, 115, 116, 121, 131, 137, 140, 141, 149, 154, 169A, 188, 191, 201, 203, 230, 253, 259, 272A, 277, 278, 287, 289, 308, 313, 338, 353, 357, 366, 368, 382, 383, 386, 387 |
| Polyvinyl acetate | 5, 7, 9, 10, 24, 29, 30, 50, 51, 71, 72, 80, 85, 94, 100, 101, 102A, 113, 116, 140, 141, 154, 155, 164, 169A, 182, 191, 193, 195, 200, 203, 214, 230, 236, 243, 248, 253, 259, 267, 269, 272A, 277, 278, 282, 299, 309, 312A, 313, 315, 319, 325, 328, 329, 338, 353, 357, 364, 367, 384, 386, 387 |
| Polyvinyl acetate/polyvinyl chloride copolymer | 7, 10, 24, 50, 71, 115, 140, 141, 169A, 195, 203, 214, 253, 259, 272A, 277, 278, 289, 309, 353, 382 |
| Polyvinyl alcohol | 5, 7, 38, 94, 116, 169A, 203, 214, 253, 259, 269, 278, 279, 289, 309, 338, 353, 367, 384, 387 |
| Polyvinyl butyral | 7, 10, 24, 50, 72, 94, 110, 113, 116, 140, 141, 169A, 203, 253, 259, 277, 278, 289, 310, 328, 338, 353 |
| Polyvinyl chloride | 7, 10, 24, 72, 94, 110, 113, 116, 140, 141, 169A, 253, 259, 277, 289, 328, 338, 353 |
| Polyvinyl ether | 7, 10, 43, 72, 115, 116, 140, 141, 154, 169A, 253, 277, 289, 353, 367 |

*continued*

**Table 10.2** *continued*

| *Basic material* | *Trade source* |
|---|---|
| Polyvinyl formal | 100, 202 |
| Resorcinol formaldehyde | 50, 71, 85, 95, 100, 101, 115, 116, 125, 133, 325, 267, 367, 387 |
| Rubber (cyclised) | 7, 140, 141, 203, 278 |
| Rubber (natural latex) | 7, 24, 51, 71, 72, 80, 94, 101, 110, 116, 131, 140, 141, 154, 155, 169A, 195, 200, 203, 214, 224, 253, 259, 272A, 277, 278, 289, 299, 305, 309, 313, 338, 353, 367, 382, 384, 386, 387 |
| Rubber (reclaim) | 5, 72, 115, 131, 140, 141, 154, 155, 169A, 171, 200, 203, 259, 277, 352, 382 |
| Rubber (synthetic latex) | 9, 22, 51, 60, 71, 72, 94, 116, 140, 141, 154, 155, 203, 214, 253, 259, 277, 289, 309, 338, 353, 384, 386 |
| Rubber (vulcanised) | 140, 141, 171, 203, 259, 338, 352, 353, 382 |
| Shellac | 188, 337 |
| Silicates | 5, 7, 20, 30, 60, 117, 164, 214, 259, 286, 309, 382, 384 |
| Silicones | 7, 72, 90, 100, 101, 116, 126, 138, 140, 141, 149, 155, 162, 188, 201, 203, 322, 322A, 336, 357, 375 |
| Sodium carboxy methyl cellulose | 193, 201, 278, 279, 297, 328 |
| Starch | 5, 6, 23, 71, 85, 101, 116, 214, 236, 253, 259, 279, 297, 309, 329, 338, 357, 359, 367, 382, 384 |
| Styrene butadiene | 5, 7, 24, 30, 51, 72, 94, 115, 116, 140, 141, 154, 155, 169A, 193, 203, 214, 243, 253, 259, 277, 278, 282, 289, 299, 305, 309, 315, 338, 352, 353, 367 |
| Urea formaldehyde | 43, 71, 76, 95, 100, 101, 115, 116, 191, 201, 325, 387 |
| Vegetable (protein soya, etc.) | 5, 71, 72, 116, 214, 259, 298, 309, 387 |
| Vinyl chloride/vinylidene copolymer | 24, 140, 141, 190, 214, 269, 274A, 277, 309, 353, 367 |
| Waxes | 29, 88, 186, 381, 382, 386 |

**Table 10.3 Industrial use of adhesives: trade sources**
Manufacturers of adhesives products employed for the applications listed below are specified by the code number system given on p. 315 (*Table 10.4*)

| *Usage* | *Trade source* |
|---|---|
| Aircraft and aerospace industries | 13, 14, 68, 71, 72, 100, 121, 132, 138, 141, 142, 157, 169, 188, 194A, 203, 207, 235, 255, 277, 285, 288, 310, 312, 312A, 350, 352, 363, 366, 378, 380 |
| Building industry | |
| general | 5, 7, 24, 65, 66, 72, 84, 98, 101, 140, 182, 220, 222, 227, 241, 253, 259, 277, 296, 326, 352, 353, 357, 367, 370 |
| external claddings | 9, 24, 26, 71, 72, 80, 140, 154, 155, 193, 285, 287, 299, 340, 352, 357, 364, 371 |
| decorative wall coverings | 9, 24, 38, 71, 72, 101, 116, 140, 154, 193, 197, 278, 279, 297, 299, 307, 326, 340, 352, 367, 371 |
| ceramic tiles | 9, 26, 38, 71, 80, 140, 154, 193, 262, 285, 287, 307, 312A, 326, 340, 367, 371 |
| ceiling finishes and skirtings | 9, 26, 38, 75, 101, 140, 154, 155, 193, 278, 299, 307, 326, 340, 367, 371 |
| ceramic floor finishes | 26, 80, 154, 307, 326, 364, 371 |
| floor coverings | 9, 38, 71, 72, 80, 101, 140, 154, 193, 243, 278, 285, 287, 299, 307, 315, 326, 335, 340, 367, 371 |
| Domestic and general purpose | 5, 9, 23, 24, 30, 71, 72, 79, 89, 90, 100, 101, 110, 113, 116, 122, 132, 140, 149, 156, 162, 169A, 171, 177, 182, 188, 192, 193, 195, 203, 204, 211, 214, 222, 227, 230, 234, 235, 237, 259, 267, 271, 279, 282, 289, 295, 296, 304, 326, 329, 338, 353, 357, 358, 364, 367, 379, 384, 386, 387 |
| Electrical components | 5, 24, 50, 71, 72, 90, 100, 125, 126, 132, 149, 162, 188, 210, 249, 277, 352, 353, 367, 383 |

*continued*

**Table 10.4** *continued*

| *Code no.* | *Manufacturer* |
|---|---|
| 30 | ATLAS PRODUCTS AND SERVICES LTD<br>Fraser Road, Erith, Kent DA8 1PN<br>Tel: Erith 32255 Telex 896176 |
| 31 | ATTWATER AND SONS LTD<br>Hopwood Street Mills, Preston, Lancs<br>Tel: 0772–4045 |
| 32 | AUSTEN CHEMICALS LTD<br>Enderby, Leicester LE9 5NH<br>Tel: 0533–862282 |
| 33 | AUTOMOBILE PLASTICS LTD<br>Autoplax House, 7 Henry Road, New Barnet, Herts EN4 8BL<br>Tel: 01–449–9147 |
| 34 | AVALON CHEMICAL CO.<br>Hitchin Lane, Shepton Mallet, Somerset BA4 5TZ<br>Tel: 0749–3061 |
| 35 | AVEBE G. A. Beneden, Oosterdiep, Veendam, Holland |
| 36 | AVON BONDING TECHNIQUES LTD<br>Haslemere Estate, Third Way, Avonmouth, Bristol BS11 9XY<br>Tel: 0272–827761 |
| 37 | BASF (UK) LTD<br>P.O. Box 4, Earl Road, Cheadle Hulme, Cheadle, Cheshire SK8 6QG<br>Tel: 061–485–622 Telex 669211 |
| 38 | BC PRODUCTS (BONDCRETE) LTD<br>Wade Road, Basingstoke, Hants RG 24 0PL<br>Tel: 0256–55511 Telex 858091 |
| 39 | BEE CHEMICAL CO. (UK) LTD<br>Kangley Bridge Road, Lower Sydenham, London SE26 5BA<br>Tel: 01–659–2141 Telex 946567 |
| 40 | BEW AUTO PRODUCTS LTD<br>Sea Plane Works Esplanade, Rochester, Kent ME1 1QP<br>Tel: 0634–403481 |
| 41 | BICC LTD<br>21 Bloomsbury Street, London WC1B 3QN<br>Tel: 01–637–1300 |
| 42 | BIFF INDUSTRIAL FASTENINGS LTD<br>Gatehouse Road, Aylesbury, Bucks HP19 3DS<br>Tel: 0296–81341 Telex 83675 |
| 43 | B.I.P. CHEMICALS INTERNATIONAL LTD<br>P.O. Box 6, Warley, West Midlands B69 4PD<br>Tel: 021–552–1551 Telex 337261 |
| 44 | B.I.P. (SHEET AND FILM DIVISION) LTD<br>Manningtree, Essex<br>Tel: Manningtree 2401 |
| 45 | BP AQUASEAL LTD<br>Kingsnorth, Hoo, Rochester, Kent ME3 9ND<br>Tel: 0634–250722 Telex 965236 |
| 46 | BP CHEMICALS LTD<br>Belgrave House, 76 Buckingham Palace Road, London SW1 0SU<br>Tel: 01–581–1388 Telex 266883 |
| 47 | BTR DEVELOPMENT SERVICES LTD<br>Hornington Road, Burton-on-Trent, Staffs DE13 0SN<br>Tel: 61611 Telex 34419 |
| 48 | BXL PLASTICS LTD<br>Buchanan House, 3 St James Square, London SW1Y 4JU<br>Tel: 01–629–8100 |
| 49 | BAGGS, C. B. LTD<br>Claremont Way Industrial Estate, Cricklewood, London NW2 1AL<br>Tel: 01–458–8011 |
| 50 | BAKELITE UK LTD<br>Thermosetting Division, Redfern Road, Tyseley, Birmingham B11 2BJ<br>Tel: 021–706–3322 |
| 51 | BALL, F. and CO. LTD<br>Barnfields, Leek, Staffs ST13 5QH<br>Tel: 0578–385612 Telex 36236 |
| 52 | BANNER AND CO. LTD SAMUEL<br>59–61 Sandhills Lane, Liverpool L5 9XL<br>Tel: 051–922–7871 Telex 627025 |
| 53 | BARDENS (BURY) LTD<br>Hollins Vale Works, Bury<br>Tel: 061–766–2804 |
| 54 | BARKER LTD., E.R.<br>Lion Works, Elliot Road, Selly Oak, Birmingham 29<br>Tel: 021–472–1929 |
| 55 | BAYER UK LTD<br>Bayer House, Paradise Road, Richmond, Surrey TW9 1S5<br>Tel: 01–940–6077 Telex 927027/8 |
| 56 | BEECHAM UHU<br>Beecham House, Great West Road, Brentford, Mddx TW8 9BD<br>Tel: 01–560–5151 Telex 935986 |
| 57 | BELL AND CO. (GRANTHAM) LTD, HENRY<br>Old Wharf Road, Grantham, Lincs<br>Tel: 0476–2281 |
| 58 | BERGER ADHESIVES<br>Portland Road, Newcastle-upon-Tyne, Tyne and Wear NE2 1BL<br>Tel: 0632–25151 Telex 35572 |
| 59 | BEROL KEMI (UK) LTD<br>55–57 Clarendon Road, Watford, Herts WD1 1SP<br>Tel: Watford 25602 Telex 923191 |
| 60 | BETTABOND ADHESIVES<br>11 Church Wood Avenue, Leeds, W. York LS16 5LF<br>Tel: 0532 785088 |
| 61 | BETTER PACKAGES (UK) LTD<br>Dartford Industrial Estate, Powder Mill Lane, Dartford, Kent DA1 1NN<br>Tel: 32–27661 Telex 896487 |

**Table 10.4** *continued*

| *Code no.* | *Manufacturer* |
|---|---|
| 62 | BITUMEN INDUSTRIES LTD<br>Ajax Avenue, Slough, Bucks<br>Tel: Slough 23274 |
| 63 | BLAGDEN CAMPBELL CHEMICALS LTD<br>AMP House, Dingwall Road, Croydon, Surrey CR9 3QU<br>Tel: 01–681–1341/0477 Telex 262388/24285 B CHEM G |
| 64 | BLOORE, G.H. LTD<br>480 Honeypot Lane, Stanmore, Mddx HA7 1JT<br>Tel: 01–952–2931 Telex 244228 |
| 65 | BLUE CIRCLE ENTERPRISES<br>Portland House, Stag Place, London SW1 5BJ<br>Tel: 01–828–3456 Telex 23701 |
| 66 | BLUNDELL PERMOGLAZE LTD<br>Building Chemicals Division, Charnley Fold Lane, Bamber Bridge, Preston<br>Tel: Preston 38331 |
| 67 | BONDING SYSTEMS LTD<br>Vines Cross Road, Horam, Heathfield, Sussex TN21 0HD<br>Tel: 04353 2226 |
| 68 | BONDMASTER LTD<br>Galvin Road, Slough, Berks SL1 4DF<br>Tel: 75–33494 |
| 69 | BONE MARKHAM LTD<br>Manor Farm Road, Wembley, Mddx HA0 1BS<br>Tel: 01–997–9555 Telex 23921 |
| 70 | BOOTS CO. LTD<br>Thane Road, Nottingham NG2 3AA<br>Tel: 0602–56111 Telex 37128/9 |
| 71 | BORDEN (UK) LTD<br>North Baddesley, Southampton, Hants SO5 9ZB<br>Tel: 0703–732131 Telex 47212 and 477649 |
| 72 | BOSTIK LTD<br>Ulverscroft Road, Leicester LE4 6BW<br>Tel: 0533–50015 Telex 34625 |
| 73 | BRIGGS AND TOWNSEND LTD<br>Station Road, Birch Vale, Stockport, Cheshire SK12 5BR<br>Tel: New Mills 46233 Telex 669238 |
| 74 | BRIGHT ENTERPRISES, R.F., LTD<br>London Road, West Kingsdown, Sevenoaks, Kent TN15 6AF<br>Tel: 047–485–2225/2852 Telex 277950 |
| 75 | BRITISH GYPSUM LTD<br>Ruddington Hall, Loughborough Road, Nottingham NG11 6LX<br>Tel: 0602–844844 |
| 76 | BRITISH INDUSTRIAL PLASTICS LTD<br>P.O. Box 6, Popes Lane, Oldbury, Warley, West Midlands B69 4PD<br>Tel: 021–552–1551»Telex 337261 |

**Table 10.4** *continued*

| *Code no.* | *Manufacturer* |
|---|---|
| 77 | BRITISH OXYGEN CHEMICALS LTD<br>Hammersmith House, Hammersmith Bridge Rd., London W6 9DX<br>Tel: 01–748–2020 Telex 934664<br>and, Vigo Lane, Chester-le-Street DH3 2RB<br>Tel: 0632–403111 Telex 53621 |
| 78 | BRITISH SISALKRAFT LTD<br>Knight Road, Strood, Rochester, Kent<br>Tel: 0634–777777 |
| 79 | BUCK AND HICKMAN LTD<br>Bank House, 100 Queen Street, Sheffield S1 2DW<br>Tel: 0742–731111 Telex 547294 |
| 80 | BUILDING ADHESIVES LTD<br>Longton Road, Trentham, Stoke-on-Trent, Staffs ST4 8JB<br>Tel: 0782–659921 Telex 36574 |
| 81 | BUNZL ADHESIVE MATERIALS LTD<br>Eastfield Industrial Estate, Scarborough, North Yorkshire YO11 3PP<br>Tel: 0732–583661 |
| 82 | BURGESS GALVIN AND CO. LTD<br>Janestown Road, Finglas, Dublin, Eire<br>Tel: 342255 Telex 24229 |
| 83 | BUSH BEACH LTD<br>175 Tottenham Court Road, London W1P 0BJ<br>Tel: 061–580–5095 Telex 8953839 BB LN C |
| 84 | CBP (UK) LTD<br>Vimy Road, Off Leighton Road, Leighton Buzzard, Beds LU7 7ER<br>Tel: 0525–3756646 Telex 82252 |
| 85 | C.P.C. (UK) Ltd<br>Trafford Park, Manchester M17 1PA<br>Tel: 061–872–5959 Telex 667022 |
| 86 | CAGO LTD<br>Alflow House, 222 Soho Hill, Handsworth, Birmingham B19 1AP<br>Tel: 021–523–3311 |
| 87 | CAMPBELL AND CO. LTD REX<br>AMP House, Dingwall Road, Croydon, Surrey<br>Tel: 01–681–1341 |
| 88 | CAMPBELL TECHNICAL WAXES LTD<br>Thames Road, Crayford, Kent<br>Tel: 252–4555 |
| 89 | CANNON B. AND CO. LTD<br>Kingston Grange, Hopwood Lane, Halifax, West Yorks 4HX1 4ET<br>Tel: 0422–50231 Telex 517322 |
| 90 | CARBORUNDUM CO. LTD, THE<br>Trafford Park, Manchester 17<br>Tel: 061–872–2381 |
| 91 | CARLESS SOLVENTS LTD<br>Hope Chemical Works, London E9<br>Tel: 01–975–5500 |
| 92 | CARLTON BROWN AND PARTNERS LTD<br>Elford Mill, Elford, Tamworth, Staffs<br>Tel: Harlaston 355 |

*continued*

**Table 10.4** *continued*

| *Code no.* | *Manufacturer* |
|---|---|
| 93 | CASEIN INDUSTRIES LTD<br>20 Linfold Street, London SW8<br>Tel: 01–622–9090 |
| 94 | CASWELL AND CO. LTD<br>Chelsea Works, St Michael's Road, Kettering, Northants NN15 6AU<br>Tel: 0536–518340 |
| 95 | CATALIN LTD<br>Farm Hill Road, Waltham Abbey, Essex EN9 1NL<br>Tel: Lea Valley 713344 Telex 23598 |
| 96 | CERTOFIX LTD<br>St Andrew's Dock, Hull, Yorks<br>Tel: 0482 27361 |
| 97 | CHAMBERLAIN AND SONS LTD, W. W.<br>Components Division, Heathfield Way, Kings Heath, Northampton<br>Tel: 0604–51761 |
| 98 | CHEMICAL BUILDING PRODUCTS LTD<br>Cleveland Road, Hemel Hempstead, Herts<br>Tel: 0442–42101 Telex 82252 |
| 99 | CHEMOX/TRADEBASE<br>6a Mead Row, Godalming, Surrey GU7 3HN<br>Tel: 04868 5661 Telex 859102 |
| 100 | CIBA-GEIGY<br>Plastics Division, Duxford, Cambridge CB2 4QA<br>Tel: 0223 832121 Telex |
| 101 | CLAM-BRUMMER LTD<br>Maxwell Road, Borehamwood, Herts WD6 1JN<br>Tel: 0223 832121 |
| 102 | COATES BROTHERS (INDUSTRIAL FINISHES) LTD<br>Sidcup-By-Pass, Sidcup, Kent<br>Tel: 01–300–1451 |
| 102A | COLAS PRODUCTS LTD<br>Galvin Road, Slough, Berks SL1 4DL<br>Tel: 0753–71711 Telex 848200 |
| 103 | COLEBRAND LTD<br>Colebrand House, 20 Warwick Street, Regent Street, London W1R 6BE<br>Tel: 01–439–9191 |
| 104 | COLOUR SPRAYS LTD<br>62 Southwark Bridge Road, London SE1<br>Tel: 01–928–7102 |
| 105 | COMPOUNDING INGREDIENTS LTD<br>Byrom House, Quay Street, Manchester M3 3HS<br>Tel: 061–834–8492 Telex 667552 CIL G |
| 106 | CONNELL LTD, R. and I<br>Ashton Road, Harold Hill, Romford, Essex<br>Tel: 04023–42893/4 Telex 897133 |
| 107 | CONREN CHEMICALS LTD<br>Silhill House, 2235 Coventry Road, Birmingham B26 3NW<br>Tel: 021–742–1433 |
| 108 | COOLAG LTD •<br>P.O. Box 3, Charlestown, Glossop, Derbyshire<br>Tel: 045–74 61611 Telex 669867 |

**Table 10.4** *continued*

| *Code no.* | *Manufacturer* |
|---|---|
| 109 | CO-OPERATIVE WHOLESALE LTD<br>Bone Products Dept., P.O. Box 3, 86 South Baileygate, Pontefract, Yorks<br>Tel: 0977–3311/2 |
| 110 | COPYDEX LTD<br>1 Torquay Street, Harrow Road, London W2 5EL<br>Tel: 01–286–7391 Telex 21729 |
| 111 | CORNELIUS CHEMICAL CO. LTD<br>Ibex House, Minories, London EC3N 1HY<br>Tel: 01–481–9331 Telex 883942 |
| 112 | COURTAULDS ACETATE LTD<br>Chemical Divisions, P.O. Box 5, Spondon, Derby DE2 7BP<br>Tel: 0332–661422 Telex 37391 |
| 113 | COW PROOFINGS LTD<br>Eastbourne Road Trading Estate, Slough, Berks SL1 4SF<br>Tel: Slough 22274 Telex 847215 |
| 114 | CRAY VALLEY PRODUCTS LTD<br>St Mary Cray, Orpington, Kent BR5 3PP<br>Tel: Orpington 32545 Telex 25898 |
| 115 | CRISPIN CHEMICAL CO. LTD<br>Coleman Road, Leicester LE5 4NQ<br>Tel: 0533–766331 |
| 116 | CRODA ADHESIVES LTD<br>Winthorpe Road, Newark, Notts NG24 2AL<br>Tel: 0636–76711 Telex 37579 |
| 117 | CROSFIELD, JOSEPH AND SONS LTD<br>P.O. Box 26, Warrington, Cheshire WA5 1AB<br>Tel: 0925–31211 Telex 627067 |
| 118 | CROXTON AND GARRY LTD<br>Curtis Road, Dorking, Surrey RH4 1XA<br>Tel: 0306–5981 Telex 859567 |
| 119 | CUMMING, J. W. LTD<br>45 Borough High Street, London SE1<br>Tel: 01–407–5967 |
| 120 | CURTIS (ONX), A. L.<br>Huntingdon Road, Lolworth, Cambridge CB3 8HB<br>Tel: 0959–80687 |
| 121 | CYANAMID OF GREAT BRITAIN LTD<br>154 Fareham Road, Gosport, Hants PO13 0AS<br>Tel: 0329–236131 Telex 86173 |
| 122 | DCMC GROUP OF COMPANIES<br>116 Brompton Road, London SW3<br>Tel: 01–584–6428 |
| 123 | DEB CHEMICAL PROPRIETARIES LTD<br>Spencer Road, Belper, Derby DE5 1JX<br>Tel: 077382–2712 Telex 377209 |
| 124 | D H INDUSTRIES LTD<br>Sullivan House, Abbey Wharf, Kingsbridge Road, Barking, Essex 1G11 0BD<br>Tel: 01–594–9826 Telex 24132 |
| 125 | DRG KWIKSEAL PRODUCTS<br>Amersham Road, Chesham, Bucks HP5 1NG<br>Tel: 02057 74141 Telex 837491 |

**Table 10.4** *continued*

| *Code no.* | *Manufacturer* |
|---|---|
| 126 | DTV GROUP LTD<br>2–12 Ernest Avenue, West Norwood, London SE27 0DJ<br>Tel: 01–670–6166/9361 Telex 262415 |
| 127 | DANCKAERTS WOODWORKING MACHINERY LTD<br>2, 4, 6, East Road, City Road, London N1 6AG<br>Tel: 01–253–7151 Telex 24374 |
| 128 | DATAC ADHESIVES LTD<br>Victoria Road, Dukinfield, Cheshire SK16 4UP<br>Tel: 061–330–8979 |
| 129 | DEGUSSA LTD<br>175 Tottenham Court Road, London W1P 0BJ<br>Tel: 01–580–8041 Telex 25433 DGLDN G |
| 130 | DENTAL FILLINGS LTD<br>49 Grayling Road, London N16<br>Tel: 01–800–7444 |
| 131 | DEVCON DIVISION ITW LTD<br>Station Road, Theale, Reading, Berks RG7 4AB<br>Tel: 0734–302304 Telex 848713 |
| 132 | DEXTER HYSOL LTD<br>Rose Industrial Estate, Bourne End, Bucks SL8 5AS<br>Tel: Bourne End 28222 Telex 84583 |
| 133 | DIAMOND SHAMROCK PROCESS CHEMICALS LTD<br>147 Kirkstall Road, Leeds LS3 1JN<br>Tel: 0532–457471 |
| 134 | DOBOY LTD<br>Sunderland Road, Sandy, Beds<br>Tel: 0767–82911 Telex 82358 |
| 135 | DOUBLE H INTERNATIONAL MARKETING LTD<br>33 Clarence Street, Staines, Mddx<br>Tel: Staines 55454 |
| 136 | DOUGLAS KANE LTD<br>Carylon Road, Atherstone CV9 1LQ<br>Tel: 08277–4511 Telex 342270 |
| 137 | DOW CHEMICAL CO. LTD<br>Heathrow House, Bath Road, Hounslow, Mddx TW5 9QY<br>Tel: 01–759–2600 |
| 138 | DOW CORNING LTD<br>Reading Bridge House, Reading, Berks RG1 8PW<br>Tel: Reading 57251 Telex 848340 |
| 139 | DUFAY TITANINE LTD<br>Darlington Road, Shildon, Co. Durham DL4 2QP<br>Tel: 038884–2541 and 2341 |
| 140 | DUNLOP SEMPTEX LTD<br>Industrial Products Division, Chester Road, Erdington, Birmingham B35 7AL<br>Tel: 021–373–8101 |
| 141 | DUNLOP LTD<br>Chemical Products Division, Chester Road, Erdington, Birmingham B35 7AL<br>Tel: 021–373–8101 |

**Table 10.4** *continued*

| *Code no.* | *Manufacturer* |
|---|---|
| 142 | DU PONT (UK) LTD<br>18 Breams Building, Fetter Lane, London EC4 1HT<br>Tel: 01–242–9044 Telex 262973 |
| 143 | DURAPLAS BUILDING PRODUCTS LTD<br>5 Elizabethan Way, Lutterworth, Leics LE17 4ND<br>Tel: 04555–5613/5 |
| 144 | DURHAM CHEMICALS LTD<br>Birtley, Chester-le-Street, Co. Durham DH3 1QX<br>Tel: 0632–402361 and 406521 Telex 53618 |
| 145 | DUSSEK CAMPBELL LTD<br>Thames Road, Crayford, Kent DA1 4QJ<br>Tel: Crayford 52966 Telex 896135 |
| 146 | DWYER PACKAGING SYSTEMS LTD<br>Unit 6, Greenwich Industrial Estate, 159 Greenwich High Road, London SE10 8JA<br>Tel: 01–858–7111 |
| 147 | EAGLE PACKAGING AND PRINTING CO. LTD<br>Compsall Mill, Compsall, Cheshire |
| 148 | EASTMAN CHEMICAL INTERNATIONAL AG<br>P.O. Box 66 Kodak House, Station Road, Hemel Hempstead, Herts HP1 1JU<br>Tel: Hemel Hempstead (0442) 62441 and 41171 Telex 825227 |
| 148A | ELECTROMAC MEASUREMENT SERVICES LTD<br>303–307 Normanton Road, Derby DE3 6UU<br>Tel: 0332–363256/7 Telex 377431 |
| 149 | EMERSON AND CUMING (UK) LTD<br>Colville Road, Acton, London W3 8BU<br>Tel: 01–571–3469 Telex 935087 |
| 150 | EMHART (UK) LTD<br>Crompton Road, Wheatley, Doncaster, Yorks<br>Tel: 0320–65226 |
| 151 | ENGELMANN AND BUCKHAM MACHINERY LTD<br>William Curtis House, Alton, Hants GV34 1HH<br>Tel: 0420–82421 Telex 858194 |
| 152 | ERLICH LTD, G. J.<br>Erli House, South Cottage, Denham Place, Bucks UB9 5BL<br>Tel: 089–583–2576 Telex 8953636 |
| 153 | ESSO CHEMICAL LTD<br>Arundel Towers, Portland Terrace, Southampton, Hants SO9 2GW<br>Tel: 0703–34191 Telex 47437 |
| 154 | EVODE LTD<br>Industrial Adhesives Division, Common Road, Stafford ST16 3EH<br>Tel: 0785–57755 Telex 36161 |
| 155 | EXPANDITE LTD<br>Chase Road, London NW10 6PS<br>Tel: 01–865–8877 Telex 25420 |

*continued*

**SAQ 2.3b**

### 2.3.3. Summary of Reference Works in General

So, do you think you now know all there is to know about chemical reference works? Can you call a title to mind to meet every problem you come up against? If you think you can, you are either blessed with a photographic memory (although you would still be wrong) or you are deluding yourself. No, of course you can't claim either of those things (no more than I could), and they do not represent what we have been trying to achieve. What I hope you can do is to realise that there might be a reference book to help you with any problem you meet; to have some idea of how to choose a reference book likely to be of use in a particular situation (assuming that you have the luxury of choice from a number of titles in any library to which you may have access); and perhaps most important of all, to be able to approach any reference book you come across with the confidence and experience to look for its strengths and weaknesses, and so to use it to maximum advantage.

This Part 2 of *Using Literature* is dominated by reference books of different kinds: this is because I think that in the many and various situations that you, the users of this Unit, will find yourselves, reference books are likely to be the most heavily consulted items of the chemical literature you will use.

So, before we leave this subject I am going to finish this Section with a final SAQ to make sure you have grasped an overall picture of what we have been doing, in addition to the specific practice I have already given you with reference works.

**SAQ 2.3c**

In a table accompanying this question I have listed all the reference works I have mentioned or discussed in Sections 2.1 to 2.3.

Keep this list; it will be a useful memory-jogger for you. Also, I shall be referring to it again for a different purpose when we reach Section 3.1.

Now, imagine you are in a well-stocked library. To help to create the illusion I have listed the books in the order in which my library shelves them, using the Universal Decimal Classification we encountered earlier (more explanations in Section 3.1). I have listed below ten questions, or types of information. For each one I want you to note the serial number from my list (ie 1–29, *not* the classification number) of the book you would go to first to try and find the information required.

There is not necessarily an absolutely correct answer to each question as the information may be found in more than one book; also, you will find that I have not referred to every book in the list.

Just note where you would go *first* for:

(*a*) a description of a process for the desulphurisation of coal;

(*b*) the title of a scientific reference book on wines and spirits;

(*c*) a list of the uses of different enzymes;

(*d*) the formula and properties of lead stannate;

(*e*) an explanation of biorheology;

(*f*) a method of identifying aspirin;

(*g*) the hazards presented by butyric acid;

(*h*) an outline description of organic dyes; ⟶

**SAQ 2.3c (cont.)**

(*i*) methods for the extraction of juices from fruit;

(*j*) a method of assay for the ammonia content of ammonium chloride.

Reference books referred to in this unit.


1. 006 *Annual book of ASTM standards.* Philadelphia: American Society for Testing and Materials, annual. 66v.

2. 006 *BSI Catalogue.* London: British Standards Institution, annual.

3. 006 *ISO Catalogue.* Geneva: International Organization for Standardization, annual.

4. 03 *Walford's guide to reference material, Vol 1: science and technology.* A. J. Walford. 4th ed. London: Library Association, 1980. 697pp.

5. 5 *McGraw-Hill encyclopedia of science and technology.* 5th ed. New York: McGraw Hill, 1982 15v.

6. 53/54 *CRC Handbook of chemistry and physics: a ready-reference book of chemical and physical data.* Robert C. Weast, ed. 63rd ed. Cleveland: CRC Press, 1982. Various pagings.

| | | |
|---|---|---|
| 7. | 54 | *The condensed chemical dictionary.* 10th ed. Revised by Gessner G. Hawley. New York; Van Nostrand Reinhold, 1981. 1135pp. |
| 8. | 54 | *Van Nostrand Reinhold encyclopedia of chemistry.* Douglas M. Considine and Glenn D. Considine, ed. 4th ed. New York: Van Nostrand Reinhold, 1984. 1082pp. |
| 9. | 542.614.8 | *Handbook of reactive chemical hazards: an indexed guide to published data.* L. Bretherick, ed. 2nd ed. London: Butterworth, 1979, 1281pp. |
| 10. | 542.614.8 | *Hazards in the chemical laboratory.* 3rd ed. L. Bretherick, ed. London: Royal Society of Chemistry. 1981p. 567pp. |
| 11. | 543 | *Comprehensive analytical chemistry.* C. L. Wilson and D. W. Wilson eds; from vol. 8 edited by G. Svehla. Amsterdam: Elsevier, 1959 to date. 20v. (in 29v to date). |
| 12. | 543 | *Handbook of analytical chemistry.* Louis Meites, ed. New York: McGraw Hill, 1963. Various pagings. |
| 13. | 543.06 | *Official methods of analysis of the Association of Official Analytical Chemists.* Sidney Williams, ed. 14th ed. Arlington: AOAC, 1984. 1141pp. |
| 14. | 543.06 | *Official, standardised and recommended methods of analysis.* 2nd ed. S. C. Jolly, ed. London: Society for Analytical Chemistry, 1974. 897pp |
| 15. | 543.544 | *Handbook of chromatography.* Gunter Zweig and Joseph Sherma, eds. Cleveland: CRC Press, 1972. 2v. |
| 16. | 577.1 | *Concise encyclopedia of biochemistry.* Thomas Scott and Mary Brewer, eds. Berlin: Walter de Gruyter, 1983. 516pp. |
| 17. | 577.1.014 | *Specifications and criteria for biochemical compounds.* 3rd ed. Washington: National Academy of Sciences, 1972. 216pp. |
| 18. | 577.6 | *Biochemical engineering and biotechnology handbook.* Bernard Atkinson and Ferda Mavituna. London: Macmillan, 1983. 119pp. |
| 19. | 615.11 | *British Pharmacopoeia.* London: HMSO, 1980. 2v. |
| 20. | 615.11 | *The Merck index: an encyclopedia of chemicals and drugs.* Martin Windholz, ed. 10th ed. Rahway; New Jersey: Merck, 1983. 1937pp. |

21. 616-074 *A–Z of clinical chemistry: a guide for the trainee.* W. Hood. Lancaster: MTP Press, 1980. 386pp.

22. 66 *Encyclopedia of chemical technology.* (Kirk-Othmer). 3rd ed. New York: Wiley, 1978–1984. 24v.

23. 66 *Encyclopedia of industrial chemical analysis.* Foster Dee Snell and Clifford L. Hilton, ed. New York: Wiley-Interscience, 1966–1975. 20v.

24. 66 *Materials and technology: a systematic encyclopedia of the technology of materials used in industry and commerce, including foodstuffs and fuels.* London: Longmans, 1968–1975. 8v.

25. 664 *Food industries manual.* M.D. Ranken, ed. 21st ed. Glasgow: L. Hill, 1984. 530pp.

26. 665.7 *IP standards for petroleum and its products, part 1: methods for analysis and testing.* Chichester: Wiley for the Institute of Petroleum, annual. 2v.

27. 668.3 *Adhesives handbook.* J. Shields. 3rd ed. London: Butterworth, 1984, 360pp.

28. 669.14 *Handbook of comparative world steel standards, Vol. 6, steel: world standards mutal speedy finder.* Rev. ed. Tokyo: International Technical Information Institute, 1982. 499pp.

29. 678 *Encyclopedia of polymer science and technology: plastics, resins, rubbers, fibers.* New York: Wiley–Interscience, 1964–1972. 16v.

**Objectives**

When you have worked through this part of the Unit you should be able to:

- recognise the wide range of subjects covered by specialist reference works;
- be aware of particular specialised titles in addition to those major titles covered in Section 2.1 and 2.2;
- find information in an even wider range of sources than you have used up to now;
- apply the advice you have received in previous Sections to selecting likely sources of information.

## 2.4. MAJOR JOURNALS

### 2.4.1. Introduction to Journals

As you can tell from the lengths (and relative complexities) of the Sections in Part 2 of this Unit, we have concentrated very heavily on reference books in our examination of the analytical literature. The reason for this, as I have already told you, is that reference works are likely to be your most constant and practical everyday source of analytical chemical information. But if you were to ask chemists what they thought were the most important items of chemical literature I am sure that most of them would reply 'papers in journals'. Over the whole field of chemistry, and within any specialised field in chemistry, there is no question that journal papers form the core, and the major substance, of the literature. When we talk about the information explosion, or the literature explosion, it is primarily the growth in journals and in the numbers of papers they contain that we are talking about.

Before we continue, here is a very brief word on terminology. To all intents and purposes, the terms 'journal' and 'periodical' can be considered interchangeable. 'Magazine' is used only for more popular and general publications (like gardening and popular hobby magazines). 'Serials' is strictly a more comprehensive term than 'periodicals' but you will find it used (although not by me) as a synonym for 'journal' or 'periodical'. The term 'paper' here will be used to describe an original contribution of research findings (or a critical review of research); an 'article' will, in this Unit, mean a commissioned piece in a commercial or news journal. Some of these definitions are rather simplistic: most of these terms are used rather loosely. But my generalisations are broadly valid and any essential differences will become apparent in the course of this Section.

Right, back to the journals. Let's see why they are considered so important.

Π Have you any idea how long it takes to write and publish a book?

There is no hard-and-fast timescale and some books can be produced very quickly when it suits the authors and publishers. In fact the timescale for producing this ACOL Unit is pretty short. But a book covering a chemical topic in some detail will take months or years rather than weeks to write. Then, even if the publisher is expecting the manuscript and has it programmed into his publishing schedule, it can still be another year or so before copies of the book are finally available for sale in the bookshops.

Modern science couldn't operate to that kind of timescale for its information could it? You are likely to need detailed up-to-date information on a specialised topic from any number of research workers throughout the world. Books are the ideal vehicles for the presentation of many kinds of information (as I hope you now realise), but *not* for concise up-to-date accounts of experiments or new techniques. This is the principal gap filled by the scientific journal. The problem was encountered right at the beginning of modern science: the Royal Society began publishing its *Philosophical Transactions* in 1667, and there has been a very great increase in the numbers of titles since then. Nobody knows exactly how many chemical journals are being published today throughout the world, but the Chemical Abstracts Service claims to monitor over 12,000 current scientific titles.

All the major research journals use many referees who read all the papers submitted in their specialist fields and advise the editor, first of all, whether the paper is worth publishing; even if the referees advise publication they might still suggest that certain alterations or additions should be made to the paper. The refereeing system has come in for some criticism in recent years: there are examples of referees not detecting fraudulent results or making some serious errors of judgment. But by-and-large the refereeing system for papers in journals works to everybody's advantage and helps to ensure that, as a reader, you are being given valid scientific data.

So far I have been describing the major scientific research journals. But, of course, there are other kinds of journals too. Soon we shall be looking at examples of all of them. Apart from the original descriptions of research work or experiments we have been talking about so far, there are other kinds of material suitable for publication in regular journals: news, and reviews of new products or techniques; news of societies, conferences and meetings; news of firms and individuals; industrial or commercial surveys; book reviews, letters and, of course, advertisements (not just for products either—where do you look for new jobs?). There is also a market for articles reviewing recent research or a range of products, techniques, or applications. We will leave review articles for discussion in our next Section.

Many journals are, as we shall see in a moment, devoted to one kind of information almost exclusively: that is research papers and related material only, or news and feature articles only. Just a few journals, mainly American ones, contain more than one basic kind of contribution. So for most of the remainder of this Section we shall simply consider examples: these will illustrate, and amplify, what I have been saying. I shall mention all the major analytical journals and introduce you to the major chemical titles. But when we come to the more specialised titles I must be rather selective. No illustrations this time, but—you know by now what I am going to say, don't you?—try and see copies of as many of these journals as you can for yourself.

Before we continue, let's make sure you know the kind of thing we are looking for.

**SAQ 2.4a**

Of the types of journal contents listed below, put a cross beside those you would expect to find in a research journal, and a tick besides those you would expect to find in a monthly news journal. Some will appear in both types of journal, of course, so put both a cross and a tick beside those.

Book reviews

Advertisements

**SAQ 2.4a (cont.)**

Feature articles

Equipment reviews

Research papers

Conference reports

Review papers

Letters to the editor

Company and personnel reviews

### 2.4.2. Major Chemical and Analytical Journals

Many of the major journals in any scientific field originate from learned or professional societies. As their subjects have grown in complexity, many of the societies have responded by making their journal publishing programme more complex, to try to deal with the problems arising from the literature explosion. Alternatively the organisation itself might have grown and amalgamated with other bodies (and its publications with theirs), resulting in an apparently rather complex pattern of publishing.

A notable example of this is the Royal Society of Chemistry (RSC). They began publishing their journal, as the Chemical Society, in 1841. But from 1972 the *Journal of the Chemical Society* has been published in sections:

(*a*) *Chemical Communications.* Twice-monthly. Publishes urgent or novel results from all branches of chemistry, with articles not normally exceeding one printed page each only.

(*b*) *Dalton Transactions.* Monthly. For inorganic chemistry.

(*c*) *Faraday Transactions, I and II.* Monthly. I, Physical chemistry, II, Chemical physics.

(*d*) *Perkin Transactions I and II.* Monthly. I, Organic and bio-organic chemistry, II, Physical organic chemistry.

Did you notice *Chemical Communications*? Here is a journal deliberately trying to circumvent the delays inevitable in traditional journal publishing (yes, even research papers can take quite a few months, often over a year, before they are published; and that delay can in some cases be vital). These 'letters journals', as they are called, are becoming increasingly popular in many scientific fields, although their unregulated growth is bringing about its own problems. The refereed research paper must remain the basic currency of scientific literature; but within closely defined limits there is room for shorter more urgent communications.

Now that is just the *Journal.* The Royal Society of Chemistry's periodical publishing doesn't stop there. Here are another six titles, all from the RSC, the first two being our first major analytical journals.

*The Analyst.* Monthly, 1875 to date. An international journal covering all branches of analytical chemistry. Publishes papers, short papers, and book reviews.

*Analytical Proceedings.* Monthly, 1963 to date. Publishes papers, news of the Analytical Division of the RSC, reports of meetings, product news, book reviews, correspondence, calendar of forthcoming events, and advertising.

*Chemical Society Reviews.* (We consider this in Section 2.5)

*Chemistry in Britain.* Monthly, 1965 to date. You might already know this. It covers the news side of chemistry, including general articles, letters, book reviews, and advertising.

*Education in Chemistry.* Every 2 months, 1964 to date. Gives background material for school and undergraduate chemical teaching. Contains news, letters, articles, book reviews, equipment notes, and advertising.

*Journal of Analytical Atomic Spectrometry.* Every 2 months, 1986 to date. This is a new international journal carrying research papers, shorter communications, and reviews on the development and application of atomic spectrometric techniques. It replaces a review publication *Annual Reports on Analytical Atomic Spectroscopy* (1970–1985).

*Journal of Chemical research.* Monthly, 1977 to date. An interesting venture of which the RSC is co-founder-publisher with the Gesellschaft Deutscher Chemiker and the Societe Chimique de France, supported by another 15 societies in Europe and America. This is published simultaneously in two different kinds of monthly parts: synopses of all papers accepted for publication, and a microfiche or miniprint version containing full texts corresponding to the synopses. We shall meet the microfiche quite often, so for those of you who might not know what they are, they consist of pages (usually 60–100) of microfilmed documents arranged in rows on an A5 (150 × 100mm) sheet, and read, of course, by means of a special viewer.

In the USA the American Chemical Society is also extensively involved in publishing. Primarily, they have their *Journal*:

*Journal of the American Chemical Society.* Fortnightly, 1879 to date. Covers all areas of chemistry and publishes only research papers and short book reviews, and letters to the editor.

Also of particular concern to us is their analytical journal:

*Analytical Chemistry.* Monthly, 1947 to date (formerly, 1929–1946, published as *Industrial and Engineering Chemistry, Analytical Edition*). This publishes research papers and review articles, as well as news items and related material: letters, notes of meetings and courses, new products, book reviews and advertising. The April issue each year is devoted to review articles covering the whole field of analytical chemistry: we shall look at these reviews again in our next Section, when we shall also mention the American Chemical Society's *Chemical Reviews.*

Altogether the American Chemical Society issues 23 different journal titles. We have looked at those most relevant to us, but to show you the scope of the Society's periodical publishing programme I have listed all their periodical titles:

Titles preceded by an asterisk are referred to in the text of the Unit.

**Journal of the American Chemical Society.* Fortnightly. 1879–.

**Accounts of Chemical Research.* Monthly. 1968–.

**Analytical Chemistry.* Monthly. 1929–.

*Biochemistry.* Fortnightly. 1962–.

*Chemical and Engineering News.* Weekly.

**Chemical Reviews.* Every two months. 1924–.

*Chemtech.* Monthly. 1909–. Articles on innovations and new ideas in chemical technology.

*Environmental Science and Technology.* Monthly.

*I and EC Fundamentals.* Quarterly. 1962–. Research papers in chemical engineering.

*I and EC Process Design and Development.* Quarterly. 1962–. Research papers and critical reviews on theoretical and experimental results relating to developments in processes and process equipment.

*I and EC Product R and D.* Quarterly. 1962–. Research papers on product-related interdisciplinary topics.

*Inorganic Chemistry.* Monthly. 1962–.

*Journal of Agricultural and Food Chemistry.* Every two months. 1952–.

*Journal of Chemical and Engineering Data.* Quarterly. 1956–.

*Journal of Chemical Education.* Monthly. 1924–.

*Journal of Chemical Information and Computer Sciences.* Quarterly. 1981–.

*Journal of Medicinal Chemistry.* Monthly. 1963–.

*Journal of Organic Chemistry.* Fortnightly. 1936–.

*Journal of Physical and Chemical Reference Data.* Quarterly. 1972–.

*Journal of Physical Chemistry.* Fortnightly. 1896–.

*Langmuir.* Every two months. 1985–. All areas of fundamental surface and colloid science.

*Macromolecules.* Monthly. 1968–.

*Organometallics.* Monthly. 1982–.

Now you are beginning to see what you are up against!

By comparison the American Institute of Chemical Engineers has just the two titles, but either or both might potentially be of considerable interest to you, particularly in the areas of applications of analytic chemistry.

> *AIChE Journal.* Monthly, 1955 to date. Publishes research and review papers, and short book reviews. Covers fundamental research and developments having immediate or potential value in chemical engineering.

> *Chemical Engineering Progress.* Monthly, 1947 to date (formerly, 1908–1946, published as *American Institute of Chemical Engineers Transactions*). This contains news and feature articles, notes on new materials and applications, letters, book reviews, and advertising.

And finally two commercial journals. Although they do not carry the prestige of the name of a professional body, some commercial journals are the leading titles in their field, and publishers such as Elsevier and Pergamon are among the most experienced scientific publishers in the world. Our last two journals here are:

> *Analytica Chimica Acta.* Monthly, Amsterdam: Elsevier, 1947 to date. This has research papers on all branches of analytical chemistry, which may be published in English, French or German (most are in fact in English: and only the journal's title is in Latin!).

> *Talanta.* Monthly, Oxford: Pergamon, 1958 to date. An international journal of pure and applied analytical chemistry, it contains research papers, short communications, analytical data, letters, and critical reviews.

### 2.4.3. Selected Specialist and Applications Journals

Here I must be selective, or we shall spend hours going through titles. You will find journal titles on an extraordinarily wide range of specialised topics, but here are notes on just a few mainly commercial titles which might be of particular interest to you. Don't forget there are many more for you to find for yourself.

1. *Journal of Chromatography.* Fortnightly (approx). Amsterdam: Elsevier, 1958 to date. Also includes *Biomedical Applications, Chromatographic Reviews*, and *Bibliography Section.* Publishes papers on all aspects of chromatography, electrophoresis, and related methods. These are mainly research papers dealing with chromatographic theory,

instrumental development, and applications. There are also shorter Notes, and a news section (new books, recent and forthcoming conferences).

2. *Atomic Spectroscopy.* Every two months. Norwalk (Conn.): Perkin–Elmer, 1980 to date (formaly, 1962–1979, Atomic Absorption Newsletter). Publishes research papers and technical notes on techniques and applications of atomic spectroscopy.

3. *Spectrochimica Acta.* Monthly. Oxford. Pergamon, 1939 to date. In two parts: A, *Molecular Spectroscopy*, and B, *Atomic Spectroscopy.* Carries research papers and occasional conference reports, book reviews, and review articles.

4. *Journal of Electroanalytical Chemistry and Interfacial Electrochemistry.* Fortnightly. Lausanne: Elsevier, 1959 to date. Again, fundamental research papers and review papers, preliminary notes, and book reviews.

And finally, just three examples of journals on specialised applications.

5. *Dyes and Pigments.* Every two months. Barking: Elsevier, 1980 to date. Publishes research and review papers on the chemistry and physics of dyes, pigments, and their intermediates.

6. *Fuel Processing Technology.* Quarterly. Amsterdam: Elsevier, 1976 to date. Publishes research and review papers on the technical and scientific aspects of processing fuels to other fuels, chemicals, and by-products.

7. *Journal of the Oil and Colour Chemists Association.* Monthly. Wembley: O.C.C.A, 1918 to date. Publishes research papers, news and book reviews.

### 2.4.4. Finding Information in Journals

Remember, I have again been very selective in my choice of examples in the previous Section. So to start with, how do you find out what journals are available? Well, as for the reference books (remember *Walford's Guide*?), there is a very useful directory of current periodicals. There are, in fact, several such directories but this one is particularly well-known, has an international coverage, and should be available in most larger libraries.

*Ulrich's international periodicals directory.* New York: Bowker, annual. 2v.

This is a classified international directory of periodicals in any language. For each publication it gives: full title, name and address of the publisher, subscription rates, name of the editor, date of first issue, and some miscellaneous information (such as where some of the journals are indexed, or audited circulation figures for a few titles). It also lists those titles which have ceased publication since the previous edition. Its arrangement is A-to-Z by subjects, with an index of titles.

You should have realised by now that you are likely to use different journals for different purposes: whether you want to keep up-to-date with news, or with research work in general, or in a specialised field. If you are looking for recent work in a particular field you might find yourself a likely journal title, but how do you proceed? That row of uniformly bound thick volumes in the library, or dusty heap of unbound issues on the floor in the corner of somebody's office or cupboard can look equally daunting. Unless you can remember something specific to look for (and *never* trust your memory—it is almost invariably faulty), forget the heap of unbound issues and go to the library. Most journals issue indexes to each complete volume, and that is the obvious place to start. And of course we know quite a bit about indexes now, don't we? Certainly we know enough to realise that the quality of the indexes might vary considerably; but at least we are not having to browse through every issue looking at the titles of papers.

Π Even assuming that the journal you have selected has an excellent, detailed subject index, which you have consulted thoroughly, are you sure you haven't missed a vital paper?

Even if you check four or five different journal titles you cannot be sure you have not missed something vital. Important work in the field might have been published in Great Britain, the USA, Japan, or Western Europe; even if you cover those countries comprehensively, what about Eastern Europe, Asia, Australia, or South America?

The indexes to each journal are fine, as long as it is *only* that journal you want to consult. If you need a wider search for papers which might have appeared in almost any chemical journal you have to use different tools: some of them you will find in the next Section, but for a complete answer (as far as that is possible) you will have to wait until Section 3.2 (and 4.1) of this Unit; then all will be revealed to you on how to find information from periodicals (as well as other publications).

I have concentrated on why journals exist, what functions the different kinds perform, eg news, research papers, and why they are so important in the process of scientific research. I have also introduced you to some useful titles in our field of analytical chemistry. But one last word before we leave the subject: in your search for the major chemical journals or for specialised titles, don't forget the major general science journals. They can also prove of great value to you. I am thinking of titles such as *Nature, Science, Scientific American*, and *New Scientist*. I shall not tell you any more about them here—perhaps you are already familiar with some or all of them. If you are not, go to a library and read a few recent issues: you might start reading them just to find out what they are like, but I think you will quickly find it a most instructive and rewarding task (sometimes even entertaining).

And just to remind you of some major titles and the main points of this Section, we finish with another SAQ.

**SAQ 2.4b**

You want to keep yourself up-to-date with what is happening in chemistry at large and in analytical chemistry in particular.

From the list below note the numbers of two journals you would choose to read regularly to try and fulfil these aims.

1. *The Analyst*
2. *Analytical Chemistry*
3. *Analytical Proceedings*
4. *Chemical Communications*
5. *Chemical Engineering Progress*
6. *Chemistry in Britain*
7. *Talanta*

**Objectives**

When you have worked through this Section you should be able to:

- recognise the role and importance of journals within the scientific literature;
- recognise the different types and functions of journals;
- recognise the different types of contents found in journals;
- appreciate the problems arising from the existence of large numbers of chemical journals;
- identify the major journals covering analytical chemistry.

## 2.5. REVIEWS, EVALUATION PAPERS

### 2.5.1. What Are Reviews?

Are you beginning to feel thoroughly bemused by the vast amounts and different types of chemical literature available? You are not alone: nearly all scientists are confronted with this problem (whether they admit it or not), many of them to a far greater extent than you are ever likely to suffer. Of course, not everything published is of equal value, but that really compounds the problem doesn't it? First you have got to find your needles in the haystack, then decide which ones are of most value.

Wouldn't it be nice if someone did it all for you, and told you both where to find particular documents and then compared them all with one another? Well, help is at hand: in fact in that rhetorical question I have given you a fairly broad, if simplistic, definition of a review. Quite simply, as its name suggests, a review usually takes a very specific topic and evaluates all the recent published literature on that topic (almost entirely journal papers); and of course all the papers referred to (sometimes running into hundreds and rarely less than 50–70) are listed with their bibliographic details so that you can, if you wish, locate those which seem most important to you.

There are in fact two sides to reviews—the literature and the subject. Our initial interest here is in the literature: convenient selected lists of recent published work, and evaluative text indicating the most important work and conclusions. This is exactly what you find in the annual reviews I shall show you in a moment. But many good review papers start as reviews of the subject, and use the published literature to demonstrate it; they indicate not only the comparative results of the various papers, but also show the trends of research and suggest areas left relatively untouched or new areas requiring further study.

We can see the roles of review papers therefore as including three important points: collation of information from different sources; summarising existing knowledge at a vital stage between initial publication of research and its ultimate assimilation into handbooks, reference works, and textbooks; and the identification of developing research fields and the indication of new and worthwhile areas.

You can therefore appreciate the potential usefulness of reviews to you, whether it be to guide you through the published literature; to keep you up-to-date with recent research in a particular field; or to give you an evaluative introduction to a research field new to you.

Over the last fifteen or so years reviews have undoubtedly developed into one of the most important features of the scientific literature. I think you can see why that should be.

So, now have a go at a question:

**SAQ 2.5a**

Look at the following uses for reviews and list them in order of likely importance to you as an analytical chemist, in relation to any specialised topic in which you might be interested.

(*a*) To find lists of papers

(*b*) To introduce an unfamiliar subject

(*c*) To give you a compact set of results from a wide range of papers.

(*d*) To evaluate published papers in comparison with one another.

(*e*) To identify areas of the subject still needing research.

### 2.5.2. Finding Analytical Chemistry Reviews

As you will have gathered, the review literature has expanded enormously over the last decade and now presents an information retrieval problem all of its own. I shall tell you about various specialist review publications in just a moment, but they are not the only sources: as you should have noticed in the previous Section, many journals publish review papers among the research papers. So a useful review paper might have been published in any of a number of different places. But again, help is at hand for you in tracing review papers. Here are two indexes to reviews, no matter where those reviews might have been published:

> *CA Reviews Index (CARI)*: Nottingham: UK chemical Information Service, 1978 to date. Every 6 months. (By the way, the 'CA' of the title stands for Chemical Abstracts.)

This is what is called a KWIC (computer-produced Keyword in Context) index to abstracts of review articles in the main *Chemical Abstracts* service. In this kind of index the keywords are listed alphabetically down the centre of each column; each entry (up to 54 characters) includes the remainder of the title. The references are to CA abstracts numbers, so you must have access to *Chemical Abstracts* to find full bibliographic details of the paper. The other index is:

> *Index to Scientific Reviews*. Philadelphia: Institute for Scientific Information, 1975 to date. Every 6 months.

This is an international interdisciplinary index to reviews in science, medicine, agriculture, technology, and behavioural sciences. It consists of five parts: (*a*) a guide and list of source publications; (*b*) a research-front speciality index: a list by code numbers of research-front specialities—active research fields identified by citation analysis; (*c*) a corporate index of institutions where the reviews have been written; (*d*) a source index: bibliographic descriptions of all items indexed. [They are listed under the name of each first author (perhaps I should have mentioned earlier that most scientific papers have more than a single author) with cross-references to secondary authors]; (*e*) a permuterm subject index to the source index and to research front numbers. This is an alphabetical list of subject terms arranged ('permuted') together.

Well, those will help you to find published reviews. But it will also be of use to you to know of some important review publications, especially those relating to analytical chemistry. I have already introduced you to the journal *Analytical Chemistry* in Section 2.4; I mentioned then its important annual review issue:

> Fundamental Reviews. *Analytical Chemistry*. The whole of the April issue each year.

This contains between 25 and 30 review articles in that single issue. These reviews cover specialised techniques and applications, and the most recent research activity over all fields of analytical chemistry, although they tend to concentrate on specific subjects (such as pharmaceuticals) in alternate years. Given the importance of reviews you will not be surprised to learn that chemical societies take a serious interest in publishing them. In the United Kingdom the Royal Society of Chemistry is responsible for many review publications, beginning with:

> *Chemical Society Reviews*. London: Royal Society of Chemistry, 1972 to date. Quarterly. (Formerly, 1947–1971, published as *Quarterly Reviews of the Chemical Society)*.

This publishes some 20 review papers (500 pages) each year. It covers the whole of chemistry and its interfaces with other disciplines, while each review paper is intended to be of interest to chemists in general.

In addition the Royal Society of Chemistry has a large programme of publishing *Specialist Periodical Reports*—volumes appearing annually or every couple of years on specific chemical topics. They provide critical coverage of major areas of research. There are usually between 6 and 10 major review articles each year in each volume in the series, on selected topics of current importance within the subject field under review. The volumes likely to be of particular interest to the analytical chemist include:

> *Electrochemistry*,
> *General and Synthetic Methods*,
> *Mass Spectrometry*,
> *Nuclear Magnetic Resonance*,
> *Spectroscopic Properties of Inorganic and Organometallic Compounds*

To give you a more detailed idea of the kind of papers they carry, the following made up the contents of the 1984 volume of *Electrochemistry*:

> Electrochemistry of porous electrodes; Flow-through and 3-phase electrodes; Semiconductor electrochemistry; Spectro-electrochemistry; Electrochemistry of transition-metal complexes; Organic electrochemistry—synthetic aspects; Solid-state gas sensors and monitors.

I list here all the *Specialist Periodical Reports* issued by the Royal Society of Chemistry.

Dates quoted indicate the years for which the literature is covered and do not necessarily correspond to years of publication.

(*a*) CURRENT TITLES;

*Amino-acids, peptides and proteins*. 1968–.
*Carbohydrate chemistry*. 1967–.
*Catalysis*. 1976–.
*Colloid science*. 1970–.

*Electrochemistry*. 1968–.
*Electron spin resonance*. 1971–.
*Electronic structure and magnetism of inorganic compounds*. 1970–.
*Enviromental chemistry*. 1975–.
*General and synthetic methods*. 1976–.
*Heterocyclic chemistry*. 1978–.
*Inorganic biochemistry*. 1977–.
*Macromolecular chemistry*. 1977–.
*Mass spectrometry*. 1968–.
*Nuclear magnetic resonance*. 1970–.
*Organic compounds of sulphur, selenium and tellurium*. 1969–.
*Organometallic chemistry*. 1971–.
*Organophosphorous chemistry*. 1968–.
*Photochemistry*. 1968–.
*Spectroscopic properties of inorganic and organometallic compounds*. 1967–.
*Theoretical chemistry*. 1972–.

(*b*) EARLIER TITLES (including those for which no further volumes are planned)

*Alicyclic chemistry*. 1970–1976.
*Aliphatic chemistry*. 1970–1975.
*Aliphatic and related natural product chemistry*. 1976–1981.
*The Alkaloids*. 1969–1982.
*Aromatic and heteroaromatic chemistry*. 1971–1978.
*Biosynthesis*. 1971–1981.
*Chemical physics of solids and their surfaces*. 1971–1978.
*Chemical thermodynamics*. 1971–1977.
*Dielectric and related molecular processes*. 1966–1976.
*Fluorocarbon and related chemistry*. 1969–1974.
*Foreign compound metabolism in mammals*. 1968–1979.
*Gas kinetics and energy transfer*. 1972–1980.
*Inorganic chemistry of the main-group elements*. 1971–1977.
*Inorganic chemistry of the transition elements*. 1970–1976.
*Inorganic reaction mechanisms*. 1969–1979.
*Molecular spectroscopy*. 1971–1976.
*Molecular structure by diffraction methods*. 1971–1977.
*Radiochemistry*. 1969–1975.
*Saturated heterocyclic chemistry*. 1970–1975
*Statistical mechanics*. 1972–1973.
*Terpenoids and steroids*. 1969–1981.

The American Chemical Society is also active in publishing reviews, through the medium of two journals:

*Chemical Reviews*. Every two months, 1924 to date.

This carries authoritative critical reviews and comprehensive summaries of recent research in all aspects of chemistry.

*Accounts of Chemical Research*. Monthly, 1968 to date.

This publishes short critical reviews.

And just to remind you, review papers are also to be found in most of the Society's specialised journals.

Let's not forget our friends the commercial publishers. Remember the CRC Press? Well, among a list of (currently) 14 quarterly critical review journals they publish you will find one entitled *Analytical Chemistry*. To demonstrate the kind of papers they publish, the titles of the papers in vol.15 of this were:

> Recent advances in air pollution analysis; particle-size measurement by using chromatography; analytical aspects of photo-acoustic spectroscopy.

Rather than fill out the text unnecessarily, again I have listed all the titles in this CRC Series of critical review journals:

*Analytical chemistry,*
*Biochemistry,*
*Biomedical engineering,*
*Clinical laboratory sciences,*
*Diagnostic imaging,*
*Environmental control,*
*Food science and nutrition,*
*Immunology,*
*Microbiology,*
*Toxicology,*
*Solid state and material science,*
*Biocompatibility,*
*Biotechnology.*

Finally, you will find many annual review publications with titles like 'Advances in …'. I have listed relevant chemical ones but take note of these three which are typical of ones likely to be of particular interest to you:

*Advances in Chromatography*. New York: Marcel Dekker, 1961 to date. Annual.

*Advances in Photochemistry*. New York: Wiley Interscience, 1969 to date. Annual.

*Vibrational Spectra and Structure: a series of advances*. Amsterdam: Elsevier, 1972 to date. Annual.

*Advances in:*
*Carbohydrate chemistry and biochemistry,*
*Clinical chemistry,*
*Drug research,*
*Electro-chemistry and electro-engineering,*
*Food research,*
*Heterocyclic chemistry,*
*Infrared and Raman spectroscopy,*
*Inorganic chemistry and radio-chemistry,*
*Lipid research,*
*Mass spectrometry,*
*Organic chemistry,*
*Organometallic chemistry,*
*Pharmaceutical sciences,*
*Physical organic chemistry,*
*Quantum chemistry,*
*X-ray analysis.*

*Annual review of:*
*Biochemistry,*
*Pharmacology and toxicology,*
*Physical chemistry.*

*Progress in:*
*Biochemical pharmacology,*
*Physical organic chemistry,*
*Rubber technology.*

So there you have the review literature: you will find it a most useful source of information (and saver of time), the only real difficulty being in tracing the existence of particular papers as their numbers grow out of proportion to the rates of growth of other forms of the literature.

But as usual, we are going to finish with a question:

**SAQ 2.5b**

You want to know about research on new instrumentation techniques in mass spectrometry over the last few years. Write down the titles of three sources of reviews where you would look for likely papers and to give yourself the best coverage you can. Then, when you have looked at these three sources, write down where you would turn next to look for further or more up-to-date reviews. ⟶

**SAQ 2.5b (cont.)**

You will be able to answer this question from the titles quoted in the text of this Section (in other words you can ignore the titles in the three lists).

**Objectives**

When you have worked through this Section you should be able to:

- recognise what review papers are and how they can benefit you;
- locate and identify likely useful sources of reviews on analytical chemistry.

## 2.6. OTHER FORMS OF LITERATURE

### 2.6.1 Introduction

I asked at the start of the previous Section if you were starting to feel bemused by the amounts and types of literature available. Well, we haven't quite finished yet! There are in fact four more kinds of literature to which I must introduce you. But what I have said already is still valid: for practical everyday purposes the standards and other reference works we have already examined will remain your major sources of information, backed up by journals. But you are likely to come across these other forms from time to time, and are certainly likely to encounter them if you carry out the kind of literature search to which I shall introduce you in Section 3.2 (and 4.1 for that matter). So this is the logical place to deal with them.

### 2.6.2. Patents

In some areas of applied chemistry, where new developments produce potentially marketable or valuable substances, techniques, or equipment, patents will be a most important source of information. Certainly, in searches I have conducted in *Chemical Abstracts* on particular industrial topics, virtually all the references found have proved to be to patents.

Π What is a patent? I expect you have some idea at least, and some of you who have experience in the chemical industry might have a very good idea. But read my response carefully: patents are very special kinds of documents.

A patent is primarily a legal contract between a national government and an inventor. This contract grants the inventor a monopoly of the commercial exploitation of his or her invention for a fixed number of years (eg, 16 years in the UK, 17 in the USA). In return, the inventor must describe his invention publicly in a patent specification. This specification must include two important features: (*a*) a full and sufficient description to enable any person skilled in the art concerned to understand and carry out the invention; (*b*) a clear statement (the 'claim') of the precise monopoly sought, and of what distinguishes the invention from what has previously been known.

So I think that from that definition you can appreciate that patents have a double importance: their legal and commercial function is obvious; but giving a full technical disclosure, the patent specification becomes a most important technical document. Because of the legal and technical importance of patents many of the larger chemical firms have special departments to deal with them, and that includes keeping a close watch on the patents taken out by rival firms. Smaller firms wanting to take out a patent would normally employ the services of a patent agent.

But what about the chemist who just needs to know what is being developed and to find out what patents have been, or are being, published? A number of journals review recent patents in their field, but this is a selective review and subject to obvious delay. Once again, there are very large numbers of patents being issued all over the world every day. I have already mentioned one source for them: *Chemical Abstracts* includes patents in its indexing of documents, and we shall be looking at how to use these services later on. If you want to survey up-to-date chemical patents on a comprehensive and regular basis, then you will probably turn to the patents services of Derwent Publications. They offer various services tailored to different requirements:

> *Central Patents Index*. An Abstracting and retrieval service for chemical patents from twelve leading industrial nations (including the UK, USA, USSR, and Japan).
>
> *Alerting Bulletins*. These are weekly bulletins listing all the chemical patents issued in a particular country, or on a specific chemical topic.
>
> *Basic Abstract Journals*. These are twelve weekly classified journals on different aspects of chemistry, giving more detailed summaries than the *Alerting Bulletins*.
>
> *Manual Code Cards*. These are cards for individual patents (or very small groups of patents) classified by very detailed codes (about 4000 specific subjects); they reduce searches to less than 60 patents for each code per year.

They also provide cumulative indexes and complete specifications on microfilm, and make available computer tapes for in-house processing or for on-line remote access (but more of that in Section 4). You can consult or buy copies of patents from the Patent Office in London, but there are also reference sets of UK, US, and other countries' patents in major provincial public libraries.

### 2.6.3. Reports

Much pure scientific and applied research is carried out under contract. As partial fulfilment of the contract the researcher has to report to his sponsors the results of the work for which they have paid. There will, of course, be a final report (or perhaps series of reports) produced at the completion of the project; but there might also be interim reports, telling how a project is progressing, or preliminary reports to explain what is to be done and why. Reports are not usually refereed and are delivered directly to the sponsors of the research. More often than not the sponsor is a government body, and many governments have established agencies to collect and issue all the unclassified reports submitted to them; that is, all those reports not affected by considerations of defence or national security secrecy. The USA is particularly prominent in this area, although other countries as well (including the UK) arrange the publication of reports to try to help in the commercial, technical, or scientific exploitation of publicly funded research. Well, just to compound our problems, if journal papers and patents can each be counted in their millions, reports can certainly be counted in their hundreds of thousands. So, again, how do we trace them? Reports can be notoriously difficult to track down, but many are listed in *Chemical Abstracts*, along with all the other literature, and we shall look at this in

Section 3.2. If you do want to trace just reports (or to try and find a particular report), here are a couple of useful sources. For the UK literature:

> *British Reports, Translations, and Theses*. Boston Spa: British Library Document Supply Centre, 1981 to date. Monthly.

This lists report literature and translations produced by British government organisations, industry, universities, and learned institutions, and most doctoral theses accepted at British universities. It covers all subjects and each issue is arranged alphabetically in subject groupings. There is an alphabetical keyword index in each issue and cumulative annual indexes of authors, report numbers, key terms, and issuing organisations.

And for US reports:

> *Government Reports Announcements and Index*. Washington: National Technical Information Service (US Dept. of Commerce), 1946 to date. Fortnightly.

This contains abstracts of publicly available US Government-sponsored research, development, and engineering reports. It too is arranged alphabetically in subject groups and is fully indexed.

Here are a couple of final points about reports (by the way, the currently fashionable term for reports and similar documents is 'grey literature'). Certain research bodies produce large amounts of report literature in series, and may also produce their own index to this material. Also, for ease of identification almost any body which issues reports gives each one some kind of identifying serial number, and this number must always be noted whenever you refer to a report (in just the same way as you refer to British Standards by their serial numbers).

### 2.6.4. Conference Papers

I shall not go into the detailed sociology of scientific conferences, but suffice it to say initially that there are surprisingly large numbers of scientific conferences held throughout the world each year: many deal with very specialised topics. Conferences range from large prestigious gatherings over several days of international or national societies, to small highly specialised day (or even half-day) meetings. By the way, I could define and draw subtle distinctions between various terms ('conference', 'symposium', 'colloquium', 'meeting', for example), but to all intents and purposes—at least as far as this Unit is concerned—you can treat them as synonyms. Anyway, conferences are obviously going to be important in their own right for those attending them, but they will extend that importance to a far wider audience if the proceedings are published. When someone reads a paper to a conference he or she is often presenting the first formal communication of the progress of a piece of research, or its results, or perhaps an overview of its significance. It is obviously frequently important that this information should be made available to an audience wider than just those who sat and listened to it (most of whom would probably welcome a printed text anyway).

(*a*) A preprint of the paper might be issued to delegates at the conference before the paper is actually read. This practice can help to stimulate discussion, especially when the delegates do not all speak the same language.

(*b*) The full proceedings of the conference might be published as a book, or in one or more issues of a journal.

(*c*) A particular paper (or papers) might be published verbatim, or with modifications, as a journal paper.

(*d*) A particular paper might be issued by its author's employing organisation as a report.

Some conference papers are never published, and sometimes (particularly for (*b*) above) there may be a considerable delay before publication.

So you might need to know about conferences, or about a particular conference. To find out what conferences are taking place you need to keep a sharp eye on the journals, and on your organisation's notice boards. More particularly, when you want to find out if a particular conference has been published, and where, you can use the following index:

> *Index of Conference Proceedings Received*. Boston Spa: British Library Document Supply Centre, 1966 to date. Monthly. There are annual cumulations from 1974 and cumulated volumes for 1964–1973 and 1974–1978.

The British Library Document Supply Centre (more of this later) tries to acquire the proceedings of any conference in any subject field. This publication is an alphabetical keyword index of the titles of conferences, including serial publications.

Also remember that the ubiquitous *Chemical Abstracts* includes conference papers.

### 2.6.5. Theses

Finally, what about all those PhD and MSc students? They all have to conduct a research programme and publish the results in a thesis which they submit to the university from which they are seeking their qualification. Unless the researcher writes up his or her results as a journal paper or in some other form of the literature, the thesis is never formally published. The nearest it gets is a copy being deposited in the university's library. This copy will normally be submitted to either or both of two major collecting agencies for academic theses: the British Library Document Supply Centre and a commercial organisation, University Microfilms International. We have already seen the British Library Document Supply Centre's announcement bulletin in Section 2.6.3, but to help you trace academic theses there are two more indexes available.

*Dissertation Abstracts International. Part B: The sciences and engineering. Part C: European abstracts.* Ann Arbor: University Microfilms International, 1969 to date. Monthly. This was preceded by the same publisher's *Comprehensive Dissertation Index 1861–1972.* 1973. 37 vols.

This contains abstracts of theses and dissertations from co-operating institutions and which are available from the publisher in microfilm form or as xerox copies. It is arranged alphabetically by broad subjects (eg Agriculture, Chemistry, Food science and technology, Pharmacology; the chemistry sub-divisions include Analytical, Pharmaceutical, etc). There is a keyword title and author index.

*For British material*:

*Index to theses accepted for higher degrees by the universities of Great Britain and Ireland and the Council for National Academic Awards.* London: Aslib, 1953 to date. Semi-annual.

This is arranged systematically under subject headings (eg Chemistry, sub-headings: Analytical Chemistry–general, analytical Chemistry–chromatography, Physical chemistry–electrochemistry, Physical chemistry–photochemistry). There are author and subject indexes.

### 2.6.6. Summary of Literature Types

So there you have the different types of literature you might come across. Before I summarise their relative importance to you, I must just mention one apparent complication—although in practice it might help you to make more sense of the results of some literature searches. None of the forms of literature to which I have introduced you is exclusive. In theory somebody could undertake a research programme and produce a number of reports from it; perhaps read one or two papers at conferences; patent any marketable products arising; write one or more journal papers; if the project is an important long-term one it might even merit a book; one of the research assistants might produce a PhD thesis on an aspect of the research; and of course if the resulting data or knowledge are of any significance they will eventually find their way into the handbooks, reference works, and textbooks.

Now, before I summarise all this, I want you to try a SAQ.

**SAQ 2.6a**

Look at the following list of literature types. Then write them down in the chronological order in which you would expect each type to be produced in the publication of any research project (eg if you think the first step in publishing new research results would be to write a reference book, put that at the top of your list). Some of the types will, in practice, overlap at least slightly, but try the exercise anyway. I shall now make one or two important points arising out of it.

Journal papers
Textbooks
Patents
Reports and conference papers
Handbooks, standards, and reference books
Review papers

You now have an idea of the pattern of publishing scientific information in its various forms. To clarify the point further, have a look at my diagram in Fig. 2.6a showing this pattern. The arrow lines show you where, approximately, within the research process you can expect each type of document to be produced. For much of our discussion here the important end of each line is its left end—that is where each type of document might start to be produced. The items above the centre line form what is called the primary literature and those below the secondary literature. I think the reasons for these names are obvious: the primary literature is giving you the researchers' own announcements of their results and findings; in the secondary literature this information is compared with and gradually assimilated into the scientific literature by other people.

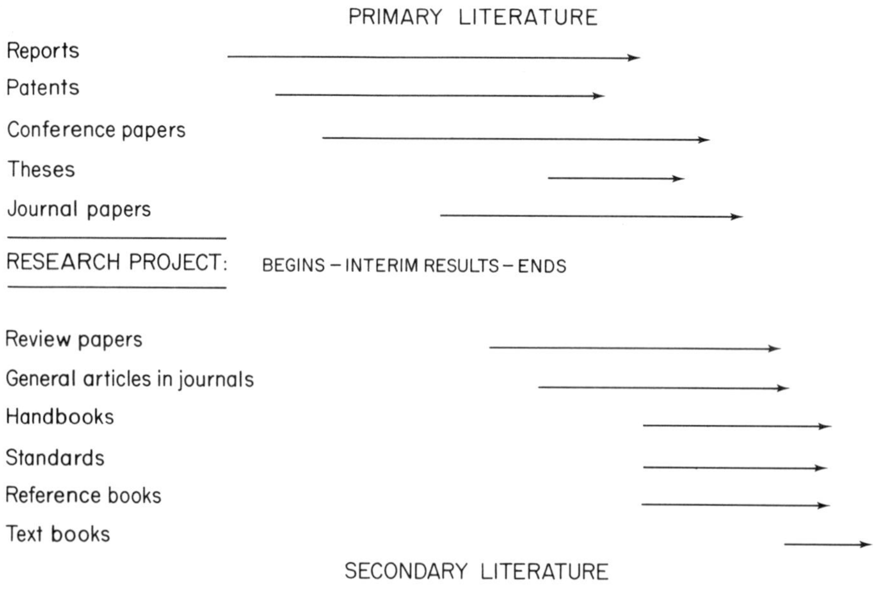


**Fig. 2.6a**

Where does all this leave you? The importance of getting at the primary literature for up-to-date information on new results and techniques is obvious. But, having said that, realistically most of your needs are going to be met by the secondary literature. When you start looking for information for yourself your normal pattern will be to work through the types of literature in my diagram in reverse order (ie from bottom to top). Start with the knowledge you have already acquired from textbooks, then turn to reference books. If you need more, then turn to handbooks and standards (or your information requirements might have been initially so precise as to make you consult these immediately). If you still need more information, or have not found what you are looking for, you are now approaching the primary literature. Your best introduction to this will be a review paper, if you can find one on the topic you are investigating. After that you are into a full search of the mass of primary literature. And at that point we look ahead to Part 3, where we shall find out where and how all this literature is kept, and then how to identify and locate it.

**Objectives**

When you have worked through this Section you should be able to:

- recognise the functions within the analytical literature of patents, reports, conference papers, and theses;
- appreciate their possible significance in relation to other forms of the literature,
- use some specialised sources of information to find these types of documents specifically.

# 3. Use of Libraries

## 3.1. LIBRARIES AND LIBRARY SERVICES

### 3.1.1. Introduction

So now you know about the vast mass of literature available and the significance of its various forms. Where do we keep it all? The answer is not quite as simple as just saying 'libraries'. So before we go on to discuss libraries, let's just put things into perspective.

Π Think of the different kinds of information you are likely to need in the course of your work. Then jot down briefly where or how you would expect to keep that information, starting with the simplest and working to most complex.

Think about it for a moment before you jot down your ideas; then go on and read my response.

I said start at the simplest, so here goes.

(*a*) Basic knowledge and information you keep in your head! This knowledge will have been built up from formal education and everyday practice. It is, of course, absolutely vital to you; but also, since human memory is fallible, it is potentially unreliable. Neither can you ever hope to remember all the details of all the information you need. So you have to aid your memory by all the other means listed here.

(*b*) Vital everyday information you will keep to hand in your laboratory—perhaps notes on paper or in a notebook. Further important information will be kept in your laboratory, perhaps in house manuals or instruction books produced by your employers.

(*c*) One or two standard reference works or books of standards might be kept in your laboratory, or in your supervisor's office. These might be supplemented by a few (often outdated!) textbooks, and perhaps some miscellaneous photocopies or a pile of issues of one or two journals. If your laboratory is well-organised all this material might be neatly arranged, recorded, and kept up-to-date. But such efficiency is not common.

(*d*) At this stage your search for information moves out of the laboratory and towards, or into, the library. We will go on to discuss all this in just a moment.

Before we do look at libraries in more detail, one last point. If you make an interesting discovery, or want to know something, what is the very first way in which you communicate your findings or requirements? Unless there are exceptional circumstances I think you would tell someone or ask someone, probably the person working next to you, or your supervisor. So before you ever approach a library you have probably already exhausted more than one avenue of investigation.

### 3.1.2. Types of Libraries

Obviously you cannot keep all the documents you are ever likely to need in your laboratory, so you must at some point rely on a library service. In this Section we are simply going to list the range of libraries to which you might have access. We will discuss their services in more detail in the next Section.

**Industrial/special libraries**. Perhaps you work in an organisation which provides an on-site technical library to house most of the documents likely to be required by its employees. This might cover one subject area in great detail or, in larger and more diverse organisations, a wider range of subjects. Either way, there are obvious advantages to such a library: it should be located conveniently for you; it is likely to contain much of the information and documentation you are expected to need; and its staff will at least be familiar with the subjects you are working in, if not experts. It is also worth noting that a few industrial libraries have been in the forefront of pioneering new documentation services, which have subsequently been followed by other libraries. So you might have an excellent in-house library to help you, or a medium-sized or a small one; or, perhaps, you have no in-house library at all. Don't worry, there are still other types of library services to help you.

**Academic libraries**. Perhaps you are pursuing your ACOL courses in conjunction with a local college or polytechnic. In that case you should already have used their library—I wonder whether you remember what I told you in all those questions in Section 2. I cannot generalise about other people's libraries, but if you are within convenient travelling distance of a polytechnic or college (especially one teaching chemistry, of course) it is well worth your while contacting the librarian and asking for permission to use the library. Many will let you use it for reference purposes. Some might even let you borrow books as well. University libraries, however, tend to be very strict about not allowing members of the public to use their services; but, again, it is worth asking, as some might let you use their books for reference. Some will do so only if you pay a fee, but if you are likely to need regular access to good collections it might be a useful investment. It is almost impossible to generalise in these matters, but you should find most, or all of the works I mention in this Unit (and more besides) in the library of any university, polytechnic, or college teaching chemistry to an undergraduate level.

**Public Libraries.** Of course, as a member of the public you have automatic access to your public library. These vary from huge municipal (or county headquarters) libraries to small rural or urban branch or mobile libraries. The large reference libraries will have bookstocks second only to those of national or university libraries—certainly they will have most or all of the books I mention in this Unit (and, don't forget, more besides). As you go down the scale in size you will find a smaller choice of books of course, and in the smallest branches will be lucky to find more than a handful of any kind of books on chemistry. *But* any branch of any public library can be an entry point to the network of the inter-library lending system (more about this in Section 3.1.3). So, if you have no other library available, your small local branch library should be able to function as an entry (and collection) point to a much wider system.

**National Libraries.** I shall mention these only briefly because I think few of you will have convenient access to a national library, or the need to consult one. But if you live in the UK and can travel easily into central London, Cardiff, or Edinburgh, there are national libraries there which you will be able to use for reference, if you cannot find what you want elsewhere (and that should very rarely, if ever, be the case). In London you should know of the Science Reference Library (a division of the British Library), which holds the largest and most comprehensive reference collection of the world's technical literature in the UK. Both its Holborn and its Bayswater branch are open freely to any adult member of the public without having to acquire a reader's ticket. We have come across the British Library Document Supply Centre (BLDSC) already as a publisher of indexes; it is primarily the focal point of inter-library lending in the United Kingdom, and I shall mention it again in Section 3.1.3.

**Other Special Libraries**. Perhaps your organisation dosn't have its own library. But there may be a research institution or similar organisation fairly near you. It is possible (but by no means certain of course) that on application to their librarian you might be able to use their library.

If you want to know what libraries exist, either nationally or locally, and what subjects they cover, a useful directory to start with is:

> *Aslib Directory. Vol l: Information sources in science, technology, and commerce.* Ellen M. Codlin, ed. 4th edn. London: Aslib, 1977. 634pp. This is an alphabetical directory (by their names) of libraries and organisations furnishing information throughout the U.K. It has a subject index.

There is also a useful series of regional library guides (*Library Resources in . . .*) published by branches of the Library Association. International directories of libraries also exist, as do national directories of libraries for most industrial countries. Before we move on, just try a short question:

**SAQ 3.1a**

Specify any *two* types of library to which you could claim automatic right of access to consult their stock.

### 3.1.3. Library Services

Well, it seems obvious doesn't it? You go to a library to read or to borrow books or journals. But there is (or should be) more to it than that. Often you are going to the library to find information: you don't necessarily know exactly where you will find it, but by the time you have finished this Unit your chances of success will have improved considerably. And remember, you might know all about your specific field in analytical chemistry (or think you do), but the librarian knows what is kept in his or her library. When you have finished this Unit you will have a sound knowledge of the kinds of documents (even of specific titles) likely to furnish the information you want, and of how to use any library to which you have access. But if you are ever uncertain, always ask a member of the library staff for help or advice—that is one of the main things they are there for. Having found your book or other source of information, you borrow it; or read it and note down the information you need; or photocopy a part of it. Most libraries have a photocopier and this provides a very convenient way of recording printed information for yourself. *But* there is a Copyright Act and there are restrictions (strictly enforced by publishers and their representatives) on what you can or cannot copy. Ask the librarian for advice, but if you are told that what you want may not be copied, don't think the librarian is being obstructive: the law does lay down rigid restrictions on the amount you may photocopy from copyright books and journals. It is also worth mentioning that photocopying restrictions are even more rigid in the USA than in the UK. One last point about photocopying: because it is so easy and convenient photocopying has a habit of replacing reading and digesting; how many people carry around piles of photocopies which they are always meaning to read but never quite getting round to?

What if the book or journal you want isn't held in the library you are using? You don't necessarily have to tramp around other libraries in the area, or give up. All the libraries in the UK, academic, public, or special, are linked in an inter-library lending system. They might use a local networking system, or the national inter-library lending system (usually both), but the result is the same: with few exceptions (current reference books in heavy demand; or valuable, rare, and antiquarian books are obvious ones) you will be able to borrow from, or read in, your local library a copy of almost any book or journal you might want. The formal national interlending system in the United Kingdom is centred on the British Library Document Supply Centre (formerly the British Library Lending Division). This is based at Boston Spa in Yorkshire and has enormous holdings of over three million volumes of books and journals, and over two million documents on microfilm or microfiche. Any of these items may be borrowed by any registered library (and that means virtually every library in the UK); if you request a journal paper it will normally be supplied as a photocopy. Often material can be borrowed with only a week or two of delay (sometimes faster than that, but sometimes also much slower); but always try to allow your library as much time as you can to obtain material for you. And try to ensure that you have full details, accurately recorded, of whatever you are seeking. Inter-library loans are not cheap services for the library obtaining material for you: taking into account staff time and postage, they will be as much as five pounds or ten dollars per loan. So there might be restrictions on the use of the services in some libraries and it is obviously up to you to use them only for essential requests. But do remember that if the only library you can use is a tiny branch of a public library or a very small works library, you can still get access to the interlending system through them: as long as you know what you want, of course, and provided you allow enough time for the material to be obtained for you. Interlending services in other countries vary considerably, but most industrial nations have established or are trying to establish systems similar to that in the UK. In the USA there is greater reliance on local or regional networking.

Finally in this Section, remember that libraries nowadays stock far more than just books and journals. They may carry all kinds of audio-visual materials, even computer programs, as well. Many use computers for various purposes, and so a modern library can be seen not just as a document repository, but also as a gateway to information of all kinds, no matter the form in which it may be stored.

Another quick question before we proceed:

**SAQ 3.1b**

(*a*) First of all note any two forms in which documents might be given you in a library, apart from books and journals.

(*b*) Then note any three other kinds of materials you might find kept in a library, apart from books, journals, and whatever you gave as your answer to (*a*).

**SAQ 3.1b**

### 3.1.4. Arrangement of Libraries

Ideally you should be able to go into your library, straight to one section of the shelves, and there find all the documents available, in whatever form, on the subject in which you are interested. In practice, for various reasons, that cannot be the case, except in the most specialist of libraries. Several practical considerations affect the layout of libraries. Their physical size and design is the first obvious one, and is unique to each individual library. The nature of the documents held and the kind of users catered for must also be taken into account. One of the first and most obvious divisions is between reference and lending stock: some libraries inter-file reference and lending books, many keep reference books separately. Some major public libraries have entirely separate reference and lending departments (even buildings). Most libraries keep materials such as microfilms, audio-visual programmes, and unbound journals in sequences separate from the main library stock. A few libraries keep much of their stock in private ('closed access') areas, so that you have to ask the staff to fetch you many of the books you need. I am not going to consider this in any more detail: the important thing is for you to realise that these kinds of arrangements exist and to find out what is shelved where in any library you use.

As far as the main public-access stock is concerned most libraries use a classified arrangement. The idea is that if you want books on a specific subject you find out the classification code for that subject and go straight to the right set of shelves. If you want to browse through related (or broader or more specific) subjects, they should all be in about the same area of the library. It sounds ideal, doesn't it? But unfortunately (perhaps inevitably) some practical considerations complicate matters. Again, except in the most specialist libraries, a fairly general classification scheme will be used. A general classification scheme tries to cover in a logical manner all fields of human knowledge. What is a peripheral subject to the analytical chemist will be a core subject to another discipline (and *vice-versa*); and what about the increasingly interdisciplinary nature of much modern science and technology? Do we put all the books on computers, for example, in one place, or scatter them among the many subjects to which they are applied?

Let's look at this in more detail by examining one classification scheme. I have chosen the Dewey Decimal Classification (no, not the Universal Decimal Classification we have already encountered—but I shall come to that soon). The Dewey Decimal is the most widely used library classification scheme in the world; if you have used a school, public, or college library at any time, you have probably already encountered it (without necessarily knowing it). It includes every field of knowledge in an encyclopaedic, logical order, and produces classification symbols (as its name tells you) of decimal numbers. To start with it divides the whole of knowledge into ten major classes:

000 Generalities
100 Philosophy and related disciplines
200 Religion
300 Social sciences
400 Language
500 Pure sciences
600 Technology (applied sciences)
700 The arts
800 Literature
900 Geography and history

These main classes are then sub-divided into ever more specific subjects, for example:

| | |
|---|---|
| 500 | Pure sciences |
| 510 | Mathematics |
| 520 | Astronomy and allied sciences |
| 530 | Physics |
| 540 | Chemistry |
| 541 | Physical and theoretical chemistry |
| 542 | Chemical laboratories, apparatus, equipment |
| 543 | Analytical chemistry |
| 543.085 | Optical methods |
| 543.085 2 | Photometric analysis |
| 543.085 3 | Refractometric and interferometric analysis |
| 543.085 6 | Polarimetric (polariscopic) analysis etc. |

So far so good. You might not agree with the way everything is arranged, but it is a reasonably clear and logical system. Provided you are looking for a precise subject as indicated in the schedules, you will not have much difficulty. But we have already seen that the applications of analytical chemistry can spread pretty widely. Can you really expect to find everything relevant within the 543 (analytical chemistry), the 544 (qualitative chemistry) or the 545 (quantitative chemistry) schedule? No, you cannot. You will have to look in other places for some applications, principally in the 600 (technology, applied sciences) schedule. Of particular concern will be the chemical technology section, the outline of which is:

600 Chemical and related technologies
661 Technology of industrial chemicals (heavy chemicals)
662 Technology of explosives, fuels, related products
663 Beverage technology
664 Food technology
665 Technology of industrial oils, fats, waxes, gases
666 Ceramic and allied technologies
667 Cleaning, colour and related technologies
668 Technology of other organic products
669 Metallurgy

This can be a bit annoying—in a large library you might have to walk a long way to get from the 540 to the 660 section (you might even have to go to another building!). But provided you realise that you might have to look in more than one place for your subject it isn't too serious. But, just suppose you are looking for a book on spectroscopic analysis in the food industry: where do you look? Section 545.83 for spectroscopic analysis or 664.07 for tests and analyses in the food industry? A book can only be shelved in one place, and the librarian has had to choose which of the two places in which he expects most of his users to look, and put the book there. Something else for you to bear in mind; but soon we shall look at library catalogues and these can help to solve some of the problems. Now I think the best way to take a closer look at the Dewey Decimal Classification is to invite you to answer a question on it, before we proceed.

**SAQ 3.1c**

I have given you in the next few pages some copies of parts of the Dewey Decimal Classification schedules and indexes. No, I am not going to be strict this time—it is not a book you are likely to consult much (but you probably will find yourself using the resulting classification numbers frequently); but if you are answering this question in a library and they have a copy of the 19th edition, ask if you may use it for a few minutes to answer this question (and have a general look at it while you have the opportunity).

Anyway, use the alphabetical index first and find classification numbers for the following:

(*a*) Spectroscopic analysis,

(*b*) Analysis of drugs, →

**SAQ 3.1c (cont.)**

(*c*) Chemical technology of soft drinks.

In each case check back to the schedules from the index to make sure you have the correct number. And ignore in the schedules any references to tables or to standard subdivisions: I am trying to keep this fairly simple for you, not teaching you to become librarians!

DEWEY

# Decimal Classification and Relative Index

*Devised by*

MELVIL DEWEY

*Edition 19*

*Edited under the direction of*
BENJAMIN A. CUSTER

Volume 2
**Schedules**

FOREST PRESS
A Division of
Lake Placid Education Foundation

ALBANY, N.Y. 12206 U.S.A.
1979

542 *Dewey Decimal Classification*

▶ **542.2–542.8 Specific types of apparatus**

Class comprehensive works in 542

.2 **Receptacles and accessory equipment**

Beakers, flasks, retorts, test tubes, funnels, crucibles, supports, stoppers, tubing

Class receptacles and accessory equipment for measuring in 542.3, for distilling in 542.4; gas apparatus in 542.7

.3 **Measuring apparatus**

Gravimetric and volumetric

Class gas measuring apparatus in 542.7

.4 **Heating and distilling apparatus**

.5 **Blowpipes**

.6 **Filters and dialyzers**

.7 **Gas apparatus**

For producing, collecting, washing, dissolving, storing, measuring, rarefying, compressing, liquefying, solidifying gases

.8 **Electrical and electronic apparatus Including computers**

543 Analytical chemistry

Use 543.001–543.009 for standard subdivisions

Class analytical chemistry of specific elements, compounds, mixtures, groupings in 546

*For qualitative chemistry, see 544; quantitative chemistry, 545*

.01 Reagents

.02 Sample preparation

.07 Instrumentation

.08 Instrumental and separation methods

Including gas analysis

SUMMARY

**543.081 Micro and semimicro methods**
**.083 Mechanical methods**
**.085 Optical methods**
**.086 Thermal methods**
**.087 Electromagnetic methods**
**.088 Radiochemical methods**
**.089 Chromatographic analysis**

.081 Micro and semimicro methods

684

.081 2 Microscopical analysis

Use of compound, electron and ultramicroscopes for identification

.081 3 Microchemical analysis

Reactions performed with small quantities, e.g., micrograms and microliters, and using small apparatus

.081 32 Systematic

.081 34 Spot tests

.081 5 Semimicro analysis

Small-scale adaptations of existing macro methods

.083 Mechanical methods

.085 Optical methods

.085 2 Photometric analysis

Colorimetric, nephelometric, turbidimetric, fluorophotometric methods

.085 3 Refractometric and interferometric analysis

.085 6 Polarimetric (Polariscopic) analysis

.085 8 Spectrochemical (Spectroscopic) analysis

Class spectroscopic interpretation of chemical structure in 541.2; each kind of spectroscopic analysis not provided for here with the subject, e.g., mass spectrographic methods 543.0873

.085 82 Microwave

.085 83 Infrared

.085 84 Visible light

Including Raman spectroscopy

.085 85 Ultraviolet

.085 86 X-ray and gamma-ray

.086 Thermal methods

Thermometric titrimetry, thermal analysis, differential thermal analysis

.087 Electromagnetic methods

.087 1 Electrical methods

*For polarographic methods, see 543.0872; coulometric methods, 543.0874*

.087 11 Conductometric methods

.087 12 Potentiometric methods

.087 2 Polarographic methods

Analysis based on current-voltage curves obtained in electrolysis with a slowly dropping mercury cathode (polarimetric titration)

.087 3 Mass spectrographic methods

.087 4 Coulometric methods

Including electrodeposition

.087 7 Magnetic methods

Including nuclear magnetic resonance

*For mass spectrographic methods, see 543.0873*

.088 Radiochemical methods

.088 2 Activation (Radioactivation) analysis

.088 4 Tracer techniques

.089 Chromatographic analysis

.089 2 Various specific types of interaction

Including adsorption, molecular sieve, partition

Class specific types of interaction applied to liquid chromatography in 543.0894, to gas chromatography in 543.0896

*For ion-exchange separations, see 543.0893*

.089 3 Ion-exchange separations

Class application to liquid chromatography in 543.0894, to gas chromatography in 543.0896

.089 4 Liquid chromatography

Including liquid-liquid, liquid-solid, column chromatography

*For paper and thin-layer chromatography, see 543.0895*

.089 5 Paper and thin-layer chromatography

.089 52 Paper

.089 56 Thin-layer

.089 6 Gas chromatography

Including gas-liquid and gas-solid chromatography

## 544 Qualitative chemistry

Systematic macro and semiquantitative methods and procedures for detecting and identifying constituents of a substance

Use 544.001–544.009 for standard subdivisions

.01 Reagents

.02 Sample preparations

.07 Instrumentation

**.1 Systematic separations**

.12 Cation separation and identification

.13 Anion separation and identification

**.2 Thermal methods**

Including pyrolysis and combustion

*For blowpipe analysis, see 544.3*

**.3 Blowpipe analysis**

Flame tests, bead tests, reactions on charcoal and plaster of paris in conjunction with blowpipe

**.4 Gas analysis**

**.5 Diffusion analysis**

Including ultrafiltration methods (dialysis)

**.6 Spectrochemical (Spectroscopic) analysis**

Add to base number 544.6 the numbers following 543.0858 in 543.08582–543.08586, e.g., microwave analysis 544.62

**.8 Micro and semimicro methods**

Add to base number 544.8 the numbers following 543.081 in 543.0812–543.0815, e.g., microscopical analysis 544.82

**.9 Other methods**

.92 Chromatographic analysis

Add to base number 544.92 the numbers following 543.089 in 543.0892–543.0896, e.g., paper chromatography 544.9252

.93 Mechanical methods

.94 Biochemical methods

Identification by means of microorganisms

.95 Optical methods

*For spectrochemical analysis, see 544.6*

.952 Photometric analysis

Colorimetric, nephelometric, turbidmetric, fluorophotometric methods

.953 Refractometric and interferometric analysis

.956 Polarimetric (Polariscopic) analysis

*For spectrochemical analysis, see 544.6*

.97 Electromagnetic methods

Add to base number 544.97 the numbers following 543.087 in 543.0871–543.0877, e.g., coulometric methods 544.974

.98 Radiochemical methods

.982 Activation (Radioactivation) analysis

.984 Tracer techniques

## 545 Quantitative chemistry

Determination of amount of a constituent in a substance

Use 545.001–545.009 for standard subdivisions

.01 Reagents

.02 Sample preparation

.07 Instrumentation

.08 Mechanical methods

**.1 Gravimetric analysis**

Including gravimetric analysis of precipitates

*For thermogravimetric methods, see 545.4*

**.2 Volumetric analysis**

Determination of elements and compounds in a substance by titration with standard solutions and indicators

.22 Neutralization methods (Alkalimetry and acidimetry)

.23 Oxidation-reduction methods (Oxidimetry and iodometry)

.24 Precipitation methods

**.3 Electromagnetic methods**

Add to base number 545.3 the numbers following 543.087 in 543.0871–543.0877, e.g., electrodeposition 545 34

**.4 Thermal methods**

Class here thermogravimetric methods

.42 Pyrolysis and combustion

.43 Blowpipe analysis

.46 Volatilization

.7 **Gas analysis**

.8 **Other methods**

.81 Optical methods

*For spectroscopic analysis, see 545.83*

.812 Photometric analysis

Colorimetric, nephelometric, turbidimetric, fluorophotometric methods

.813 Refractometric and interferometric analysis

.816 Polarimetric analysis (Polariscopic analysis)

.82 Radiochemical methods

.822 Activation analysis (Radioactivation analysis)

.824 Tracer techniques

.83 Spectroscopic analysis

Add to base number 545.83 the numbers following 543.0858 in 543.08582–543.08586, e.g., microwave analysis 545.832

.84 Micro and semimicro methods

Add to base number 545.84 the numbers following 543.081 in 543.0812–543.0815, e.g., semimicro quantitative analysis 545.845

.89 Chromatographic analysis

Add to base number 545.89 the numbers following 543.089 in 543.0892–543.0896, e.g., gas chromatography 545.896

.728 *Polonium

.73 Halogen group (Group 7A)

.731 *Fluorine

.732 *Chlorine

.733 *Bromine

.734 *Iodine

.735 *Astatine

.75 Rare (Inert, Noble) gases

Group O

.751 *Helium

.752 *Neon

.753 *Argon

.754 *Krypton

.755 *Xenon

.756 *Radon (Niton)

**.8 Periods of periodic table**

Class hydrogen and its compounds in 546.2, other specific elements and their compounds in 546.38–546.75

.81 Period 1

Hydrogen and helium

.82 Period 2

Lithium, beryllium, boron, carbon, nitrogen, oxygen, fluorine, neon

.83 Period 3

Sodium, magnesium, aluminum, silicon, phosphorus, sulfur, chlorine, argon

.84 Period 4

Potassium, calcium, scandium, titanium, vanadium, chromium, manganese, iron, cobalt, nickel, copper, zinc, gallium, germanium, arsenic, selenium, bromine, krypton

.85 Period 5

Rubidium, strontium, yttrium, zirconium, niobium, molybdenum, technetium, ruthenium, rhodium, palladium, silver, cadmium, indium, tin, antimony, tellurium, iodine, xenon

*Add as instructed under 546

.86 Period 6

Cesium, barium, lanthanide series, hafnium, tantalum, tungsten, rhenium, osmium, iridium, platinum, gold, mercury, thallium, lead, bismuth, polonium, astatine, radon

.87 Period 7

Francium, radium, actinide series

## 547 Organic chemistry

Add to notation for each term identified by * as follows:

04 Special topics of general applicability
044 Theoretical chemistry
Add to 044 the numbers following 541.2 in 541.22–541.28, e.g., molecular structure 0442
045 Physical chemistry
Add to 045 the numbers following 541.3 in 541.34–541.39, e.g., radiochemistry 0458
046 Analytical chemistry
0464 Qualitative
0465 Quantitative

*For biochemistry, see 574.192*

.001 Philosophy and theory

Class theoretical chemistry in 547.12

.002–.009 Standard subdivisions

Notations from Table 1

---

▶ 547.01–547.08 General groupings of compounds

Unless other instructions are given, class compounds with components in two or more subdivisions of this schedule in the number coming last in the schedule, e.g., sulfonamides 547.067 (*not* 547.042)

Class comprehensive works in 547, natural compounds in 547.7–547.8

### SUMMARY

**547.01 Hydrocarbons**
**.02 Halogenated compounds**
**.03 Oxy and hydroxy compounds**
**.04 Nitrogen compounds**
**.05 Organometallic compounds**
**.06 Sulfur compounds**
**.07 Phosphorus compounds**
**.08 Silicon compounds**

.01 *Hydrocarbons

*Add as instructed under 547

.065 Oxy derivatives of thioethers

Sulfones, sulfoxides, thioaldehydes, thioketones

.066 *Sulfinic acids

.067 *Sulfonic acids

.07 *Phosphorus compounds

.071 Phosphonium compounds, phosphines

.073 *Phosphoalcohols

.074 *Phosphoacids

.075 Phosphoaldehydes, phosphoketones

.076 *Phosphinic acids

.077 *Phosphonic acids

.08 *Silicon compounds

SUMMARY

**547.1 Physical and theoretical chemistry**
**.2 Synthesis and named reactions**
**.3 Analytical chemistry**
**.4 Aliphatic compounds**
**.5 Cyclic compounds**
**.6 Aromatic compounds**
**.7 Macromolecular and related compounds**
**.8 Other organic substances**

.1 **Physical and theoretical chemistry**

Add to base number 547.1 the numbers following 541 in 541.2–541.7, e.g., electrochemistry 547.137; however, class synthesis and named reactions in 547.2

Class physical and theoretical chemistry of specific compounds and groups of compounds in 547.01–547.08, 547.4–547.8

.2 **Synthesis and named reactions**

Class chain, reversible, irreversible, homogeneous, heterogeneous reactions in 547.1393

.21 Alkylation, acylation, aromatization

Examples: Friedel-Crafts, Würtz-Fittig, Wittig reactions

.22 Halogen and hydroxy addition and substitution

.223 Halogenation

.225 Hydrolysis and saponification

***Add as instructed under 547**

.23 Oxidation and reduction

Hydrogenation, dehydrogenation, peroxidation, quinonization

.24 Esterification

.25 Amination and diazotization

.26 Nitration and nitrosation

.27 Sulfonation

.28 Polymerization and condensation

Polycondensation, copolymerization, addition and condensation polymerization

.29 Fermentation processes

.3 **Analytical chemistry**

Use 547.3001–547.3009 for standard subdivisions

.301–.308 General principles

Add to base number 547.30 the numbers following 543.0 in 543.01–543.08, e.g., reagents 547.301

.34 Qualitative chemistry

Use 547.34001–547.34009 for standard subdivisions

Add to base number 547.34 the numbers following 544 in 544.01–544.98, e.g., microscopical analysis 547.3482

.35 Quantitative chemistry

Use 547.35001–547.35009 for standard subdivisions

Add to base number 547.35 the numbers following 545 in 545.01–545.89, e.g., volumetric analysis 547.352

---

▶ **547.4–547.6 Aliphatic and cyclic compounds**

Class comprehensive works in 547, macromolecular compounds in 547.7

.4 ***Aliphatic compounds**

Observe order of precedence in note under 547.01–547.08

.41 *Hydrocarbons

.411 *Paraffins (Alkanes)

.412 *Olefins (Alkenes)

.413 *Acetylenics (Alkynes)

*Add as instructed under 547

.42–.48 Other compounds

Add to base number 547.4 the numbers following 547.0 in 547.02–547.08, e.g., carboxylic acids 547.437

*For proteins, see 547.75*

.5 ***Cyclic compounds**

Class here alicyclic compounds

*For aromatic compounds, see 547.6*

---

▶ 547.51–547.58 Alicyclic compounds

Nonaromatic compounds with ring structure

Observe order of precedence in note under 547.01–547.08

Class comprehensive works in 547.5

.51 *Alicyclic hydrocarbons

Add to base number 547.51 the numbers following 547.41 in 547.411–547.413, e.g., cycloparaffins 547.511

.52–.58 Other alicyclic compounds

Add to base number 547.5 the numbers following 547.0 in 547.02–547.08, e.g., alicyclic acids 547.537

.59 *Heterocyclic compounds

Cyclic compounds containing other than carbon atoms in the ring

.592 *With hetero oxygen atoms

Furans, pyrans, oxazoles

.593 *With hetero nitrogen atoms

Pyrroles, porphyrins, chlorophylls, pyridines, pyrazoles, imidazoles, diazines

.594 *With hetero sulfur atoms

Thiophenes, thiazoles

.595 With several different hetero atoms

Oxazines, oxdiazines, oxdiazoles

.596 *With fused hetero rings

Quinolines, purines

Class nucleic acids [*formerly* 547.596] in 547.79

.6 ***Aromatic compounds**

Observe order of precedence in note under 547.01–547.08

*Add as instructed under 547

.02 *Halogenated compounds

.03 *Oxy and hydroxy compounds

.031 *Alcohols

.035 *Ethers

.036 Aldehydes and ketones

.037 *Acids

.038 *Esters

.04 Nitrogen compounds

.041 *Nitro and nitroso compounds

.042 Amines and amides

.043 *Azo compounds

.044 *Nitriles and isonitriles

.05 *Organometallic compounds

.053–.056 Specific organometallic compounds

Add to base number 547.05 the numbers following 546 in 546.38–546.68, e.g., organoaluminum compounds 547.05673; then add as instructed under 547, e.g., analytical chemistry of organoaluminum compounds 547.05673046; however, class organosilicon compounds in 547.08

.057 Other specific organometallic compounds

.057 1 *Organometallic compounds of nitrogen group

.057 15 *Organoarsenic

.057 16 *Organoantimony

.057 18 *Organobismuth

.057 2 *Organometallic compounds of oxygen group

.057 24 *Organoselenium

.057 26 *Organotellurium

.057 28 *Organopolonium

.06 *Sulfur compounds

.061 *Sulfites (Thioethers)

.063 *Hydrosulfites (Thioalcohols, mercaptans)

.064 *Thioacids

*Add as instructed under 547

615 *Dewey Decimal Classification*

**SUMMARY**

**615.1 Drugs (Materia medica)**
**.2 Inorganic drugs**
**.3 Organic drugs**
**.4 Practical pharmacy**
**.5 Therapeutics**
**.6 Methods of medication**
**.7 Pharmacodynamics**
**.8 Physical and other therapies**
**.9 Toxicology (Poisons and poisoning)**

.1 Drugs (Materia medica)

Class here pharmacology

Class drug therapy in 615.58

*For specific drugs and groups of drugs, see 615.2–615.3; practical pharmacy, 615.4; physiological and therapeutic action of drugs, 615.7*

.11 Pharmacopeias

Add "Areas" notation 4–9 from Table 2 to base number 615.11

.12 Dispensatories

Add "Areas" notation 4–9 from Table 2 to base number 615.12

.13 Formularies (Collected prescriptions)

Add "Areas" notation 3–9 from Table 2 to base number 615.13

.14 Posology

Prescription writing, dosage determination, incompatibilities

.18 Drug preservation technique

Proper packaging and other methods of preserving drug quality and potency

.19 Pharmaceutical chemistry

Manufacture, preparation, analysis of drugs

Use 615.19001–615.19009 for standard subdivisions

.190 1 Analysis

.190 15 Chemical analysis

.190 18 Assay methods

.191 Manufacture and preparation

856

▶ **615.2–615.3 Specific drugs and groups of drugs**

Pharmaceutical chemistry, preservation, general therapeutics

Class comprehensive works in 615.1, a specific drug or group of drugs affecting a specific system in 615.7

.2 **Inorganic drugs**

Add to base number 615.2 the numbers following 546 in 546.2–546.7, e.g., calomel 615.2663

.3 Organic drugs

.31 Synthetic drugs

Add to base number 615.31 the numbers following 547.0 in 547.01–547.08, e.g., sulfonamides 615.3167

Class a specific synthetic drug not provided for here with the drug, e.g., synthetic vitamins 615.328

.32 Drugs of vegetable origin

Class enzymes of vegetable origin in 615.35

.321 Pharmacognosy

Crude drugs and simples, alkaloids, herbals

*For drugs derived from specific plants, see 615.323–615.327*

.322 Drugs derived from bryophytes

Add to base number 615.322 the numbers following 588 in 588.1–588.3, e.g., drugs derived from Musci 615.3222

.323–.327 Drugs derived from specific plants

Add to base number 615.32 the numbers following 58 in 583–587, e.g., belladonna 615.32379

*For drugs devived from bryophytes, see 615.322; drugs devived from thallophytes, 615.329*

.328 Vitamins

.329 Drugs derived from thallophytes

Class here antibiotics

Add to base number 615.329 the numbers following 589 in 589.1–589.9, e.g., streptomycin 615.32992

.34 Fish-liver oils

.35 Enzymes

Pepsin, diastase, trypsin, chymotrypsin, papain

**641 Food and drink**

*For meals and table service, see 642*

.01 Philosophy and theory

.013 Gastronomy and pleasures of eating

SUMMARY

641.1 Applied nutrition
.2 Beverages (Drinks)
.3 Foods and foodstuffs
.4 Preservation and storage
.5 Cookery
.6 Cookery of and with specific materials
.7 Specific cookery processes and techniques
.8 Cookery of various specific kinds of composite dishes

.1 **Applied nutrition**

Study of nutrient constituents in foods in relation to body needs

Class nutritive values of specific foods and foodstuffs in 641.33–641.39, role of food in personal health in 613.2

.104 Special topics of general applicability

.104 2 Calories

Including calorie counters

.12 Proteins

.13 Carbohydrates

.14 Fats and oils

.16 Water

.17 Minerals

Elements and compounds

.18 Vitamins

.2 **Beverages (Drinks)**

Class here interdisciplinary works on production, manufacture, preservation, preparation, use

Class a specific aspect with the subject, e.g., manufacture (commercial preparation) 663

.21 Alcoholic beverages

*For wines, see 641.22; brewed and malted beverages, 641.23; distilled liquors, 641.25*

.22 Wines

.222 Grape wines

Add to base number 641.222 the numbers following 663.22 in 663.2204–663.224, e.g., champagne 641.2224

.23 Brewed and malted beverages

Class malt whiskies in 641.252

.25 Distilled liquors

.252 Grain whiskies

.253 Brandies

.255 Compound liquors

Distilled spirits flavored with various seeds, roots, leaves, flowers, fruits, e.g., cordials (liqueurs), gin

.259 Other

Mescal, tequila, rum, vodka, potato whisky

.26 Nonalcoholic beverages [*formerly* 641.3]

Class specific nonalcoholic beverages and kinds of beverages in 641.3

.3 **Foods and foodstuffs**

Class here interdisciplinary works on production, manufacture, preservation, preparation, use; comprehensive works on foods and beverages

Use 641.3001–641.3009 for standard subdivisions

Class comprehensive works on nonalcoholic beverages [*formerly* 641.3] in 641.26

Class a specific aspect with the subject, e.g., manufacture (commercial preparation and preservation) 663–664

*For beverages, see 641.2*

.302 Health foods

Including organically grown foods

▶ 641.303–641.309 Classes of foodstuffs by origin

Class comprehensive works in 641.3

.303 Vegetable

.306 Animal

.309 Mineral

.31 Selection and purchase

Principles of evaluation and guides to procurement of foods and foodstuffs to ensure nutritive value, quality, safety, economy

Class nutrition in 641.1, selection and purchase of specific foods and foodstuffs in 641.33–641.39

▶ 641.33–641.39 Specific foods and foodstuffs

Class comprehensive works in 641.3

.33 Foods and foodstuffs derived from field crops

.331 Cereals

Add to base number 641.331 the numbers following 633.1 in 633.11–633.18, e.g., rice and rice products 641.3318

.336 Sugars and syrups

.337 Alkaloidal products

Add to base number 641.337 the numbers following 633.7 in 633.72–633.78, e.g., chocolate 641.3374

.338 Seasonings and flavorings

Add to base number 641.338 the numbers following 633.8 in 633.82–633.84, e.g., vanilla 641.3382

.34 Fruits and nuts

Add to base number 641.34 the numbers following 634 in 634.1–634.8, e.g., apples 641.3411

.35 Vegetables

Add to base number 641.35 the numbers following 635 in 635.1–635.8, e.g., potatoes 641.3521

.36 Meats

Class here comprehensive works on meats and seafood

Add to base number 641.36 the numbers following 636 in 636.1–636.9, e.g., turkey 641.36592

Class game and seafood in 641.39

# 660 Chemical and related technologies

*For pharmaceutical chemistry, see 615.19; elastomers and elastomer products, 678*

.01–.03 Standard subdivisions

Notations from Table 1

.04 Chemical technologies of specific states of matter

Add to base number 660.04 the numbers following 530.4 in 530.41–530.44, e.g., plasma technology 660.044

Class industrial gases in 665.7

.05–.09 Standard subdivisions

Notations from Table 1

**.2 Chemical engineering**

*For industrial stoichiometry, see 660.7*

.28 The chemical plant and its work

.280 01 Philosophy and theory

.280 02 Miscellany

.280 028 Techniques, procedures, apparatus, equipment, materials

Class raw and auxiliary materials in 660.282, process equipment in 660.283

[.280 028 9] Safety measures

Do not use; class in 660.2804

.280 03–.280 09 Standard subdivisions

Notations from Table 1

.280 4 Safety technology

Safe engineering procedures; prevention and alleviation of explosions and other hazards

.280 7 Specific types of chemical plant

.280 71 Bench-scale

.280 72 Pilot

.280 73 Full-scale

---

► 660.281–660.283 General considerations

Class comprehensive works in 660.28, applications to specific processes in 660.284

.281 Process design, assembly, automation

Including process control

.282 Raw and auxiliary materials

.283 Process equipment

Instruments, apparatus, machinery

.284 Specific processes

.284 2 Unit operations Transport phenomena engineering

Unit operations: operations basically physical

Class here separation processes

.284 22 Crushing, grinding, screening

.284 23 Mass transfer

Including absorption, adsorption, gas chromatography

*For precipitation, filtration, solvent extraction, see 660.28424; fractional distillation, 660.28425*

.284 24 Precipitation, filtration, solvent extraction

.284 245 Filtration

.284 248 Solvent extraction

.284 25 Fractional distillation

.284 26 Evaporative and drying processes

.284 27 Heat transfer

Class a specific heat transfer process with the process, e.g., melting 660.284296

.284 29 Other

.284 292 Momentum transfer and fluidization

Including mixing

.284 293 Humidification

.284 296 Melting

.284 298 Crystallization

.284 4 Unit processes

Operations basically chemical

Add to base number 660.2844 the numbers following 547.2 in 547.21–547.29, e.g., hydrogenation 660.28443

.29 Applied physical chemistry

Add to base number 660.29 the numbers following 541.3 in 541.34–541.39, e.g., electrochemistry 660.297

.6 **Industrial biology**

.62 Industrial microbiology

.63 Industrial biochemistry

.7 **Industrial stoichiometry**

## 661 Technology of industrial chemicals (heavy chemicals)

Large-scale production of chemicals used as raw materials or reagents in manufacture of other products

Use 661.001–661.009 for standard subdivisions

Class industrial gases in 665.7

---

► 661.03–661.08 Compounds

Class comprehensive works in 661; acids, bases, salts in 661.2–661.6; organic compounds in 661.8

.03 Metallic compounds

Class metallic compounds other than those of alkali and alkaline earth metals in 661.04–661.07

.038–.039 Alkali and alkaline earth

Add to base number 661.03 the numbers following 546.3 in 546.38–546.39, e.g., sodium compounds 661.0382

.04–.07 Other compounds

Add to base number 661.0 the numbers following 546 in 546.4–546.7, e.g., sulfur compounds 661.0723

*For hydrogen compounds, see 661.08*

.622 Properties, tests, analysis

.622 09 Historical treatment

Class properties, tests, analysis of coal from specific places in 662.6229

.622 1–.622 5 Of specific types

Add to base number 662.622 the numbers following 553.2 in 553.21–553.25, e.g., analysis of bituminous coal 662.6224

.622 9 Of coal from specific places

Add "Areas" notation 1–9 from Table 2 to base number 662.6229

Class specific types of coal regardless of place in 662.6221–662.6225

.623 Treatment

Washing, coloring, drying, grinding, sizing, desulfuration

.624 Storage, transportation, distribution

.625 Uses

As a fuel, as a raw material

Class a specific use with the subject, e.g., metallurgical use 669.81

.65 Bagasse, briquettes, wood, sawdust

*For charcoal, see 662.74*

.66 Synthetic fuels

.662 Synthetic petroleum

.662 2 Bergius process

Production through hydrogenation and liquefaction of coal

.662 3 Fischer-Tropsch processes

Production through hydrogenation of carbonaceous gases

.666 Rocket fuels (Rocket propellants)

Liquid and solid

.669 Other liquid fuels

Fuel alcohols, benzenes from waste products

**.7 Coke and charcoal**

.72 Coke

.74 Charcoal

**.8 Other fuels**

*1148*

.82 Colloidal and mud fuels

.86 High-energy boron fuels

.9 **Nonfuel carbons**

.92 Graphite and graphite products

.93 Adsorbent carbons

Activated carbons, animal black, bone char, carbon black, adsorbent charcoals, decolorizing carbons, lampblack

**663 Beverage technology**

Manufacture (commercial preparation and preservation), packaging

Class household preparation of beverages in 641.87

SUMMARY



.1 **Alcoholic beverages**

*For wines and wine making, see 663.2; brewed and malted beverages, 663.3; distilled liquors, 663.5*

.11 Raw and auxiliary materials

.12 Preliminary preparations

.13 Fermentation

.14 Packing

.15 Refrigeration and pasteurization

.16 Distillation

.17 Aging

.19 Bottling

.2 **Wines and wine making**

Use 663.2001–663.2009 for standard subdivisions

.201–.209 General principles

Add to base number 663.20 the numbers following 663.1 in 663.11–663.19, e.g., fermentation 663.203

.22 Grape wines

Natural and fortified

.220 4 Special topics of general applicability

Add to base number 663.2204 the numbers following 663.1 in 663.11–663.19, e.g., fermentation of grape wine 663.22043

.222 White

.223 Red

Including rosé

.224 Sparkling

White and red

[.29] Wines from other fruits

Number discontinued; class in 663.2

**.3 Brewed and malted beverages**

Add to base number 663.3 the numbers following 663.1 in 663.11–663.19, e.g., fermentation 663.33

*For specific kinds of brewed and malted beverages, see 663.4*

**.4 Specific kinds of brewed and malted beverages**

Class malt whiskies in 663.52

.42 Beers and ales

.49 Sake and pulque

**.5 Distilled liquors**

Use 663.5001–663.5009 for standard subdivisions

.501–.509 General principles

Add to base number 663.50 the numbers following 663.1 in 663.11–663.19, e.g., distillation 663.506

.52 Grain whiskies

.53 Brandies

.55 Compound liquors

Distilled spirits flavored with various seeds, roots, leaves, flowers, fruits, e.g., gin, cordials (liqueurs)

.59 Other

Mescal, potato whisky, tequila, vodka, rum

**.6 Nonalcoholic beverages**

*For nonalcoholic brewed beverages, see 663.9; milk, 637.1*

.61 Potable water

Bottled and canned

.62 Mineralized and carbonated beverages

Artificial and natural

Class mineralized and carbonated water in 663.61

.63 Fruit and vegetable juices

Class fermented cider in 663.2

.64 Milk substitutes

Beverages and beverage constituents from soybeans, coconuts, malted cereals

.9 **Nonalcoholic brewed beverages**

Add to each subdivision identified by * as follows:
1 Raw and auxiliary materials
2 Preliminary preparations
3 Fermentation and oxidation
4 Firing, roasting, curing
5 Blending
7 Specific varieties
8 Concentrates
9 Packaging

.92 Cacao, cocoa, chocolate

.93 *Coffee

.94 *Tea

.96 Herb teas

Catnip, sassafras, maté, other aromatic and medicinal teas

.97 Coffee substitutes

Chicory, acorns, cereal preparations

**664 Food technology**

Manufacture (commercial preparation and preservation), packaging of edible products for human and animal consumption

Class here comprehensive works on processing food and beverages

Class household preservation, storage, cookery in 641.4–641.8

*For dairy and related technologies, see 637; beverage technology, 663*

.001 Philosophy and theory

.002 Miscellany

*Add as instructed under 663.9

[.002 8] Techniques, procedures, apparatus, equipment, materials

Do not use; class in 664.01–664.09

.003–.009 Standard subdivisions

Notations from Table 1

.01 Raw and auxiliary materials

.02 Processes

.022 Extraction

.023 Refining

.024 Manufacture

.028 Preservation techniques

.028 1 Preliminary treatment

.028 2 Canning

.028 4 Drying and dehydrating

.028 42 By slow, thermal processes

.028 43 Through pulverization and flaking

.028 45 By freeze-drying

.028 5 Low-temperature techniques

*For freeze-drying, see 664.02845*

.028 52 Cold storage

.028 53 Deep freezing

.028 6 Chemical preservation

Including brining, pickling, smoking

*For chemical preservation by use of additives, see 664.0287*

.028 7 Chemical preservation by use of additives

.028 8 Treatment with electromagnetic radiations (Irradiation)

.06 Additives

Production, properties, use

Including food colors

*For chemical preservation by use of additives, see 664.0287*

.07 Tests, analyses, quality controls

For texture, taste, odor, color, contaminants

Class tests, analyses, quality controls of additives in 664.06

.08 By-products

.09 Packaging and waste control

.092 Packaging

.096 Waste control

Class here pollution control

**SUMMARY**

**664.1 Sugars, syrups, their derived products**
**.2 Starches and jellying agents**
**.3 Fats and oils**
**.4 Food salts**
**.5 Other flavoring aids**
**.6 Special-purpose foods and aids**
**.7 Grains, other seeds, their derived products**
**.8 Fruits and vegetables**
**.9 Meats and allied foods**

**.1 Sugars, syrups, their derived products**

[.102 8] Techniques, procedures, apparatus, equipment, materials

Do not use; class in 664.11

---

► 664.11–664.13 Sugars and syrups

Class comprehensive works in 664.1

.11 General principles of sugars and syrups

Class general principles applied to specific sugars and syrups in 664.12–664.13

.111 Raw and auxiliary materials

.112 Preliminary preparations

.113 Extraction and purification

.114 Concentration

Production of syrups

.115 Crystallization

Production of sugars

.116 Additives

.117 Tests, analyses, quality controls

For texture, taste, color, contaminants

Class tests, analyses, quality controls of additives in 664.116

DEWEY

# Decimal Classification and Relative Index

*Devised by*

MELVIL DEWEY

*Edition 19*

*Edited under the direction of*
BENJAMIN A. CUSTER

Volume 3
**Relative Index**

FOREST PRESS
A Division of
Lake Placid Education Foundation

ALBANY, N.Y. 12206 U.S.A.
1979

*Dewey Decimal Classification*

| | |
|---|---|
| Amusement | |
| park bldgs. (continued) | |
| *other aspects see* Public structures | |
| parks | |
| landscape design | 712.5 |
| recreation | 791.068 |
| Amusements *see* Recreation | |
| Amvets | 369.186 2 |
| Amylases *see* Saccharolytic enzymes | |
| Ana *see* Quotations | |
| Anabantoidea *see* Acanthopterygii | |
| Anabaptist | |
| church bldgs. | |
| architecture | 726.584 3 |
| building | 690.658 43 |
| *other aspects see* Religious-purpose bldgs. | |
| churches | 284.3 |
| Christian life guides | 248.484 3 |
| doctrines | 230.43 |
| creeds | 238.43 |
| general councils | 262.543 |
| govt. & admin. | 262.043 |
| parishes | 254.043 |
| missions | 266.43 |
| moral theology | 241.044 3 |
| private prayers for | 242.804 3 |
| pub. worship | 264.043 |
| rel. associations | |
| adults | 267.184 3 |
| men | 267.244 3 |
| women | 267.444 3 |
| young adults | 267.624 3 |
| rel. instruction | 268.843 |
| rel. law | 262.984 3 |
| schools | 377.843 |
| secondary training | 207.124 3 |
| sermons | 252.043 |
| theological seminaries | 207.114 3 |
| *s.a. other spec. aspects* | |
| Anabaptists | |
| biog. & work | 284.3 |
| *s.a.* | *pers.*–243 |
| Anabolism *see* Metabolism | |
| Anacanthini *see* Acanthopterygii | |
| Anacardiaceae *see* Sapindales | |
| Anacardiaceous fruits | |
| agriculture | 634.44 |
| soc. & econ. aspects *see* Primary industries | |
| foods | 641.344 4 |
| preparation | |
| commercial | 664.804 44 |
| domestic | 641.644 4 |
| *other aspects see* Sapindales; *also* Fruits | |
| Anacardium occidentale *see* Cashew nuts | |
| Anaconda Mont. | *area*–786 87 |
| Anaerobic | |
| digestion sewage trmt. | |
| technology | 628.354 |
| *other aspects see* Sewage treatment | |
| respiration | |
| pathology | 574.212 8 |
| animals | 591.212 8 |
| plants | 581.212 8 |
| *s.a. spec. organisms* | |
| physiology | 574.128 |
| animals | 591.128 |
| microorganisms | 576.112 8 |
| plants | 581.128 |
| *s.a. spec. organisms* | |
| Anaesthetics | |
| pharmacodynamics | 615.781 |
| *other aspects see* Drugs | |
| Anagrams | |
| recreation | 793.73 |
| Analgesics | |
| pharmacodynamics | 615.783 |
| *other aspects see* Drugs | |
| Analogue | |
| computers | |
| electronic data proc. | 001.64 |
| spec. subj. | *s.s.*–028 54 |
| electronic eng. | 621.381 957 |
| instruments | |
| manufacturing | |
| soc. & econ. aspects | |
| *see* Secondary industries | |
| technology | 681.1 |
| *other aspects see* Computers | |
| Analogue-to-digital converters | |
| electronic eng. | 621.381 959 6 |
| *other aspects see* Computers | |
| Analogy | |
| logic | 169 |
| nat. rel. | 219 |
| Analysis | |
| commodities | |
| foods | |
| comm. proc. | 664.07 |
| *other aspects see* Foods | |
| *s.a. spec. foods e.g.* Meats | |
| minerals | 549.1 |
| spec. minerals | 549.2–.7 |
| pharmaceuticals | 615.190 1 |
| *misc. aspects see* Pharmacology | |
| *s.a. spec. drugs* | |
| engineering | 620.004 22 |
| aircraft eng. | 629.134 1 |
| parts | 629.134 3 |

44

*Relative Index*

## *Relative Index*

*Dewey Decimal Classification*

*Dewey Decimal Classification*

Petroleum (continued)
processing
soc. & econ. aspects *see* Secondary industries
technology 665.5
radiesthesia
parapsychology 133.323 7
resources
economics 333.823 2
govt. control
law 346.046 823 2
spec. jur. 346.3–.9
pub. admin. *see* Subsurface resources govt. control pub. admin.
*→ supply
law 343.092 6
spec. jur. 343.3–.9
synthetic
manufacturing
soc. & econ. aspects *see* Secondary industries
technology 662.662
tech. instruments & machinery
manufacturing
soc. & econ. aspects *see* Secondary industries
technology 681.766 5
*s.a. spec. kinds*
transportation 665.543
*s.a.* Oils; *also spec. uses*
Petroleum coke
chem. tech. 665.538 8
*other aspects see* Petroleum
Petroleum Co. Mont. *area*–786 28
Petroleum gas
chem. tech. 665.773
soc. & econ. aspects *see* Secondary industries
Petroleum-derived chemicals
chem. tech. 661.804
*other aspects see* Organic chemicals
Petrologists
biog. & work 552.009 2
*other aspects see* Scientists
*s.a.* *pers.*–552
Petrology 552
*misc. aspects see* Earth sciences
Petrosaviaceae *see* Alismatales
Pets
animal husbandry for 636.088 7
*s.a. spec. animals*
Petticoats *see* Underwear
Pettis Co. Mo. *area*–778 48
Petty officers
naval forces 359.338
*s.a. spec. naval forces*
Petunias *see* Solanales
Petuntse *see* Clays

Petworth West Sussex Eng. *area*–422 62
Pews *see* Seats ecclesiastical furniture
Pewsey Vale Wiltshire Eng. *area*–423 17
Pewter
arts decorative 739.533
*other aspects see* Tin
Phaeophyta
botany 589.45
med. aspects
gen. pharm. 615.329 45
toxicology 615.952 945
vet. pharm. 636.089 532 945
toxicology 636.089 595 294 5
paleobotany 561.93
*other aspects see* Plants
Phaëthontes *see* Pelecaniformes
Phalangers *see* Marsupialia
Phalanges
feet *see* Lower extremities bones
hands *see* Upper extremities bones
Phalangida
zoology 595.43
*other aspects see* Arachnida
Phalansterianism
socialist school
economics 335.23
Phalanxes
mil. organization 355.31
*s.a. spec. mil. branches*
Phalarideae *see* Pooideae
Phalaropes *see* Charadriiformes
Phallales *see* Lycoperdales
Phanerogamia *see* Spermatophyta
Phanerozonea *see* Asteroidea
Phantasms
occultism 133.14
Phantoms *see* Ghosts
Phareae *see* Pooideae
Pharisees
Judaism 296.812
*misc. aspects see* Judaism
Pharmaceutical
chemistry 615.19
*misc. aspects see* Pharmacology
*s.a. spec. drugs*
services
soc. welfare 362.1782
*misc. aspects see* Medical services
Pharmaceuticals *see* Drugs
Pharmacists
biog. & work 615.409 2
prof. duties & characteristics
law 344.041 6
spec. jur. 344.3–.9

*802*

* reserves
in nature 553.282
in storage 333.823211

Relative Index

Now you know something about how and why librarians use classification schemes to arrange their books. As I said above, the Dewey Decimal Classification we have been examining is the scheme in commonest use in many kinds of libraries throughout the world. But before we leave the subject I shall just introduce a couple more that you might come across.

If you discover that a library classifies its chemistry books with the symbol 'QD', then that library is using the Library of Congress Classification. This is used by a number of university libraries in the UK and, of course, more widely in the USA. All I am going to say about it here is that it is built up on the same encyclopaedic principles as the Dewey Decimal Classification, but it uses a mixture of letters and numerals for its more detailed subject classification; for example, QD 341.H9 signifies hydrocarbons, eg benzene, etc. If you do find yourself using a library which employs this scheme then you may look at it in greater detail in that library.

I have left the Universal Decimal Classification (UDC) till last. This is based initially on exactly the same scheme as the Dewey Decimal Classification (except that there is no class 400 in the UDC—language is treated with literature in class 8). But the UDC was devised for specialist collections and it covers science and technology especially in far greater detail. It is published in Europe, and is sponsored and published in the United Kingdom by the British Standards Institution. The UDC is used by a fair number of libraries in Europe; a very few academic libraries in the United Kingdom employ it (including my own college's library), but it is principally used in specialised libraries, and for arranging specialist indexes or lists. While its outline is more-or-less identical with the Dewey Decimal Classification, there are considerable differences in detail. I have included a few pages of the schedules to illustrate it: you may compare these with the Dewey Decimal Classification schedules I have shown you.

**539.389.3** MATHEMATICS AND NATURAL SCIENCES **54.08**

| | |
|---|---|
| 539.389.3 | After-effects. Ageing. Recovery. Accommodation |
| .4 | Hysteresis |
| .4 | **Strength. Resistance to stress** |
| | ·620.1 |
| **Special auxiliary subdivisions** | |
| 539.4.01 | Theory. Various influences on strength |
| .011 | Theory of stress resistance. Cohesion. Extension. Stretching. Elastic limit. Breaking point |
| .012 | Design theory. Permissible stresses |
| .013 | Influence of shape of a body on its strength. Notches. Corners. Waists, constrictions |
| .014 | Mechanical influences on strength. Tension. Initial stress. Inherent stress |
| .015 | Influence of structure on stress |
| .016 | Influence of treatment, working. Cold working. Heat treatment |
| .019 | Other influences on strength. Time. Dimensions |
| **Principal divisions** | |
| 539.41 | Resistance to elastic deformation. Cohesive strength |
| .411 | Resistance to compression. Crush resistance. Buckling resistance |
| .412 | Resistance to tension. Tear resistance |
| .413 | Resistance to flexion, bending |
| .414 | Resistance to torsion, twisting |
| .415 | Resistance to shear. Resistance to punching |
| .416 | Resistance to creasing, folding, bending |
| .417 | Resistance to drawing. Resistance to rolling |
| .419 | Resistance to complex stresses. Resistance to dropping. Resistance to tumbling |
| .42 | Breakage. Structural fracture. Tensile strength. Breaking strength |
| .43 | Resistance to repeated stresses |
| .431 | Resistance to alternating stresses. Fatigue resistance |
| .432 | Resistance to repeated complex stresses, e.g. in transmission shafts |
| .433 | Resistance to vibration, oscillation |
| .434 | Creep resistance. Long-time creep strength. Heat resistance. Resistance to quenching |
| .435 | Natural stress limit. Endurance strength |
| .5 | **Properties of materials affecting deformability** |
| | ·532.13; 539.214 |
| .51 | Malleability. Weldability |
| .52 | Ductility |
| .53 | Hardss. Softness. Hardness scales. Wear resistance |
| .54 | Hardenability |
| .55 | Toughness. Cleavability. Flexibility |
| .56 | Brittleness. Friability. Fragility |
| .57 | Consistency. Flow characteristics |
| .58 | Compressibility of solids |
| .6 | **Intermolecular forces** |
| | ·539.196 |
| .61 | Cohesion. Adhesion |
| .62 | Friction |
| | ·531.43 |
| .621 | Sliding friction |
| .622 | Rolling friction |
| .63 | Impact effects |
| .67 | Internal friction. Attenuability |
| .8 | **Other physico-mechanical effects** |
| .87 | Mechanical effects of physical phenomena |
| .89 | Effects of mechanical processes. Effects of pressure |
| .893 | Production of very high pressures. Hypercompression |
| 54 | **CHEMISTRY. MINERALOGICAL SCIENCES** |
| | ·550.4; 66 |
| **Special auxiliary subdivisions** | |
| 54-1 | State of substance |
| | ·62-403/-405 |
| -11 | State as regards occurrence and treatment |
| -112 | Natural. Native. Raw. Untreated |
| -114 | Man-made. Synthetic. Artificial. Prepared |
| 54-116 | Processed. Treated. Converted. Modified |
| -12 | State as regards atomic or molecular structure |
| -13 | Gaseous state |
| -134 | Vapour state |
| -138 | Colloidal state with gaseous continuous phase. Mists. Smokes. Aerosols |
| -139 | Supercritical fluid state |
| -14 | Liquid state |
| -145 | Solutions. Homogeneous liquid phase |
| -148 | Colloidal state with liquid continuous phase. Suspensions. Emulsions. Sols. Gels. Foams |
| -149 | Coarse suspensions |
| -16 | Solid phase |
| -161 | Amorphous state |
| -162 | Crystalline state |
| -165 | Solid solutions. Homogeneous solid phase |
| -168 | Colloidal state in solid medium |
| -18 | Special states as regards properties |
| -182 | Colloidal state in general |
| | ·541.182 |
| -183 | Adsorbed |
| | ·541.183 |
| -185 | Mixtures. Heterogeneous systems in more than one phase |
| -188 | Activated |
| -19 | Alloys in general |
| | ·54-165; 54-185 |
| -3 | Particular kinds of compound |
| -31 | Oxides |
| -32 | Acids |
| -36 | Bases. Hydroxides |
| -38 | Salts. Analogous compounds |
| -381 | Basic salts |
| -383 | Neutral salts |
| -384 | Acid salts |
| .1 | Primary acid salts |
| .2 | Secondary acid salts |
| .3 | Tertiary acid salts |
| -39 | Peroxy compounds |
| -4 | Chemicals. Reagents |
| -41 | Reagents in general |
| -42 | Standard solutions |
| -43 | Indicators |
| -44 | Catalysts |
| -45 | Buffers. Buffer solutions |
| -48 | Degree of purity of chemicals |
| 54.01 | Chemical substances and systems. Origin. Occurrence. Phases |
| .02 | Composition. Structure. Isotopes |
| .021 | Composition and formulas in general. Molecular formulas. Symbols etc. |
| .022 | Structure and arrangement of atoms in the molecule. Structural formulas |
| .024 | Radicals |
| .027 | Isotopes |
| | *For individual isotopes, see note 'Isotopes' under 546* |
| .03 | Physical properties and constants. Mechanical and physical effects |
| .04 | Chemical properties and constants |
| .05 | Production. Preparation. Isolation. Purification etc. |
| .057 | Synthesis |
| .06 | Analysis, investigation and handling in general |
| .061 | Qualitative investigation. Evidence. Identification experiments |
| .062 | Quantitative investigation. Determination of quantity |
| .063 | Micro-analysis. Semimicro-analysis |
| .064 | Trace analysis. Detection and determination of impurities |
| .066 | Investigation by means of specific reactions |
| .07 | Apparatus and equipment for preparation, investigation and analysis |
| .08 | Measurement principles, methods, techniques. Instrumentation |
| | 54.08 ≈ 53.08, e.g. |

165

| | |
|---|---|
| 542.9 (cont'd) | 542.9 $\simeq$ 66.09 |
| 543 | ANALYTICAL CHEMISTRY<br>*Chemical analysis of particular substances or classes of substance is denoted by colon, e.g.*<br>543:546.13 Analytical determination of chlorine<br>:637.12 Analysis of milk<br>→620.1 |
| **Special auxiliary subdivisions** | |
| 543.05 | Sampling. Preparation for analysis |
| .052 | Sampling procedure. Selection of sampling points. Sampling errors etc. |
| .053 | Technique and apparatus for sample collection. Sampling tubes. Sampling vessels, bottles etc. |
| .055 | Preparation of samples by dry treatment. Grinding etc. |
| .056 | Preparation of samples by wet treatment. Dissolving etc. |
| .06 | Analytical procedure<br>543.06 $\simeq$ 54.06 |
| .08 | Analytical measurement<br>543.08 $\simeq$ 53.08 |
| **Principal divisions** | |
| 543.2 | Special chemical methods of analysis |
| .21 | Separation and determination by wet methods. Wet assay. Gravimetric analysis |
| .211/.215 | Cation separation and determination |
| .217 | Anion separation and determination |
| .219 | Spot tests. Spot reactions |
| .22 | Separation and determination by dry methods |
| .224 | Bead reactions. Bead tests |
| .225 | Blowpipe analysis |
| .226 | Thermal analysis |
| .24 | Volumetric analysis. Titration. Titrimetric analysis |
| .241 | Alkalimetry. Acidimetry |
| .242 | Oxidation analysis. Reduction analysis |
| .243 | Precipitation using silver salts: silver nitrate etc. |
| .244 | Precipation using salts other than silver salts |
| .25 | Electroanalysis. Electrolytic methods. Electrometric methods |
| .251 | Electrolytic precipitation<br>→541.135; 621.3.035.2 |
| .252 | Electrolytic gas liberation. Measurement of gas released |
| .253 | Polarographic analysis |
| .257 | Electrometric titrations |
| .1 | Potentiometric analysis. Determination of hydrogen ion concentration. pH value<br>→541.132.3; 543.241; 631.415 |
| .2 | Standard electrodes. Quinhydrone. Calomel. Hydrogen. Oxygen. Glass etc.<br>→541.135.5; 542.8; 621.3.035.2 |
| .27 | Sampling and analysis of gas, including air |
| .271 | Apparatus and technique in general. Combined apparatus, trains. Gas detection papers |
| .272 | Selective absorption of gases. Determination of specific gaseous constituents. Absorption trains and units |
| .273 | Analysis by combustion methods |
| .274 | Other methods for determination of gaseous constituents. Gravimetric, titrimetric measurement of properties. |
| .275 | Determination and properties of non-gaseous constituents in suspensions, aerosols, smokes etc. Water content, moisture. Solid content, dust. Grit etc.<br>→533.275 |
| .278 | Titrimetric or gravimetric determination of gaseous decomposition products in analysis of non-gaseous substances |
| .279 | Gasometric analysis. Volumetric measurement of gaseous decomposition products of substance analysed |
| 543.3 | Water sampling and analysis<br>→628.1; 663.6 |
| .31 | Inorganic substances in water. Inorganic impurities in general |
| .32 | Hardness of water |
| .37 | Dissolved gas content of water |
| .38 | Organic substances in water. Organic impurities |
| .39 | Microbiological and bacteriological testing and analysis of water<br>→576.8 |
| .4 | Optical methods of analysis<br>→681.785 |
| .41 | Simple visual methods. Colour reactions. Flame coloration |
| .42 | Spectrum analysis. Spectroscopy. Spectrography. Spectrometry. Spectrophotometry. Fluorescence analysis |
| .43 | Colorimetric analysis. Nephelometric analysis |
| .45 | Refractometric analysis. Refractometry |
| .46 | Interferometric analysis. Interferometry |
| .47 | Polarimetric analysis. Polarimetry |
| .5 | Physicochemical methods of analysis<br>*except optical methods,* → *543.4* |
| .51 | Positive ray analysis. Mass spectrometry |
| .52 | Radiometric analysis |
| .53 | Radiochemical methods of analysis. Activation analysis (analysis with radiations exciting radioactivity or nuclear reactions) |
| .54 | Capillary analysis. Adsorption analysis. Stalagmometry. Chromatography. Electrophoresis |
| .7 | Analysis of inorganic substances |
| .71 | Water content, moisture. Volatile constituents |
| .72 | Soluble constituents. Insoluble constituents |
| .73 | Inorganic impurities |
| .74 | Organic impurities |
| .8 | Analysis of organic substances |
| .81 | Water content, moisture. Volatile constituents |
| .82 | Analysis by heating to high temperatures |
| .83 | Soluble constituents |
| .84 | Ultimate analysis |
| .85 | Determination of fat content. Ester value. Saponification value. Determination of oxygen compounds |
| .86 | Determination of content of other substances. Determination of proteins, enzymes, vitamins etc. |
| .87 | Oxidizability. Saturation. Explosiveness. Stability. Reactivity |
| .9 | Analysis by means of biological reactions |
| .92 | Sensory reactions. Organoleptic testing. Characteristic smell, taste, texture<br>→541.69 |

546 INORGANIC CHEMISTRY

**Classification of compounds**

Most chemical compounds are to be considered as the result of an actual or hypothetical reaction of two oxides with each other, or of an oxide with a hydride. In this way, a chemical compound may be designated by the oxide or hydride radicals from which it is derived.

The chemical compound is denoted by means of the apostrophe. A binary compound is denoted by citing the number for the metal radical (cation), and adding the number for the acid radical (anion). In the second number, the digits 546 are replaced by the apostrophe. Thus:

546.561'131 Cupric chloride CuCl

from

546.561 Monovalent copper
546.131 Chlorides

Chemical compounds containing several metallic radicals (cations) or acid radicals (anions) are denoted by citing the numerically highest of the numbers for the metals, and adding the numbers for the other metals and the numbers for the acids in reverse numerical order. Thus:

546.623'32'226 Potassium aluminium sulphate

from

546.623 Trivalent aluminium

# 66 CHEMICAL TECHNOLOGY. CHEMICAL AND RELATED INDUSTRIES

→54; 550.4; 550.84; 577.1; 621.794; 623.459; 631.4

**Special auxiliary subdivisions**

*The special auxiliaries -2/-9 (not -1) listed under 62 are also applicable in 66, except for -3 under 661.185 and 662.2. -9 is further developed here. For -1, see specific sections 661.725-11/-12; 661.8-11/-18; 664.71-11/-12; 666.3-1; 666.9-1; 667-12; 669-1; 669.15-192/-198; 669.35-191/-193*

| | |
|---|---|
| 66-2 | Fixed and movable components of machines etc. in chemical technology |
| -4 | State, condition, form of materials etc. in chemical technology |
| -5 | Operation and control of machines and processes in chemical technology |
| -6 | Chemical machinery etc. according to fuels, heat sources |
| -7 | Servicing, maintenance, protection of chemical machinery etc. |
| -8 | Machines etc. according to motive power, propulsive force. Source of energy of chemical machinery |
| -9 | Process and plant operating characteristics |
| -91 | State of material (characteristics and variables) |
| -911 | State of aggregation<br>→62-403/-405<br>66-911.3/.6 ≃ 54-13/-16, e.g.<br>66-911.3 Gaseous state (from 54-13)<br>.4 Liquid (from 54-14)<br>.45 Solutions<br>.48 Colloidal state with liquid continuous phase. Suspensions. Emulsions. Sols. Gels. Foams<br>.6 Solid (from 54-16)<br>.61 Amorphous<br>.62 Crystalline |
| -912 | Processes in the fluidized state |
| -913 | State of humidity |
| .1 | Dry |
| .2 | Damp. Humid. Moist |
| .3 | Wet |
| -914 | Concentration characteristics<br>→66.082 |
| .3 | Low concentration. Dilute |
| .5 | Medium concentration |
| .7 | High concentration. Concentrated |
| -915 | pH value<br>→541.132.3 |
| .1 | Acid |
| .2 | Neutral |
| .3 | Alkaline |
| -916 | rH value |
| .1 | Oxidizing |
| .2 | Reducing |
| -93 | Processes, machines and equipment according to mode of working, continuity, associations, media |
| -932 | According to continuity |
| .2 | Continuous |
| .3 | Semi-continuous |
| .4 | Discontinuous |
| -936 | Processes according to ambient or phase conditions |
| .1 | Processes in the dry |
| .2 | Processes in the wet |
| .3 | Reactions between gases and vapours |
| .4 | Reactions between gases and liquids |
| .5 | Reactions between liquids |
| .6 | Reactions between gases and solids |
| .7 | Reactions between liquids and solids |
| .8 | Reactions between solids |
| 66-94 | Direction, velocity, rate, duration of processes etc. |
| -942 | Parallel processes |
| -944 | Opposed, reversed processes. Counterflow processes |
| -946.1 | Processes without agitation |
| .3 | Processes with agitation |
| -947.1 | Very rapid |
| .2 | Rapid |
| .3 | Accelerated |
| .5 | Brief. Short duration |
| -948.1 | Slow |
| .2 | Prolonged. Extended |
| .3 | Retarded. Delayed |
| .5 | Lengthy. Long-lasting |
| -95 | Simple and multiple processes. Single-stage and multi-stage processes |
| -951 | Simple, simplified processes |
| -952 | Multiple, complex, repeated processes |
| -953 | Single-stage processes |
| -954 | Multi-stage processes |
| -956 | Single-bath processes |
| -957 | Multi-bath processes |
| -96 | Processes associated with purification and refining |
| -962 | Processes without purification or refining |
| -963 | Processes with purification or refining |
| -965 | Particulars of processes according to the form or condition of the material to be treated |
| .1 | Loose. Flocculent. In the form of flakes |
| .2 | Coiled. In a coil |
| .3 | In a hank or skein |
| .4 | As ribbon or tape |
| .61 | Stressed. Under tension |
| .62 | Not stressed. Not under tension. Slack. Limp |
| .81 | Open. Uncovered |
| .82 | Closed. Covered |
| -966 | Processes according to durability (permanence, fastness) |
| .1 | Durable (e.g. wash-resistant) |
| .2 | Not durable |
| .5 | Fast (colourfast) |
| -967 | Processes according to liquor ratio |
| .1 | Low liquor ratio |
| .2 | Medium liquor ratio |
| .3 | High liquor ratio |
| -97 | Thermal characteristics. Temperature. Temperature range<br>*Denote specific scale and value by abbreviation and figure, following an asterisk (→ Table I(h)), e.g.*<br>66-97*C0 *F32 *K273 Freezing point of water = 0°C = 32°F = 273K<br>66-97*C100 *F212 *K373 Boiling point of water = 100°C = 212°F = 373K |
| -971 | Thermodynamic characteristics |
| .2 | Exothermic |
| .4 | Endothermic |
| -973 | Very low temperature |
| -974 | Low temperature |
| -975 | Room temperature |
| -976 | Medium, moderate temperature |
| -977 | High temperature |
| -978 | Very high, extremely high temperature |
| -98 | Pressure. Pressure range<br>→66.083 |
| -982 | Vacuum. In vacuo |
| -983 | Very low pressure |
| -984 | Low pressure |
| -985 | Atmospheric pressure. Normal pressure |
| -986 | Medium, moderate pressure |
| -987 | High pressure. Hyperbaric |
| -988 | Very high, extremely high pressure |

662.959.2 Furnaces for combustion of both gaseous and liquid fuels
.3 Furnaces for combustion of both gaseous and solid fuels
.4 Special devices for burners in steam generators with internal combustion
→621.181.2
.5 Gaseous-fuel furnaces with flue gas recirculation. Drum gas heaters
.6 Gaseous-fuel furnaces for flameless surface combustion
.96 Smoke consumption. Smoke abatement. Treatment of combustion products
→628.511 and references; 662.613.5; 662.8.056; 662.842
.963 Purification of emissions by washing, scrubbing
→66.074.5
.964 Devices for separating and solid and incandescent particles in smoke
.965 Precipitation from smoke
.966 Methods of smoke prevention (smoke abatement). Smoke combustion. Smoke consumption (in the strict sense). Smoke dispellents
→662.892
.98 Heating of apparatus and utensils. Various heating methods
→542.4; 621.365; 621.745; 644.1; 683.9
.981 Solid-fuel heating appliances. Solid-fuel ovens, stoves, hotplates etc.
.982 Liquid-fuel heating appliances. Liquid-fuel burners, stoves etc. Spirit burners
.983 Gas-fired heating appliances. Gas burners, rings etc. Bunsen burners
.986 Heating by direct contact. Heating by injection of hot gas, air, steam. Heating by contact with hot liquids or solids
.987 Indirect heating. Heat transfer media. Heat exchangers. Baths of hot water, molten salts, molten metals etc.
→536.24
.989 Heating by means of chemical reactions (heat of reaction)
.99 Heat recovery. Use of natural heat. Prevention of heat loss. Insulation
→62-68; 620.97; 621.186.8; 621.577; 662.957
.992.8 Combustion control and testing
.82 Control and testing in the combustion chamber
.84 Control and testing of combustion gas, flue gas
.994 Utilization of waste gas
.995 Heat storage in general
→662.925; 662.957
.997 Utilization of heat from natural phenomena. Use of geothermal energy. Volcanic heat. Hot springs, geysers. Solar heat
→620.91; 621.472; 697.329
.998 Industrial heat insulation and insulators. Heat loss
→66.043.2; 662.614; 662.926

663 **INDUSTRIAL MICROBIOLOGY. INDUSTRIAL MYCOLOGY. ZYMURGY, FERMENTATION INDUSTRY. BEVERAGE INDUSTRY. STIMULANT INDUSTRY**
→178; 351.761; 579; 613.3; 613.8; 641.87

**Special auxiliary subdivisions**

*The .03 auxiliaries are applicable only in 663.1/.5*

663.031 Raw materials, secondary materials and their preparation. Milling. Grinding. Crushing. Water treatment. Steam treatment
.1 Ferments
663.031.1 ≃ 663.1
.2/.4 Raw materials
663.031.2/.4 ≃ 663.14.031.2/.4
663.031.8 Liquors forming bases of blended drinks, e.g. sugar syrups
→663.812; 664.14
.032 Equipment for preparation of mash and wort (for yeast processes and other processes)
.033 Fermentation vessels and accessories
.035 Extraction and pressing of yeast
.036 Preservation and storage of yeast
.05 Additions. Preservation. Aftertreatment
*The .05 auxiliaries are not applicable at 663.91*
.051 Additions for improvement of flavour. Aromatic supplements
.052 Additions for improvement of appearance or physical properties (e.g. viscosity, foaming capacity)
→667.777
.053 Preservation methods for beverages
666.053 ≃ 664.8.03
.054 Diluents for beverages
.057 Methods and equipment for infusing beverages with carbon dioxide. Carbonation, aeration of beverages
.058 Various aftertreatments
.2 Natural maturing, ageing of beverages. Cellaring. Means of preservation in cellars
.3 Artificial maturing, ageing of beverages
.4 Mixing of beverages
.059 Bottling and final treatment before consumption
→683.5

**Principal divisions**

663.1 MICROBIOLOGICAL INDUSTRIES. SCIENCE AND TECHNIQUE OF APPLIED MICROBIOLOGY. APPLIED MYCOLOGY
→577.15; 579.6; 582.28; 637.146; 66.098
.12 Yeasts in general. Brewing yeast. Ascending and descending yeasts
→664.872
.13 Pure yeast. Equipment for yeast culture. Pure culture apparatus
.14 Fermentation of yeast. Baker's yeast. Pressed yeast. Nutritional yeast
663.14:636.087.2 Yeast as animal fodder

**Special auxiliary subdivisions**

663.14.031 Raw materials (and their treatment)
.2 Carbon compounds
.3 Nitrogen compounds
.4 Nutrient salts (except nitrogen salts) and acids (either as additives or produced by fermentation)
.036 Preservation of yeast and fermentation liquors
.038 Processes and equipment for treatment of yeast
.3 Aftertreatment to increase fermentative capacity and other properties
.5 Defects of yeast, and means of improvement. Flocculation
.7 Conversion of brewer's and distiller's yeast into baker's yeast
.039.3 Factors affecting fermentation (e.g. concentration, acidity, pressure, temperature
.4 Foam, froth formation (and measures affecting it)

**Principal divisions**

663.142 Fermentation processes without aeration. Viennese method
.143 Fermentation processes with aeration
.15 Industrial fermentation processes. Production of enzymes (other than by yeast)
663.15 ≃ 577.15
663.15:661.715 Production of hydrocarbons by fermentation
.16 Vitamin production by fermentation
663.16 ≃ 577.16

To start with, the UDC does not use zeros in the same way (so science is 5, not 500, and chemistry 54, not 540). But the major difference which will strike you is the use of various typographical symbols for different purposes. Do you remember the list of reference books at the end of Section 2.3? Have another look at it now and you can see a list of typical, fairly simple, UDC classification numbers. Notice the solidus (/) in item 6, the dash (–) in 21, and, most important, the colons (:) in 9 and 10. These are not all the symbols used, but to keep this as simple as I can, I am going to tell you only about the use of the colon, which is the commonest and most important of these symbols in the UDC, and then just mention some of the others. If you remember, I pointed out that one of the problems with the Dewey Decimal Classification is that the same subject has to be accomodated in different areas for various applications. A solution to this would be to express each subject once only and allow the librarian to bring numbers together when necessary from any part of the scheme: this is what UDC does, mainly with the colon. Look again at items 9 and 10 in that figure: 542 expresses 'practical laboratory chemistry' and 614.8 expresses 'hazards, accident prevention; etc. Bring the two numbers together with a colon and we have neatly expressed the subject of these two books (hazards in the chemical laboratory). We have chosen to shelve the books in the chemistry section, but inverted entries (614.8:542) in the classified catalogue (see the next Section 3.1.5 for a more detailed examination of these) will bring together at 614.8: in the medicine section all the works on occupational hazards, whether in chemical, medical, engineering, or any other kind of laboratory or environment. And we can do the same with any other subjects: computers in chemistry would be 54:681.3; spectroscopic analysis in the food industry would be 543.42:664 (or 664:543.42 if you preferred to shelve the item under food technology). This is of course, particulary suited to more specialised libraries. If you have only a few chemical books in a general collection they can all be quickly found and examined; but if you have a lot of documents on practical laboratory chemistry then you don't want to have to sift through too many at the simple number 542.

I said I would just mention the other symbols used by the UDC. Some of these are (in their arbitrary filing order):

+ joining two discrete subjects (eg 543.42 + 543.52, 'spectroscopy and radiometric analysis'),

/ joining consecutive numbers (eg 53/54, 'physics and chemistry'),

: we have looked at this above, expressing a relationship between subjects,

(O) documents in a particular form (eg 543(03), 'reference work on analytical chemistry'),

(1/9) documents relating to a particular place (eg 543(430), 'analytical chemistry in Germany'),

“..” documents relating to a particular time (eg 543“18”, ‘analytical chemistry in the 19th Century’),

– special auxiliary sub-divisions as indicated in the tables at particular subjects.

So the advantage of the UDC for specialist collections is fairly obvious: the librarian can sub-divide any part of the collection in a high degree of detail to cope with large amounts of material on particular subjects. There is a price to be paid, of course. The numbers can become very long and excessively complex, but used with discretion this can be a very effective classification scheme to meet particular requirements.

Let’s just complete this Section with a simple question.

**SAQ 3.1d**

Look at the following three classification numbers and write down the classification scheme to which each one belongs:

(*a*) 543.089 4

(*b*) QD 330

(*c*) 663.14.031.3:543.053

**SAQ 3.1d**

### 3.1.5. Use of Library Catalogues

Whenever you use a lending library you must remember that a significant proportion of the books will be out on loan at any one time. We have already seen another problem in relying on what you find on the shelf when looking for books: each book can be shelved in only one place, but might have relevance to more than one subject. One answer to these problems is to consult the library catalogue: and it is usually the simplest and most effective answer. If a library can be compared with a reference book, then the catalogue is its index. Unless you are very experienced and know exactly what you want and where it is, never look up anything in a reference book without starting at the index; likewise never look for books or information in a library without starting at the catalogue.

Library catalogues are changing. Until recently you would have found most catalogues in the form of card indexes, and these remain a very common form of library catalogue. But as computers are used more and more commonly to produce library catalogues, the common form of print-out is increasingly the microfiche. Now an increasing number of libraries (of all types) are installing public computer terminals for their users to gain direct on-line access in order to interrogate their catalogue data bases. Whatever form of catalogue you find, the library concerned should give you instruction in how to handle any machinery you might need (microfiche readers or computer terminals); and the principles of using a library catalogue remain the same, no matter what form the catalogue takes.

Let's start with the easy bit. You want to know if the library has a copy of the second edition of Lever's *Inorganic electronic spectroscopy*. You go to the author catalogue and look up 'Lever' and find, in a card catalogue, an entry looking something like this:

LEVER, A. B. P.
Inorganic electronic spectroscopy.
2nd ed. Elsevier, 1984. 863pp.

Shelved at: 543.42

Some of the details and presentation will vary from library to library, but that is the kind of entry you will find in most card catalogues. If you were checking a computer-produced catalogue on a microfiche, your entry would look much like this:

LEVER, A. B. P
Inorganic electronic spectroscopy /
A.B.P Lever –2nd ed. –
Amsterdam; Oxford: Elsevier, 1984-xvi, 863pp –
(Studies in physical and theoretical
chemistry; 33)
Previous ed.: 1968 0-444-452389-3
Shelved at: 543.42

As you can see, this carries much more information about the book, but there are still some twenty entries to most frames on each microfiche in the catalogue. This makes browsing in the catalogue much easier than with a card catalogue. The computer takes a standard centrally produced description of the book and inserts it into the library catalogue data base under any headings selected by the librarian. The information about the book in this entry tells you the title, author, edition, publication details, number of pages, the name of any publisher's series in which the book is issued, a note of when the first edition was published, an international standard book number (ISBN, used as a control number within the data base) and whereabouts in the library the book is shelved. A few card catalogues have entries under the titles of the books as well as their authors; the majority of microfiche catalogues will have these title entries in addition to the author entries. So, if you did not know the author's name you could look under 'Inorganic electronic spectroscopy' and still find the entry.

That was a fairly straightforward example. Inevitably, though, some complications arise. What if there is more than one author? Usually you will find a second entry under the name of a second, or even third author; after three authors for one book librarians usually stop putting in any more entries. In some card catalogues you might find a reference from a second author to the first. In a microfiche catalogue a standard descriptive entry for the book will be found under whatever headings are used. For example:

> FRITZ, James S.
> Quantitative analytical chemistry /
> James S. Fritz, George H. Schenk.
> - 3rd ed. Boston; London: Allyn
> and Bacon, 1974. - xiv, 689p.
> Previous ed.: 1969 0-205-04203-1
> Shelved at: 543.062

There will be an identical entry with the heading 'Fritz, James S.' replaced by 'Schenk, George H.' and another under the title as well. What if there is no author indicated? There will then be an entry under the title, and there might be an entry under the name of an organisation responsible for the document (provided it is *not* just a commercial publishing firm). So, if you want to check whether the library has a copy of any particular book you are seeking, then look in the alphabetical author catalogue under the author's name, or under the title if the library has an alphabetical title catalogue as well.

Before moving on, try this question:

**SAQ 3.1e**

Write down the name words you would look up in the catalogue to find works by the following authors or for which the following bodies are responsible:

(*a*) Knut Schmidt-Nielsen,

(*b*) The Chemical Society's Continuing Education Commitee,

(*c*) Richard T. O'Connell,

(*d*) Anthony Hartley De Borde,

(*e*) The United Kingdom Department of Industry's Microprocessor Applications Project,

(*f*) Glasgow University's Department of Chemistry.

**SAQ 3.1e**

That is fine if you know what book you are looking for. But what if you just want information on mass spectrometry, or any other subject in which you might be interested? You will want to know what books the library has (many of which are likely to be missing from the shelves in a busy lending library), and perhaps there might be a tape-slide or videocassette lecture programme, if your library keeps that kind of material. You might want to know what books the library has on the applications of mass spectrometry, and which might, of course, be shelved under these applications in different sections of the library. Well, as I said before, a book can be shelved in only one place, but we can put as many entries as we like in a catalogue.

Subject catalogues are found in one of two types: alphabetical or classified. Alphabetical subject catalogues use subject headings often similar to those you would find in book indexes, and with all their good and bad points. You can look directly under a subject you want, but related subjects will be scattered by the alphabet. Also, if the subject you want is not listed you might have to think of a more general heading (although a good alphabetical subject catalogue should contain plenty of cross-references to guide you from one term to another). The other form of catalogue is to have subject entries arranged by their classification numbers. The advantage is that related subjects will generally be brought together by the scheme and the catalogue will largely (but not completely due to the extra entries) follow the shelf arrangement of the library. The disadvantage is that you have to convert your subject from words into unfamiliar classification numbers before you can look it up. Even with good indexes to the classification scheme (and by no means all libraries have these) this can sometimes be a complex process. But there can be considerable rewards in terms of flexibility in retrieving information. This is particularly true when one uses a classification scheme like the Universal Decimal Classification.

Take the book by Lever which I gave as an example earlier. Well, in my library we have classified it at 543.42 : 546 (spectroscopy: inorganic chemistry) with another entry at 535.33 (emission spectra). The book is shelved at 543.42. But because of the extra entries in the classified catalogue at 546.543.42 and 535.33, the book should also be found in the catalogue by those interested in inorganic chemistry or the physics of emission spectra.

So another final moral. Get to know the layout, the strengths and the weaknesses of any library you are going to use. Whenever you come to look for documents within that library always start at the catalogue, just as you would start at the index to a reference book. That way you will be able to find out whether the library holds any particular item you are seeking, or all that it holds on the subject you are interested in, irrespective of what does or does not happen to be on the shelves at the time of your visit.

Now you know how to find what is in your particular library, let's see where we have got to before moving on to the final step of discovering all that has been published by a particular author or by any authors on your particular subject. Move on to section 3.2.1 and we will summarise the impressive progress you have already made.

**Objectives**

When you have worked through this Section you should be able to:

- recognise the different types of libraries and how they might be of use to you;
- make the best use of the different services libraries can offer you;
- request a book (or other document) from another library through the inter-library lending system with the best chance of success;
- find your way around any library to which you have access;
- find the books you want in a library by using the author/title or subject catalogue.

## 3.2. ABSTRACTING/INDEXING SERVICES

### 3.2.1. Introduction

Let's just pause and take our breath a moment. What have we achieved so far in this Unit? What do we know and what can we do? Then, what do we still need to know and do? Well, we know that there is an enormous amount of potentially relevant literature being published. This literature appears in different forms and we can now recognise these forms and appreciate their importance in relation to one another. We know that we can obtain a lot of useful information quite conveniently by consulting published standards and major reference books; we know some useful titles and have practised using them. In particular, we know how important indexes are and have some useful experience in their good and bad points. We recognise what kind of information and publications are likely to be kept and where, and we can now use with confidence any library to which we have access. So, what is left?

If you check a library catalogue you will find that it lists all the documents that library holds. But libraries do not include individual journal papers in their catalogues (except for a very few very specialised collections), and no library (except a very few major national libraries) can afford to buy all the books and other documents of potential usefulness to their customers. That is why a sophisticated inter-library lending network has been set up. So, when you have exhausted the resources of your own library, or if its collection cannot cater for the details you need in your specific subject, there is one more huge step for you to take: quite simply to try to find everything published on your special subject. You have to know what has been published before you can start to select material. It is this step that we are now going to take.

### 3.2.2. Bibliographies, Guides, and General Indexes

Although I could overwhelm you with precise definitions and myriads of examples, for our present purposes we shall take a very generalised and simplistic definition of a bibliography. Quite simply, let us assume that a bibliography is a list of references to documents (of any form—journal papers, books, reports, etc) compiled for a specific purpose. A bibliography might therefore appear as a list of recommended reading items in a textbook, a list of references in a review paper, a regular index to new publications (although we shall deal with these more as indexing services soon) or a separately published book or report. Bibliographies might include material selected as recommended reading, or material selected as illustrating a particular subject; or they might aim to be comprehensive uncritical lists of all available relevant material. The obvious advantage of a comprehensive bibliography on your subject is that it will save you a lot of search-time and effort. But bibliographies in analytical chemistry are not very common (except as lists of references to review papers) and they can quickly go out of date. On a very general level, however, there is a regular national bibliography of all the books published in the UK which you might sometimes need to use, if you are trying to trace details of any particular book, or of books on any particular subject. This is:

> *British National Bibliography.* London:
> British Library, Bibliographic Services Division,
> 1950 to date. Weekly, with quarterly and annual cumulations.

This is a list of all the new British books and pamphlets received in the Copyright Office of the British Library (so effectively a list of all new British books). It is arranged by use of the Dewey Decimal Classification (so you would know how to look up analytical chemistry, wouldn't you?) and has an index of authors, titles, and publishers' series. There is a monthly alphabetical subject index and, of course, all the indexes cumulate into the annual volumes.

Guides to the literature are usually published in book form, although some libraries issue short publications giving guides to particular areas of or items within their stock, which can also be considered as literature guides. They give selective descriptions and/or evaluations of documents of different kinds on fairly broad subjects. They might be descriptive texts with full references to sources in notes or separate lists; or they might be systematic lists, similar to bibliographies, but with more detailed descriptive or evaluative notes; or they might combine both approaches. I have already mentioned three on chemistry in my introduction to this Unit. Do you remember the books by Mellon, by Bottle, and by Maizell? These cover all the issues I deal with in this Unit, but over the whole field of chemistry. Obviously they don't go into the same detail that we have here, but for a wider view of chemical literature or an introduction to documents on non-analytical subjects you will find them most useful. If you want to look at an even wider field you could turn to either or both of the following which cover the major literature of the whole of science and engineering:

> Chen, Ching-Chih. *Scientific and technical information sources.* Cambridge (Mass): MIT Press, 1977, 519pp.
>
> Malinowsky, H. Robert and Richardson, Jeane M. *Science and engineering literature: a guide to reference sources.* 3rd ed. Littleton (Col): Libraries Unlimited, 1980. 342pp.

All this can be very useful in its own way, but what about systematic searches of all the available chemical or analytical literature? I shall come to that in just a moment; but before I do, here are a couple of general indexing services which cover journal articles in all fields of science and technology. The idea of these is to provide a comprehensive index to articles appearing in a wide range of journals; obviously they will save you a lot of time and effort in searching through journals and can indicate useful material in sources which might never have occurred to you. The major British service is:

> *Current Technology Index.* London: Library Association, 1981 to date. Monthly, with annual cumulations. From 1962 to 1980 published as *British Technology Index*.

This is an alphabetical subject index of the main articles appearing in British technical journals. It covers all branches of engineering and chemical technology, and includes material on the pure science of man-made objects and industrial processes, the chemistry of individual substances, and instruments. There is no evaluation in the selection of articles to be indexed. It is intended as an index for tracing journal articles on highly specific topics: the headings are therefore very detailed and are made co-extensive with the subjects of the articles. Let's just take a closer look at this: it will give us an interesting and useful introduction to indexing services and emphasise one or two worthwhile points. So try this question.

**SAQ 3.2a**

When I set you some of the first of these practical SAQs I promised you that we would find ourselves in many varied laboratories and types of employment. Well, let's go the whole hog and find ourselves analysing extra-terrestrial materials.

You will be working eventually on the analysis of samples returned from the planet Mars. You decide that initially you want a fairly broad introduction to the obviously special problems involved, and so you turn to *Current Technology Index*. Quite simply, go to the 1983 annual volume and see if you can find a likely paper. When you find one write down its details as a standard bibliographic reference as I showed you in the Introduction (ie Author, title of paper, title of journal, volume, date and pages).

I haven't had to say this for a little while, have I? But here we go again: *Current Technology Index* is widely available in libraries of all kinds in the United Kingdom, so try to use a copy of the book. Use the copies on the next few pages only if you are quite unable to find the book itself.

# CURRENT TECHNOLOGY INDEX

ANNUAL VOLUME 1983

LIBRARY ASSOCIATION PUBLISHING
7 RIDGMOUNT STREET, LONDON WC1E 7AE
1984

**AMPLITUDE** (Contd.)
*See*
Amplifiers, Gyrotron, Travelling wave : Conical—Cylindrical junctions : Amplitude
Amplifiers, Transistor : Distortion, Non-linear : Effect of maximum output amplitude
Film : Acrylic plastics, Thermosetting : Mechanical properties : Measurement : Vibrations : Amplitude
Shift work : Circadian rhythms : Amplitude
Ultrasonics, Industrial : Equipment : Testing : Vibrations : Amplitude
Vibrations, Non-linear, Damped : Amplitude
Waveforms, Sinusoidal, Variable period : Amplitude

AMPLITUDE CONSTRAINED DISCONTINUOUS MODEL REFERENCE ADAPTIVE CONTROL SYSTEMS
*See*
Control systems, Adaptive, Model reference, Discontinuous, Amplitude constrained

AMPLITUDE DISTORTION
*See*
Filters (Digital), Recursive : Digital to analogue converters : Distortion, Amplitude

AMPLITUDE MODULATED HOMODYNE NETWORK ANALYSERS
*See*
Network analysers, Homodyne, Amplitude modulated

AMPLITUDE MODULATED STRIPE CONTACT GALLIUM ARSENIDE—GALLIUM ALUMINIUM ARSENIDE LASERS
*See*
Lasers : Gallium arsenide—Gallium aluminium arsenide, Stripe contact, Amplitude modulated

AMPLITUDE MODULATED SYNCHRONOUS DATA TRANSMISSION
*See*
Data transmission, Synchronous, Amplitude modulated

AMPLITUDE MODULATION DATA TRANSMISSION
*See*
Data transmission, Amplitude modulation

AMPOULES
*See*
Parenteral drugs : Ampoules

AMPUTEES, Upper limb
*See*
Work, Manual : Amputees, Upper limb

α-AMYLASE
*See also*
Brewing : Mashing : α-Amylase
Malt : Barley : α-Amylase

**β-AMYLASE**
**Production : Bacillus subtilis**
Production and purification of a maltose-producing amylase from *Bacillus subtilis* IMD 198. W.M. Fogarty & E.J. Bourke. *J. Chem. Technol. Biotechnol.*, 33B (Sep 83) p.145-54

AMYLASES
*See*
Pork : Luncheon meat : Amylase activity

ANAEROBIC CURING
*See*
Adhesives : Acrylic plastics : Curing, Anaerobic

ANALOGUE FROM DIGITAL CONVERTERS
*See*
Digital to analogue converters

**ANALOGUE TO DIGITAL CONVERTERS**
*See also*
Computers : Analogue to digital converters
Control systems, Pulse code modulation, Differential : Analogue to digital converters
Microprocessors : Analogue to digital converters
Multimeters, Digital read out : Analogue to digital converters
Waveforms, Transient : Recorders : Analogue to digital converters

Enhancing the dynamic range of an a/d converter. J. Lacy & F. McMullin. *Electron. Engng.*, 55 (Aug 83) p.49+
**Amplifiers, Differential**
Instrumentation amplifier design. [Burr-Brown OPA 101 BM operational amplifier] *Measmt. Control*, 16 (Aug 83) p.312

**ANALOGUE TO DIGITAL CONVERTERS** (Contd.)
**Flash, Metal—Oxide—Semiconductor, Complementary, Sapphire substrate**
Eight-bit flash A/D converter. [RCA Somerville CA3300 analogue to digital converter] V. Zazzu. *Electron. Engng.*, 54 (Dec 82) p.51+
**Microcomputers, Single board : Teaching**
Analog to digital conversion using a microprocessor. A.M. Chadwick & W. Herdman. *Int. J. Elect. Engng. Educ.*, 19 (Oct 82) p.299-306

**ANALOGUE TO TERNARY CONVERTERS**
**Pipeline**
Pipeline analogue to ternary converter. H.N. Shivashankar & A.P. Shivaprasad. *Int. J. Electron.*, 53 (Oct 82) p.357-61

**ANALYSERS, Multichannel**
*See also*
X-ray astronomy : Satellites, Artificial : Detectors : Analysers, Multichannel
**Microcomputers : Programs : Languages : Forth**
Writing a FORTH discompiler and using FORTH in digital filters and multichannel analysis. G. Filbey. *Microprocessors Microsysts.*, 7 (Jun 83) p.223-7

ANALYSIS, Chemical
*See*
Apples : Determination of trace elements
Arsenic sulphides, Amorphous : Analysis
Astronautics flights : Mars : Samples : Analysis
Bitumen, Athabasca : Asphaltenes : Clay : Analysis
Blood : Analysis
Boilers : Flue gas : Analysis
Brassica : Determination of glucosinolates
Brewing : Analysis
Buses : Diesel engines : Wear : Lubricating oil analysis
Ceramics : Analysis
Clay, Sedimentary : Analysis
Coal : Analysis
Coal, Bituminous : Hydrogenation : Catalysts : Zinc chloride : Products : Analysis
Coal, Bituminous : Liquefaction : Products : Analysis
Coal : Freeze conditioning agents : Glycols : Analysis
Coal : Gasification : Steam : Catalysts : Minerals : Reaction products : Analysis
Coal : Liquefaction : Products : Hydrocracking : Products : Analysis
Coal : Prospecting : Boreholes : Drilling, Core : Samples : Analysis
Coal : Solvent extraction : Hydrogen peroxide—Trifluoroacetic acid : Products : Analysis
Concrete : Analysis
Cosmetics : Analysis
Desulfovibrio : Corrosion agents : Hydrogen sulphide evolution : Determination
Drugs, Illicit : Analysis
Flour : Analysis
Food : Analysis
Furnaces, Blast : Top gases : Analysis
Gas oil : Determination of sulphur
Gears, Spur, Carburised : Wear : Lubricating oil analysis
Glass : Melting : Tanks : Fumes : Analysis
Iron—Titanium : Oxidation : Electrolysis : Films : Analysis
Leaves : Protein : Concentrates : Analysis
Lubricating oil analysis
Machinery : Wear : Lubricating oil analysis
Milk : Products : Analysis
Oil shale : Analysis
Pathogens : Analysis
Pesticides : Analysis
Phosphoric acid : Determination of trace metals
Shale oil : Hydrogenation : Products : Nitrogen compounds, Organic : Analysis
Sludge : Determination of metals
Steel : Production : Analysis
Sugars : Analysis
Tomatoes, Frozen : Analysis
Tops : Wool, Shrink resistant : Finishing : Chlorine—Resin process : Studies : Amino acids analysis
Water : Analysis
Water : Determination of bromides

**ANALYSIS, Chemical** (Cont.)
*See*
Water : Pollution : Analysis
Water : Purification : Analysis
Wines : Analysis
*Related headings*
Amperometric titrations
Auger emission spectroscopy
Chromatography
Colorimetry
Complexometric analysis
Coulometry
Electrodes, Ion selective
Gas analysers
Gravimetry
Immunoassaying
Karl Fischer method
Microanalysis
Muon spin rotation
Nuclear magnetic resonance
Oxygen flask analysis
Polarography
Potentiometric stripping analysis
Potentiometric titrations
Radioactivation analysis
Spectrofluorimeters
Spectrofluorimetry
Spectrophotometers
Spectrophotometry
Spectroscopy, Atomic
Spectroscopy, Emission
Spot tests
Thermogravimetry
Titrations
Turbidimetry
X-ray analysis
X-ray diffraction analysis
X-ray fluorescence spectroscopy
**Calculations**
Chemometrics—a buzz word explained. D. Betteridge. *Lab. Pract.*, 32 (Oct 83) p.13+
**Instruments**
Analytical instruments in process management. J.R.P. Clarke. *Processing*, 29 (Jun 83) p.25+
Detection, measurement and analytical equipment. *Process Biochem.*, 18 (Jan/Feb 83) suppt. p.xi-xv
**Instruments : Output : Data processing : Computers**
Multiple launch of analytical instruments by Philips at CSI Conference. [Philips PV8020] [Philips Automated Laboratory Management System] *Found. Trade J.*, 155 (11/25 Aug 83) p.166-7

ANALYSIS, Thermal
*See*
Thermal analysis

ANALYTIC SERIES EXPANSION
*See*
Particles, Spherical : Diffusion : Boundaries, Moving : Analysis : Differential equations, Partial : Solution : Series expansion, Analytic

ANALYTICAL ESTIMATING
*See*
Incentives : Work study : Estimating, Analytical

ANATASE
*See*
Coal, Bituminous : Hydrodesulphurisation : Catalysts : Cobalt—Molybdenum, Alumina supported : Deactivation : Anatase

ANCHORS
*See*
Dams, Rockfill : Anchors
Structures : Concrete : Post-tensioning : Anchors

ANDERSON EFFECT
*See*
Metal—Insulator transition

**ANECHOIC CHAMBERS**
**Performance : Noise, Low level : Measurement : Transducers**
Measurement of background noise in sound-insulated rooms. P.K. Moller. *J. Sound Vib.*, 85 (22 Nov 82) p.143-50

ANEMOMETERS, Hot film
*See*
Arteries, Stenosed : Flow : Waveforms : Studies : Anemometers, Hot film

*30* *CURRENT TECHNOLOGY INDEX*

SOURCE—SINK FLOW
*See*
Nuclear reactors : Fuels : Uranium : Isotope separation : Centrifuges, Gas : Flow, Source—Sink

**SOURSOP**

**Juices : Aroma compounds : Gas chromatography : Adsorbents : Polymers, Porous**
Trapping of soursop (*Annona muricata*) juice volatiles on Porapak Q by suction. M.R.B. Franco & D.B. Rodriguez-Amaya. *J. Sci. Fd. Agric.*, 34 (Mar 83) p.293-9

SOUTH AFRICA
*See*
Asbestos, Crocidolite : Mining : Pomfret mine
Brewing : South Africa
Buses : Transport : Operation : South Africa
Food : Research : South Africa
Pesticides : Analysis : Inter-laboratory calibration : South Africa
Ports : South Africa
Railways : South Africa
Shipping : Industry : South Africa
Ships : Repairs : South Africa
Sugar, Refined : Conditioning : South Africa
Water : Engineering : South Africa

SOUTH AMERICA
*See*
Animal feedingstuffs : Industry : South America

SOUTH EAST ASIA
*See*
Containers : Glass : Industry : South East Asia
Shipbuilding : South East Asia

SOUTH KOREA
*See*
Nuclear energy : South Korea
Shipbuilding : South Korea
Shipping : Industry : South Korea

SOUTH PACIFIC
*See*
Technology, Small scale : South Pacific

SOUTH WEST BRITAIN
*See*
Foundry practice : South West Britain

SOUTH WEST ENGLAND
*See*
Gas, Natural : Utilisation : South West England
Gas, Town : South West England
Water : Engineering : South West England

SOUTH YORKSHIRE
*See*
Railways : South Yorkshire
Refuse disposal : South Yorkshire

SOUTHERN SPAIN
*See*
Copper : Mining : Southern Spain
Silver : Mining : Southern Spain

SOUTHWARK
*See*
Flats : London : Southwark
Housing : Town planning : London : Southwark

SOUTHWELL PLOT METHOD
*See*
Shells, Cylindrical, Stringer stiffened : Epoxy resins : Buckling : Critical load : Determination : Southwell plot method

**SOYA BEAN OIL**

**Refining**
Quality characteristics of physically refined soyabean oil: effects of pre-treatment and processing time and temperature. I.M. Jawad, S.P. Kochhar & B.J.F. Hudson. *J. Fd. Technol.*, 18 (Jun 83) p.353-60

SOYA BEANS : Flour
*See*
Maize : Dough : Fermentation : Products : Supplementation : Soya flour

**SOYA BEANS : Globulin**

**Food products : Heating : Effect of organosulphur compounds**
Effect of sulphydryl and disulphide compounds on the formation and quality of thermal aggregates of soya bean 11S globulin. T. Yamagishi, F. Yamauchi & K. Shibasaki. *J. Sci. Fd. Agric.*, 33 (Nov 82) p.1092-1100

**SOYA BEANS : Protein**

**Food products**
Make the most of soya. L. Schutte. *Fd. Mf.*, 58 (May 83) p.21+
Preparation of protein hydrolysate from defatted coconut and soybean meals. Pt.1: effect of process variables on the amino nitrogen released and flavour development. C.B. Pham & R.R. Del Rosario. *J. Fd. Technol.*, 18 (Feb 83) p.21-34
Textured vegetable protein comes of age. J.H. Landau. *Fd. Flavs. Ingredients Process. Packag.*, 5 (Sep 83) p.33+

**Food products : Sensory testing**
Preparation of protein hydrolysate from defatted coconut and soybean meals. Pt.2: quality and sensory evaluation of products. C.B. Pham & R.R. Del Rosario. *J. Fd. Technol.*, 18 (Apr 83) p.163-70

**SOYA BEANS : Protein—Lipid film**

**Food products**
Development of an improved soya protein-lipid film. E.C. Chuah, A.Z. Idrus, C.L. Lim & C.C. Seow. *J. Fd. Technol.*, 18 (Oct 83) p.619-27

SPACE CHARGE
*See*
Air gaps : Electrodes, Plane—Point : Streamer-to-spark transition : Space charge
Diodes, Aluminium—Alumina—Tellurium : Current—Voltage relationships : Effect of space charge
Power transmission lines, Overhead : Conductors, Bundle : Corona : Models : Space charge
Vacuum : Gaps : Breakdown : Initiation : Space charge

SPACE CHARGE LIMITED CURRENT
*See*
Current, Space charge limited

SPACE CHARGE LIMITED SOLID STATE DIODES
*See*
Diodes, Single carrier injection

SPACE FRAMES
*See*
Frames, Space

SPACE SHUTTLE LAUNCHED ARTIFICIAL SATELLITES
*See*
Satellites, Artificial, Space shuttle launched

**SPACE SHUTTLES**

*See also*
Satellites, Artificial : Transport : Space shuttles

Maiden voyage of 'Columbia', pt.3. J.A. Pfannerstill. *Spaceflight*, 24 (Dec 82) p.460-5
Meeting the challenge. D.F. Robertson. *Spaceflight*, 25 (Sep/Oct 83) p.366-7
Soviet space shuttle. G. Borrowman. *Spaceflight*, 25 (Apr 83) p.149-51
STS-5 mission. *Spaceflight*, 24 (Dec 82) p.450-2

**Boosters : Nozzles : Bearings : Laminates : Rubber—Steel**
America's space shuttle rides into orbit on natural rubber. D.M. Cox. *Rubb. Dev.*, 35 no.4 (1982) p.84-7

**Experiments : Columns, Stayed, Self-expanding : Vibrations : Analysis**
Vibration characteristics of self-expanding stayed columns for use in space. J.R. Banerjee & F.W. Williams. *J. Sound Vib.*, 90 (22 Sep 83) p.245-61

**Heat shields : Tiles : Fibres : Silica**
Insulating the space shuttle. [Lockheed Missiles and Space Co.] [Manville Products Corporation: Q-Fiber] S.J. Previte. *Insulation J.*, 27 (Aug 83) p.14-16

**Laboratories, In-orbit**
Countdown for *Columbia*. [Spacelab 1] *Spaceflight*, 25 (Jul/Aug 83) p.292-3
Scientists gain a foothold in space. [Spacelab 1] P. Marsh. *New Scientist*, 99 (22 Sep 83) p.874-9
Spacelab crew prepares for launch. J. Bird. *Spaceflight*, 25 (Jul/Aug 83) p.290-1
Spacelab: v.i.p. among Shuttle passengers. D. Velupillai. *Flight Int.*, 124 (22 Oct 83) p.1106-8

**SPACE SHUTTLES** (Contd.)

**Laboratories, In-orbit : Experiments : Payloads**
First Spacelab payload (f.s.l.p.). D.J. Shapland. *Aeronaut. J.*, 87 (Feb 83) p.39-51

**Laboratories, In-orbit : Photography**
Photography aboard Spacelab. H.J.P. Arnold. *Br. J. Photogr.*, 130 (25 Nov 83) p.1246-8

**Laboratories, In-orbit : Vacuum chambers : Pipes : Degassing : Pumps, Turbomolecular**
Analytical approach for calculating vacuum properties of outgassing tubing systems. [Spacelab 1] H. Hamacher. *Vacuum*, 32 (Dec 82) p.729-33

**Life support systems**
Shuttle service. P. Kelsey. *Bldg. Serv.*, 5 (Jul 83) p.48-9

**Panels, Honeycomb, Sandwich : Vibrations : Modal density**
Modal density of honeycomb plates. B.L. Clarkson & M.F. Ranky. *J. Sound Vib.*, 91 (8 Nov 83) p.103-18

**Photography**
First for the shuttle—and Pioneer. H.J.P. Arnold. *Br. J. Photogr.*, 130 (17 Jun 83) p.622-3

SPAIN
*See*
Copper : Mining : Southern Spain
Fishing : Industry : Spain
Hydroelectric power : Spain
Industrial research : Spain
Observatories : La Palma (Canary Islands)
Offshore drilling : Spain
Papermaking : Industry : Spain
Shipbuilding : Spain
Shipping : Industry : Spain
Silver : Mining : Southern Spain
Water : Supplies : Inter-river transfer : Spain

SPALLING
*See*
Ball mills : Balls : Spalling
Gears, Carburised : Spalling

SPANGLER'S FORMULA
*See*
Pipes : Plastics, Underground : Loading : Spangler's formula

SPARE PARTS
*See*
Motor cars : Spare parts
Motor vehicles : Spare parts

SPARK ARRESTERS
*See*
Cupolas : Spark arresters

SPARK DISCHARGE
*See*
Air gaps : Electrodes, Plane—Point : Streamer-to-spark transition
Air gaps : Electrodes, Plane—Point : Transition to spark discharge
Lasers : Carbon dioxide, Pulsed, Transversely excited : Preionisation : Spark discharge
Sulphur hexafluoride : Gas discharge : Electrodes, Plane—Point : Transition to spark discharge

SPARK ERODED PHOSPHOR BRONZE
*See*
Diaphragms : Phosphor bronze, Spark eroded

SPARK ERODED STEEL
*See*
Steel, Spark eroded

**SPARK EROSION**

*See also*
Steel : Coating : Metals : Spark erosion
Wires : Drawing : Dies : Diamond, Synthetic : Manufactures : Spark erosion
Wrenches, Power : Sockets, Impact : Spark erosion

Seminar reveals some new ideas. A. Astrop. *Machinery Prod. Engng.*, 141 (15 Jun 83) p.21-2

**Surface roughness : Calculation**
Analysis of surface erosion in electrical discharge machining. S.T. Jilani & P.C. Pandey. *Wear*, 84 (1 Feb 83) p.275-84

**SPARK EROSION MACHINES**

*See also*
Dies : Manufactures : Spark erosion machines
Moulds : Manufactures : Spark erosion machines

Now you know more about this service, let me just mention finally its American equivalent:

> *Applied Science and Technology Index*. New York: H.W. Wilson, 1958 to date. Monthly, with quarterly and annual cumulations. From 1913 to 1957 published as *Industrial Arts Index*.

This is arranged alphabetically by subject headings, although they are not as detailed or complex as those in *Current Technology Index*. It covers all subjects from aeronautics to transport and, of course, includes chemistry.

### 3.2.3. Chemical Abstracts

Now we get to the core of this Section. When chemists talk about literature searches, this is usually the service they mean. There are other indexing and abstracting services available, and I shall introduce a few in Section 3.2.4; but the supreme source of information on chemical literature must be *Chemical Abstracts*.

Π What's the difference between an abstracting and an indexing publication?

You have seen a couple of general indexing publications above and know what they are like. Well, an abstracting journal is just like an indexing journal, except that in addition to giving you the bibliographic details of the journal paper or whatever document it is indexing, it also gives you a brief summary (an abstract) of the document.

Just to keep things tidy and consistent let me give you the proper details of the publication:

> *Chemical Abstracts*. Columbus (Ohio): American Chemical Society, 1907 to date. Weekly.

Since its foundation over seventy-five years ago *Chemical Abstracts* has grown into the world's largest and most comprehensive index of the chemical literature. In addition to indexing all the articles published in some 14,000 current journals, it also covers patents, reports, monographs (ie specialist books), and conference proceedings over the entire fields of analytical and applied chemistry, biochemistry, chemical engineering, macromolecular, organic, inorganic, and physical chemistry. Each document included during the period covered by each issue or volume of *Chemical Abstracts* is listed as a bibliographic citation, and usually an informative summary (the abstract) of its contents is also published. The references and their abstracts are arranged in subject sections and are numbered consecutively to facilitate the production of indexes to them. *Chemical Abstracts* is published weekly and every 26 weeks the issues are cumulated into a volume, with the indexes to that whole volume issued a few months later. The weekly issues are published on a fortnightly cycle, with material on biochemistry and organic chemistry published one week, and with the material on macromolecular and applied chemistry, chemical engineering, and physical, inorganic and analytical chemistry published the next; any one subject area is thus covered every fortnight.

You have to search the weekly issues since the last cumulated volume individually to bring a search up-to-date. Since the issues are arranged in broad subject sections you can also scan newly arrived issues on a regular basis to try and keep up-to-date with information in your field. Each weekly issue has an author and keyword index, and a patent index.

But the real value of *Chemical Abstracts* lies in the six monthly volumes and their comprehensive indexes. The following four different indexes are the ones you will find of most value, so I am confining my descriptions to them.

The *Author Index* obviously lists the references under the names of authors; there are full cross-references to and from all the joint authors of any item (no matter how many). Brief titles are quoted for each document indexed to help you to find your way to the particular reference you require.

The *General Subject Index* and the *Chemical Substance Index* are probably the trickiest indexes to use. But once you have mastered their peculiarities they can be immensely useful to you. One thing you must remember about them both is that before you ever look anything up in them you must always first consult the index guide. This indicates the preferred subject headings or chemical compound names, and suggests possible synonymous terms you could also use. It also includes scope notes and, where relevant, illustrative structural diagrams. You will get to use these indexes in just a moment.

Finally the *Formula Index*. All the compounds referrred to in the literature are indexed under their molecular formulae and arranged by the Hill system. Each formula is presented with its elements listed in alphabetical order; except when carbon is present, the formula is written with carbon first and hydrogen second, followed by the other elements in alphabetical order. These re-arranged formulae are then listed in alphabetical (and numerical) order.

You will come across one or two further specialised indexes to *Chemical Abstracts* (an *Index of Ring Systems; Registry Number Index*; and *Patents Indexes*); but I advise you to concentrate on those I have dealt with here. Enough description—the best way for you to find out what *Chemical Abstracts* can do for you is to use it. So, off to the library (if you aren't there already) and tackle this exercise.

**SAQ 3.2b**

I have tried to make my descriptions and explanations of *Chemical Abstracts* as clear and simple as possible. But given the enormous amount of material indexed there are bound to be complexities. These are best confronted by actually using the service, so here are a few questions to make you use the different indexes I described. All the questions are set from one volume (Vol. 100) only, and so remember that you are covering only six-months chemical literature. And let me remind you to try to use the books themselves if you can get access to a set of *Chemical Abstracts*: believe me, you will get a much greater sense of achievement by finding these needles in what will at first seem a dauntingly large haystack, than if you rely on the copies on the following pages. I must also warn you that, due to difficulties in printing this Unit, the copies of *Chemical Abstracts* we have given you here look simpler than the real thing. You are only seeing about two thirds of a single column each time—there are in fact three columns to each page of the indexes and two columns per page in the abstracts section in the original publication. But sets of *Chemical Abstracts* are not all that widespread, so use the photocopies if you must; they will serve to demonstrate my points and give you useful initial experience. In each case refer to the abstracts and write your reference out as a full bibliographic citation, as I showed you in Section 1.7 (but note that you will have to edit out, in doing so, some of the data provided with the abstracts and references).

We are going to start with the easiest search, by authors, and then proceed into gradually more complex subject searches.

(*a*) You have been told that Richard A. Dluhy has published papers on applications of Fourier-transform infrared spectroscopy. Find two papers, of which he is one of the authors, on this subject, by using the author index to volume 100 of *Chemical Abstracts.* I should warn you that reference 98632p is not one of the two papers you are seeking.

(*b*) Let's go back to our space question. Having read the paper you found in *Current Technology Index*, you want more detailed information. In particular you want information on the water content of the atmosphere of Mars.

⟶

**SAQ 3.2b (cont.)**

Use the General Subject Index to volume 100 of *Chemical Abstracts* to find a paper on this subject. And meanwhile, back on Earth, you are also involved in stratospheric analysis. You have been told of a new balloon-carried quadrupole mass spectrometer linked to a microprocessor and developed in Europe. Use the same General Subject Index to find a paper describing this.

(*c*) Do you remember a question about Isoniazid earlier in this Unit? Following up what you learned in the *British Pharmacopoeia* you want to know whether more recent work has been done on its determination in tablets. Use the Chemical Substance Index to volume 100 of *Chemical Abstracts* to find a relevant paper.

(*d*) To cross-check on this and related substances, look in the Chemical formula Index and note just the reference number and the CA Registry Number (you don't have to look the full reference up this time) of a paper on Isoniazid mixed with another compound. You need to find the formula first so you could check it quickly and easily in either the *British Pharmacopoeia* or *Merck Index* (or look back to the photocopy with SAQ 2.1e if you are stuck).

CODEN: CHABA8 100 1A–1534A (1984)
ISSN: 0009–2258 **100 A 1**

# Chemical Abstracts

Published by the

American Chemical Society

Volume 100

Author Index, A–Lea

(Part 1 of 2 Parts)

January–June, 1984

*A publication of the*
CHEMICAL ABSTRACTS SERVICE
P. O. Box 3012, Columbus, Ohio 43210

CODEN: CHABA8 100 1535A–3087A (1984)
ISSN: 0009-2258 **100 A 2**

# Chemical Abstracts

Published by the

American Chemical Society

## Volume 100

## Author Index, Leb–Z

(Part 2 of 2 Parts)

## January–June, 1984

*A publication of the*
CHEMICAL ABSTRACTS SERVICE
P. O. Box 3012, Columbus, Ohio 43210

Mathematical simulation of certain biosynthesis phenomena in bacterial batch culture, 207855m
**Mencke, Detlev**
Electronic components of flowmeters., 123226u
**Mencl, Frantisek** See Nemcova, Jitka
**Menczel, E.**
—; Bucks, D. A. W.; Maibach, H. I.; Wester, R. C.
Malathion binding to sections of human skin: skin capacity and isotherm determinations, 30691b
**Menczel, Gyorgy** See Jover, Bela
**Menczel, Jozsef**
—; Varga, J.
Influence of nucleating agents on crystallization of polypropylene. I. Talc as a nucleating agent, 210890z
**Mendau** See also Rihter-Mendau.
**Mende, Dierk**
—; Maroti, P.; Wiessner, W.
Energy distribution between two photosystems during the life-cycle of synchronized cultures of Chlorella fusca, 3698c
**Mende, Guenter**
—; Von Borany, J.; Deutscher, M.; Fenske, F.; Kuester, G.; Schmidt, B. R.
Anodic silicon oxidation as a method for passivating electronic devices, 43746y
**Mende, Hans Horst** See Hartmann, U.
**Mende, S. B.**
—; Eather, R. H.; Rees, M. H.; Vondrak, R. R.; Robinson, R. M.
Optical mapping of ionospheric conductance, 148069h
**Mende, Volker** See Bucka, Hartmut
**Mendel, Eric** See Basi, Jagtar S.
**Mendel, Ralf R.**
Release of molybdenum cofactor from nitrate reductase and xanthine oxidase by heat treatment, 31326y
**Mendel, Verne E.** See Aponte, Gregory William; Rabinowitz, L.
**Mendeleeva, Z. G.** See Gromov, L. A.
**Mendelevich, A. I.** See Shkoropad, D. E.
**Mendel-Hartvig, J.** See Lewis, James Wyatt
**Mendelovici, E.**
—; Nadiv, S.; Lin, I. J.
Morphological and magnetic changes during mechano-chemical transformation of lepidocrocite to hematite, 220449b
**Mendelovici, L.**
—; Tzehoval, H.; Steinberg, M.
The adsorption of oxygen and nitrous oxide on platinum-ceria catalyst, 74650s
**Mendelovitz, Simona** See Aharonowitz, Yair
**Mendelsohn, A. J.** See Harris, Stephen Ernest; Wisoff, P. J. K.
**Mendelsohn, David H.** See Goldner, Ronald B.
—; Goldner, R. B.
Ellipsometry measurements as direct evidence of the Drude model for polycrystalline electrochromic tungsten(VI) oxide films, 200829x
**Mendelsohn, D. L.** See Henderson, Jack Bradford
**Mendelsohn, Ella** See Manor, Haim
**Mendelsohn, Frederick A. O.** See Aguilera, Greti; Nichols, Nancy R.; Quirion, Remi
—; Aguilera, G.; Saavedra, J. M.; Quirion, R.; Catt, K. J.
Characteristics and regulation of angiotensin II receptors in pituitary, circumventricular organs and kidney, 18397c
—; Quirion, R.; Saavedra, J. M.; Aguilera, G.; Catt, K. J.
Autoradiographic localization of angiotensin II receptors in rat brain, 169054q
**Mendelsohn, Geoffrey**
—; Baylin, S. B.
Ectopic hormone production - biological and clinical implications, 185822u
**Mendelsohn, John M.** See Masui, Hideo; Rice, Carol
**Mendelsohn, Morris A.** See Luck, Russell M.
—; Navish, F. W., Jr.; Luck, R. M.; Yeoman, F. A.
Effects of thermal, oxidative and hydrolytic degradation on physical properties of solar collector sealants, 24435e
**Mendelsohn, Marshall H.**
—; Gruen, D. M.; Dwight, A. E.
Aluminum lanthanum nickel hydride, 150103h
**Mendelsohn, M. L.** See Ehling, U. H.
**Mendelsohn, Richard** See Dluhy, Richard A.
—; Dluhy, R. A.; Crawford, T.; Mantsch, H. H.
Interaction of glycophorin with phosphatidylserine: a Fourier transform infrared investigation, 116761m
**Mendelson, Carole R.** See Cleland, William H.; Ng, Valerie L.
**Mendelson, Dan**
—; Tas, J.; James, Jan
Cuprolinic Blue: a specific dye for single-stranded RNA in the presence of magnesium chloride. II. Practical applications for light microscopy, 3063s
**Mendel'son, E. L.**
—; Gol'din, V. A.; Breger, A. Kh.
The effect of water-equivalent flow structure on the capacity of radiation-chemical apparatuses, 59487a
**Mendelson, Haim** See Leibowitz, Elia M.
**Mendelson, Jack H.** See Mello, Nancy K.; Mellow, N. K.
Chronic effects of cannabis on human brain function and behavior, 46392r
—; Ellingboe, J.; Mello, N. K.
Acute effects of natural and synthetic cannabis compounds on prolactin levels in human males, 79731d
**Mendelson, N. H.** See Saxe, Charles L., III
**Mendelson, S.**
Theory and mechanisms for martensitic transformations in ferroelectrics. Part I. Proper transitions, 201730v
Theory and mechanisms for martensitic transformations in ferroelectrics. Part II. Improper transitions, 201731w
Theory and mechanisms for martensitic transformations in ferroelectrics. Part III. Successive transitions, 201732x
**Mendelson, Wallace B.** See Weissman, Ben Avi
**Mendelson, Wilford L.** See Chodosh, Daniel F.
**Mendelssohn, A. S.**
Coal-oil mixture test at Florida Power and Light Company-Sanford No. 4, 36775d
**Mendenhall, Charles L.** See Chedid, A.; Roselle, Gary A.
—; Chedid, A.; Kromme, C.
Altered proline uptake by mouse liver cells after chronic exposure to ethanol and its metabolites, 98109y
**Mendenhall, J. V.** See Eng-Wilmot, D. L.
**Mendenhall, Marcus Holden**
High-energy heavy-ion-induced enhanced adhesion, 60220q
**Mendes, Carmen Lucia de Oliveira**
—; Qassim, R. Y.
Application of the Davies-Taylor equation to a large bubble rise in liquid-fluidized beds, 194273t
**Mendes, Geraldo F.** See Braga, E. S.
—; Cescato, L.; Frejlich, J.
Gratings for metrology and process control. 2: Thin film thickness measurement, 148267w
**Mendes, Nelson F.** See Musatti, Chloe C.
**Mendes, P. J.** See Peichl, R.
—; Gil, J. M.; De Campos, N. A.; Peichl, R.; Weidinger, A.
Phase transitions in the tantalum-hydrogen system observed by PAC, 54530z
**Mendes e Silva, Leda M.** See Srivastava, Rajendra M.
**Mendes Filho, J.** See Melo, F. E. A.
**Mendez** See also Gonzalez-Mendez, Hernandez Mendez, Morales Mendez, Perez Mendez.
**Mendez, A.** See Ferrando, A.
**Mendez, Bernardita** See Cathala, Guy
**Mendez, Bernardo**
—; Martin, P.; Rojas, A. C.
Carbon-13 NMR spectra of untreated natural samples. II. Germination of some common seeds, 32303a
**Mendez, C.** See Juhan-Vague, Irene
**Mendez, Enrique** See Gavilanes, Jose G.; Lopez-Otin, Carlos; Lopez Otin, Carlos
**Mendez, Emilio E.** See Heiblum, Mordehai; Robbins, David John; Sun, Ya Li; Washburn, Sean
**Mendez, Frances** See Bases, Robert
**Mendez, Jesus**
—; Castro-Poceiro, J.
Furocoumarins from Angelica pachycarpa, 206459y
**Mendez, J. C.** See Garcia-Rejon, A.
**Mendez, J. E.** See Sylvester-Bradley, R.
**Mendez, L.** See Errea, L. F.
**Mendez, Louis** See Bases, Robert
**Mendez, M.**
—; Duprez, D.; Apesteguia, C. R.; Barbier, J.
Characterization of a sulfided platinum-aluminum oxide catalyst by chemisorption and titrations: comparison of the dynamic and static volumetric methods, 180704x
**Mendez, Q. Jose D.**
—; Hicks, J. J.
Metabolism and function of polyamines in plant cells, 188694q
**Mendez, Roberto H.**
—; Verga, A. D.; Kriner, A.
The photometric and radial velocity variations of the central star of the planetary nebula IC 418, 148062a
**Mendez, Sixto R.** See Kim, Yoon Shin
**Mendez, Vicente**
Geology and mineral manifestations on the Tabarin Peninsula (Argentine sector, Antarctica), 71551z
**Mendez, Victor M.** See Geller, Irving
**Mendez del Castillo, G.** See Varela Julia, M.
**Mendez Jorrin, Gladys** See Mesa Crespo, Dora L.
**Mendez Llorens, A.** See Thwayeb, Y.
**Mendez Marco, Maria M.** See Mendez Marco, M. T.; Sanz Munoz, M.
**Mendez Marco, M. T.** See Sanz Munoz, M.
—; Mendez Marco, M. M.; Sanz Munoz, M.
Inhibitors of porphobilinogenase of Pinus pinea seeds in germination, 64018c
**Mendez-Moreno, R. M.**
—; Calles, A.; Yepez, E.; Ortiz, M. A.
Electron gas with screened Coulombic interactions, 26249w
**Mendham, J.** See Chambers, D.
**Mendieta, F. J.**
Phase noise of semiconductor lasers in single mode optical fiber heterodyne systems, 14992v
The phase noise of semiconductor lasers in single-mode optical fiber heterodyne systems, 182897m
**Mendieta, H. B.**
—; Pate, D. L.
Water quality of Belton Lake, central Texas, 108725n
**Mendiev, T. R.** See Mamedov, A. M.
**Mendillo, Michael** See Wand, Ronald H.
**Mendiola, A.** See Garcia Casado, P.
**Mendiola, Jesus** See Alemany, C.; Duran, Pedro
—; Alemany, C.; Jimenez, B.; Maurer, E.
Poling strategy of PLZT ceramics, 201704q
—; Jimenez, B.
Review of recent work on ferroelectric composite systems, 183837d
—; Pardo, L.
An XRD study of 90° domains in tetragonal PLZT under poling, 201705r
**Mendiola-Morgenthaler, Leticia** See Leu, Stefan
**Mendioroz, S.** See Gil Llambias, Francisco Javier
**Mendiratta, Madan G.**
—; Ehlers, S. K.
$DO_3$-domain growth kinetics in the iron-20 aluminum-5 at.% silicon alloy, 10706g
—; Kim, Hak Min; Lipsitt, H. A.
Slip directions in B2 iron-aluminum alloys, 107438j
**Mendiratta, Ranjana** See Malaviya, Brajesh
—; Malaviya, B.; Kamboj, Ved P.
Immuno-interception of HCG through local immunization of rabbit fallopian tubes, 45471d
**Mendiratta, R. G.** See Kishore, N.; Puri, Rajinde
**Mendis, Aruna H. W.** See Gibson, J. M.
—; Rees, Huw H.; Goodwin, T. W.
The occurrence of ecdysteroids in the cestode Moniezia expansa, 83051z
**Menditto, Peter** See Sears, Alan R.
**Mendivil, S.** See Archer, Thomas E.
**Mendizabal, R.** See Macias, Antonio
**Mendkovich, A. S.** See Churilina, A. P.; Gultya... P.
**Mendla, Klaus**
—; Cantz, M.
Specificity studies on the oligosaccharide neuraminidase of human fibroblasts, 117082c
**Mendlein, John** See Cuppoletti, E. Rabon J.; In Wha Bin
**Mendler, Nikolaus** See Hagl, S.
—; Struck, E.; Sebening, F.
Orthotopic canine heart transplantation after 2... hours cold perfusion, 136816m
**Mendlewicz, Julien** See Copinschi, Georges
**Mendonca, A. H.**
—; Nobrega, A. W.; Mulder, R. Udo; Vianna, M... C.; Almeida, C. E.; Winter, M.
Preoperational environmental monitoring of the Angra reactor site, 27341g
**Mendonca, A. L.** See Ritz, K. N.
**Mendonca, M. T.** See Licht, Paul
**Mendoza** See also Sevilla-Mendoza.
**Mendoza, Claudio** See Butler, K.
**Mendoza, Celia G.** See Gersich, Francis M.
**Mendoza, Eugenio E.** See Canto, Jorge
**Mendoza, Evelyn Mae T.** See Hernandez, Vic...
—; Mujer, C. V.; Rodriguez, F. M.; Ramirez, D...
Ontogeny of some chemical constituents in mak... and normal coconut endosperms, 20559a
**Mendoza, G.** See Maxwell, W. M. C.
**Mendoza, Hector**
An evaluation of the use of Mexico City wastew... on the irrigation of crops, 25680t
**Mendoza, Luis A.** See Hernandez, Jose O.
**Mendoza, L. G.** See Phillips, Robert Dixon
**Mendoza, Natividad S.**
—; Amemura, A.
(1→2)-β-D-Glucan-hydrolyzing enzymes in Cytophaga arvensicola: partial purification an... properties of endo-(1→2)-β-D-glucanase and β-D-glucosidase specific for (1→2)- and (1→3)-linkages, 19682d
**Mendoza, R.** See Laine, J.
**Mendoza, Stanley A.** See Reznik, Vivian M.
**Mendoza, S. P.** See Coe, Christopher L.
**Mendoza-Alvarez, Maria Eugenia** See Setter...
—; Schmid, H.; Rivera, J. P.
Spontaneous birefringence and spontaneous Fa... effect in cobalt bromine boracite $Co_3B_7O_{13}Br$ 202057t
**Mendoza Gamez, Gaston**
—; Flores Herrera, F.
Mexico City's master plan for reuse, 39130a
**Mendoza-Hernandez, G.**
—; Calcagno, M.; Sanchez-Nuncio, H. R.; Diaz-Zagoya, J. C.
Dehydroepiandrosterone is a substrate for estra... 17β-dehydrogenase from human placenta. 1...
**Mendoza Villela, Elba** See Blade Font, Artur
**Mendoza-Zelis, L.** See Bibiloni, A. G.; Desim...
**Mendrela, E. A.**
—; Mendrela, E. M.; Osadnik, Z.; Staszak, J.
High capacity magnetic separator for volatile ... 108473d
**Mendrela, E. M.** See Mendrela, E. A.
**Mendte, K.** See Maus, W.
**Menduina, C.** See Aicart, E.
**Mendyk, Ewaryst** See Leboda, Roman
**Mendzheritskii, A. M.**
—; Vovchenko, I. B.; Kozlova, L. S.
Effect of antibodies to myelin basic protein on proteolytic activity in the brain and the kin... system of cerebrospinal fluid and blood. 465...
**Mene, P.** See Cinotti, Giulio Alberto
**Menear, S.**
—; Bradbury, A.; Chantrell, R. W.
Ordering temperatures in ferrofluids, 6072...
**Menegatti, Enea** See Antonini, Eraldo; Ascen... Paolo; Scatturin, Angelo
—; Guarneri, M.; Bolognesi, M. C.; Ascenzi, P.; Antonini, E.
Structure-function relationship for Kunitz- and Kazal-type trypsin inhibitors, 47696e
**Menegatto, Guido**
A modification of the method of Mathur and ... for the determination of oxytocinase in the p... of pregnant women, 98772j
**Meneghelli, Ulysses G.**
—; Godoy, R. A.; Oliveira, R. B.; Santos, J. C...; Dantas, R. O.; Troncon, L. E. A.
Effect of pentagastrin on the motor activity ... dilated and nondilated sigmoid and rectum ... Chagas' disease, 1084a
**Meneghelli, Virgilio** See Mazzocchi, Giuseppe
**Meneghin, Mariano** See Giordano, Claudio
—; Piccinelli, P.; Giordano, C.
2',4'-Difluoro-4-hydroxy-(1,1'-diphenyl)-3-carboxylic acid, P 209399q
**Meneghini, G.** See Lucci, Alberto
**Meneghini, Rogerio** See Hoffmann, Maria Ed...; Mello Filho, Alberto C.
**Meneguzzi, Gim**
—; Binetruy, B.; Grisoni, M.; Cuzin, F.
Plasmidial maintenance in rodent fibroblasts ... BPV1-pBR322 shuttle vector without immed... apparent oncogenic transformation of the re... cells, 115824x
**Menendez, Jose**
—; Cardona, M.
Temperature dependence of the first-order Ra... scattering by phonons in silicon, germanium, ... α-tin: Anharmonic effects, 111372a
—; Hoechst, H.
Study of the phase transition in heteroepitaxi... grown films of α-tin by Raman spectroscop...

CODEN: CHABA8 100 4769CS–6466CS (1984)
ISSN: 0009-2258 **100 CS 4**

# Chemical Abstracts

Published by the

American Chemical Society

Volume 100

Chemical Substance Index, O–P

(Part 4 of 5 Parts)

January–June, 1984

*A publication of the*
CHEMICAL ABSTRACTS SERVICE
P. O. BOX 3012, COLUMBUS, OHIO 43210

6295CS *VOL. 100, 1984 - CHEM. SUBSTANCE INDEX* **4-Pyridinecarboxylic acid,**

GABA uptake inhibition by, 114852t
—, **1,4,5,6-tetrahydro-4,6-dioxo-**
**methyl ester** *[88499-68-7]*, prepn. and transesterification of, 51409f
**octyl ester** *[88499-70-1]*, prepn. and nitrosation of, 51409f
**pentyl ester** *[88499-69-8]*, prepn. and nitrosation of, 51409f
—, **1,2,3,4-tetrahydro-4-(4-methoxypheny=l)-6-[2-(4-methoxyphenyl)ethenyl]-2-oxo-** *[89055-89-0]*, 103134j, 103135k
—, **1,4,5,6-tetrahydro-2-(4-methoxypheny=l)-1-methyl-**
**ethyl ester** *[88653-03-6]*, 68344d
—, **1,2,5,6-tetrahydro-1-methyl-**
**methyl ester** *(arecoline) [63-75-2]*
acute dyskinesia from, 61674c
analogs, muscarinic receptor response to, in heart and ileum, receptor subtypes in relation to, 204102c
binding of, to muscarinic receptors of brain, 96165q
cardiovascular response to intracerebroventricular injection of, muscarinic receptors in relation to, 150822a
glycogen of organs of turtle in response to, 154177h
GTPase of striatum stimulation by, 62306w
learning and memory response to, cholinergic nervous system in relation to, 62368t
memory response to cholinomimetics and, senile dementia treatment in relation to, 114871y
muscarinic receptor binding of, in pheochromocytoma cells, 151485w
muscarinic receptor response to, in heart and ileum, receptor subtypes in relation to, 204102c
neuronal response to, in posterior hypothalamus, in fever, 45197u
prepn. and reaction with acetoacetate ester, P 68189g
sister chromatid exchange induction by, in bone-marrow cells, caffeine effect on, 98022q
**methyl ester, hydrobromide** *[300-08-3]*
conversion of, to base, P 68189g
memory retention response to, 79734g
**methyl ester, hydrochloride** *[61-94-9]*, rectum of lizard and tortoise contraction response to, 61733w
**2-propynyl ester** *[35516-99-5]*, muscarinic receptor response to, in heart and ileum, receptor subtypes in relation to, 204102c
—, **1,2,3,4-tetrahydro-6-methyl-5-nitro-2,=4-dioxo-**
**ethyl ester** *[88499-60-9]*, sapon. and decarboxylation of, 51409f
—, **1,4,5,6-tetrahydro-1-methyl-2-phenyl-**
**ethyl ester** *[88653-02-5]*, 68344d
—, **1,2,5,6-tetrahydro-1-nitroso-**
**methyl ester** *[55557-02-3]*, mutagenicity of, liver microsome of hamster activation of, 19036w
—, **1,4,5,6-tetrahydro-5-nitroso-4,6-dioxo-**
**octyl ester** *[88499-72-3]*, 51409f
**pentyl ester** *[88499-71-2]*, 51409f
—, **1,2,5,6-tetrahydro-1-(phenylmethyl)-**
**methyl ester** *[88928-69-2]*, 103128k
—, **1,4,5,6-tetrahydro-1-(phenylmethyl)-**
**methyl ester** *[75532-94-4]*, 103128k
—, **6-[2-(4,5,6,7-tetrahydro-4,4,7,7-=tetramethylbenzo[b]thien-2-yl)-1-=propenyl]-**
**ethyl ester, (E)-** *[90103-50-7]*, P 191728j
**ethyl ester, (Z)-** *[90103-49-4]*, P 191728j
—, **6-[2-(5,6,7,8-tetrahydro-5,5,8,8-=tetramethyl-2-naphthalenyl)-1-=propenyl]-**
**ethyl ester, (E)-** *[90103-36-9]*, P 191728j
—, **2-[[3-(trifluoromethyl)phenyl]amino]-** *(niflumic acid) [4394-00-7]*
antiulcer compns. contg. sodium salicylate and, P 39626a
chromatog. sepn. of, from fifteen-component mixt., high-performance liq., optimization of, 216177u
diffusion of, in simulated lipid membrane of intestine and stomach, other physicochem. properties in relation to, 44826m
macrophage migration and proliferation response to, 185487p
pharmacol. activity of, caffeine enhancement of, P 197818k
sepn. of, from fifteen-component mixt., by over pressurized thin-layer chromatog., 216176t
**2-(4-morpholinyl)ethyl ester** *(morniflumate) [65847-85-0]*, absorption of, after rectal administration to humans, 132057a
**3-Pyridine-4-$^{13}$C-carboxylic acid** *[89945-68-6]* 187885r
**4-Pyridinecarboxylic acid** *[55-22-1]*, **analysis**
chromatog. sepn. of, from fifteen-component mixt., high-performance liq., optimization of, 216177u
in cyanide detn. in waters by spectrophotometry, 179854q
sepn. of, from fifteen-component mixt., by over pressurized thin-layer chromatog., 216176t
**4-Pyridinecarboxylic acid** *[55-22-1]*, **biological studies**
as isoniazid metabolite, ascorbate effect on, 61401m
prolyl 4-hydroxylase inhibition by, kinetics of, oxoglutarate-binding site structure in relation to, 64081t
**4-Pyridinecarboxylic acid** *[55-22-1]*, **preparation**
prepn. of, by air oxidn. of picoline, 34383a
tungsten oxide hydroxide aqua complex, prepn., IR spectrum and thermal stability of, 131396q
**4-Pyridinecarboxylic acid** *[55-22-1]*, **reactions**
amidation of, with amino sugar, 51958j
chlorination of, by phenylphosphinic dichloride and phosphorus pentachloride, trichloromethyl deriv. from, P 85412a
cyclization of, with phenylenediamine, benzimidazole deriv. from, 103290g
reaction of, with diphenylcyclopropenone, P 122772g
**4-Pyridinecarboxylic acid** *[55-22-1]*, **compounds**
metal complexes, 44420z
**ammonium** *(OC-6-11)*-**hexachloromolybdate(3-) (2:1:1)** *[88437-45-0]*
prepn., IR spectrum and thermal decompn. of, 150008f
thermal rearrangement of, 131324q
**boron complex** *[88503-48-4]*, 60906t
**cobalt-ruthenium complex** *[88547-64-2]*, intramol. electron transfer properties of, 121600n
**di-μ-chlorooctachlorodimolybdate(4-) (4:1)** *[88746-01-4]*
formation of, from ammonium bis(pyridinium=carboxylic acid) hexachloromolybdate(3-), 131324q
formation of, from ammonium protonated isonicotinic acid hexachloromolybdate, 150008f
**hexachloro-μ-oxodioxobis(4-pyridinecarboxy=lato-$N^1$)dimolybdate(4-) (2:1)** *[89697-23-4]*, prepn. and IR spectrum of, 167051n
*(OC-6-11)*-**hexachlororhenate(2-) (2:1)** *[88383-54-4]*, prepn. and thermal decompn. of, 60841t
**molybdenum complex, homopolymer** *[89030-78-4]*, prepn., IR spectrum and thermal stability of, 131396q
**octachloro-μ-oxodioxodimolybdate(4-) (4:1), tetrahydrate** *[89697-21-2]*, prepn., IR spectrum and thermal decompn. of, 167051n
**rhenium complex** *[88383-58-8]*, 60841t
**ruthenium complex** *[69010-44-2]*
elec. potential of, in soln., incorporated couple in chlorosulfonated polystyrene coated electrode in relation to, 217713c
reaction products with chlorosulfonated polystyrene, electrodes coated with, electrochem. properties of, 217713c
**trifluoromethanesulfonate** *[88496-42-8]*, prepn. and reaction of, with cobalt amino acid complexes, 121600n
**vanadium complex** *[89528-76-7]*, prepn. and IR spectrum of, 150046a
**4-Pyridinecarboxylic acid, esters**
**2-aminoethyl ester, cobalt complex** *[88130-08-9]*, prepn. and redn. kinetics of, 113859p
**3-aminopropyl ester, cobalt complex** *[88130-11-4]*, prepn. and redn. kinetics of, 113859p
**ethyl ester** *[1570-45-2]*
acylation by, of alkanoate esters, P 209808x
basicity of, based on visible transition blue shift in formation of iodine or methanol complexes, 129023x
partition coeff. of, structure in relation to, 167721f
reaction of, with Grignard salts, 209592x
**methyl ester** *[2459-09-8]*
basicity of, based on visible transition blue shift in formation of iodine or methanol complexes, 129023x
partition coeff. of, structure in relation to, 167721f
reaction of, with bis(butylphenyl)cyclopropenone, P 122772g
reaction of, with iodomethylpenicillanate dioxide, P 103062j
thromboxane synthetase inhibition by, mol. structure in relation to, 153003m
**methyl ester, vanadium complex** *[89716-13-2]*, prepn. and hydroxylation by, of hydrocarbons, P 174641h
**phenyl ester** *[94-00-8]*, thromboxane synthetase inhibition by, mol. structure in relation to, 153003m
**3-pyridinylmethyl ester, cobalt complex** *[88130-14-7]*, prepn. and redn. kinetics of, 113859p
**4-pyridinylmethyl ester, cobalt complex** *[88130-20-5]*, prepn. and redn. kinetics of, 113859p
**3-(3-pyridinyl)propyl ester, cobalt complex** *[88130-17-0]*, prepn. and redn. kinetics of, 113859p
**1,7,7-trimethylbicyclo[2.2.1]hept-2-yl ester** *[88382-45-0]*, prepn. of, by direct esterification under shear and high pressure, 51848y
**4-Pyridinecarboxylic acid, hydrazides**
**2-acetylhydrazide** *[1078-38-2]*
detn of, in blood by HPLC, 185218b
as isoniazid metabolite, ascorbate effect on, 61401m
of urine of humans, acetylator phenotype evaluation in relation to, 44768u
**2-(aminothioxomethyl)-2-methylhydrazide** *[88317-38-8]*, prepn. and cyclization of, 34482g
**2-benz[cd]indol-2-ylhydrazide, copper complex** *[89420-16-6]*, pigment, for coatings and plastics, manuf. of, P 140821b
**2-[bis(4-chlorophenyl)hydroxyacetyl]-1-=phenylhydrazide** *[89562-91-4]*, prepn. and analgesic activity of, 209592x
**[(5-carboxy-2-hydroxy-3-iodophenyl)=methylene]hydrazide** *[89342-84-7]*, prepn. and antimicrobial activity of, 135676d
**[(5-carboxy-2-hydroxyphenyl)methylene]=hydrazide** *[89342-70-1]*, prepn. and antimicrobial activity of, 135676d
**[(5-chloro-2-hydroxyphenyl)methylene]=hydrazide** *[732-92-3]*, conformation of, dipole moment in relation to, 155959b
**[(4-chlorophenyl)cyanomethylene]hydrazide** *[90017-04-2]*, cyclocondensation of, of oxadiazole deriv. from, 191799h

**Consult 1984 Index Guide for Cross-references and Indexing Policies**

**2-(dithiocarboxy)hydrazide, monopotassium salt** *[61019-32-7]*, cyclization of, 34482g
**2-(ethoxythioxomethyl)hydrazide** *[90053-23-9]*, prepn. and IR spectra of, 202406f
**2-(2-ethyl-2-hydroxy-1-oxobutyl)-1-=phenylhydrazide** *[89562-84-5]*, prepn. and analgesic activity of, 209592x
**(1-ethyl-3-phenyl-2-propynylidene)hydrazide** *[88467-09-8]*, 51416f
**hydrazide** *(isoniazid) [54-85-3]*
acetylation of, in humans, acetylation phenotype evaluation in relation to, 44768u
adrenaline autoxidn. in presence of, superoxide dismutase detn. in relation to, 47506t
aminoacetone urinary excretion response to, 79809k
amino acids of brain regions response to, 96597g
ammonia formation response to, in amaranth and mung bean leaves, photorespiration in relation to, 188886d
*Bacillus subtilis* strains from Enterogermina susceptibility to, 206393x
as candidate, for in vitro teratogenesis evaluation, 30789q
central nervous system response to, R 95903a
chlorophyll fluorescence induction by oxidn. of, 100111p
chromosomes damage from aminosalicylic acid and, in human lymphocytes of tuberculosis patients, 61437c
clotiazepam pharmacokinetics response to, 167695a
compd. with methylbenzoquinoneimine, spectrum of, 220920y
condensation of, with alkynones, isonicotinoyl hydrazones from, 51416f
conjugative and oxidative metab. of, by liver, enzymes in, species in relation to, 152063u
convulsions from, amino acids in nerve endings in relation to, 45926f
convulsions from, ethyl-β-carboline carboxylate potentiation of, GABA receptors of brain in mediation of, 189769e
cyclocondensation of, with β-diketones, 103240r
detn of, in blood by HPLC, 185218b
detn. of, by bromatometry and iodometry, 12758z
detn. of, in formulations by oxidn. and titrn., 145077y
detn. of, Metol and tetravalent cerium in spectrophotometric, 220920y
detn. of, phenyliodosoacetate in titrimetric, 95844y
detn. of, in tablets by spectrophotometry, 126991a
GABA of brain response to, hydrazine metabolite in relation to, 96607k
glycolate pathway inhibition by, in bundle sheath strands of corn and *Panicum maximum*, 117903w
glycollate metab. inhibition by, in cyanobacteria, glycollate excretion response to, 153899h
heart toxicity from, ionol reversal of, 291d
hepatotoxicity of, mechanism of, in humans and lab. animals, R 131872a
interaction of, with ethanol, hepatotoxicity in relation to, 96185w
interaction of, with oxazepam and triazolam in humans, 61396p
liver damage from, nootropil inhibition of, 44944y
lymphocyte transformation and adrenoceptor binding response to, 114976m
metab. of, ascorbate effect on, 61401m
metabolites, ascorbate effect on, 61401m
mutagenicity of, in lymphoma assay, 134046m
*Mycobacterium* susceptibility to, vs. tri-alanine deriv., 99720c
nerve toxicity of, nerve conduction velocity in, 203193w
*Nocardia asteroides* susceptibility to, 117732q
oxidn. by *Arthrobacter* inhibition by, P 101632w
partition coeff. of, structure in relation to, 167721f
pharmacokinetics of, in children with tuberculosis, therapeutic blood concns. in relation to, 167708g
pharmacokinetics of, in humans, ethambutol and rifampicin effect on, 150588v
polymorphic metab. of, genetic variation in, statistical anal. of, 44871x
reaction of, with benzothiazinedithiones, 51539y
reaction of, with formylrifamycin, P 138862x
resistance to, in *Mycobacterium tuberculosis*, 48436g
sensitivity to, of *Mycobacterium tuberculosis* complex, detn. of, conventional and radiometric tests comparison for, 48037c
toxicity of, to central nervous system, acetylator phenotype in relation to, 167773z
toxicity of, to liver, bromsulphalein test in relation to, 61420a
transition metal selenocyanate complexes, 44421a
tuberculosis therapy with, phytohemagglutinin as immune adjuvant in, 79593k
uptake of, by phagocytes of humans and lab. animals, 17168a
**hydrazide, cobalt complex** *[88873-28-3]*, prepn. and cytotoxic inactivity of, 131330p
**hydrazide, cobalt complex, homopolymer** *[90169-09-8]*, prepn. and thermal dehydration of, 220542b
**hydrazide, cobalt complex, homopolymer** *[90169-11-2]*, 220542b

CODEN: CHABA8 100 1F–1468F (1984)
ISSN: 0009-2258 **100 F 1**

# Chemical Abstracts

Published by the

American Chemical Society

Volume 100

Formula Index, A–$C_{15}$

(Part 1 of 2 Parts)

January–June, 1984

*A publication of the*

CHEMICAL ABSTRACTS SERVICE

P. O. Box 3012, Columbus, Ohio 43210

2-Furancarboximidic acid
methyl ester [51282-48-5], 120794a
4-Isoxazolecarboxaldehyde, 3,5-dimethyl- [54593-26-9], 6280w
Methanamine, N-(2-furanylmethylene)-
N-oxide [41106-11-0], 139036t
Phenol, 4-(hydroxyamino)- [3505-87-1], P 85377j
2-Propenamide, N-(1-oxo-2-propenyl)- [20602-80-6]. For general derivs. see *Chemical Substance Index*
polymer with fibroin and 2-propenamide [88583-07-7], P 56676a
2-Propenenitrile, 3-(acetyloxy)-2-methyl- [76651-75-7], P 102778k
——, 3-ethoxy-2-formyl- [88223-52-3], P 22328y
2-Propenoic acid, 2-cyano-
ethyl ester [7085-85-0], 7215r, P 12704d, P 12705e, P 23098k, P 23329m, 34900a, P 35479y, P 52753a, P 69326m, P 122408r, P 211127t, R 211183h, P 211222v
ethyl ester, homopolymer [25067-30-5], 54481j, 103969k, P 176749y
ethyl ester, polymer with bis(2,2,3,3,4,4,5,5,6,6,7,7-dodecafluoroheptyl) methylenepropanedioate [89174-22-1], 103969k
ethyl ester, polymer with bis(2,2,3,3,4,4,5,5-octafluoropentyl) methylenepropanedioate [89174-20-9], 103969k
ethyl ester, polymer with bis(2,2,3,3-tetrafluoropropyl) methylenepropanedioate [89174-18-5], 103969k
ethyl ester, polymer with butyl 2-methyl-2-propenoate [90216-37-8], P 193173y
ethyl ester, polymer with dibutyl methylenepropanedioate [89174-14-1], 103969k
ethyl ester, polymer with diethyl methylenepropanedioate [89174-11-8], 103969k
ethyl ester, polymer with diheptyl methylenepropanedioate [89174-16-3], 103969k
ethyl ester, polymer with dipropyl methylenepropanedioate [89174-13-0], 103969k
ethyl ester, polymer with ethyl 2-propenoate [90216-36-7], P 193173y
ethyl ester, polymer with methyl 2-methyl-2-propenoate [27072-86-2], P 193173y
ethyl ester, polymer with methyl 2-propenoate [27072-83-9], P 193173y
——, 3-cyano-
ethyl ester, (Z)- [40594-97-6], 173938y
——, 2-methyl-
cyanomethyl ester [7726-87-6]. For general derivs. see *Chemical Substance Index*
cyanomethyl ester, homopolymer [66218-08-4], 200817a, 218931j
2-Propynoic acid, 3-(1-aziridinyl)-
methyl ester [88474-06-0], 51377u
Pyridine, 3-methoxy-
1-oxide [14906-61-7], 174614b
——, 4-methoxy-
1-oxide, radical ion(1+) [88058-00-8], 5731g
1-oxide, trichloroacetate [90254-28-7], 209012h
1-oxide, trifluoroacetate [76187-47-8], 209012h
2-Pyridinemethanol, 3-hydroxy- [14047-53-1], P 51574f
——, 5-hydroxy- [40222-77-3], 103139q, 174616d
2(1H)-Pyridinone, 4-methoxy- [52545-13-8], 34379d, 68134k
1H-Pyrrole-1-acetic acid [19167-98-7], 103113b
1H-Pyrrole-2-carboxylic acid
methyl ester [1193-62-0], P 6326r, 67911z
——, 1-methyl- [6973-60-0], P 6313j, P 85595d
——, 5-methyl- [3757-53-7], 208894k
1H-Pyrrole-3-carboxylic acid, 4-methyl- [64276-66-0], 208894k
1H-Pyrrole-2,5-dione, 3,4-dimethyl- [17825-86-4], P 53366p, 120917j
——, 1-ethyl- [128-53-0]. See *Chemical Substance Index*
2,5-Pyrrolidinedione, 1-ethenyl- [2372-96-5], 51413c, 121860x
polymer with 1-ethenyl-2-pyrrolidinone [83017-09-8]. For general derivs. see *Chemical Substance Index*
polymer with octyl 2-methyl-2-propenoate and 2-propenenitrile [88677-51-4]. For general derivs. see *Chemical Substance Index*
——, 1-methyl-3-methylene- [85639-10-7], 174631e

**$C_6H_7NO_2S$**

Acetic acid, (1H-pyrrol-2-ylthio)- [89597-78-4], 209570p
——, (1H-pyrrol-3-ylthio)- [89597-76-2], 209570p
Benzenesulfonamide [98-10-2], 6245p, P 36367x, P 104462b, 133827e, 139049z, P 149944b, 167726m, 182600j, 209139e. For general derivs. see *Chemical Substance Index*
monopotassium salt [29290-80-0], 182600j
monosodium salt [18522-93-5], 51178e
1,4-Oxathiin, 2,3-dihydro-5-isocyanato-6-methyl- [88258-66-6], 22624e
Pyridine, 2-(methylsulfinyl)-
1-oxide [75853-85-9], 103123e
——, 2-(methylsulfonyl)- [17075-14-8], 167760t
2H-1,3-Thiazin-2-one, 5-acetyl-3,6-dihydro- [88406-78-4], 34478k
5-Thiazoleacetic acid, 4-methyl- [5255-33-4], 206216e
2-Thiopheneacetic acid, α-amino-
(S)- [43189-45-3], P 85502w
2-Thiophenecarboxylic acid, 3-amino-
methyl ester [22288-78-4], 67961r, 209221a
——, 4-amino-
methyl ester [89499-43-4], 156449z
3-Thiophenecarboxylic acid, 4-amino-5-methyl- [66319-09-3], P 174651m

**$C_6H_7NO_2S_2$**

5-Thiazoleacetic acid, 2,3-dihydro-4-methyl-2-thioxo-
disodium salt [88585-70-0], P 103052f
Thiophene, 4-methyl-2-(methylthio)-3-nitro- [30129-92-1], P 34136x

**$C_6H_7NO_2S_3$**

1,3-Dithiole-4-carboxylic acid, 5-amino-2-thioxo-
ethyl ester [88714-58-3], 138996u

**$C_6H_7NO_3$**

Acetamide, N-(2,5-dihydro-2-oxo-3-furanyl)- [90237-92-6], 210388a
2-Furancarboxamide, 5-(hydroxymethyl)- [89149-72-4], 120794a
2(5H)-Furanone, 5-(2-aminoethylidene)-4-hydroxy-
hydrobromide, (Z)- [89004-85-3], 103092u
3-Isoxazolecarboxylic acid
ethyl ester [3209-70-9], P 120778q
4-Isoxazolecarboxylic acid, 3,5-dimethyl- [2510-36-3], 6280w
7-Oxabicyclo[4.1.0]hept-3-en-2-one, 3-amino-5-hydroxy-
[1S-(1α,5β,6α)]- [89020-30-4], 99624e
4-Oxazolecarboxylic acid, 2-ethyl- [75395-42-5], 34311a
2-Propenoic acid, 2-cyano-
methoxymethyl ester, polymer with methyl 2-methyl-2-propenoate [90235-74-8], P 193173y
——, 2-cyano-3-methoxy-
methyl ester [13974-74-8], P 102778k
2H-Pyran-2-one, 3-amino-4-hydroxy-6-methyl- [83432-20-6], 67687f
1H,3H-Pyrrolo[1,2-c]oxazole-1,3-dione, tetrahydro-
(S)- [45736-33-2], 175432c

**$C_6H_7NO_3S$**

Benzenesulfonamide, N-hydroxy- [599-71-3], 182600j
monosodium salt [39510-76-4], 182600j
——, 2-hydroxy- [3724-14-9], P 139141y, P 156634d
Benzenesulfonic acid, 2-amino- [88-21-1], P 8502u, 50925j, P 102981w, P 174434t, P 176455z, P 211643h
——, 3-amino- [121-47-1], 35813c, 35827k, P 85398a, P 87268y, P 102981w, P 105102w, P 140815c, 147298v, 173991k, P 211656q. For general derivs. see *Chemical Substance Index*
calcium salt (2:1) [88884-87-1], P 85398a
monoammonium salt [88884-86-0], P 85398a
monosodium salt [1126-34-7], P 85398a, P 87657z, P 211655p
——, 4-amino- [121-57-3]. See *Chemical Substance Index*
compd. with L-lysine [24735-59-9], 175237t
compd. with (±)-α-methylbenzenemethanamine (1:1) [89326-50-1], 138268h
monosodium salt [515-74-2], P 48096w, P 138755q, P 176669x
polymer with 5-amino-4-hydroxy-1,7-naphthalenedisulfonic acid, formaldehyde and 1,3,5-triazine-2,4,6-triamine [89655-58-3], P 161077d
polymer with formaldehyde [67471-65-2], P 193065q
polymer with formaldehyde and 1,3,5-triazine-2,4,6-triamine [89655-59-4], P 161077d
2-Pyridinol
methanesulfonate (ester) [25795-97-5], 117903w, 188886d
1H-1,2-Thiazine-4-carboxylic acid, 1-methyl-
1-oxide [88628-87-9], 68243v
5-Thiazoleacetic acid, 2,3-dihydro-4-methyl-2-oxo- [18600-98-1], 44801z
——, α-hydroxy-4-methyl- [87764-53-2], 44801z
2,4-Thiazolidinedione, 3-(2-oxopropyl)- [88419-03-8], 103228t
2-Thiopheneacetamide, N,α-dihydroxy- [89008-60-6], 103097z
3-Thiopheneacetamide, N,α-dihydroxy- [89008-63-9], 103097z
3-Thiophenecarboxylic acid, tetrahydro-2-imino-4-oxo-
methyl ester [90312-12-2], P 211664r

**$C_6H_7NO_3S_3$**

1,2,4-Dithiazolidine-4-acetic acid, 3-oxo-5-thioxo-
ethyl ester [89570-24-1], P 209838g

**$C_6H_7NO_4$**

Acetamide, N-(tetrahydro-2,5-dioxo-3-furanyl)-
(S)- [41148-79-2], P 109128g
2H-Azirine-2,3-dicarboxylic acid
dimethyl ester [16504-44-2], 156453u
2,5-Pyrrolidinedione, 1-(acetyloxy)- [14464-29-0], 81839p

**$C_6H_7NO_4S$**

Benzenesulfonic acid, 3-amino-4-hydroxy- [98-37-3], P 8502u, P 23557j, P 149944b, P 193523n, P 193525q
monosodium salt [83266-79-9], P 35853r
2-Pyridinemethanesulfonic acid, α-hydroxy- [3343-41-7], 153899h
2H-1,4-Thiazine-3,5-dicarboxylic acid, 3,6-dihydro-
(R)- [83923-11-9], 47637m

**$C_6H_7NO_5$**

2-Butenoic acid, 4-[(carboxymethyl)amino]-4-oxo-
(Z)- [54930-24-4]. For general derivs. see *Chemical Substance Index*
2,4-Pyrrolidinedicarboxylic acid, 5-oxo- [89267-01-6], 116607r

**$C_6H_7NO_5S$**

3-Thia-1-azabicyclo[3.2.0]heptane-2-carboxylic acid, 7-oxo-
3,3-dioxide, sodium salt, cis-(±)- [88586-38-3], 68052g

**$C_6H_7NO_6S_2$**

1,3-Benzenedisulfonic acid, 4-amino- [137-51-9], P 140815c
1,4-Benzenedisulfonic acid, 2-amino- [98-44-2], P 8504w, P 53189h, P 53199m, P 105086u, P 193522m
monosodium salt [24605-36-5], P 87247r

**$C_6H_7NS$**

Benzenethiol, 2-amino- [137-07-5]. See *Chemical Substance Index*
compd. with 1-butanamine (1:1) [88860-17-7], 85824c
compd. with N-ethylethanamine (1:1) [88860-14-4], 85824c
——, 3-amino- [22948-02-3], 68660d
——, 4-amino- [1193-02-8], P 12651j, 41688g, P 53366p, 55603n, 67511u, 68660d, 152952h, 156414g, P 210350y
monosodium salt [6976-04-1], P 12651j
Cyclopentene, 1-isothiocyanato- [87656-51-7], 5905a
——, 3-isothiocyanato- [52566-12-8], 5905a
Methanamine, N-(2-thienylmethylene)- [54433-71-5], P 120704n
2,4-Pentadienenitrile, 2-(methylthio)-
(E)- [90231-47-3], P 209190q
(Z)- [90231-48-4], P 209190q
Pyridine, 2-(methylthio)- [18438-38-5], P 6201w, 103123e, 167721f, 208632y
hydriodide [6313-75-3], 51804f
——, 3-(methylthio)- [18794-33-7], P 68079w
——, 4-(methylthio)- [22581-72-2], P 68079w
2-Pyridinemethanethiol [2044-73-7], 174609d
2(1H)-Pyridinethione, 4-methyl- [18368-65-5], 116349h

**$C_6H_7N_2O_3$**

Pyridinium, 1-methoxy-3-nitro-
perchlorate [76856-83-2], 208925w

**$C_6H_7N_2Tl$**

Thallium, [N-(1H-pyrrol-2-ylmethylene)methanaminato-N,N']- [88266-51-7], 22779w

**$C_6H_7N_3$**

1H-Imidazole-1-carbonitrile, 2,5-dimethyl- [83505-97-9], 34467f
Propanedinitrile, [(dimethylamino)methylene]- [16849-88-0], 50999m
2-Pyridinecarboxaldehyde
hydrazone [25976-65-2], P 128600w
2-Pyridinecarboximidamide [52313-50-5], 34507u, 51541t
monohydrochloride [51285-26-8], 51542u
3-Pyridinecarboximidamide [23255-20-1], 51541t
4-Pyridinecarboximidamide [33278-46-5], 51541t
monohydrochloride [6345-27-3], P 85725w
1H-Pyrrole-3-carbonitrile, 2-amino-4-methyl- [59146-60-0], 34498a

**$C_6H_7N_3O$**

Acetamide, N-2-pyrimidinyl- [13053-88-8], 167721f
2,4-Cyclohexadien-1-ol, 6-azido-
trans-(±)- [90057-19-5], 210307q
Methanimidamide, N-hydroxy-N'-2-pyridinyl- [69512-30-7], 51517q
1H-Pyrazole-4-carbonitrile, 4,5-dihydro-1,4-dimethyl-5-oxo- [89209-28-9], 120951r
1H-Pyrazolium, 4-cyano-5-hydroxy-1,1-dimethyl-
hydroxide, inner salt [89209-26-7], 120951r
2-Pyridinamine, N-methyl-N-nitroso- [16219-98-0], 98024a
3-Pyridinamine, N-methyl-N-nitroso- [69658-91-9], 98024a
4-Pyridinamine, N-methyl-N-nitroso- [16219-99-1], 98024a
2-Pyridinecarboxamide, 3-amino- [50608-99-6], 103298r
3-Pyridinecarboxamide, 2-amino- [13438-65-8], 34513t
——, 6-amino- [329-89-5], P 6343u, 30789q, 100784k, 116241a, P 120901z
2-Pyridinecarboximidamide, N-hydroxy- [1772-01-6]. For general derivs. see *Chemical Substance Index*
2-Pyridinecarboxylic acid
hydrazide [1452-63-7]. For general derivs. see *Chemical Substance Index*
3-Pyridinecarboxylic acid
hydrazide [553-53-7], 34482g, P 138862x, P 156609z, 191800b, 220920y. For general derivs. see *Chemical Substance Index*
4-Pyridinecarboxylic acid
hydrazide [54-85-3]. See *Chemical Substance Index*
hydrazide, mixt. with N-[4-[[(aminothioxomethyl)hydrazono]methyl]phenyl]acetamide [62682-46-6], 271x
5-Pyrimidinecarboxaldehyde, 4-amino-2-methyl- [73-68-7], P 103380m
Urea, 3-pyridinyl- [13114-65-3], 34384b

**$(C_6H_7N_3O)_n$**

Poly[imino[1-(1H-imidazol-4-ylmethyl)-2-oxo-1,2-ethanediyl]]
(S)- [26854-81-9], 48402t, 86116k, 103863w

**$C_6H_7N_3OS$**

5-Thiazolecarbonitrile, 4-amino-2-ethoxy- [29422-49-9], 174706h
Urea, (1,2-dihydro-2-thioxo-3-pyridinyl)- [89498-85-1], P 156624a

**$C_6H_7N_3OS_2$**

Acetamide, N-[(2-thiazolylamino)thioxomethyl]- [89012-12-4], 103294m

**$C_6H_7N_3O_2$**

Acetamide, 2-cyano-N-cyclopropyl-2-(hydroxyimino)- [70792-01-7], P 2188z
——, 2-cyano-2-[(2-propenyloxy)imino]- [50833-90-4], P 2188z
——, N-(1,6-dihydro-6-oxo-3-pyridazinyl)- [88259-94-3], P 22803z
——, N-(1,2-dihydro-2-oxo-4-pyrimidinyl)- [14631-20-0], 203559v
1,2-Benzenediamine, 3-nitro- [3694-52-8], 134046m, 162725a
mono(octyl phosphonate) [83316-34-1], P 47061u

CODEN: CHABA8 100 1537GS–3173GS (1984)
ISSN: 0009–2258 **100 GS 2**

# Chemical Abstracts

Published by the

American Chemical Society

Volume 100

General Subject Index, K–Z

(Part 2 of 2 Parts)

January–June, 1984

*A publication of the*
CHEMICAL ABSTRACTS SERVICE
P. O. BOX 3012, COLUMBUS, OHIO 43210

**Mass spectra** — *CHEMICAL ABSTRACTS - VOL. 100, 1984* — **1780GS**

methane reaction with platinum in relation to, 22256y
of methanoazaannulenecarboxylates, 120859e
of methoxybenzenepolycarboxylates, water elimination in, 208928z
of methoxybenzenepolycarboxylic acid Me esters, 22255x
of methoxybutanol or butanediol, 174113n
of methoxycinnamic acid, 208938c
of methoxycyclohexane, Pr-radical elimination paths in, 67668a
of methoxypyridinium perchlorates, 208925w
of *O*-methylaldose isopropylidene derivs., 64441k
of methylcyclohexenol, 116443j
of methylcyclohexyl acetates, 191183c
of methylenecyclopropanecarboxylate esters, 85285c
of methylglucose polysaccharide, 22925r
of methylphosphines, 139197w
of methyltolyloxadiazoles, 33999u
of middle mols., 208014z
MIKE, unimol. and collision-activated, of (trimethylsilyl)methyl-substituted chlorosilanes, 103439n
MIKES, chem.-ionization, of pyrocatechol, dimethylpentadiene, and their cycloaddn. product, 33899m
of mol. clusters generated by supersonic expansion, size dependent intensity fluctuations in, 148036v
of monobactams, 67693e
of multiply charged ions in hydrogen-contg. laser-induced multielement plasma, 218547p
neg.-ion
of carbon dioxide, in glow discharge, 27804k
chem.-ionization, of derivatized phenylalanine and tyrosine, 210366h
chem.-ionization, of histamine pentafluorobenzyl deriv., 96822b
chem.-ionization, of nucleoside, 175187b
chem.-ionization, of oxygen-18 labeled pentafluorobenzyltrimethylsilyl hydroxyeicosatetraenoate and leukotriene $B_4$, 188253v
of digitonin, by fast-atom bombardment, 170985a
fast-atom-bombardment, of boronate complexes of triols, sugars, and nucleosides, 121506m
fast-atom-bombardment, of glycosphingolipids, 188275d
of pentafluorophenyl derivs., 138429m
of sulfur dioxide, metastable transitions in, ion-mol. reactions in relation to, 129361f
of thiabicycloalkanes, 138434j
of neg. ions formed over chromium trichloride at high temps., 27802h
neutralization-reionization, of org. compds., 5647j
of neutral species, from electron interactions with gases, 129363h
of nickel complexes, 5733j
of nitrogen-contg. arom. compds., 42387v
of nitrogen oxide clusters in nozzle flow, 215865e
of nitrogen-sulfur heterocycles, 67790j
of nitrogen trifluoride discharge, 166345z
of nitrous oxide, metastable decay process of excited oxodinitrogen ion produced in, 127460p
of nonadiyne, 67683b
of nucleobase-functionalizing *β*-cyclodextrins, 121485d
of nucleoside phosphite ester derivs., 47965e
of nucleotides, 7050h
of octane derivs. at low temps. and voltages, 138455e
of octene, 50923g
of oligodeoxyribonucleotides and oligoribonucleotides, P 135436a
of oligomeric tannins, 175141g
of oligosaccharide and sugars, 3022c
of oligosaccharides, of salivary mucin of nest of Chinese swallow, 170300y
optimization of information selection from spectral library to establish useful training sets in pattern recognition or retrieval methods in, 184878e
of organoaluminum compd. decompn. products, 127399a
of organotin carbonyl compds. and their oxazolidine and dioxolane derivs., 85838k
of organotin compds., 85839m
of org. azides and halides, R 155902c
of org. compds. using surface ionization on hot rhenium filament, 218548q
of org. peroxides, R 155897e
of oxacephems, 208924v
of oxazolidines, amino alcs. and hydroxyalkyl lactams, 102546h
of oxo- and thioxoimidazolidinedimethanol, 156081w
of (oxopropyl)cycloalkanones, 156085a
of oxygen(1+) in topside nighttime ionosphere, satellite observations of, 182685r
of oxygen-18-labeled adenosine phosphorothioates, from fast-atom bombardment, 153029z
of ozonolysis products from erucic acid, 51059e
of penta-*O*-benzoylgalactofuranoses, 139493q
of penta-*tert*-butylcyclopentastibine, 6709t
of pentamethylpyrimidodiindole, 34495p
PEPICO, of hexamethyldisilane, 103436j
of peptides
19949w
deuteration in relation to, 872u
metastable mapping in, 192251k
of peptides and low-mol.-wt. proteinase inhibitors, 205399e
of peptides and polymers with grand-scale tandem instrument, R 199925e
of paracetylated alditolnitrile of muramic acid, 2922j
of peralkylcyclopolysilanes, 51684e
of (perfluoroalkylthio)benzene and -butane, 191478c
of perfluorodecalin and perfluoromethyldecalin isomers, 120334e
of permethylated peptides, capillary gas chromatog. columns in relation to, 2927q
of $PGF_{2α}$ metabolite derivs., 79972h
of phencyclidine analogs, 18823p
of phenolic compds., of plant cell walls, 48623r
of phenols, anisoles, and benzyl alcs., 138437n
of phenothiazine antipsychotic agent, 156082z
of phenyldioxaphospholane oxides, 34613a
of phenyloxetane, carbon exchange in, 50922f
of phenylthiohydantoin derivs. of amino acids, 103846t
of phosphorylated and thiophosphorylated aminobenzothiazine, 6672a
of phosphorylated mercaptobutenoates, 34618f
of photocycloadduct of cyclohexenedicarboxylic anhydride and its double bond positional isomer, 50815y
of photodecompn. products of benzoylaminothiazoline derivs., 77221p
photodissocn.
of alkanes, 120340r
of isomeric $[C_4H_5N]^{+}$ ions, 156092a
photoelectron-photoion, of acetone enol, 67664w
photoelectron-photoion coincidence, of cyclopentane isomers, 208952c
photoionization
of aniline van der Waals complexes with helium and methane, 94070f
of halogenated methane, 218554p
of hydrogen chloride, abs. cross-sections for, 218102q
laser-induced, of rare gas and iodine ions, 182619x
mol. beam, of oxidized metal clusters, 111221a
of oxides, cluster ion distributions and correlations in, 200393a
of vinyl chloride from reaction of chlorine atoms with vinyl bromide, chlorine-atom measurements by, 50777n
of photoionized isobutene, 120442a
of piperazines, 147m
of pirisudanol, 18821m
of plant hormones of indole type, 205917j
of plasmas used in deposition of materials, 147799j
of platelet-activating factor of human neutrophil, 49705f
of polyethers of thio- and dithioethylene glycol, 67678d
of polymers, data bank of, 23171d
of polymethoxy flavones, 207947t
of polymn. and sublimation of cesium iodide, 145957k
of polyoxymethylene glycol trimethylsilyl deriv., 95840u
of polystyrene plasma-induced sheaths, 192809y
of polyunsatd. fatty acid methoxy derivs., 117376b
of porphyrins from Boscan crude oil, R 153191w
pos.-ion, fast-atom-bombardment, of glycosphingolipids, 188275d
of potassium dicyanoiodate di-Et diiminooxalate, 95494j
of primary laser desorption products, in gas desorption study from stainless steel surface, 59176y
of procyanidins $B_3$ and $B_4$, 120725v
of protonated acetone, propanal, propylene oxide, oxetane, and allyl alc., 174096j
of protriptyline photolysis products, 156339m
of pteridine trimethylsilylated derivs., 206072s
of pyrazolones and pyrazolethiones, 138443m
of pyridotriazines and their dihydro analogs, 5618a
of pyridoxamine phosphate hydrochloride, 138442k
of (pyrimidinylthio)acetic acids, 85086p
of pyrocarboxyglutamic acid, 116607r
pyrolysis, of oligodeoxyribonucleotides, 117367z
of pyrrolopyridinones, 209754b
of pyrrolopyrimidinones, 68258d
of pyrylium and pyridinium salts, 102591u
of quaternary ammonium salts, energy deposition in desorption-ionization in relation to, 14833u
of rare earth *β*-diketone complexes, 200379a
of rare-gas cluster ions, 182668n
retro-Diel-Alder reaction in, R 173907n
of rotenoid derivs., 50902z
of scandium(III) and yttrium(III) complexes of trifluoromethyl-*β*-diketones, 200395c
of Schiff bases derived from dimethylvaleraldehyde, acetylacetone, and benzoylacetone, 14725k
secondary-ion
of acidic glycosphingolipids and permethylated derivs., 117365x
of alkali metal halides in cluster ion formation, 129353e
of aminoglycoside antibiotics, 156914v
of beryllium-doped aluminum gallium arsenide double heterostructure lasers, reduced threshold current temp. dependence in relation to, 15039b
of coatings, R 16774z
of deuterated glycerol, 50907e
of gallium indium arsenide contg. germanium and manganese, redistribution in relation to, 111654u
of Hb F Izumi trypsin digest, 153283c
of hydrogen and deuterium in solid state, 111657x
of metal oxides, cation formation in relation to, 59180v
of metals bombarded by bromine ions, 164866w
of oligoribonucleotides, 210331t
of oligosaccharides in amide matrixes, 68623u
of oligosaccharides of sialoglycoproteins of egg of salmon, 47285p
of peptides, sequencing by, 139611b
pos., metal hydride peaks in, 114174y
in tokamak limiter surface study, 163641g
of silane glow discharge plasma during hydrogenated amorphous silicon film deposition, 78317m
of silatranes, 22685n, 209958w
of silicon-contg. cyclic polysulfides, 121164y
of silver-thiophosphoryl chloride reaction, 78915e
of silylated $C_{6-10}$ fatty acids, 20014u
of silylquinolizines, 6609k
of sodium cluster, 147998y
of sodium clusters under equil. conditions, 200398f
of sodium sulfate and sulfite and thiosulfate, 167225x
spark-source, of mercuric iodide, purity in relation to, 113837e
of spirobicyclic sulfuranes, 164870t
of stearic acid-based vinyl monomers, 210473r
of steroidal glycosides, of starfish, 48917q
of steroids, R 85978f
of sterols, 103744h
of stilbenols and stilbenediols, 208926x
of sublimation and polymn. of sodium iodide, 145958m
tandem
of cationic iron and technetium complexes, 94067k
data systems for evaluation of, R 202769b
ionic reaction mechanism from, R 190979e
of radicals, R 190981z
structural information from, R 190978d
of tetrahydroisoquinolines, mechanism of substituent loss in, 156099h
of tetrahydrooxazaphosphorin oxides, 138438p
of tetrahydroquinolines, 138498b
of tetrahydrotrimethylquinolinol and its oxidn. product, 102550e
of tetraiododiborane, 78916f
of tetramethyldiphosphine, MO calcns. in relation to, 121190d
of thiabicycloalkanes, 138434j
of thienylsilanes, 22694q
of (thioacyl)hydrazone tautomers, 191254b
of thiobenzanilides, 208927y
of thiophenecarboxylic acids, 67670v
of thiophenemercuric compds., fragmentation patterns in, 121265g
of tin compds. found in household products, field ionization in, 59188d
of tin oxide-tungsten oxide system, heavy mol. forms in, 111650q
of TLC spots, fast-atom bombardment for, 29107j
of toxic substances, for data base, 126043t
of transition metal diethylselenothiocarbamates, 167076z
of transition metal trialkylbismuthine carbonyl complexes, 202448w
of trialkyl- and triphenyldiazepines, 86079p
of trichloroacetonitrile oligomerization products, 157044y
of trichlorobiphenyl metabolites, 133709t
of trifluoropentanedione bis-Schiff bases and their copper(II) and nickel(II) complexes, 5628d
of trimethylammonium compds. related to choline and carnitine, 208936a
of trimethylsilyl or acetyl derivs. of Me hydroxypalmitate, 5625a
of trinitrobenzene derivs., 191181a
of trinitrotoluene and its isotopically labeled analogs, 33980f
of trioxaphosphabicyclodecane, stereospecificity of hydrogen rearrangement in, 209985c
of trioxaphosphathionobicyclodecane diastereomers, 209984b
of triphenyltin carboxylates, 5633b
triple quadrupole, of furan and benzene and thiophene mixt., 129355g
of tris(acetylacetonato)lanthanide complexs with *o*-phenanthroline and *α,α'*-bipyridine, 218546n
of tris(acetylacetonato)lanthanide reaction products with nitrogen-contg. crown ethers, 150002z
of tungsten carbonyl thiolato dinuclear complexes, 44412y
of two-layer systems contg. tellurium and metals or oxides, 149395y
of uranium, using intracavity laser ionization, 111647u
of versicoside A, 83014q
of vitamin $B_{12}$ analogs, 129362g
of weakly coordinated cations and cation radicals, 208930u
of xylidines, 85138g

**Mass spectrometers**
P 59464r, 111659z, 182908r, R 219156x
accelerator, for detn. of aluminum-26, 167257j
for angle-resolved ion kinetics energy spectra of collisionally scattered ion beams, 165162a
app. for, for anal. of gas discharge plasmas, 200636g
on Atm. Explorer, expts. using, 42759t
automation of, system for, 200635f
automation of scan control of, for isotopic anal. of light gases, 220733q
Bennett, performance of, evaluation of, by computer simulation, 114075e
for biochem. anal., 64267h
calibration in, in temp.-programmed desorption evolved gas anal., 114068p
in capacity and condition study for condensation for mol. with fragment, P 15180r
charged-particle analyzer for, P 129697v
for charged-particles, 199490w
collision-activated dissocn., with high resoln., 59374m
with combination of time resoln. and mass dispersive techniques, P 59475v
computerized magnetic field regulation for, P 59425d
for consecutive ion dissocn. studies, P 200739t
detectors, P 129702t
for detn. of elements, with inductively coupled plasma source and continuum flow sampling, 28932n
digital scanner for, P 218886y
direct sample in chamber for, P 129661d
double-beam, in detn. of isotopes, effect of peripheral magnetic field on, 16815p

**Consult 1984 Index Guide for Cross-references and Indexing Policies**

**CODEN: CHABA8 1I–4I, 1G–1507G, 5I–231I (1984)**
**ISSN: 0009-2258**

# Chemical Abstracts

**Published by the**

## American Chemical Society

## Index Guide

## 1984

This *Index Guide* outdates the 1982 *Index Guide* which can now be discarded.
The *Index Guide* will be reissued in September 1985

*A publication of the*
CHEMICAL ABSTRACTS SERVICE
P. O. BOX 3012, COLUMBUS, OHIO 43210

**α-Amyrenone**
See *Urs-12-en-3-one [638-96-0]*
**β-Amyrenone**
See *Olean-12-en-3-one [638-97-1]*
**δ-Amyrenone**
See *Olean-13(18)-en-3-one [20248-08-2]*
**β-Amyrenonol**
See *Olean-12-en-11-one, 3-hydroxy-, (3β)- [38242-02-3]*
**β-Amyrilene II**
See *Oleana-2,12-diene [640-25-5]*
**α-Amyrin**
See *Urs-12-en-3-ol, (3β)- [638-95-9]*
**β-Amyrin**
See *Olean-12-en-3-ol, (3β)- [559-70-6]*
**——, 30-hydroxy-**
See *Olean-12-ene-3,29-diol, (3β,20β)- [6813-59-8]*
**γ-Amyrin**
See *Oleanan-3-ol, 12,13-epoxy-, (3β,12β)- [28413-82-3]*
**δ-Amyrin**
See *Olean-13(18)-en-3-ol, (3β)- [508-04-3]*
**Amyrolin**
See *2H,8H-Benzo[1,2-b:3,4-b′]dipyran-2-one, 8,8-dimethyl- [523-59-1]*
**α-Amyrone**
See *Urs-12-en-3-one [638-96-0]*
**β-Amyrone**
See *Olean-12-en-3-one [638-97-1]*
**δ-Amyrone**
See *Olean-13(18)-en-3-one [20248-08-2]*
**Amytal**
See *2,4,6(1H,3H,5H)-Pyrimidinetrione, 5-ethyl-5-(3-methylbutyl)- [57-43-2]*
**Amyzyl**
See *Benzeneacetic acid, α-hydroxy-α-phenyl-, esters, 2-(diethylamino)ethyl ester, hydrochloride [57-37-4]*
**Amzyme TX 8**
See *Amylase [9000-92-4]*
**AN (polyester)**
See also *Polyesters*
**AN 1 (pharmaceutical)**
See *Benzeneacetonitrile, α-[(1-methyl-2-= phenylethyl)amino]- [17590-01-1]*
**AN 3**
See *Iron alloy, base, Fe 57-60,Ni 24-25,Al 13-14,Cu 3-4,Ti 0.2-0.3 (YuND4) [39360-80-0]*
**AN7 (steel)**
See *Steel, (PP-AN7) [50948-27-1]*
**AN 41**
See *Pyridine, 2-ethenyl-5-ethyl-, homopolymer [27755-56-2]*
**AN101**
See *Iron alloy, base, Fe 58-72,Cr 22-28,Si 2.5-5,Ni 1.3-3.6,C 2.2-3,Mn 0-2.5 (PL AN101) [58411-34-0]*
**AN 105**
See *Iron alloy, base, Fe 80-82,Mn 14-15,Ni 3-4,C 0.8-0.9,Si 0.2-0.4 (PP-AN105) [57890-01-4]*
**AN106**
See *Iron alloy, base, Fe 83-87,Cr 12-15,Mn 0.2-0.8,Si 0.3-0.6,C 0.1-0.2,Ti 0.1-0.2 (PP-AN106) [57687-83-9]*
**AN111**
See *Chromium alloy, base, Cr 35-40,Ni 28-35,Fe 14-30,C 4.5-5.5,Si 1.5-3.5,Mn 1-2,B 0.2-0.4 (PL AN111) [63414-40-4]*
**AN 162**
See *Benzenemethanol, α-[[(2,5-dichloropheny= l)[2-(diethylamino)ethyl]amino]methyl]- [33189-65-0]*
**AN 278**
See *1,4-Benzenediamine, N,N′-dibutyl-N,N′-= dinitroso- [19433-82-0]*
**AN 448**
See *3H-Imidazo[2,1-a]isoindol-5-ol, 5-(4-chlorophenyl)-2,5-dihydro- [22232-71-9]*
**AN 600**
See *2-Propene-1-sulfonic acid, 2-methyl-, sodium salt, polymer with 2-propenenitrile [27103-76-0]*
**AN 1200**
See *2-Propene-1-sulfonic acid, 2-methyl-, sodium salt, polymer with 2-propenenitrile [27103-76-0]*
**AN 56477**
See *Benzenamine, N,N-bis(2-chloroethyl)-4-= methyl-2,6-dinitro- [26389-78-6]*
**Anabactyl**
See *4-Thia-1-azabicyclo[3.2.0]heptane-2-= carboxylic acid, 6-[(carboxyphenylacetyl)= amino]-3,3-dimethyl-7-oxo-, disodium salt, [2S-(2α,5α,6β)]- [4800-94-6]*
**Anabasamine**
See *2,3′-Bipyridine, 5-(1-methyl-2-piperidinyl)-, (+)- [20410-87-1]*
**Anabaseine**
See *2,3′-Bipyridine, 3,4,5,6-tetrahydro- [3471-05-4]*
**Anabasine**
See *Pyridine, 3-(2-piperidinyl)-, (S)- [494-52-0]*
**Anabellamide**
See *L-Phenylalanine, N-benzoyl-, 2-(benzoylamino)-3-phenylpropyl ester, (S)- [63631-36-7]*
**Anabilysine**
See *Pyridinium, 1-(1-amino-1-carboxypenty= l)-3-[1-(1-amino-1-carboxypentyl)-2-= piperidinyl]-, chloride [66517-63-3]*
**Anabiol**
See *Estr-4-ene-3,17-diol, dipropanoate, (3β,17β)- [1986-53-4]*
**Anabion**
See *Benzenesulfonamide, 4-amino-N-[4-= [(methylamino)sulfonyl]phenyl]- [547-53-5]*
**Anabolin**
See *Androsta-1,4-dien-3-one, 17-hydroxy-17-= methyl-, (17β)- [72-63-9]*
**Anabolism**
See *Metabolism*, anabolic
**Anaboliberin**
See *L-Valine, N-[N-[N²-[N-[N-(5-oxo-L-= prolyl)-L-seryl]-L-histidyl]-L-arginyl]-L-= leucyl]- [68107-09-5]*
**Anabsin**
See *2H-8,15-Epoxy-7,13b-ethanopentaleno[1″,= 2″:6,7;5″,4″:6′,7′]dicyclohepta[1,2-b:1′,2′-= b′]difuran-2,12(11H)-dione, 3,3a,4,5,6,6a,6b,7,7a,8,9,10,10a,13a,13c,14b-= hexadecahydro-6,16-dihydroxy-3,6,8,11,14,= 15-hexamethyl-, [3S-(3α,3aα,6β,6aα,6bβ,= 7α,7aβ,8α,10aβ,11β,13aα,13bα,13cβ,14bβ,= 15S*,16R*)]- [72542-39-3]*
**Anabsinthin**
See *2H-8,15-Epoxy-7,13b-ethanopentaleno[1″,= 2″:6,7;5″,4″:6′,7′]dicyclohepta[1,2-b:1′,2′-= b′]difuran-2,12(11H)-dione, 3,3a,4,5,6,6a,6b,7,7a,8,9,10,10a,13a,13c,14b-= hexadecahydro-6-hydroxy-3,6,8,11,14,15-= hexamethyl-, [3S-(3α,3aα,6β,6aα,6bβ,7α,= 7aβ,8α,10aβ,11β,13aα,13bα,13cβ,14bβ,15R*)]- [6903-12-4]*
**(15:0)-Anacardic acid**
See *Benzoic acid, 2-hydroxy-6-pentadecyl- [16611-84-0]*
**Anacardic materials**
See *Cashew*, nutshell liq.
**Anacardiol**
See *Benzamide, 3-ethoxy-N,N-diethyl-4-hydroxy- [13898-68-5]*
**Anacardium occidentale**
See *Cashew*
**Anacardium orientale**
See *Semecarpus anacardium*
**Anacharis**
See *Elodea*
**Anacin**
See *Benzoic acid, 2-(acetyloxy)-, mixt. with 3,7-dihydro-1,3,7-trimethyl-1H-purine-2,= 6-dione [53908-20-6]*
**Anacrotine**
See *Senecionan-11,16-dione, 6,12-dihydroxy-, (6β)- [5096-49-1]*
**Anacyclin (contraceptive)**
See *19-Norpregna-1,3,5(10)-trien-20-yn-17-ol, 3-methoxy-, (17α)-, mixt. with (17α)-19-norpregn-4-en-20-yn-17-ol [8015-14-3]*
**Anacyclin (natural product)**
**——, 12,13-didehydro-**
See *2,4,12-Tetradecatriene-8,10-diynamide, N-(2-methylpropyl)- [29428-83-9]*
**Anacystis**
See also *Synechococcus*
**Anadensin**
See *Dicyclopenta[a,d]cycloocten-2(1H)-one, 4,5,6,6a,7,8,9,9a,10,10a-decahydro-10a-= hydroxy-1,4,9a-trimethyl-7-(1-= methylethyl)-, [1S-(1α,4β,6aα,7β,9aβ,10aα)]- [58740-18-4]*
**Anadol**
See *4-Piperidinol, 1,3-dimethyl-4-phenyl-, propanoate (ester), hydrochloride [14405-05-1]*
**Anadoline**
See *2-Butenoic acid, 2-methyl-, 2-hydroxy-1,= 3-dimethyl-2-[[(2,3,5,7a-tetrahydro-1-= hydroxy-1H-pyrrolizin-7-yl)methoxy]= carbonyl]butyl ester, N-oxide, [1R-[1α,7[1R*(E),2S*],7aβ]]- [28513-29-3]*
**Anadrol**
See *Androstan-3-one, 17-hydroxy-2-= (hydroxymethylene)-17-methyl-, (5α,17β)- [434-07-1]*
**Anadur**
See *Estr-4-en-3-one, 17-[3-[4-(hexyloxy)= phenyl]-1-oxopropoxy]-, (17β)- [52279-57-9]*
**Anaferine**
L-(+)——see *2-Propanone, 1,3-di-2-= piperidinyl-, [S-(R*,R*)]- [19519-53-0]*
**Anaflex**
See *Urea, polymers, polymer with formaldehyde [9011-05-6]*
**Anaflon**
See *Acetamide, N-(4-hydroxyphenyl)- [103-90-2]*
**Anafranil**
See *5H-Dibenz[b,f]azepine-5-propanamine, 3-chloro-10,11-dihydro-N,N-dimethyl-, monohydrochloride [17321-77-6]*
**Anagadiol**
See *Olean-18-ene-1,3-diol, (1β,3β)- [33600-94-1]*
**Anagalligenin B**
See *Oleanane-3,16,23-triol, 13,28-epoxy-, (3β,4α,16α)- [33722-92-8]*
**Anagalligenone B**
See *Oleanan-16-one, 13,28-epoxy-3,23-= dihydroxy-, (3β,4α)- [33809-48-2]*
**Anagalligenone lactone B**
See *Oleanan-28-oic acid, 3,13,23-trihydroxy-= 16-oxo-, γ-lactone, (3β,4α)- [35061-05-3]*
**Anagasta kuehniella**
Mediterranean flour moth is also indexed at this heading
**Anagestone acetate**
See *Pregn-4-en-20-one, 17-(acetyloxy)-6-= methyl-, (6α)- [3137-73-3]*
**Anagrelide**
See *Imidazo[2,1-b]quinazolin-2(3H)-one, 6,7-dichloro-1,5-dihydro- [68475-42-3]*
**Anagyrine**
See *7,14-Methano-4H,6H-dipyrido[1,2-a:1′,= 2′-e][1,5]diazocin-4-one, 7,7a,8,9,10,11,= 13,14-octahydro-, [7R-(7α,7aβ,14α)]- [486-89-5]*
**Analcime ($AlNa(SiO_3)_2.H_2O$)** *[1318-10-1]*
The naturally occurring mineral only is indexed at this heading. Synthetic analcimes are indexed at *Silicic acid ($H_2SiO_3$), aluminum sodium salt (2:1:1), monohydrate* and(or) at *Zeolites*
**Analcite**
See *Analcime ($AlNa(SiO_3)_2.H_2O$) [1318-10-1]*
**Analeptics**
See *Central nervous system stimulants*
**Analergine**
See *1H-Imidazole-2-methanamine, 4,5-dihydro-N-phenyl-N-(phenylmethyl)- [91-75-8]*
**Analgesia**
See also *Anesthesia*
acupuncture——see *Acupuncture*
agents inducing (new or class)——see *Analgesics*
**Analgesics**
Studies of analgesics as a class as well as studies of new analgesics are indexed at this heading. Studies of known analgesics are indexed at those specific headings and at *Analgesia*
See also
*Anesthetics*
*Narcotics*
**Analgin**
See *Methanesulfonic acid, [(2,3-dihydro-1,5-= dimethyl-3-oxo-2-phenyl-1H-pyrazol-4-= yl)methylamino]-, sodium salt [68-89-3]*
**Analgocain**
See *Methanesulfonic acid, [(2,3-dihydro-1,5-= dimethyl-3-oxo-2-phenyl-1H-pyrazol-4-= yl)methylamino]-, sodium salt, mixt. with 2-(diethylamino)-N-(2,6-dimethylphenyl)= acetamide monohydrochloride [65168-73-2]*
**Analog computers**
See
*Computer application*
*Computers*
analog
**Analogy theory**
See *Mathematics*
**Analysis**
The methodologies of chemical analysis are indexed at this heading. Specific analytical numbers are indexed at such headings as *Acetyl number, Iodine number, Saponification number.* Analytical methods for specific substances and for classes of substances (excepting elements, anions, cations, metals in combined form, biological materials, and organic compounds in the broadest meanings of these terms) are indexed at headings for the specific substances and the classes
See also *Separation*
calibration in——see *Calibration*
chromatog.——see *Chromatography* headings
colorimetric——see *Spectrochemical analysis*
by distn.
see
*Distillation*
*Distillation apparatus*
electron microprobe——see *Electron microprobe analysis*
electrophoretic——see *Electrophoresis and Ionophoresis*
ESCA——see *Photoelectron spectroscopy*
forensic——see *Legal chemistry and medicine*
immunochem.——see *Immunochemical analysis*
lab. ware
see
*Crucibles*
*Laboratory ware*
masking in——see *Masking, analytical*
mass spectrometric——see *Mass spectroscopy*
for metals of anal. Group II——see *Hydrogen sulfide group*
for metals of anal. Group III——see *Ammonium sulfide group*
nephelometric——see *Nephelometry*
polarimetric——see *Polarimetry*
radiochem. or radioactivation——see *Radiochemical analysis*
sampling in——see *Sampling*
spectrochem.——see *Spectrochemical analysis*
standardization in
see
*Standardization*
*Standard solutions*
*Standard substances*
thermal
see
*Thermal analysis*
*Thermogravimetric analysis*
titrimetric
see such headings as
*Argentometry*
*Chelatometry*
*Iodometry*
*Titration*
of trace elements——see *Trace elements*
turbidimetric——see *Turbidimetry*
voltammetric
see such headings as
*Chronopotentiometry*
*Polarography*
*Voltammetry*
x-ray——see *X-ray analysis*
**Anamarin**
See *2H-Pyran-2-one, 5,6-dihydro-6-[3,4,5,6-= tetrakis(acetyloxy)-1-heptenyl]-, [6R-[6R*(1E,3R*,4S*,5S*,6S*)]]- [73413-69-1]*
**Anamnestic reaction**
See *Antibodies*
**Ananas comosus**
See *Pineapple*
**Ananasic acid**
See *9,19-Cyclolanost-24-en-26-oic acid, 3,11,15-trihydroxy-, (3β,11α,15α,24E)- [60877-02-3]*
**Ananas sativus**
See *Pineapple*
**Ananolide**
See *Cyclohexanepropanoic acid, 2-propenyl ester [2705-87-5]*

**Atisinone, dihydro–** *CHEMICAL ABSTRACTS 1984* **118G**

**——, dihydro-**
See *Atidan-15-one, 16,17-didehydro-21-(2-hydroxyethyl)-4-methyl- [74260-93-8]*
**Atisirene**
See *Atis-16-ene, (5β,8α,9β,10α,12α)- [20230-48-2]*
**Atisirenic acid**
See *Atis-16-en-18-oic acid, (4α,5β,8α,9β,10α,12α)- [75678-89-6]*
**Ativan**
See *2H-1,4-Benzodiazepin-2-one, 7-chloro-5-(2-chlorophenyl)-1,3-dihydro-3-hydroxy- [846-49-1]*
**ATJ-S**
See *Graphite [7782-42-5]*
**Atka mackerel**
See
*Pleurogrammus azonus*
*Pleurogrammus monopterygius*
**Atlac**
See also *Polyesters*
**Atlac 382-2**
See *Poly[oxy(1,4-dioxo-2-butene-1,4-diyl)oxy-1,4-phenylene(1-methylethylidene)-1,4-phenylene], (E)- [27879-05-6]*
**Atlac 382-05**
See *Poly[oxy(1,4-dioxo-2-butene-1,4-diyl)oxy-1,4-phenylene(1-methylethylidene)-1,4-phenylene], (E)- [27879-05-6]*
**Atlac 382E**
See *2-Butenedioic acid (E)-, polymers, polymer with α,α′-[(1-methylethylidene)di-4,1-phenylene]bis[ω-hydroxypoly[oxy(methyl-1,2-ethanediyl)]] [39382-25-7]*
**Atlac 382ES**
See *2-Butenedioic acid (E)-, polymers, polymer with α,α′-[(1-methylethylidene)di-4,1-phenylene]bis[ω-hydroxypoly[oxy(methyl-1,2-ethanediyl)]] [39382-25-7]*
**Atlac G 382**
See *Poly[oxy(1,4-dioxo-2-butene-1,4-diyl)oxy-1,4-phenylene(1-methylethylidene)-1,4-phenylene], (E)- [27879-05-6]*
**Atlac 4010L**
See *Poly[oxy(1,4-dioxo-2-butene-1,4-diyl)oxy-1,4-phenylene(1-methylethylidene)-1,4-phenylene], (E)- [27879-05-6]*
**Atlac 382-05L**
See *Poly[oxy(1,4-dioxo-2-butene-1,4-diyl)oxy-1,4-phenylene(1-methylethylidene)-1,4-phenylene], (E)- [27879-05-6]*
**Atlac 580-05L**
See *Poly[oxy(1,4-dioxo-2-butene-1,4-diyl)oxy-1,4-phenylene(1-methylethylidene)-1,4-phenylene], (E)- [27879-05-6]*
**Atlac T 500**
See *2-Butenedioic acid (E)-, polymers, polymer with α,α′-[(1-methylethylidene)di-4,1-phenylene]bis[ω-hydroxypoly[oxy(methyl-1,2-ethanediyl)]] [39382-25-7]*
**Atlantia**
See *Atalantia*
**Atlantic Acid Fast Blue B**
See *Benzenesulfonic acid, 2-[(4-amino-3-bromo-9,10-dihydro-9,10-dioxo-1-anthracenyl)amino]-5-methyl-, monosodium salt [6424-75-5]*
**Atlantic Acid Violet 4BNS**
See *Benzenemethanaminium, N-[4-[[4-(dimethylamino)phenyl][4-[ethyl[(3-sulfophenyl)methyl]amino]phenyl]methylene]-2,5-cyclohexadien-1-ylidene]-N-ethyl-3-sulfo-, hydroxide, inner salt, sodium salt [1694-09-3]*
**Atlantic Alizarine Milling Blue RB**
See *Benzenesulfonic acid, 3,3′-[(9,10-dihydro-9,10-dioxo-1,4-anthracenediyl)diimino]bis[2,4,6-trimethyl-, disodium salt [4474-24-2]*
**Atlantic bonito**
See *Sarda sarda*
**Atlantic Chrysoidine Y**
See *1,3-Benzenediamine, 4-(phenylazo)-, monohydrochloride [532-82-1]*
**Atlantic cod**
See *Cod*
**Atlantic croaker**
See *Micropogon undulatus*
**Atlantic cutlassfish**
See *Trichiurus lepturus*
**Atlantic Direct Black E-NB**
See *C.I. Direct Black 170 [86167-73-9]*
**Atlantic Direct Blue 3R-NB**
See *C.I. Direct Blue 285 [86167-90-0]*
**Atlantic Direct Brown 3GN-NB**
See *C.I. Direct Brown 232 [86167-93-3]*
**Atlantic Direct Brown 3RB-NB**
See *C.I. Direct Brown 231 [86167-92-2]*
**Atlantic Direct Fast Blue 2RL-NB**
See *C.I. Direct Blue 286 [86167-91-1]*
**Atlantic Direct Fast Brown BCW-NB**
See *C.I. Direct Brown 238 [86167-76-2]*
**Atlantic Direct Green 2B-NB**
See *C.I. Direct Green 92 [83929-82-2]*
**Atlantic Direct Navy Blue BH-NB**
See *C.I. Direct Blue 165 [12222-03-6]*
**Atlantic Fast Turquoise LGA**
See *Cuprate(2-), [29H,31H-phthalocyaninedisulfonato(4-)-$N^{29}$,$N^{30}$,$N^{31}$,$N^{32}$]-, disodium [1330-38-7]*
**Atlantic halibut**
See *Hippoglossus hippoglossus*
**Atlantic herring**
See *Herring*
**Atlantic mackerel**
*Scomber scombrus* is also indexed at this heading
**Atlantic Malachite Green**
See *Methanaminium, N-[4-[[4-(dimethylamino)phenyl]phenylmethylene]-2,5-cyclohexadien-1-ylidene]-N-methyl-, chloride [569-64-2]*

**Atlantic Milling Red B**
See *C.I. Acid Red 167 [61901-41-5]*
**Atlantic Resin Fast Black LW-NB**
See *C.I. Direct Black 170 [86167-73-9]*
**Atlantic salmon**
See *Salmo salar*
**Atlantic threadfin**
See *Polydactylus octonemus*
**Atlantic wolffish**
See *Anarhichas lupus*
**α-Atlantone**
See *2,5-Heptadien-4-one, 2-methyl-6-(4-methyl-3-cyclohexen-1-yl)-, [R-(E)] [26294-59-7]*
**β-Atlantone**
See *1,5-Heptadien-4-one, 6-methyl-2-(4-methyl-3-cyclohexen-1-yl)-, (R)- [38331-79-2]*
**Atlas 1086**
See *2-Butenedioic acid (E)-, polymers, polymer with (chloromethyl)oxirane, 2,2-dimethyl-1,3-propanediyl bis(2-methyl-2-propenoate) and 4,4′-(1-methylethylidene)bis[phenol] [56166-18-8]*
**Atlas Ago 500**
See also *Urethane polymers*
**Atlasetox**
See *Phosphorothioic acid, esters, O,O-dimethyl O-[2-(methylthio)ethyl] ester, mixt. with O,O-dimethyl S-[2-(methylthio)ethyl] phosphorothioate [8065-62-1]*
**Atlas G numbers**
See *G Numbers*
**Atlox 1087**
See *Sorbitan, mono-9-octadecenoate, poly(oxy-1,2-ethanediyl) derivs., (Z)- [9005-65-6]*
**Atlox 1045A**
See *Poly(oxy-1,2-ethanediyl), α-hydro-ω-hydroxy-, ether with D-glucitol (6:1), dodecanoate (Z)-9-octadecenoate [55069-68-6]*
**Atlox 8916TF**
See *Sorbitan, mono-9-octadecenoate, poly(oxy-1,2-ethanediyl) derivs., (Z)- [9005-65-6]*
**ATM**
See *Quaternary ammonium compounds,* $C_{17-20}$-alkyl
**ATM 1**
See *Phenol, polymers, polymer with formaldehyde [9003-35-4]*, graphite-modified
**ATM 2 (nylon)**
See *Poly[imino(1-oxo-1,6-hexanediyl)] [25038-54-4]*, graphite fiber-reinforced
**ATM 2 (phenolic resin)**
See *Phenol, polymers, polymer with formaldehyde [9003-35-4]*
**ATM 10**
See *Phenol, polymers, polymer with formaldehyde [9003-35-4]*, graphite-filled
**ATM 11**
See *2-Propenoic acid, 2-methyl-, esters, 2-ethyl-2-[[(2-methyl-1-oxo-2-propenyl)oxy]methyl]-1,3-propanediyl ester [3290-92-4]*
**Atmer 122**
See *Octadecanoic acid, esters, ester with 1,2,3-propanetriol [11099-07-3]*
**Atmer 645**
See *Sorbitan, monododecanoate, mixt. with sorbitan monohexadecanoate, sorbitan monooctadecanoate and (Z)-sorbitan mono-9-octadecenoate [76011-54-6]*
**Atmolysis**
See
*Isotope separation*
*Permeability and Permeation*
*Separation*
atmolysis
**Atmos 300**
See *9-Octadecenoic acid (Z)-, esters, ester with 1,2,3-propanetriol [37220-82-9]*
**Atmosphere**
The higher levels of the atmosphere are indexed at this heading, while those close to the earth's surface are indexed at *Air*. However, the effects of all levels on biological systems are indexed at *Atmosphere, environmental*. Atmospheres of celestial bodies are indexed at such headings as *Moon, Planets, Stars, Sun*. Protective and inert atmospheres are indexed at *Controlled atmospheres*
See also *Sky*
air- or afterglow of——see *Airglow*
aurora of——see *Aurora*
clouds in——see *Clouds*
cosmic rays in——see *Cosmic ray*
humidity of——see *Humidity*
ionosphere——see *Ionosphere*
lightning discharge in——see *Lightning*
magnetosphere——see *Magnetosphere*
nitrogen fixation in
see such headings as
*Nitrogen fixation*
*Ammonia [7664-41-7], preparation*
*Nitrogen oxide (NO) [10102-43-9], preparation*
pollution of——see *Air pollution*
pptn. from
see such headings as
*Dew*
*Fog*
*Hail*
*Rain*
*Snow*
*Waters, natural*
radioactive fallout in——see *Radioactive fallout*
smoke in——see *Smoke*

**Atmosphere, environmental**
Effects on biological systems of artificial and natural atmosphere with respect to composition and pressure are indexed at this heading
See also
*Life support systems*
*Oxygen [7782-44-7], biological studies*
**ATM 1T**
See *Phenol, polymers, polymer with formaldehyde [9003-35-4]*
**Atmul 84**
See *9-Octadecenoic acid (Z)-, esters, ester with 1,2,3-propanetriol [37220-82-9]*
**Atmul 122**
See *Glycerides,* $C_{18}$ mono-
**Atmul 124**
See *Octadecanoic acid, esters, ester with 1,2,3-propanetriol [11099-07-3]*
**Atmul P 40S**
See *Octadecanoic acid, esters, ester with 1,2,3-propanetriol [11099-07-3]*
**Atmul T 95**
See *Glycerides,* mixed mono- and di-
**ATNB 1300X16**
See also *Epoxy resins*
**ATO 3030BN1**
See *1-Propene, polymers, homopolymer, isotactic [25085-53-4]*
**ATO 1002CN22**
See *Ethene, polymers, homopolymer [9002-88-4]*
**Atolide**
See *Benzamide, 2-amino-N-[4-(diethylamino)-2-methylphenyl]- [16231-75-7]*
**Atomaric acid**
See *1-Naphthalenepropanoic acid, decahydro-5-[(2-hydroxy-5-methoxy-3-methylphenyl)methyl]-5,6,8a-trimethyl-2-(1-methylethylidene)-, [1S-(1α,4aβ,5β,6β,8aα)]- [55907-34-1]*
**Atomic absorption spectra**
See such headings as
*Helium-group gases, properties*
spectra of
*Spectrochemical analysis*
at. absorption
*Transition metals, properties*
spectra of
*Hydrogen [1333-74-0], properties*
spectrum of
*Sodium [7440-23-5], properties*
spectrum of
**Atomic batteries**
See *Nuclear batteries*
**Atomic beams**
Studies of atomic beams themselves, or of beams of classes of elements, are indexed at this heading. For beams of specific elements, see headings for those elements
accommodation coeffs. for——see *Accommodation coefficient*
ionic——see *Ion beams*
magnetic resonance of——see *Atomic beam magnetic resonance*
**Atomic bombs and weapons**
See *Nuclear bombs*
**Atomic emission spectroscopy (AES)**
See such headings as
*Spectra*
*Spectrochemical analysis*
*Spectrometry*
*Ultraviolet and visible spectrometry*
**Atomic energy**
See
*Nuclear energy*
*Power*
nuclear
**Atomic form factors**
See *Atomic scattering factor*
**Atomic heat**
See *Heat capacity*
**Atomic integral**
exchange——see *Exchange, quantum mechanical, integrals for*
**Atomic nuclei**
Atomic nuclei themselves, and nuclei of classes of elements, are indexed at this heading. For studies of nuclei of specific elements, see those specific headings. For interactions with particles and waves and for emission of radiation, see those specific and class headings.
For sources of numerical data on atomic nuclei see the list of Data Collection and Analysis Centers, Index Guide, Appendix II, ¶15
See also *Radioelements*
acoustic resonance of——see *Nuclear acoustic resonance*
alignment, orientation or polarization of——see *Nuclear polarization*
angular momentum of
see
*Nuclear energy level*
*Spin, nuclear*
anti——see *Antielements*
binding energy of——see *Nuclear binding energy*
effective elec. charge of——see *Nuclear charge*
elec. moments of
see
*Multipole moment*
*Quadrupole moment*
energy from
see
*Nuclear energy*
*Power*
nuclear
energy levels of——see *Nuclear energy level*
entropy of——see *Entropy, nuclear*
forces in——see *Nuclear force*
formation of, in astrophys. processes——see *Nucleosynthesis*

*[1358-11-8]*

**Isonaphthazarin**
See *1,4-Naphthalenedione, 2,3-dihydroxy- [605-37-8]*

**1,3-ββ-Isonaphthazoledione**
See *1H-Benz[f]isoindole-1,3(2H)-dione [4379-54-8]*

**γ-Isonaphthocyclinone**
See *8,16-Methano-2H-furo[2''',3''':5'',6'']= pyrano[3'',4'':6',7']naphtho[2',3':5,6]= cyclohepta[1,2-g][2]benzopyran-11-acetic acid, 8-(acetyloxy)-3,3a,5,6,8,10,11,13,15,= 16,18,18b-dodecahydro-7,14,17-trihydroxy-= 5,13-dimethyl-2,6,15,18-tetraoxo-, methyl ester, [3aS-(3aα,5α,8α,11β,13α,16β,18bα)]- [85425-04-3]*

**αβ-Isonaphthofuran**
See *Naphtho[1,2-c]furan [232-74-6]*

**ββ-Isonaphthofuran**
See *Naphtho[2,3-c]furan [268-51-9]*

**1,3-αβ-Isonaphthofurandione**
See *Naphtho[1,2-c]furan-1,3-dione [5343-99-7]*

**1,3-ββ-Isonaphthofurandione**
See *Naphtho[2,3-c]furan-1,3-dione [716-39-2]*

**Isonarciclasine**
See *[1,3]Dioxolo[4,5-j]phenanthridin-6(2H)-one, 1,3,4,5-tetrahydro-2,3,4,7-tetrahydroxy-, [2S-(2α,3β,4β)]- [40041-93-8]*

**Isonarciprimine**
See *[1,3]Dioxolo[4,5-j]phenanthridin-6(5H)-one, 2,7-dihydroxy- [25540-29-8]*

**Isonarcpropyline**
See *Benzoic acid, 6-[[6-[2-(dimethylamino)= ethyl]-2-(1-methylethoxy)-1,3-= benzodioxol-5-yl]acetyl]-2,3-dimethoxy- [16010-52-9]*

**Isonardosinone**
See *Naphth[2,3-b]oxiren-2(1aH)-one, 4,5,6,6a,7,7a-hexahydro-7-(1-hydroxy-1-= methylethyl)-6,6a-dimethyl-, [1aS-(1aα,6α,6aα,7β,7aα)]- [27062-01-7]*

**Isonaringenin**
See *4H-1-Benzopyran-4-one, 7-[[6-O-(6-= deoxy-α-L-mannopyranosyl)-β-D-= glucopyranosyl]oxy]-2,3-dihydro-5-= hydroxy-2-(4-hydroxyphenyl)-, (S)- [14259-46-2]*

**Isonarthogenin**
See *Spirost-5-ene-3,27-diol, (3β,25S)- [7050-40-0]*

**Isonate**
See also *Urethane polymers*

**Isonate 191**
See also *Benzene, 1,1'-methylenebis[4-isocyanato- [101-68-8]*

**Isonate 143L**
See also *Benzene, 1,1'-methylenebis[4-isocyanato- [101-68-8]*

**Isonate 125M**
See also *Benzene, 1,1'-methylenebis[4-isocyanato- [101-68-8]*

**Isonate 390P**
See *Isocyanic acid, polymethylenepolyphenylene ester [9016-87-9]*

**Isonate 136T**
See also
*1,1'-Biphenyl, 3,3'-diisocyanato-4,4'-= dimethyl- [21633-66-9]*
*1,1'-Biphenyl, 4,4'-diisocyanato-3,3'-= dimethyl- [91-97-4]*

**Isonate 148T**
See also *1,1'-Biphenyl, 4,4'-diisocyanato-3,3'-= dimethoxy- [91-93-0]*

**Isonauclefine**
See *Indolo[2',3':3,4]pyrido[1,2-g]-1,6-= naphthyridin-5(7H)-one, 8,13-dihydro- [62298-28-6]*

**Isonebularine**
See *7H-Purine, 7-β-D-ribofuranosyl- [2149-71-5]*

**Isonel 200**
See *1,4-Benzenedicarboxylic acid, polymers, polymer with 1,2-ethanediol, 1,2,3-propanetriol and 1,3,5-tris(2-= hydroxyethyl)-1,3,5-triazine-2,4,6(1H,3H,= 5H)-trione [36620-86-7]*

**Isonel 678**
See *1,4-Benzenedicarboxylic acid, polymers, polymer with 1,2-ethanediol and 1,3,5-tris(2-hydroxyethyl)-1,3,5-triazine-2,= 4,6(1H,3H,5H)-trione [26061-97-2]*

**Isonemotinic acid**
See *6,8,10-Undecatriynoic acid, 4-hydroxy- [556-35-4]*

**Isoneoambrosin**
See *Azuleno[4,5-b]furan-2,9-dione, 4,5,6,8,9a,9b-hexahydro-3,6,9a-trimethyl-, [6S-(6α,9aα,9bβ)]- [20555-07-1]*

**Isoneoavroside**
See *4H-1-Benzopyran-4-one, 6-α-D-glucopyr= anosyl-5,7-dihydroxy-2-(4-hydroxyphenyl)- [53152-14-0]*

**Isoneobavachalcone**
See *Benzaldehyde, 2-hydroxy-5-[3-(4-= hydroxyphenyl)-1-oxo-2-propenyl]-4-= methoxy-, (E)- [76444-57-0]*

**Isoneobavaisoflavone**
See *[3,6'-Bi-4H-1-benzopyran]-4-one, 2',3'-dihydro-7-hydroxy-2',2'-dimethyl- [40357-43-5]*

**Isoneobilirubinic acid**
See *1H-Pyrrole-3-propanoic acid, 5-[(4-ethyl-2,5-dihydro-3-methyl-5-oxo-= 1H-pyrrol-2-yl)methyl]-4-methyl- [13129-10-7]*

**Isoneocembrene A**
See *1,3,7,11-Cyclotetradecatetraene, 1,7,11-trimethyl-4-(1-methylethyl)-, (all-E)- [64363-64-0]*

**Isoneomatatabiol**
See *Cyclopenta[c]pyran-1-ol, 1,3,4,4a,5,6,7,7a-= octahydro-4,7-dimethyl-, (1α,4β,4aβ,7β,7aβ)- [34258-02-1]*

**Isoneomorphine**
See *Morphinan-3,6-diol, 8,14-didehydro-4,5-= epoxy-17-methyl-, (5α,6β)- [2291-70-5]*

**Isoneonepetalactone**
See *Cyclopenta[c]pyran-1(3H)-one, 4,4a,5,6-tetrahydro-4,7-dimethyl-, (4R-cis)- [76549-18-3]*

**Isoneopine**
See *Morphinan-6-ol, 8,14-didehydro-4,5-= epoxy-3-methoxy-17-methyl- (5α,6β)- [16008-33-6]*

**Isoneotutinone**
See *Spiro[2,5-ethanooxireno[3,4]cyclopenta= [1,2-c]pyran-6(3H),2'-oxirane]-3,7-dione, hexahydro-1b-hydroxy-5a-methyl-8-(1-= methylethylidene)-, [1aS-(1aα,1bβ,2β,5β,= 5aβ,6β,6aα)]- [54355-79-2]*

**Isoneovitexin**
See *4H-1-Benzopyran-4-one, 8-α-D-glucopyr= anosyl-5,7-dihydroxy-2-(4-hydroxyphenyl)- [53152-15-1]*

**Isonepetefuran**
See *7H-4,7a-Propano-5H-oxireno[h][2]= benzopyran-5-one, 3-(acetyloxy)-7b-[2-= (3-furanyl)ethyl]hexahydro-1a-= (hydroxymethyl)-4-methyl-, [1aS-(1aα,3α,3aβ,4β,7aβ,7bα)]- [37759-57-2]*

**Isonephrosterinic acid**
See *3-Furancarboxylic acid, 2,5-dihydro-4-= methyl-5-oxo-2-undecyl- [75232-40-5]*

**Isoniazid**
See *4-Pyridinecarboxylic acid, hydrazides, hydrazide [54-85-3]*

**Isoniazid methanesulfonate calcium**
See *4-Pyridinecarboxylic acid, hydrazides, 2-(sulfomethyl)hydrazide calcium salt (2:1) [6059-26-3]*

**Isonicotinamidine**
See *4-Pyridinecarboximidamide [33278-46-5]*

**Isonicotinamidoxime**
See *4-Pyridinecarboximidamide, N-hydroxy- [1594-57-6]*

**Isonicotinic acid**
See *4-Pyridinecarboxylic acid [55-22-1]*

**Isonicotinic acid hydrazide**
See *4-Pyridinecarboxylic acid, hydrazides, hydrazide [54-85-3]*

**Isonicotinic acid hydrazide-adenine dinucleotide phosphate**
See *Adenosine 5'-(trihydrogen diphosphate), esters, 2'-(dihydrogen phosphate), 5'→5'-ester with 4-(hydrazinocarbonyl)-1-β-D-= ribofuranosylpyridinium hydroxide, inner salt [3891-19-8]*

**Isonicotinic anhydride**
See *4-Pyridinecarboxylic acid, anhydrides, anhydride [7082-71-5]*

**Isonicotinohydroxamic acid**
See *4-Pyridinecarboxamide, N-hydroxy- [4427-22-9]*

**Isonicotinohydroxamoyl chloride**
See *4-Pyridinecarboximidoyl chloride, N-hydroxy- [4185-99-3]*

**Isonicotinonitrile**
See *4-Pyridinecarbonitrile [100-48-1]*

**Isonicotinuric acid**
See *Glycine, N-(4-pyridinylcarbonyl)- [2015-20-5]*

**Isonidorellalactone**
See *2(5H)-Furanone, 4-[2-[1,2,3,4,4a,7,8,8a-= octahydro-4a-(hydroxymethyl)-1,2,5-= trimethyl-1-naphthalenyl]ethyl]-, [1R-(1α,2α,4aβ,8aα)]- [70387-33-6]*

**Isonigerone**
See *4H-Naphtho[1,2-b]pyran-4-one, 5-hydroxy-6-(5-hydroxy-6,8-dimethoxy-= 2-methyl-4-oxo-4H-naphtho[2,3-b]= pyran-10-yl)-8,10-dimethoxy-2-methyl-, (S)- [76045-72-2]*

**Isonipecaine**
See *4-Piperidinecarboxylic acid, 1-methyl-4-= phenyl-, ethyl ester [57-42-1]*

**Isonipecotic acid**
See *4-Piperidinecarboxylic acid [498-94-2]*

**Isonipecotonitrile**
See *4-Piperidinecarbonitrile [4395-98-6]*

**Isonitramine**
See *2-Azaspiro[5.5]undecan-7-ol, (6S-trans)- [65620-67-9]*

**Isonitrarine**
See *Nitrarine, (3α)- [57969-03-6]*

**Isonitriles**
See *Isocyanides*

**Isonitrin A**
See *Spiro[6-oxabicyclo[3.1.0]hex-3-ene-2,2'-= oxirane], 4-isocyano-3'-methyl-, [1α,2β(S*),5α]-(+)- [83016-48-2]*

**Isonitrin B**
See *6-Oxabicyclo[3.1.0]hex-3-ene-2-methanol, 2-hydroxy-4-isocyano-α-methyl-, [1α,2β,2(R*),5α]-(-)- [83058-95-1]*

**Isonitrin C**
See *3,7-Dioxatricyclo[4.1.0.0^{2,4}]heptane-5-= methanol, 5-hydroxy-1-isocyano-α-methyl-, [1α,2β,4β,5α,5(S*),6α]-(-)- [83016-49-3]*

**Isonitrin D**
See *2-Cyclopenten-1-one, 2-ethenyl-5-= hydroxy-5-isocyano-, (+)- [83016-50-6]*

**Isonitrinic acid E**
See *2-Propenoic acid, 3-(4-isocyano-6-= oxabicyclo[3.1.0]hex-3-en-1-yl)-, (+)- [83016-51-7]*

**Isonitro compounds**
See *aci-Nitro compounds*

**Isonitrosine**
See *2,3-Butanedione, 1-(dimethylamino)-, 2-oxime, monohydrochloride [17224-46-3]*

**Isonitrosoacetone**
See *Propanal, 2-oxo-, 1-oxime [306-44-5]*

**Isonitroso compounds**
See *Oximes*

**Isonixin**
See *3-Pyridinecarboxamide, N-(2,6-dimethylp= henyl)-1,2-dihydro-2-oxo- [57021-61-1]*

**Isonkolbisine**
See *Furo[2,3-b]quinolin-4(9H)-one, 6-(2,3-dihydroxy-3-methylbutoxy)-7-= methoxy-9-methyl-, (-)- [82513-82-4]*

**Isonobilin**
——, **hydroxy-**
See *2-Butenoic acid, 2-methyl-, 2,3,3a,4,5,6,7,8,= 9,11a-decahydro-7,9-dihydroxy-10-= methyl-3,6-bis(methylene)-2-= oxocyclodeca[b]furan-4-yl ester [63549-62-2]*

**Isonocardicin A**
See *1-Azetidineacetic acid, 3-[[[4-(3-amino-= 3-carboxypropoxy)phenyl](hydroxyimino)= acetyl]amino]-α-(4-hydroxyphenyl)-2-oxo-, [3S-[1(S*),3R*[Z(R*)]]]- [88198-78-1]*

**Isonocardicin D**
See *1-Azetidineacetic acid, 3-[[[4-(3-amino-= 3-carboxypropoxy)phenyl]oxoacetyl]= amino]-α-(4-hydroxyphenyl)-2-oxo-, [3S-[1(S*),3R*(R*)]]- [88198-77-0]*

**Isonol C 100**
See *2-Propanol, 1,1'-(phenylimino)bis- [3077-13-2]*

**Isononanoic acid**
See also *Octanoic acid, 7-methyl- [693-19-6]*

**Isonootkatone**
See *2(3H)-Naphthalenone, 4,4a,5,6,7,8-= hexahydro-4,4a-dimethyl-6-(1-= methylethylidene)-, (4R-cis)- [15764-04-2]*

**Isonopaline**
See *L-Glutamic acid, N-[4-[(aminoiminometh= yl)amino]-1-carboxybutyl]-, (S)- [64199-70-8]*

**Isonopinol**
See *Bicyclo[3.1.1]heptan-3-ol, 6,6-dimethyl- [28664-09-7]*

**Isonopinone**
See *Bicyclo[3.1.1]heptan-3-one, 6,6-dimethyl- [4722-54-7]*

**8-Iso-19-noranthratestosterone**
See *8H-Cyclopent[a]anthracen-8-one, 1,2,3,3a,4,5,5a,6,9,10,10a,11,11a,11b-= tetradecahydro-3-hydroxy-3a-methyl-, [3S-(3α,3aα,5aβ,10aβ,11aβ,11bβ)]- [24887-38-5]*

**Isonorargemonine**
See *Dibenzo[a,e]cycloocten-5,11-imin-2-ol, 5,6,11,12-tetrahydro-3,8,9-trimethoxy-13-= methyl-, (5S)- [18826-67-0]*

**Isonorcedrenol**
See *3H-3a,7-Methanoazulen-6-ol, 4,5,6,7,8,8a-hexahydro-6,8,8-trimethyl-, [3aR-(3aα,6β,7α,8aβ)]- [65669-74-1]*

**Isonordrimenone**
See *2(1H)-Naphthalenone, 4a,5,6,7,8,8a-= hexahydro-3,4a,8,8-tetramethyl-, (4aR-trans)- [51020-10-1]*

**Isonorechinocystenedione**
(+)——see *28-Norolean-12-ene-3,16-dione, (17α)- [17020-10-9]*

**Isonormetanephrine**
See *Benzenemethanol, α-(aminomethyl)-3-= hydroxy-4-methoxy- [6924-27-2]*

**Isonortropinone**
See *2-Azabicyclo[2.2.2]octan-3-one [3306-69-2]*

**Isonoruron**
See *Urea, N,N-dimethyl-N'-[octahydro-4,7-= methano-1H-inden-1(or 2)-yl]- [28805-78-9]*

**Isonovobiocin**
See *Benzamide, N-[7-[[2-O-(aminocarbony= l)-6-deoxy-5-C-methyl-4-O-methyl-α-L-= lyxo-hexopyranosyl]oxy]-4-hydroxy-8-= methyl-2-oxo-2H-1-benzopyran-3-yl]-4-= hydroxy-3-(3-methyl-2-butenyl)- [10543-94-9]*

**Isonox**
See *4(3H)-Quinazolinone, 2-methyl-3-(2-= methylphenyl)-, mixt. with 2-[2-[2-[4-= [(4-chlorophenyl)phenylmethyl]-1-= piperazinyl]ethoxy]ethoxy]ethanol (Z)-2-butenedioate (1:2) (salt) [56335-22-9]*

**Isonuatigenin**
See *Spirost-5-ene-3,25-diol, (3β,25S)- [7050-41-1]*

**Isoobacunoic acid**
See *Limonoic acid, 19-deoxy-, δ-lactone [751-28-0]*

**Isoobacunoic acid diosphenol**
See *Limonoic acid, 5,6-didehydro-19-deoxy-= 6-hydroxy-, δ-lactone, (1ξ)- [74806-50-1]*

**Isoobtusilactone**
See *2(3H)-Furanone, 3-(11-dodecenylidene)= dihydro-4-hydroxy-5-methylene-, [S-(E)]- [56522-14-6]*

**Isoobtusilactone A**
See *2(3H)-Furanone, dihydro-4-hydroxy-5-= methylene-3-tetradecylidene-, [S-(E)]- [56522-16-8]*

**Isoobtusilactone B**
See *2(3H)-Furanone, 3-(7-hexadecenylidene)= dihydro-4-hydroxy-5-methylene-, [S-(E,Z)]- [58940-66-2]*

**Isoobtusol**
See *Spiro[5.5]undecan-3-ol, 2,9-dibromo-8-= chloro-1,1,9-trimethyl-5-methylene-, [2R-[2α,3α,6α(8S*,9S*)]]- [73494-23-2]*

**Isoochracinic acid**
See *1-Isobenzofuranacetic acid, 1,3-dihydro-= 4-hydroxy-3-oxo- [66964-78-1]*

**Isooconovine**
See *4H-Dibenzo[de,g]quinolin-1-ol, 5,6,6a,7-tetrahydro-2,3,10,11-= tetramethoxy-6-methyl-, (S)- [72170-11-7]*

**Mass spectrometers** *CHEMICAL ABSTRACTS 1984* **854G**

**Mass spectrometers**
ion sources—see *Ion sources*
for subat. particles—see *Nuclear spectrometers*
**Mass spectrometry**
See *Mass spectroscopy*
**Mass spectroscopy**
The methodology of mass analysis of atoms, molecules, and molecular fragments, and as it relates to classes of substances, is indexed at this heading.
app.
see
*Ion sources*
*Mass spectrometers*
of subat. particles—see *Nuclear spectrometry* and headings for particle classes and specific particles
**Mass synchrometers**
See *Mass spectrometers*, synchrometers
**Mass transfer**
in crystn. and crystal growth
see
*Crystal growth*
*Crystallization*
*Crystal nucleation*
*Epitaxy*
dialysis
see
*Dialysis*
*Electrodialysis*
diffusion
see
*Diffusion*
*Electrodiffusion*
distillation—see *Distillation*
electrophoresis—see *Electrophoresis and Ionophoresis*
iontophoresis—see *Iontophoresis*
Marangoni effect in—see *Marangoni effect*
osmosis
see
*Electroosmosis*
*Osmosis*
photophoresis—see *Photophoresis*
sorption
see
*Absorption*
*Adsorption*
*Chemisorption*
*Sorption*
thermophoresis—see *Thermophoresis*
**Mast cell degranulating peptide (Apis mellifera reduced)**
**cyclic (3→15),(5→19)-bis(disulfide)**—see *Mast cell degranulating peptide (Apis mellifera) [32908-73-9]*
**Mast cell degranulating peptide (Vespa xanthoptera)**
See *Mast cell degranulating peptide (Vespula lewisii), 3-L-tryptophan-5-glycine-6-L-= isoleucine-9-L-methionine-13-L-leucine- [72093-22-2]*
**Mast cell degranulating peptide HR I (Vespa orientalis)**
See *Mast cell degranulating peptide (Vespula lewisii), 6-L-isoleucine-10-L-valine-13-L-= valine- [80533-94-4]*
**Masticadienolic acid**
See *Lanosta-7,24-dien-26-oic acid, 3-hydroxy-, (3β,13α,14β,17α,20S,24Z)- [472-30-0]*
**Masticadienonic acid**
See *Lanosta-7,24-dien-26-oic acid, 3-oxo-, (13α,14β,17α,20S,24Z)- [514-49-8]*
**Mastication, biological**
See *Digestion, biological*, mastication
**Masticort**
See *Pregna-1,4-diene-3,20-dione, 11,17,21-trihydroxy-, (11β)-, mixt. with 4-amino-N-[(butylamino)carbonyl]= benzenesulfonamide and 2-(2-furanylmet= hylene)hydrazinecarboxamide [8067-02-5]*
**Masticort-sulfa**
See *Tylosin, [R-(R*,R*)]-2,3-dihydroxybutane= dioate (1:1) (salt), mixt. with 4-amino-N-(2,6-dimethoxy-4-= pyrimidinyl)benzenesulfonamide, (11β)-11,17,21-trihydroxypregna-1,4-= diene-3,20-dione and 5-[(3,4,5-= trimethoxyphenyl)methyl]-2,4-= pyrimidinediamine [65608-65-3]*
**Mastics**
pavement—see *Pavements and Roads*
**Mastic tree**
See also *Pistacia*
**Mastisan B**
See *Neomycin, mixt. with 4-amino-N-(4,6-= dimethyl-2-pyrimidinyl)benzenesulfonamide and 5-methyl-2,4(1H,3H)-pyrimidinedione [60294-71-5]*
**Mastisan E**
See *Erythromycin, mixt. with 4-amino-N-(4,= 6-dimethyl-2-pyrimidinyl)= benzenesulfonamide [60294-70-4]*
**Mastocytoma**
See
*Mast cell*
neoplasm
*Neoplasm*
mastocytoma
**Mastocytosis**
See *Urticaria*, pigmentosa
**Mastoiditis**
See *Sinus*, disease, mastoiditis
**Mastoparan**
See *Mast cell degranulating peptide (Vespula lewisii) [72093-21-1]*
**Mastoparan II**
See *Mast cell degranulating peptide (Vespula lewisii), 6-L-isoleucine-10-L-valine-13-L-= valine- [80533-94-4]*

**Mastoparan M**
See *Mast cell degranulating peptide (Vespula lewisii), 6-L-isoleucine-13-L-leucine- [79396-77-3]*
**Mastoparan X**
See *Mast cell degranulating peptide (Vespula lewisii), 3-L-tryptophan-5-glycine-6-L-= isoleucine-9-L-methionine-13-L-leucine- [72093-22-2]*
**Masurium**
See *Technetium [7440-26-8]*
**Masu salmon**
See *Oncorhynchus masou*
**Maszil**
See *Aluminum alloy, base. Al 95-98,Si 0.7-1.3,Mg 0.6-1.2,Mn 0.4-1,Fe 0-0.5,Cr 0-0.2,Zn 0-0.2,Cu 0-0.1,Ti 0-0.1 (DIN 3.2315) [12732-13-7]*
**Matacil**
See *Phenol, 4-(dimethylamino)-3-methyl-, methylcarbamate (ester) [2032-59-9]*
**Matairesinol**
See *2(3H)-Furanone, dihydro-3,4-bis[(4-= hydroxy-3-methoxyphenyl)methyl]-, (3R-trans)- [580-72-3]*
**Matairesinoside**
See *2(3H)-Furanone, 3-[[4-(β-D-glucopyranos= yloxy)-3-methoxyphenyl]methyl]dihydro-= 4-[(4-hydroxy-3-methoxyphenyl)methyl]-, (3R-trans)- [23202-85-9]*
**Matatabidiether**
See *2,4-Methano-4H-furo[2,3-b]pyran, hexahydro-3,5-dimethyl- [75019-51-1]*
**Matatabiether**
(+)—see *2-Oxabicyclo[3.2.1]octane, 1,4-dimethyl-8-methylene-, (1S-endo)- [24189-95-5]*
**Matatabiol**
See *6aH-Cyclopenta[b]furan-6a-methanol, 2,3,3a,4-tetrahydro-3,6-dimethyl-, (3α,3aβ,6aβ)-(-)- [34257-97-1]*
**1,9-Matatabioxid-7-al**
See *2-Oxabicyclo[3.2.1]octane-8-carboxaldehyde, 1,4-dimethyl-, [1R-(exo,syn)]- [75019-50-0]*
**Mataven**
See *DL-Alanine, N-benzoyl-N-(3-chloro-4-= fluorophenyl)-, methyl ester [52756-25-9]*
**Mat epox**
See also *Epoxy resins*
**Materials**
Studies relating to various aspects of materials are indexed at more specific headings when possible. This heading is reserved for studies too broad or diffuse to be classified more specifically
building—see *Building materials*
composite—see *Composites*
handling of—see *Handling of materials*
raw—see *Raw materials*
strength of—see *Strength*
testing of—see *Testing of materials*
**Matexil DA-AC**
See *2-Naphthalenesulfonic acid, polymer with formaldehyde, sodium salt [36290-04-7]*
**Mathematics**
Studies of new and novel mathematical methods are indexed at this heading. Studies utilizing established mathematical methods are indexed only at property, process, etc. headings
See also *Calculation*
group theory
see also
*Crystal field theory and Ligand field theory*
*Field theory*
probability—see *Probability*
process simulation by—see *Process simulation*
in quantum mechanics
see
*Quantum mechanics*
*Statistical mechanics*
quantum
space in—see *Space, topological*
statistics
see
*Probability*
*Statistical mechanics*
*Statistics and Statistical analysis*
*Stochastic process*
*Thermodynamics*
statistical
**Mathieu equation**
See *Mathematics*
**Mathiola**
See *Matthiola*
**Matricaria**
See also *Chamomile*
**Matricaria ester**
See *2,8-Decadiene-4,6-diynoic acid, methyl ester, (Z,Z)- [928-36-9]*
**Matricaria inodora**
See *Tripleurospermum maritimum inodorum*
**Matricaria lactone**
See *2(5H)-Furanone, 5-(4-hexen-2-ynylidene)- [23251-68-5]*
**Matricarianol**
See *2,8-Decadiene-4,6-diyn-1-ol [505-98-6]*
**Matricaric acid**
See *2,8-Decadiene-4,6-diynoic acid [505-91-9]*
**Matricarin**
See *Azuleno[4,5-b]furan-2,7-dione, 4-(acetyloxy)-3,3a,4,5,9a,9b-hexahydro-3,= 6,9-trimethyl-, [3S-(3α,3aα,4α,9aα,9bβ)]- [5989-43-5]*
**—, deacetoxy-**
See *Azuleno[4,5-b]furan-2,7-dione, 3,3a,4,5,9a,9b-hexahydro-3,6,9-trimethyl-, [3S-(3α,3aα,9aα,9bβ)]- [17946-87-1]*
**Matricin**
See *Azuleno[4,5-b]furan-2(3H)-one, 4-(acetyloxy)-3a,4,5,9,9a,9b-hexahydro-9-= hydroxy-3,6,9-trimethyl-, [3S-(3α,3aα,4α,= 9α,9aα,9bβ)]- [29041-35-8]*

**Matricins**
See *Proteins*, matricins
**Matrigon**
See *2-Pyridinecarboxylic acid, 3,6-dichloro- [1702-17-6]*
**Matrine**
See *Matridin-15-one [519-02-8]*
**Matrix**
See such headings as
*Embedding materials*
*Mathematics*
*Matrix media*
*Molds (forms)*
*Phase*
*Quantum mechanics*
**Matrix calibration technique**
See *Calibration*
**Matrix media**
Viscous and vitreous media for dispersing, trapping, etc. are indexed at this heading
for embedding—see *Embedding materials*
for encapsulation
see
*Encapsulation*
*Potting*
**Matrol**
See *Pregna-4,6-diene-3,20-dione, 17-(acetyloxy)-6-chloro- [302-22-7]*
**Matromycin**
See *Oleandomycin, phosphate (1:1) (salt) [7060-74-4]*
**Matrosite**
See *Coal components*, matrosite
**Mat rush**
See *Scirpus lacustris*
**Matsu-take**
See *Tricholoma matsutake*
**Matsutake alcohol**
See *1-Octen-3-ol, (R)- [3687-48-7]*
**Matter**
Studies relating to various aspects of matter are indexed at more specific headings when possible. This heading is reserved for studies too broad or diffuse to be classified more specifically
anti—see *Antimatter*
disintegration of
see such headings as
*Alpha decay*
*Fission*
*Nuclear reaction*
*Radioactivity*
*Radioelements*
nuclear—see *Nuclear matter*
states of
see such headings as
*Colloids*
*Crystals*
*Gases*
*Glassy state*
*Liquids*
*States of matter*
subnuclear—see *Nuclear matter*, sub-
**Matteucci effect**
See *Electric potential*
**Matteucin**
See *4H-1-Benzopyran-4-one, 2,3-dihydro-5,7-= dihydroxy-2-(2-hydroxyphenyl)-6,8-= dimethyl-, (S)- [77744-53-7]*
**Matteucinol**
See *4H-1-Benzopyran-4-one, 2,3-dihydro-5,7-= dihydroxy-2-(4-methoxyphenyl)-6,8-= dimethyl-, (S)- [489-38-3]*
**Matthiola**
Stocks are also indexed at this heading
**Matting agents**
See *Luster*, lowering of, agents for
**Mattox rearrangement**
See *Rearrangement*, Mattox
**Matulane**
See *Benzamide, N-(1-methylethyl)-4-[(2-= methylhydrazino)methyl]-, monohydrochl= oride [366-70-1]*
**Maturase**
See *Nuclease, messenger ribonucleate maturation endoribo- [69494-91-3]*
**cytochrome b**—see *Nuclease, cytochrome b messenger ribonucleate maturation endoribo- [77750-81-3]*
**Maturation**
See
*Animal growth*
*Development, mammalian*
*Development, nonmammalian*
*Plant growth and development*
*Puberty*
**Maturin**
See *Naphtho[2,3-b]furan-4-carboxaldehyde, 3-(hydroxymethyl)-9-methoxy-5-methyl- [62706-43-8]*
**Maturinone**
See *Naphtho[2,3-b]furan-4,9-dione, 3,5-dimethyl- [22985-02-0]*
**—, tetrahydro-**
See *Naphtho[2,3-b]furan-4,9-dione, 5,6,7,8-tetrahydro-3,5-dimethyl-, (R)- [24393-84-8]*
**Maturone**
See *Naphtho[2,3-b]furan-4,9-dione, 3-(hydroxymethyl)-5-methyl- [29100-83-2]*
**Mauiensine**
See *Ajmalan-17-ol, 19,20-didehydro-, (17S,19E)- [6883-73-4]*
**Mauritine A**
See *Propanamide, 2-(dimethylamino)-N-[2-= methyl-1-[[3,3a,11,12,13,14,15,15a-= octahydro-12,15-dioxo-13-(phenylmethyl)-= 5,8-ethenopyrrolo[3,2-b][1,5,8]= oxadiazacyclotetradecin-1(2H)-yl]= carbonyl]propyl]-, [3aS-[1[R*(R*)],3a= R*,13R*,15aR*]]- [38478-72-7]*

**Placcel H 7** *CHEMICAL ABSTRACTS 1984* **1098G**

**Placcel H 7**
See *Poly[oxy(1-oxo-1,6-hexanediyl)] [25248-42-4]*
**Placcel M**
See *Hexanoic acid, 6-hydroxy-, 5-hydroxypentyl ester [84228-89-7]*
**Placenta**
See also
*Trophoblast*
*Umbilical cord*
**Placental extracts**
See
*Placenta*
ext.
*Placental hormones*
**Placental hormones**
corticotropic——see *Corticotropin, placental [55127-96-3]*
estrogens——See *Estrogens*
gonadotropic——see *Gonadotropin, chorionic [9002-61-3]*
lactogenic——see *Lactogen, placental [9035-54-5]*
progestogens——see *Progestogens*
thyrotropic——see *Thyrotropin, placental [39316-88-6]*
**Placental lactogen**
See *Lactogen, placental [9035-54-5]*
**Placidyl**
See *1-Penten-4-yn-3-ol, 1-chloro-3-ethyl- [113-18-8]*
**Placodine**
See *Maucherite [12044-65-4]*
**Placodiolic acid**
(−)-——see *3(4H)-Dibenzofuranone, 2,8-diacetyl-4a,9b-dihydro-1,7,9-= trihydroxy-4a-methoxy-6,9b-dimethyl-, trans-(-)- [38965-63-8]*
**Plafibride**
See *Propanamide, 2-(4-chlorophenoxy)-2-= methyl-N-[[(4-morpholinylmethyl)= amino]carbonyl]- [63394-05-8]*
**Plafos PF**
See *Poly[oxy[(chloromethyl)-1,2-ethanediyl]], α,α′,α″-phosphinylidynetris[ω-[2-chloro= (chloromethyl)ethoxy]- [72536-93-7]*
**Plagiochenedial**
See *Bicyclo[5.1.0]octane-2-acetaldehyde, 3-formyl-8,8-dimethyl-α,4-bis(methylene)-, [1S-(1α,2α,3α,7α)]- [73407-82-6]*
**Plagiochilal A**
See *Bicyclo[5.1.0]octane-2-acetaldehyde, 3-formyl-8,8-dimethyl-α,4-bis(methylene)-, [1S-(1α,2α,3α,7α)]- [73407-82-6]*
**Plagiochilide**
See *4H-Cyclopropa[3,4]cyclohepta[1,2-c]= pyran-4-one, 4a,5,6,7,7a,8,8a,8b-= octahydro-1,8,8-trimethyl-5-methylene-, [4aR-(4aα,7aβ,8aβ,8bα)]- [67779-72-0]*
**Plagiochilin A**
See *Spiro[5H-cyclopropa[3,4]cyclohepta[1,= 2-c]pyran-5,2′-oxirane]-1-methanol, 4-(acetyloxy)-4,4a,6,7,7a,8,8a,8b-= octahydro-8,8-dimethyl-, acetate, [4R-(4α,4aα,5β,7aβ,8aβ,8bα)]- [67779-73-1]*
**Plagiochilin B**
See *Spiro[5H-cyclopropa[3,4]cyclohepta[1,= 2-c]pyran-5,2′-oxirane]-1,8-dimethanol, 4-(acetyloxy)-4,4a,6,7,7a,8,8a,8b-= octahydro-8-methyl-, diacetate, [4R-(4α,4aα,5β,7aβ,8α,8aβ,8bα)]- [70447-97-1]*
**Plagiochilin C**
See *4H-Cyclopropa[3,4]cyclohepta[1,2-c]= pyran-1-methanol, 4-(acetyloxy)-4a,5,6,7,= 7a,8,8a,8b-octahydro-8,8-dimethyl-5-= methylene-, acetate, [4R-(4α,4aα,7aβ,= 8aβ,8bα)]- [73100-51-3]*
**Plagiochilin D**
See *Spiro[5H-cyclopropa[3,4]cyclohepta[1,= 2-c]pyran-5,2′-oxirane]-1,8,8(4H)-= trimethanol, 4-(acetyloxy)-4a,6,7,7a,8a,= 8b-hexahydro-, triacetate, [4R-(4α,4aα,= 5β,7aβ,8aβ,8bα)]- [73100-52-4]*
**Plagiochilin E**
See *Spiro[5H-cyclopropa[3,4]cyclohepta[1,= 2-c]pyran-5,2′-oxirane]-1,8,8(4H)-= trimethanol, 4-(acetyloxy)-4a,6,7,7a,8a,= 8b-hexahydro-, triacetate, [4R-(4α,4aα,= 5α,7aβ,8aβ,8bα)]- [73137-02-7]*
**Plagiochilin F**
See *4H-2,5-Dioxabenzo[ef]cyclopropa[kl]= heptalene-1,9-diol, 6a-[(acetyloxy)= methyl]-1,6,6a,6b,7,8,9,9a,9b,9c-= decahydro-9-methyl-, 1-acetate [73100-53-5]*
**Plagiochilin G**
See *Spiro[5H-cyclopropa[3,4]cyclohepta[1,= 2-c]pyran-5,2′-oxirane]-1,8,8(4H)-= trimethanol, 4-(acetyloxy)-4a,6,7,7a,8a,= 8b-hexahydro-, α¹,α⁸-diacetate, [4R-(4α,4aα,5β,7aβ,8β,8aβ,8bα)]- [73039-18-6]*
**Plagiochilin H**
See *4H-Cyclopropa[3,4]cyclohepta[1,2-c]= pyran-4-ol, 4a,5,6,7,7a,8,8a,8b-octahydro-= 1,8,8-trimethyl-5-methylene-, acetate, [4R-(4α,4aα,7aβ,8aβ,8bα)]- [74320-19-7]*
**Plagiochilin I**
See *Spiro[5H-cyclopropa[3,4]cyclohepta[1,= 2-c]pyran-5,2′-oxirane]-1-methanol, 4-(acetyloxy)-4,4a,6,7,7a,8,8a,8b-= octahydro-8,8-dimethyl-, [4R-(4α,4aα,5β,= 7aβ,8aβ,8bα)]- [74320-18-6]*
**Plagioclase**
See *Plagioclase-group minerals*
**Plagioclase amphibolite**
See *Amphibolite*, plagioclase
**Plagioclase granite**
See *Monzonite*, quartz
**Plagioclase-group minerals**
The mineral class is indexed at this heading. Specific plagioclases are indexed at such headings as *Albite* (*$Na(AlSi_3O_8)$*), *Anorthite* (*$Al_2Ca(SiO_4)_2$*)
**Plagioclase porphyrite**
See *Porphyrite*, plagioclase
**Plagioclase rhyolite**
See *Rhyolite*, plagioclase
**Plagioclasite**
See *Anorthosite*
**Plagiodera versicolora**
Imported willow leaf beetle is also indexed at this heading
**Plagiodial**
See *3-Cyclopentene-1-acetaldehyde, 2-formyl-α,3-dimethyl-, [1S-(1α,2α)] [73532-70-4]*
**Plagiogranite**
See *Monzonite*, quartz
**Plagiolactone**
See *Cyclopenta[c]pyran-3(4H)-one, 4a,5-dihydro-4,7-dimethyl-, (4S-cis)- [63808-10-6]*
**Plague**
See also *Yersinia pestis*
**Plaice**
Studies of plaice as a group, and of species not further identified, are indexed at this heading. Studies of identified species are indexed at the scientific names
**Plainact 55**
See *Titanate(2-), tetrakis[2,2-bis[(2-= propenyloxy)methyl]-1-butanolato-O¹]bis= (ditridecyl phosphito-O″)-, dihydrogen [64157-14-8]*
**Plainact 41B**
See *Titanate(2-), bis[bis(2-ethylhexyl) phosphito-O″]tetrakis(2-propanolato)-, dihydrogen [65460-52-8]*
**Plainact 46B**
See *Titanate(2-), bis(ditridecyl phosphito-O″)= tetrakis(2-ethyl-1-hexanolato)-, dihydrogen [65460-53-9]*
**Plainact 9S**
See *Titanium, tris(dodecylbenzenesulfonato-O)= (2-propanolato)- [61417-55-8]*
**Plainact 138S**
See *Titanate(2-), bis[P,P-bis(2-ethylhexyl) diphosphato(2-)-O″,O‴][hydroxyacetato= (2-)-O¹,O²]-, dihydrogen [67729-57-1]*
**Plainact 238S**
See *Titanate(2-), bis[P,P-bis(2-ethylhexyl) diphosphato(2-)-O″][1,2-ethanediolato(= 2-)-O,O′]-, dihydrogen, (T-4)- [65467-75-6]*
**Plainact TTS**
See *Titanium, tris(isooctadecanoato-O)(2-= propanolato)- [61417-49-0]*
**Plait point**
See *Critical solution state and phenomena*
**Plakalbumins**
See *Ovalbumins*, plakalbumins
**Plakinic acid A**
See *1,2-Dioxolane-3-acetic acid, 5-(2,6-dimethyl-8-phenyl-3,7-octadienyl)-= 3,5-dimethyl- [87803-05-2]*
**Plakinic acid B**
See *1,2-Dioxane-3-acetic acid, 6-(2,6-dimethy= l-8-phenyl-3,7-octadienyl)-4,6-dimethyl- [87803-06-3]*
**Plakins**
See
*Bacteriolysins*
of blood platelets
*Blood platelet*
bacteriolysins of
**Plakortic acid**
See *1,2-Dioxane-3-acetic acid, 4-ethyl-6-(2-= ethyl-3-hexenyl)-6-methyl- [87803-10-9]*
**Plakortin**
See *1,2-Dioxane-3-acetic acid, 4-ethyl-6-(2-= ethyl-3-hexenyl)-6-methyl-, methyl ester [66940-35-0]*
**Planac BT 1000**
See *Poly(oxy-1,4-butanediyloxycarbonyl-1,= 4-phenylenecarbonyl) [24968-12-5]*
**Planac ST 2030S**
See *Poly(oxy-1,4-butanediyloxycarbonyl-1,= 4-phenylenecarbonyl) [24968-12-5]*
**Planaic acid**
See *Benzoic acid, 2,4-dimethoxy-6-pentyl-, 4-carboxy-3-methoxy-5-pentylphenyl ester [5366-07-4]*
**Planarian**
Studies of planarians as a group are indexed at this heading. More specific studies are indexed at the genus-species names
**Planatol HM 93/94**
See *Acetic acid ethenyl ester, polymers, polymer with ethene [24937-78-8]*
**Planavin**
See *Benzenamine, 4-(methylsulfonyl)-2,6-= dinitro-N,N-dipropyl- [4726-14-1]*
**Plancher rearrangement**
See *Rearrangement*, Plancher
**Planckeon**
See *Gravitation*, Planckeons
**Planck's free-energy function**
See *Free energy function*
**Planets**
Earth——see *Earth*
**Planipart**
See *Benzenemethanol, 4-amino-3,5-dichloro-= α-[[(1,1-dimethylethyl)amino]methyl]- [37148-27-9]*
**Planium (polymer)**
See *Ethene, polymers, homopolymer [9002-88-4]*, aluminum-coated
**Plankton**
See also *Neuston*
**Planocaine**
See *Benzoic acid, 4-amino-, 2-(diethylamino)ethyl ester, monohydrochloride [51-05-8]*
**Planococcus citri**
Citrus mealybug is also indexed at this heading
**Planofixe**
See *1-Naphthaleneacetic acid [86-87-3]*
**Planography**
See
*Lithography*
*Printing*
**Planotox**
See *Acetic acid, (2,4-dichlorophenoxy)-, esters, 2-butoxyethyl ester [1929-73-3]*
**Planovin**
See *Pregna-4,6-diene-3,20-dione, 17-(acetyloxy)-6-methyl-, mixt. with (17α)-3-methoxy-19-norpregna-1,3,5(10)-= trien-20-yn-17-ol [8064-51-5]*
**Plant**
Studies relating to various aspects of plants are indexed at more specific headings when possible. For specific plants and groups of plants, see those specific headings. This heading is reserved for studies too broad or diffuse to be classified more specifically
See also
*Forest*
*Plant tissue*
*Vegetable materials*
*Weed*
absorption by
see
*Leaf absorption*
*Root absorption*
*Stem absorption*
anal. of——see *Plant analysis*
breeding of——see *Plant breeding and selection*
cells of——see *Plant cell*
cell wall of——see *Cell wall*
classification of——see *Taxonomy*
development of——see *Plant growth and development*
disease——see *Plant disease*
drying agents for——see *Plant desiccants*
fertilizer expts. with——see *Fertilizer experiment*
fertilizers for——see *Fertilizers*
flowers of——see *Flower*
forage
see also
*Clover*
*Feed*
*Grass*
*Grassland*
fossil——see *Fossils*, plant
genetics of——see *Genetics*
graft——see *Graft and Grafting, plant*
growth of——see *Plant growth and development*
growth substances, hormones or regulators for——see *Plant hormones and regulators*
herbicides for——see *Herbicides*
hydroponic culture of——see *Hydroponics*
insecticides for——see *Insecticides*
juices——see *Sap*
leaves of——see *Leaf*
medicinal——see also *Pharmaceuticals*, Oriental
metab. by——see *Plant metabolism*
mulches for——see *Mulches*
nectary of——see *Nectary*
nutrition of——see *Plant nutrition*
parthenocarpy——see *Parthenocarpy*
pasture——see *Grassland*
pest control on and protective agents for
see
*Fungicides and Fungistats*
*Sprays*
pesticides for——see *Pesticides*
photoperiodism of——see *Photoperiodism*
pollens——see *Pollen*
reprodn. by——see *Plant reproduction*
respiration by——see *Plant respiration*
retting of——see *Retting*
rooting of cuttings of——see *Root*
roots of——see *Root*
rotated——see *Crop rotation*
sap——see *Sap*
seeds of——see *Seed*
selection of——see *Plant breeding and selection*
senescence of——see *Plant senescence*
silage from——see *Silage*
stems of——see *Stem*
taxonomy of——see *Taxonomy*
tumor——see *Tumor, plant*
weed control in——see *Weed control*
**Plantacupreins**
See *Cupreins*, planta-
**Plantacyanins**
See *Proteins*, cusacyanins
**Plantaginin**
See *4H-1-Benzopyran-4-one, 7-(β-D-glucopyr= anosyloxy)-5,6-dihydroxy-2-(4-= hydroxyphenyl)- [26046-94-6]*
**Plantaglucide**
See *Psyllium gum [8063-16-9]*
**Plantago**
See *Plantain*
**Plantagonine**
See *5H-2-Pyrindine-4-carboxylic acid, 6,7-dihydro-7-methyl-, (S)- [21913-34-8]*
**Plantagoside**
See *4H-1-Benzopyran-4-one, 2-[3-(β-D-= glucopyranosyloxy)-4,5-dihydroxyphenyl]-= 2,3-dihydro-5,7-dihydroxy-, (S)- [78708-33-5]*
**Plantain**
For bananas see *Banana*
*P. psyllium*, seed gum——see *Psyllium gum [8063-16-9]*
**Plantarenaloside**
See *Cyclopenta[c]pyran-4-carboxaldehyde, 1-(β-D-glucopyranosyloxy)-1,4a,5,6,7,7a-= hexahydro-4a-hydroxy-7-methyl-, [1S-(1α,4aα,7β,7aα)]- [72396-01-1]*

**Q 53**
See *Benzenemethanaminium, 3,4-dichloro-N-= [2-hydroxy-2-methyl-5-(1-methylethyl)= cyclohexyl]-N,N-dimethyl-, chloride [29379-98-4]*
**Q 56**
See *Benzenemethanaminium, N-[2-hydroxy-= 2-methyl-5-(1-methylethyl)cyclohexyl]-= N,N,4-trimethyl-, chloride [29379-99-5]*
**Q 58**
See *Cyclohexanaminium, N-heptyl-2-= hydroxy-N,N,2-trimethyl-5-(1-= methylethyl)-, bromide [20091-61-6]*
**Q 64**
See *Cyclohexanaminium, 2-hydroxy-N,N,2-= trimethyl-5-(1-methylethyl)-N-octyl-, bromide [20091-62-7]*
**75Q**
See *Naphthenic oils*
**Q 174**
See *2-Propenoic acid, 2-methyl-, esters, 3-(trimethoxysilyl)propyl ester [2530-85-0]*
**Q 225**
See *Ethene, polymers, homopolymer [9002-88-4]*
**Q 254**
See *2,5-Cyclohexadiene-1,4-dione, 2,3-dimethyl-5-(3,7,11,15,19,23,27,31,35-= nonamethyl-2,6,10,14,18,22,26,30,34-= hexatriacontanonaenyl)-, (all-E)- [4299-57-4]*
**Q 332**
See *Phosphonic acid, [[(3-methyl-2-= pyridinyl)amino]methylene]bis- [70010-76-3]*
**Q 3103**
See *Acetic acid ethenyl ester, polymers, polymer with ethene [24937-78-8]*
**Q 6203**
See also *Polyesters*
**Q 6602**
See also *Polyesters*
**Q 8349**
See also *Polyesters*
**Q 19262**
See *1-Propanamine, 3,3'-[oxybis(2,1-= ethanediyloxy)]bis- [4246-51-9]*
**Q 2-3026**
See *Siloxanes and Silicones*
**Q 2-5125**
See *Siloxanes and Silicones*
**Q 3-6527**
See *Siloxanes and Silicones*
**Q 4-6558**
See *Siloxanes and Silicones*
**Q 4-3559**
See *Siloxanes and Silicones*
**Q 5-0161**
See *Siloxanes and Silicones*, di-Me, Me 3,3,3-trifluoropropyl
**Q 5-0167**
See *Siloxanes and Silicones*, Me trifluoropropyl
**Q 9-5700**
See *1-Octadecanaminium, N,N-dimethyl-N-= [3-(trimethoxysilyl)propyl]-, chloride [27668-52-6]*
**Q 2A**
See *8H-1,3-Dioxolo[4,5-h]isoindolo[1,2-b]= [3]benzazepin-8-one, 9,10,14-trimethoxy- [75617-78-6]*
**QA-CLS**
See *Starch, ethers, 2-hydroxy-3-= (trimethylammonio)propyl ether, chloride [56780-58-6]*
**QB 1**
See *2H-Quinolizine, 3-[(4-chlorophenyl)= methyl]octahydro-, [R-(R*,R*)]-2,3-= dihydroxybutanedioate (1:1) [34255-03-3]*
**QB 200**
See also *Epoxy resins*
**Q 2-4011B**
See *Siloxanes and Silicones*
**Q-Broxin**
See *Lignosulfonic acid, chromium iron salt [8075-74-9]*
**Q-C 50**
See *Quaternary ammonium compounds*, compounds, coco alkylbis(hydroxyethyl)= methyl, ethoxylated
**QCD**
See *Quantum chromodynamics*
**Q-Cel 300**
See *Silicates*, microspheres
**Q-D 86P**
See *1-Octadecanaminium, N,N-dimethyl-N-= octadecyl-, chloride [107-64-2]*
**QDX**
See *1,3,5,7-Tetrazocine, 1-acetyloctahydro-3,5,= 7-trinitro- [13980-00-2]*
**QE 1044**
See *2-Propenenitrile, polymers, polymer with ethenylbenzene [9003-54-7]*
**Q-e equation**
See *Q-e value in polymerization*
**Q-Enzyme**
See *Glycosyltransferase, α-glucan-branching [9001-97-2]*
**Q-e value in polymerization**
See also *Reactivity ratio in polymerization*
**QF 1 (pharmaceutical)**
See *8H-1,3-Dioxolo[4,5-h]isoindolo[1,2-b]= [3]benzazepin-8-one, 6-(dimethylamino)-= 5,6-dihydro-9,10,14-trimethoxy- [75617-77-5]*
**QF 1 (silicone)**
See *Siloxanes and Silicones*, fluorinated
**Q 6F**
See *7H-Pyrrolo[3,2-g]quinolin-7-one, 1,2,3,8-tetrahydro-1,2,3,3,8-pentamethyl-= 5-(trifluoromethyl)- [58721-74-7]*
**Q-factor**
Quality factors for electric circuits and devices are indexed only at such headings as *Electric circuits, Oscillators and Resonators*, and at headings for specific devices
**Q-Film**
See *Poly(oxy-1,2-ethanediyloxycarbonyl-2,= 6-naphthalenediylcarbonyl) [24968-11-4]*
**QG 100**
See *Silica, vitreous [60676-86-0]*
**QH 25**
See *Benzenemethanol, α-[[(1,1-dimethylethy= l)amino]methyl]-4-hydroxy-3-[[(4-= methoxyphenyl)methyl]amino]-, monohydrochloride [60853-38-5]*
**QI 3037**
See *Siloxanes and Silicones*
**Qiana**
See *Polyamide fibers*
**Qianhu**
See *Pharmaceuticals*, Oriental
**Qikron**
See *Benzenemethanol, 4-chloro-α-(4-= chlorophenyl)-α-methyl- [80-06-8]*
**Qikron VP-JET**
See *Phosphoric acid, esters, 2,2-dichloroethenyl dimethyl ester, mixt. with 4-chloro-α-(4-= chlorophenyl)-α-methylbenzenemethanol [76188-65-3]*
**QINA**
See *Pyridinium, 4-carboxy-1-methyl-, chloride [5746-18-9]*
**Qing Hau acid**
See *1-Naphthaleneacetic acid, 1,2,3,4,4a,5,6,= 8a-octahydro-4,7-dimethyl-α-methylene-, [1R-(1α,4β,4aβ,8aβ)]- [80286-58-4]*
**Qing Hau Sau**
See *3,12-Epoxy-12H-pyrano[4,3-j]-1,2-= benzodioxepin-10(3H)-one, octahydro-3,6,= 9-trimethyl-, [3R-(3α,5aβ,6β,8aβ,9α,12β,= 12aR*)]- [63968-64-9]*
**Qing Hau Sau I**
See *2H-Naphtho[1,8-bc]furan-2-one, 2a,3,4,5,5a,6,8a,8b-octahydro-5,8-dimethyl- [82442-48-6]*
**Qing Hau Sau II**
See *3H-Oxireno[7,8]naphtho[8a,1-b]furan-3-one, decahydro-7,9a-dimethyl-4-methylene-, [1aR-(1aα,1bR*,4aβ,7β,7aβ,9aα)]- [50906-56-4]*
**Qing Hau Sau III**
See *10aH-9,10b-Epoxypyrano[4,3,2-jk][2]= benzoxepin-2(3H)-one, octahydro-3,6,9-= trimethyl-, [3R-(3α,3aβ,6β,6aβ,9β,10aα,= 10bβ)]- [72826-63-2]*
**Qing Hau Sau IV**
See *10aH-9,10b-Epoxypyrano[4,3,2-jk][2]= benzoxepin-2(3H)-one, octahydro-8-= hydroxy-3,6,9-trimethyl-, [3R-(3α,3aβ,= 6β,6aβ,8β,9β,10aα,10bβ)]- [82003-85-8]*
**Qing Hau Sau V**
See *Naphtho[1,8-bc]pyran-2(3H)-one, decahydro-9-hydroxy-6,9-dimethyl-3-= methylene-, [3aR-(3aα,6α,6aα,9α,9aα,9bα)]- [82003-84-7]*
**Qingyanshengenin**
See *Pregn-5-en-20-one, 3,8,14,17-tetrahydrox= y-12-[(4-hydroxybenzoyl)oxy]-, (3β,12β,14β,17α)- [84745-94-8]*
**QL 7014**
See also *Urethane polymers*
**QLAR (Quantitative localization activity relationship)**
See such headings as
*Pharmaceuticals*
pharmacokinetics of
*Pharmacology*
**Q-loid A 30**
See *Aluminum oxide* ($Al_2O_3$) *[1344-28-1]*
**Q-lure**
See *2-Butanone, 4-[4-(acetyloxy)phenyl]- [3572-06-3]*
**QM 589**
See *2-Propenoic acid, esters, 1,7,7-trimethylbic= yclo[2.2.1]hept-2-yl ester, exo- [5888-33-5]*
**QM 657**
See *2-Propenoic acid, 2-methyl-, esters, 2-[(3a,4,5,6,7,7a-hexahydro-4,7-methano-= 1H-inden-5-yl)oxy]ethyl ester [75662-22-5]*
**QM 672**
See *2-Propenoic acid, esters, 2-[(3a,4,5,6,7,7a-= hexahydro-4,7-methano-1H-inden-5-yl)= oxy]ethyl ester [79638-11-2]*
**QM 4119**
See *2-Thiophenemethanol, 5-chloro-α-[[(1,1-= dimethylethyl)amino]methyl]- [51452-63-2]*
**QM 4229**
See *2-Thiophenemethanol, 5-chloro-α-[[(1-= methylethyl)amino]methyl]- [51452-62-1]*
**QM 5119**
See *2-Thiophenemethanol, 4,5-dichloro-α-= [[(1,1-dimethylethyl)amino]methyl]- [38450-46-3]*
**QM 5229**
See *2-Thiophenemethanol, 4,5-dichloro-α-= [[(1-methylethyl)amino]methyl]- [38450-47-4]*
**QM 6008**
See *2H-[1]Benzothieno[2,3-e]-1,4-diazepin-= 2-one, 1,3,6,7,8,9-hexahydro-5-phenyl- [29462-18-8]*
**QM 6011**
See *2-Thiophenemethanol, 3,4-dichloro-α-= [[(1-methylethyl)amino]methyl]- [54954-71-1]*
**QM 6013**
See *2-Thiophenemethanol, 3,4-dichloro-α-= [[(1,1-dimethylethyl)amino]methyl]- [54954-70-0]*
**$QMB_3$**
See *Methanaminium, N,N,N-trimethyl-, octahydrotriborate(1-) [12386-10-6]*
**QMB 1020**
See *2,5-Furandione, polymers, polymer with 1,3-pentadiene [27322-62-9]*
**Q1N**
See *Iron alloy, base, Fe 93-95,Ni 2.5-3.2,Cr 1.4-1.6,Mn 0.6-0.8,Mo 0.3-0.6,Si 0-0.5,C 0-0.2 (HY 80) [54824-12-3]*
**2Q4N**
See *4-Quinolinamine, 2-[2-(5-nitro-2-= furanyl)ethenyl]- [1036-81-3]*
**QN 0301WAX**
See *1,3-Cyclopentadiene, polymers, polymer with 1,3-pentadiene [51877-34-0]*
**QN 0330WAX**
See *1,3-Cyclopentadiene, polymers, polymer with 1,3-pentadiene [51877-34-0]*
**QQM-44**
See *Magnesium alloy, base, Mg 94-97,Al 2.5-3.5,Zn 0.6-1.4,Mn 0.2-0.7 (ASTM AZ31B) [12634-55-8]*
**QR 451**
See also *Acrylic polymers*
**QR 711**
See *2,4,11,13-Tetraazatetradecanediimidamide, N,N''-bis(2-ethylhexyl)-3,12-diimino- [22573-93-9]*
**Qβ-replicase**
See *Nucleotidyltransferase, ribonucleate, RNA-dependent [9026-28-2]*
**QSAR (quantitative structure activity relationships)**
See such headings as
*Linear free energy relationship*
*Molecular orbital*
*Molecular structure-biological activity relationship*
*Molecular structure-property relationship*
**QSS**
See *Rubber, silicone*
**Q-Switch I**
See *1-Benzopyrylium, 8-[5-(6,7-dihydro-6-= methyl-2,4-diphenyl-5H-1-benzopyran-8-= yl)-2,4-pentadienylidene]-5,6,7,8-= tetrahydro-6-methyl-2,4-diphenyl-, perchlorate [21016-19-3]*
**Q-Switch II**
See *1-Benzopyrylium, 8-[5-[6,7-dihydro-2,4-= bis[4-(pentyloxy)phenyl]-5H-1-= benzopyran-8-yl]-2,4-pentadienylidene]-= 5,6,7,8-tetrahydro-2,4-diphenyl-, perchlorate [21016-22-8]*
**QT**
See *Ethanaminium, N,N,N-triethyl-2-oxo-2-= [(2,4,6-trimethylphenyl)amino]- [59333-81-2]*
**Q-Thane P 250-1**
See also *Urethane polymers*
**Q-Thane PA 10**
See *Rubber, urethane*, polyester-
**Q-Thane PA 30**
See *Rubber, urethane*, polyester-
**Q-Thane PH 56**
See also *Urethane polymers*
**Q-Thane PS 80**
See *Rubber, urethane*, polyester-
**Q-Thane PS 94M**
See also *Urethane polymers*
**Q 58u**
See *Cyclohexanaminium, N-heptyl-2-= hydroxy-N,N,2-trimethyl-5-(1-= methylethenyl)-, bromide [29380-01-6]*
**Quaalude**
See *4(3H)-Quinazolinone, 2-methyl-3-(2-= methylphenyl)- [72-44-6]*
**Quaccor 1001**
See *2-Furanmethanol, homopolymer [25212-86-6]*
**Quack grass**
See *Agropyron repens*
**Quacorr 1001**
See *2-Furanmethanol, homopolymer [25212-86-6]*
**Quadramer 5**
See *2-Propenoic acid, polymers, polymer with N-(1,1-dimethylethyl)-2-propenamide, 1-ethenyl-2-pyrrolidinone and 2-propenamide [34459-55-7]*
**Quadratic acid**
See *3-Cyclobutene-1,2-dione, 3,4-dihydroxy- [2892-51-5]*
**Quadricyclane**
See *Tetracyclo[3.2.0.0$^{2,7}$.0$^{4,6}$]heptane [278-06-8]*
**Quadricyclanone**
See *Tetracyclo[3.2.0.0$^{2,7}$.0$^{4,6}$]heptanone [1072-92-0]*
**Quadriphenyl**
See *1,1':4',1'':4'',1'''-Quaterphenyl [135-70-6]*
**Quadriplegia**
See *Paralysis*, quadriplegia
**Quadriprismane**
See *Pentacyclo[4.2.0.0$^{2,5}$.0$^{3,8}$.0$^{4,7}$]octane [277-10-1]*
**Quadrol**
See *2-Propanol, 1,1',1'',1'''-(1,2-ethanediyldinit= rilo)tetrakis- [102-60-3]*
**Quadron**
See *Quasiparticles and Excitations*, quadrons
**Quadrone**
See *6,8b-Ethano-8bH-cyclopenta[de]-2-= benzopyran-1,4-dione, octahydro-10,10-= dimethyl-, (3aα,5aβ,6α,8aα,8bα)- [66550-08-1]*
**Quadrosilan**
See *Cyclotetrasiloxane, 2,2,4,6,6,8-hexamethyl-= 4,8-diphenyl-, cis- [33204-76-1]*
**Quadrupole**
coupling and interactions of
see
*Anisotropic coupling constant*
*Quadrupole coupling*
elec. moment of—see *Quadrupole moment*
magnetic moment of
see
*Magnetic moment*

*Nuclear magnetic moment*
**Quadrupole coupling**
See also
*Nuclear magnetic resonance*
*Nuclear quadrupole resonance*
anisotropic const. for——see *Anisotropic coupling constant*
**Quadrupole moment**
Electric quadrupole moments only are indexed at this heading. Magnetic quadrupole moments are indexed at *Magnetic moment* and *Nuclear magnetic moment*
coupling and interactions of
see
*Anisotropic coupling constant*
*Quadrupole coupling*
of energy level transitions
see
*Energy level transition*
*Nuclear energy level*
**Quadrupole relaxation**
See also *Nuclear quadrupole resonance*
**Quadrupole resonance**
See *Nuclear quadrupole resonance*
**Quadrupole shielding**
See *Screening, electronic and nuclear*, quadrupole
**Quadrupole splitting**
See such headings as
*Electron spin resonance*
*Energy level, hyperfine structure*
*Moessbauer effect*
*Nuclear magnetic resonance*
**Quail**
See also *Coturnix*
**Qualidil**
See *1-Azoniabicyclo[2.2.2]octane, 1,1'-(1,6-hexanediyl)bis[3-(phenylmethyl)-, dichloride [3563-63-1]*
**Qualitex**
See *Rubber, natural*
**Quality control**
Studies of quality control itself, or as it relates to classes of substances, are indexed at this heading. For quality control involving specific substances and processes, see those headings
chem. anal. for——see *Analysis* and related headings
**Quamoclit**
See *Ipomoea*
**Quanta**
See such headings as
*Graviton*
*Photon*
*Planck's constant*
*Quantization*
*Quasiparticles and Excitations*
**Quantacure 597**
See *9H-Thioxanthen-9-one, 2-(1-methylethyl)- [5495-84-1]*
**Quantacure 659**
See *9H-Thioxanthen-9-one, 2-(1-methylethyl)- [5495-84-1]*
**Quantacure ITX**
See *9H-Thioxanthen-9-one, 2-(1-methylethyl)- [5495-84-1]*
**Quantacure PDO**
See *1,2-Propanedione, 1-phenyl-, 2-[O-(ethoxycarbonyl)oxime] [65894-76-0]*
**Quantacure SKS**
See *Benzenemethanesulfonic acid, 4-(oxophenylacetyl)-, sodium salt [77076-13-2]*
**Quanta II**
See *Barium chloride fluoride (BaClF) [13718-55-3]*
**Quanta III**
See *Lanthanum bromide oxide (LaBrO) [13875-40-6]*
**Quantalan**
See *Cholestyramine [11041-12-6]*
**Quantitative localization activity relationship (QLAR)**
See such headings as
*Pharmaceuticals*
pharmacokinetics of
*Pharmacology*
**Quantitative structure activity relationships (QSAR)**
See such headings as
*Linear free energy relationship*
*Molecular orbital*
*Molecular structure-biological activity relationship*
*Molecular structure-property relationship*
**Quantization**
See also *Quasiparticles and Excitations*
**Quantometers**
See *Spectrometers*
**Quantril**
See *2H-Benzo[a]quinolizine-3-carboxamide, 2-(acetyloxy)-N,N-diethyl-1,3,4,6,7,11b-= hexahydro-9,10-dimethoxy-, (2α,3β,11bα)- [23844-24-8]*
**Quantum amplification**
of light——see *Lasers*
of microwaves——see *Masers*
**Quantum chaos**
See *Quantum mechanics*
**Quantum chromodynamics**
Studies of quantum chromodynamics itself and as it pertains to classes of atomic nuclei and elementary particles are indexed at this heading. Quantum chromodynamics as it pertains to specific nuclei and particles is indexed at headings for those elements and particles
**Quantum fluids**
See
*Bose fluids*
*Fermi fluids*
*Fluids*
quantum
*Superfluids*
**Quantum hydrodynamics**
See
*Flow*
quantum hydrodynamic
*Fluid mechanics*
hydrodynamics, quantum
**Quantum mechanical amplifiers**
See
*Lasers*
*Masers*
*Quantum amplification*
**Quantum mechanics**
The methodology of quantum mechanical calculations is indexed at this heading. Calculations pertaining to specific atoms, molecules and particles are indexed at headings for those specific substances
See also *Energy level* headings
See also *Field theory*
of electromagnetic radiation——see *Quantum electrodynamics*
Hamiltonian, for electronic spin——see also *Electron spin resonance*
orbital integrals for
see
*Atomic integral*
*Molecular integral*
statistical——see *Statistical mechanics*, quantum
uncertainty principle in——see *Uncertainty principle*
valence-bond method——see *Valence bond theory*
wave functions for
see
*Atomic orbital*
*Molecular orbital*
*Nuclear wave function*
*Wave function*
**Quantum optics**
See
*Laser radiation*
*Lasers*
*Optics*
quantum
*Quantum amplification*
**Quantum state**
See *Energy level* headings
See
*Nuclear energy level*
*Quantum number*
**Quantum statistics**
See *Statistical mechanics*, quantum
**Quarelin**
See *Methanesulfonic acid, [(2,3-dihydro-1,5-= dimethyl-3-oxo-2-phenyl-1H-pyrazol-4-= yl)methylamino]-, sodium salt, mixt. with 1-[(3,4-diethoxyphenyl)methylene]-6,7-= diethoxy-1,2,3,4-tetrahydroisoquinoline hydrochloride and 3,7-dihydro-1,3,7-= trimethyl-1H-purine-2,6-dione [62201-28-9]*
**Quark**
See also
*Antiquark [12585-47-6]*
*Hadron [12585-58-9]*
in elementary-particle structure model——see *Quark model*
**Quark model**
Studies of the quark model itself, and as it pertains to classes of elementary particles are indexed at this heading. The quark model as it pertains to specific particles is indexed at headings for those particles, and as it pertains to atomic nuclei at *Nuclear model*
**Quartamin 24P**
See *1-Dodecanaminium, N,N,N-trimethyl-, chloride [112-00-5]*
**Quartolan**
See *Benzenemethanaminium, N-[2-(dodecyla= mino)-2-oxoethyl]-N,N-dimethyl-, chloride [100-95-8]*
**Quartz**
fused or vitreous——see *Silica, vitreous [60676-86-0]*
**Quartz, properties**
fluid inclusions in——see *Geological liquid inclusions*
fluid inculsions in——see *Geological gas inclusions*
**Quartz** *[14808-60-7]*, **uses and miscellaneous**
Products described as quartz but produced from the fused form are indexed at *Silica, vitreous*
**Quartz diorite**
See *Diorite*, quartz
**Quartz dolerite**
See *Dolerite*, quartz
**Quartz keratophyre**
See *Keratophyre*, quartz
**Quartz monzonite**
See *Monzonite*, quartz
**Quartz porphyry**
See *Porphyry*, quartz
**Quasiparticles and Excitations**
Quasiparticles or quantized elementary excitations are indexed at this heading. The process of exciting to higher energy levels is indexed at such headings as *Energy level excitation* and *Optical pumping*, or at headings for resulting emissions, such as *Fluorescence*, *Laser radiation*, *Microwave*, *Ultraviolet radiation*
demons——see *Plasmon*, demon
dielectrons
see
*Electron, solvated*
di-
*Polaron*
dielectron
ferrons——see *Polaron*, magnetic
helicons——see *Helicon*
holes——see *Hole*
magnons——see *Spin wave and Magnon*
muron——see *Phonon*, muron
phonons——see *Phonon*
plasmons——see *Plasmon*
polaritons——see *Polariton*
polarons
see
*Electron, solvated*
*Polaron*
rotons——see *Roton*
solidons——see *Phonon*, solidon
**Quasisynergism**
See *Synergism*, quasi-
**Quasi-valence number**
See *Electron configuration*
**Quasoft HS 60**
See *Ethene, polymers, homopolymer [9002-88-4]*
**Quassimarin**
See *Picras-3-ene-2,16-dione, 15-[2-(acetylox= y)-2-methyl-1-oxobutoxy]-13,20-epoxy-1,= 11,12-trihydroxy-, (1β,11β,12α,15β)- [59938-97-5]*
**Quassin**
See *Picrasa-2,12-diene-1,11,16-trione, 2,12-dimethoxy- [76-78-8]*
**Quassinol**
See *Picrasa-12,14-diene-2,16-dione, 1,11-epoxy-1,12-dihydroxy-, (1α,11α)- [21018-87-1]*
**Quat 188**
See *1-Propanaminium, 3-chloro-2-hydroxy-= N,N,N-trimethyl-, chloride [3327-22-8]*
**Quatacaine**
See *Propanamide, 2-methyl-N-(2-= methylphenyl)-2-(propylamino)-, monohydrochloride [6205-09-0]*
**Quatal**
See *Pentanedial, mixt. with benzyl-$C_{12-16}$-= alkyldimethylammonium chlorides [84503-59-3]*
**Quatamine**
See *Benzeneethanaminium, 4-amino-N,N,N-= trimethyl-β-oxo-, chloride [24293-73-0]*
**Quateleron**
See *1-Butanaminium, 3-[(4-butoxybenzoyl)= oxy]-N,N,N-triethyl-2-methyl-, iodide [3818-40-4]*
**Quaterbenzene**
See *Quaterphenyl [29036-02-0]*
**Quaterene**
See *Pentacyclo[16.2.1.$1^{3,6}$.$1^{8,11}$.$1^{13,16}$]tetracosa-= 3,5,8,10,13,15,18,20-octaene [35007-03-5]*
**Quaternary ammonium compounds**
Studies of completely substituted ammonium compounds (i.e., compounds containing a positively charged nitrogen atom to which no hydrogen is directly attached) as a class are indexed at this heading. Compounds in which one or more of the hydrogen atoms of the $NH_4$ group remain unsubstituted are indexed as addition compounds of the uncharged nitrogen compounds (see note at *Ammonium, compounds*). Specific ammonium compounds are indexed at "aminium" names, usually derived from the name of the preferred primary amine (see *Amines*), the other 3 atoms or groups being expressed as *N*-substituents; e.g., *2-Butanaminium, 3-chloro-N,N-bis= (2-chloro-1-methylpropyl)-N-methyl-*, perchlorate. Multiplicative names are employed under the same circumstances as for other classes of compounds, e.g., *Benzeneethanaminium, 4,4',4''-nitrilotris= [N,N,N-triethyl-*. When the quaternized nitrogen atom is a member of a ring system, the final "e" of the ring name is dropped and "ium" substituted, e.g., *Morpholinium, 4,4,-diethyl-*, chloride; *Oxazolium, 3-(4-chlorophenyl)-2-hydroxy-*, hydroxide, inner salt. When a nitrogen heterocycle is attached directly or indirectly to an amine group or alkanamine, as in *3-Morpholinee= thanamine*, and both nitrogen atoms are quaternized, the compound is named on the basis of the cyclic parent, e.g., *Morpholinium, 4,4-dimethyl-3-[2-(trimethylammonio)= ethyl]-*, dichloride
betaines——see *Betaines*
iminium——see *Iminium compounds*
**Quaternary ammonium compounds, polymers**
See also *Ionene polymers*
**Quaternatine**
See *Yohimban-16-carboxylic acid, 17-hydroxy-11-methoxy-, methyl ester, (3β,16β,17α,20α)- [57499-04-4]*
**Quaternine**
See *Akuammilan-17-oic acid, 2,5-epoxy-1,2-= dihydro-10,11-dimethoxy-1-methyl-, methyl ester, (2α,5α)- [57499-02-2]*
**Quaterninol**
See *Akuammilan-17-ol, 1,2-dihydro-10,11-= dimethoxy-1-methyl-, (2β)- [57499-03-3]*
**Quaternium 15**
See *3,5,7-Triaza-1-azoniatricyclo[3.3.1.$1^{3,7}$]= decane, 1-(3-chloro-2-propenyl)-, chloride [4080-31-3]*
**Quaternium 22**
See *1-Propanaminium, 3-(D-gluconoylamin= o)-N-(2-hydroxyethyl)-N,N-dimethyl-, chloride [51812-80-7]*
**Quaternium 34**
See *Quaternary ammonium compounds*, dicoco alkyldimethyl, chlorides
**Quaternium 40**
See *2-Propen-1-aminium, N,N-dimethyl-N-= 2-propenyl-, chloride, homopolymer [26062-79-3]*
**Quaternization**
catalysts——see *Quaternization catalysts*

# Chemical Abstracts

Published by the

American Chemical Society

Volume 100

January 2–January 16

(Abstracts 1–23018)

1984

CHABA 8

P. O. BOX 3012, COLUMBUS, OHIO 43210

(6 μg/min) on the renal vascular response produced by two 10 min i.v. amphotericin B (0.35 mg/kg) infusions were examd. in rats. In the control group, amphotericin B decreased renal blood flow 1.7 mL/min (22%) and 3.5 mL/min (44%) during the 1st and 2nd amphotericin B infusions, resp. In animals pretreated with aminophylline the decrease in RBF produced by amphotericin B was only 0.4 mL/min (5.5%) and 1.3 mL/min (15%) during the 1st and 2nd amphotericin B infusions, resp. In contrast, neither saralasin nor the direct vasodilator sodium nitroprusside (0.4-2 μg/min) influenced the renal vascular response to amphotericin B. Apparently, the renal vascular response to amphotericin B is not linked to the formation of angiotensin II, but rather might be mediated by increases in renal adenosine levels.

100: 270w **Activity of methisoprinol against rabies virus. In vitro evaluation.** D'Arca, S. U.; Pana, A.; Grassi, M. (Ist. Ig. G. Sanarelli, Univ. Stud. Roma, Italy). *EOS--Riv. Immunol. Immuno= farmacol.* **1983,** 3(1), 8-10 (Ital). *Methisoprinol* **[36703-88-5]** at 250 μg/mL inhibited the replication of rabies virus in BHK-21 cell culture. Methisoprinol was not toxic to the cells at concns. of ≤1000 μg/mL.

100: 271x **Activity of several chemopreparations in relation to transmissive viral gastroenteritis and enterovirus of swine.** Potopal'skii, A. I.; Spivak, N. Ya.; Shved, A. D.; Mel'nichenko, V. S.; Krasnova, E. F. (Inst. Mol. Biol. Genet., Kiev, USSR). *Mikrobiol. Zh. (Kiev)* **1983,** 45(5), 75-8 (Russ). *Amitosine* **[58318-41-5]** (a product of the condensation of *Chelidonium* alkaloids with thiophosphamide), *isathizon* **[62682-46-6]**, *OL-56* **[52233-35-9]** (a thiazoline deriv.), and various modified nucleic acid prepns. were tested for antivirus activity against transmissive viral gastroenteritis and swine enterovirus growing in porcine kidney cell cultures. Isathizon inhibited the reprodn. of gastroenteritis virus by 2.5 log units, and prepn. PP (an unspecified nucleic acid modified chem. in an unspecified manner) inhibited the enterovirus.

100: 272y **Treatment of immature and mature Fasciola hepatica infections in sheep with triclabendazole.** Boray, J. C.; Crowfoot, P. D.; Strong, M. B.; Allison, J. R.; Schellenbaum, M.; Von Orelli, M.; Sarasin, G. (Res. Cent., Ciba-Geigy Australia Ltd., Kemps Creek, Australia). *Vet. Rec.* **1983,** 113(14), 315-17 (Eng).

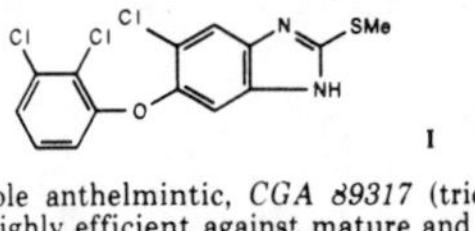

I

A new benzimidazole anthelmintic, *CGA 89317* (triclabendazole)(I) **[68786-66-3]** was highly efficient against mature and early immature *F. hepatica* infections in sheep. At 2.5 mg/kg the efficiency was 90 and 98% against flukes aged 8 and 12 wk, resp. At 5 mg/kg the drug was 92 and 98% efficient against flukes aged 4 and 8 wk, resp., and 100% against 12-wk-old flukes. An efficiency of 93 and 98% was achieved against 1-wk-old flukes and 99 to 100% against flukes aged 2 to 4 wk at 10 mg/kg. At this dose rate the drug had 100% efficiency against 6-wk-old flukes. If the dose was increased to 15 mg/kg, 98% efficiency was achieved one day after infection. Triclabendazole was equally efficient when administered orally or by intraruminal or intra-abomasal injection. A max. tolerated dose of 200 mg/kg was established.

100: 273z **Nephrotoxicity of cefotiam in rats.** Nakai, Y.; Chiba, S.; Suhara, I.; Miyajima, H.; Yamazaki, M.; Takano, K. (Drug Saf. Eval. Lab., Takeda Chem. Ind., Ltd., Osaka, Japan 569). *Infection (Munich)* **1983,** 11(5), 283-5 (Eng). Eight-week-old Wistar female

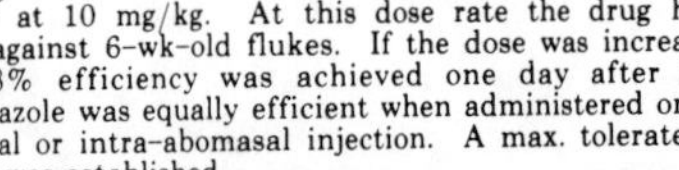

I

rats were treated i.m. with 300, 1,000 and 3,000 mg/kg/day of *cefotiam* (I) **[61622-34-2]** for 5 days. The nephrotoxicity of cefotiam was detd. on the basis of the no. of tubular epithelial cells excreted, the malate dehydrogenase activity in the urine and the hisotol. examn. of the kidneys. Cephalothin (3,000 mg/kg/day) was used as a ref. compd. There were slight increases in the no. of tubular epithelial cells excreted and in the malate dehydrogenase activity in the urine of some animals receiving 3,000 mg/kg/day of cefotiam or cephalothin. All of these changes disappeared within a few days after the dosing period. The histol. examn. of the kidneys at the end of both the dosing (5 days) and the recovery (7 days) periods revealed no pathol. changes indicating nephrotoxicity. Apparently, the nephrotoxicity of cefotiam was comparable to that of cephalothin; the urinary changes in animals receiving the highest dose were considered to be an early sign of nephrotoxicity. The max. non-toxic dose of cefotiam was 1,000 mg/kg/day under the present exptl. conditions.

100: 274a **Antiviral activity in vitro of new synthetic compound G502 against DNA and RNA viruses.** Bonina, L.; Arena, A.; Merendino, R. A.; Mastroeni, P. (Microbiol. Inst., Messina Univ., Messina, Italy). *Drugs Exp. Clin. Res.* **1983,** 9(8-9), 577-84 (Eng).

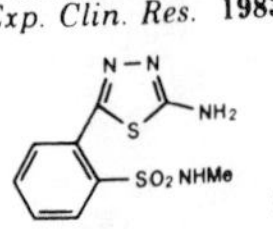

I

In order to evaluate the antiviral effects of *G502* (I) **[84597-71-7]**, expts. were carried out in HEp-2 cells where virus replication was inhibited. An attempt was made to understand its mechanism of action by carrying out expts. with DNA and RNA viruses. The precise stage of viral replication that is blocked by G502 is as yet unknown. G502 possesses properties that make it an excellent candidate for clin. evaluation.

100: 275b **Trypanosoma cruzi: 4-aminopyrazolopyrimidine in the treatment of experimental Chagas' disease.** Avila, Jose Luis; Avila, Angela; Munoz, Edgar; Monzon, Hector (Inst. Nac. Dermatol., Caracas, Venez. 1010). *Exp. Parasitol.* **1983,** 56(2), 236-40 (Eng). An allopurinol metabolite, *4-aminopyrazolopyrimidine*

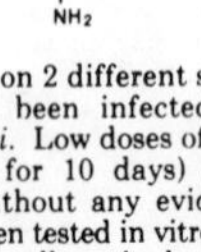

I

(I) **[2380-63-4]** was tested on 2 different strains of mice (NMRI-IVIC and C57B1/6J) that had been infected 4 days earlier with the virulent Ya strain of *T. cruzi.* Low doses of 4-aminopyrazolopyrimidine (0.125-0.500 mg/kg daily for 10 days) decreased parasitemia and increased survival time, without any evidence of toxicity, compared with untreated animals. When tested in vitro, 4-aminopyrazolopyrimidine was 6-fold more active than allopurinol as a trypanostatic drug. The low therapeutic doses of 4-aminopyrazolopyrimidine suggest that this drug may be useful in the treatment of acute Chagas' disease.

100: 276c **Etiotropic action of rifadin in herpetic keratitis.** Semenova, E. N.; Terskikh, I. I.; Kudoyarov, R. G. (USSR). *Deposited Doc.* **1982,** VINITI 4570-82, 10 pp. (Russ). Avail. VINITI. *Rifadin* (I) **[13292-46-1]** was much more effective than interferon in bringing about epithelialization and recovery in rabbits with herpetic keratitis. In clin. studies, patients with xyloid herpetic keratitis recovered 6 days earlier when treated with I than when given interferon. In deep opthalmic herpes, I was more effective when used in conjunction with other antiviral drugs. I was well tolerated by the patients.

100: 277d **Antibiotic therapy for pancreatic sepsis: differences in bioactive blood and tissue levels.** Trudel, Judith L.; Mutch, David O.; Brown, Phillip R.; Richards, Geoffrey K.; Brown, Rea A. (Surg. Clin., McGill Univ., Montreal, PQ Can.). *Surg. Forum* **1982,** 33, 26-8 (Eng). In dogs with pancreatitis, i.v. administration of *gentamicin* **[1403-66-3]** (5 mg/kg), *cefazolin* **[25953-19-9]** (12.5 mg/kg), and *tetracycline* **[60-54-8]** (12.5 mg/kg) failed to achieve therapeutic pancreatic levels despite the fact that therapeutic blood levels were achieved, whereas administration of *clindamycin* **[18323-44-9]** (15 mg/kg), *metronidazole* **[443-48-1]** (10 mg/kg), and *chloramphenicol* **[56-75-7]** (50 mg/kg) achieved therapeutic tissue levels regardless of the blood levels measured. For pancreatic sepsis, the antibiotics of choice should be evaluated on the basis of their tissue bioavailability rather than on blood levels alone.

100: 278e **Effects of topical disinfectants and antibiotics on human fibroblasts.** Lineaweaver, William; McMorris, Sally; Howard, Richard (Coll. Med., Univ. Florida, Gainesville, FL USA). *Surg. Forum* **1982,** 33, 37-9 (Eng). In human fibroblast cultures, the antibiotics *neomycin* **[1404-04-2]** (1%), *kanamycin* **[8063-07-8]** (2%), and *bacitracin* **[1405-87-4]** 50 units/mL were nontoxic, whereas the disinfectants *povidone-iodine* **[25655-41-8]** (1%), HOAc **[64-19-7]** (0.25%), $H_2O_2$ **[7722-84-1]** (3%), and NaOCl **[7681-52-9]** (3%) killed 100% of the exposed cells. Correlation of this model with bactericidal and wound strength data could provide specific criteria for evaluating the effects of topical agents.

100: 279f **Effectiveness of modified steroid-antibiotic therapies for lethal sepsis in the dog.** Beller, Beverly K.; Archer, Linda T.; Passey, Richard B.; Flournoy, Dayl J.; Hinshaw, Lerner B. (Veterans Adm. Med. Cent., Oklahoma City, OK 73104 USA). *Arch. Surg. (Chicago)* **1983,** 118(11), 1293-9 (Eng). Dogs permanently recover (survive at least seven days) from LDs of *Escherichia coli* when treated early with i.v. intermittent infusions of *methylprednisolone sodium succinate* **[2375-03-3]** and *gentamicin* **[1403-66-3]**. The therapeutic effectiveness of const. or bolus i.v. infusion of methyl= prednisolone combined with gentamicin or *netilmicin* **[56391-56-1]** was evaluated.. Four groups of anesthetized dogs were infused for one hour with *E. coli* and treated as follows (% survival indicated): no treatment (0%); const. infusion of methylprednisolone and gentamicin (100%); bolus infusion of methylprednisolone and gentamicin (57%); and const. infusion of methylprednisolone and netilmicin (83%). Const. or bolus infusion of methylprednisolone was begun 15 min after *E. coli* infusion was started. Gentamicin or netilmicin administration was begun when all organisms had been infused. The probability of recovery from shock was significantly increased when dogs were treated with const. infusion of methyl= prednisolone and intermittent infusions of gentamicin or netilmicin, but was only moderately increased when treated with intermittent bolus infusions of methylprednisolone and intermittent infusions of gentamicin.

100: 280z **Effect of magnesium ions, DEAE-dextran, and dextran sulfate on plaque formation in IB-RS-2 swine cells by attenuated Newcastle disease virus.** Kubrusly, Flavia Saldanha; Koseki, Ignez (Sec. Biol. Cel., Inst. Biol., Brazil). *Arq. Inst. Biol., Sao Paulo* **1982,** 49(1-4), 71-4 (Port). In monolayer cultures of swine IB-RS-2 cells infected with attenuated Newcastle disease virus, treatment with $MgSO_4$ and *DEAE-dextran* **[9015-73-0]** in the absence of trypsin resulted in the formation of sparse, small plaques. Addn. of trypsin to the agar overlay resulted in the formation of is minimized. A generally applicable guideline that emerged from this study is that detergents should be used at approx. their cmc which should not be exceeded by the concn. of membrane. Similar considerations should apply to the use of detergents in purifying and reconstituting intrinsic membrane proteins.

100: 2444c **Interactions of benzoquinone with a model membrane bilayer as reported by a positronium probe.** Wang, Y. Y.; Hancock, A. J.; Jean, Y. C. (Dep. Phys., Univ. Missouri, Kansas City, MO 64110 USA). *J. Am. Chem. Soc.* **1983,** 105(25), 7272-6 (Eng). Techniques of positronium chem. were used to examine the properties of model bilayer membranes in the presence and absence of *p*-benzoquinone at temps. below and above the transition temp. of the phospholipid aggregate. The chem. reactivity between positronium atoms and *p*-benzoquinone mols. in the bilayer systems (water-DL-α-dipalmitoylphosphatidylcholine) was found to vary as a function of lipid concn. By comparison of the rate consts. of the reaction with the rate consts. for *p*-benzoquinone in various pure solvents, it is suggested that the location of the *p*-benzoquinone mols. in lipid bilayers is in a region of high polarity. Positronium lifetimes and formation probabilities were measured at 25° and 45° for a variety of lipid concns. The variation of *o*-positronium lifetime as a function of lipid concn. was found to be greater for aggregates in the liq. cryst. phase than for those in the gel phase. The *o*-positronium formation probabilities were found to increase as a function of lipid concn. in the liq. cryst. phase, but to decrease in the gel phase. The implications of these aspects of positronium annihilation in this bilayer membrane system are discussed in relation to the charge transfer properties of *p*-benzoquinone.

100: 2445f **Order-disorder phase transition and lipid dynamics in rabbit small intestinal brush border membranes. Effect of proteins.** Muetsch, Beat; Gains, Nigel; Hauser, Helmut (Lab. Biochem., Eidg. Tech. Hochsch. Zurich, CH 8092 Zurich, Switz.). *Biochemistry* **1983,** 22(26), 6326-33 (Eng). The total lipids extd. from rabbit intestinal brush border membranes formed smectic lamellar phases when dispersed in water. ESR and NMR spectra were obtained with various hydrophobic probes incorporated in either brush border vesicle membranes or their extd. lipids. By use of a variety of chem. different spin-labels, the temp. dependence of brush border membranes and their extd. lipids was probed. The temp. dependence of various ESR spectral parameters showed discontinuities that, by comparison with DSC, were assigned to a lipid thermotropic phase transition. DSC showed that the lipid in brush border membranes undergoes a broad, reversible phase transition of low enthalpy between 10 and 30°, with a peak temp. of ~25°. Hence, the brush border membrane of rabbit small intestine functions in the liq.-cryst. state, well above the peak temp. and also above the upper limit of the lipid phase transition. Therefore, in itself, the thermotropic lipid phase transition is unlikely to play a physiol. role. The low enthalpy of the lipid phase transition, indicative of a lack of cooperativity, was primarily attributed to the relatively high cholesterol content and to heterogeneity in the lipid compn. of this membrane. Lipid-protein interactions appeared to play a minor role in this context. Papain treatment and also alk. treatment of brush border vesicles produced significant protein losses. Papain digestion solubilized ~65% of the total protein; alk. treatment at pH 8.8 and 37° and at pH 11 and 4° led to protein losses of 60 and 58%, resp. The alk. treatment at pH 11 led to the quant. removal of the cytoskeletal proteins. Removal of these membrane-bound enzymes and of the cytoskeletal proteins apparently had no effect on the fluidity and mol. packing of brush border vesicles. It also did not affect the reversible thermal behavior of these vesicles.

100: 2446g **Differential scanning calorimetric studies of aqueous dispersions of mixtures of cholesterol with some mixed-acid and single-acid phosphatidylcholines.** Davis, P. J.; Keough, K. M. W. (Dep. Biochem., Mem. Univ. Newfoundland, St. John's, NF Can. A1B 3X9). *Biochemistry* **1983,** 22(26), 6334-40 (Eng). DSC studies were performed on aq. dispersions of mixts. of cholesterol with either 1,2-distearoyl-*sn*-glycero-3-phosphocholine, 1,2-dioleoyl-*sn*-glycero-3-phosphocholine, 1-stearoyl-2-oleoyl-*sn*-glycero-3-phosphocholine, or 1-oleoyl-2-stearoyl-*sn*-glycero-3-phosphocholine. At low concns. of cholesterol, the endotherms obtained on heating dispersions of phosphatidylcholine (PC)-cholesterol mixts. had different shapes for each PC at a fixed cholesterol concn. The endotherms could be resolved into broad and narrow components. The $\Delta H$ and the temps. of max. excess heat capacity ($T_{max}$) for both narrow and broad components of each type of mixt. responded in different ways to increasing cholesterol concn. At high cholesterol concns., all mixts. gave 1 broad endotherm, but the concn. of cholesterol at which the endotherms could not be resolved from base-lines was different for each lipid. The results were consistent with there being at least a quant. difference in the interaction between cholesterol and different PCs in the gel phase.

100: 2447h **BLM destruction as a result of electrical breakdown.** Sukharev, S. I.; Arakelyan, V. B.; Abidor, I. G.; Chernomordik, L. V.; Pastushenko, V. F. (Inst. Electrochem., Moscow, USSR). *Biofizika* **1983,** 28(5), 756-60 (Russ). At the initial stage of bilayer membrane (BLM) rupture due to elec. breakdown, the change of current, I, in time, *t*, is of an exponential pattern and is practically independent of BLM voltage. This agrees with the idea of evolution of sufficiently large (supercrit.) pores by the energetic profile of the system corresponding to zero voltage in the pore region. The calcd. coeffs. of pore diffusion and of lateral viscosity of BLM are close to the coeff. values of lipid self-diffusion and lateral viscosity of lipid bilayers found by other methods.

100: 2448j **Mode of proteoliposome association with the pla[illegible] bilayer phospholipid membrane.** Severina, I. I. (Fac. Biol., M[illegible] Lomonosov Moscow State Univ., Moscow, USSR). *Biokhim[illegible]* (*Moscow*) **1983,** 48(9), 1522-9 (Russ). Proteoliposomes [illegible] reconstituted from *Halobacterium halobium* membrane fragm[illegible] azolectin, and cholesterol with or without nystatin. The bacter[illegible] rhodopsin-mediated electrogenesis was monitored with a proteolipo[illegible] suspension and phenyldicarbaundecaborane (PCB⁻) probe or prot[illegible] liposomes assocd. with a planar bilayer membrane. In light, PC[illegible] was absorbed by proteoliposomes. The PCB⁻ uptake was inhibi[illegible] by nystatin added to the incubation mixt. contg. proteolipos[illegible] when the latter were reconstituted in the presence of nystatin. [illegible] nystatin-contg. proteoliposomes were assocd. with a planar bil[illegible] azolectin membrane in the presence of $Ca^{2+}$. In such a syst[illegible] bacteriorhodopsin generates a photocurrent which charges [illegible] proteoliposome-contg. (cis-side) compartment neg. and the trans[illegible] compartment pos. The photoresponse increased severalfold [illegible] nystatin was added to the trans-side soln. Nystatin addn. [illegible] ineffective when proteoliposomes were reconstituted without nysta[illegible] Taking into account the fact that nystatin forms ion-permeab[illegible] pores in a membrane only when present on both sides of [illegible] membrane and when the membrane is bilayer, one can explain [illegible] above data assuming that (*1*) the intraproteoliposomal soln. does [illegible] mix with the extraproteoliposomal one when the proteoliposomes [illegible] attached to a planar black membrane, and (*2*) the attach[illegible] proteoliposomes are sepd. from the trans-side bathing soln. with [illegible] bimol. membrane. If this is the case, nystatin in the trans[illegible] bathing soln. and inside the attached proteoliposomes can form p[illegible] across the part of the planar membrane which separates [illegible] proteoliposome interior from the trans-side soln. Through [illegible] pores, $H^+$ (pumped by bacteriorhodopsin from the cis-side soln. [illegible] the proteoliposomes interior) or some other intraproteoliposomal [illegible] can be equilibrated with those in the trans-side soln. As a result, [illegible] bacteriorhodopsin-generated photocurrent increases.

100: 2449k **Association of blood coagulation factors V and [illegible] with phospholipid monolayers.** Mayer, Lawrence D.; Pusey, M[illegible] Lee; Griep, Mark A.; Nelsestuen, Gary L. (Dep. Biochem., Un[illegible] Minnesota, St. Paul, MN 55108 USA). *Biochemistry* **1983,** 22(2[illegible] 6226-32 (Eng). Blood-coagulation factors X and V were found [illegible] adsorb to phospholipid monolayers and to induce surface press[illegible] changes. These proteins also adsorbed to the air-water interface [illegible] formed protein surface films. Plots of surface pressure change ([illegible] vs. initial monolayer surface pressure ($\pi_0$) appeared biphasic [illegible] factors X and V, indicating 2 distinct adsorption processes. Surf[illegible] pressure changes in monolayers spread below the collapse pressure [illegible] the resp. proteins were characteristic of protein adsorption to [illegible] air-water interface, whereas those obsd. above the protein collap[illegible] pressures were consistent with specific protein-acidic phospholip[illegible] interactions. Phosphatidylserine-dependent surface pressure chang[illegible] were very small for both proteins. Factor X-induced surfac[illegible] pressure changes required the presence of $Ca^{2+}$, whereas facto[illegible] V-induced changes occurred in the presence or absence of Ca[illegible] Protein-monolayer binding characteristics were comparable to tho[illegible] obtained with bilayer vesicles of similar compn. and indicated th[illegible] absence of significant membrane surface curvature effects. The max[illegible] surface concn. corresponded to 1 bound factor X mol./1400 Å[illegible] Comparison of surface pressure changes induced by factor V with [illegible] those induced by myelin basic protein suggested that the membrane-[illegible] binding processes of the 2 proteins involves similar but small degree[illegible] of acyl chain perturbation. Thrombin digestion of factor V had no [illegible] effect on surface pressure change and the isolated 80,000-dalton [illegible] peptide of factor $V_a$ also showed approx. similar surface pressure [illegible] effects. The vitamin K-dependent proteins caused a smaller surface [illegible] pressure change per bound protein mol. The results indicated that [illegible] the prothrombinase proteins assoc. primarily, if not exclusively, with [illegible] the head groups of the phospholipids.

100: 2450d **Fourier transform infrared spectroscopic studies of the effect of calcium ions on phosphatidylserine.** Dluhy, Richard; Cameron, David G.; Mantsch, Henry H.; Mendelsohn, Richard (Div. Chem., Natl. Res. Counc. Canada, Ottawa, ON Can. K1A 0R6). *Biochemistry* **1983,** 22(26), 6318-25 (Eng). Fourier transform IR (FT-IR) spectroscopy was used to investigate the complex conformational changes that occur as phosphatidylserine (PS) binds $Ca^{2+}$. The spectra confirmed the isothermal crystn. of the hydrocarbon chains in the PS-$Ca^{2+}$ complex. However, in contrast with DSC, which detects no phase transitions below 100° in liposomal complexes, several FT-IR parameters detected structural changes at 30-40° in these complexes analogous to those obsd. in solid-solid phase transitions of alkanes. Site symmetry splitting obsd. in the $PO_2^-$ bands suggested that $Ca^{2+}$ binds to the PS phosphate as a bidentate ligand; in addn., $Ca^{2+}$ caused a dehydration of the phosphate ester. No evidence was found for the specific chelation of $Ca^{2+}$ by the ionized carboxylate group or the dehydration of this group; instead, the carboxylate existed in an immobilized conformation in the presence of $Ca^{2+}$. Splitting of the degenerate vibrations of the CO group at the interfacial region suggested different rotational chain isomers in the $Ca^{2+}$ complex and the possibility of H-bonding with trapped interstitial water.

100: 2451e **Investigations on the order-disorder behavior of various phospholipids of natural and synthetic origin by optical and calorimetric techniques.** Brandenburg, K.; Seydel, U. (Div. Biophys., Res. Inst. Borstel, D-2061 Borstel, Fed. Rep. Ger.). *Thermochim. Acta* **1983,** 69(1-2), 71-102 (Eng). The thermodn. states and the thermotropic gel-to-liq. cryst. phase transition of liposomes prepd. from physiol. relevant phospholipids and phospholipid mixts. were investigated with fluorescence and

100: 10334j **Modeling of vertical water distribution in the atmosphere of Mars.** Kulikov, Yu. N.; Rykhletskii, M. V. (Inst. Prikl. Mat. im. Keldysha, Moscow, USSR). *Astron. Vestn.* 1983, 17(3), 144-52 (Russ). A numerical study was done of the phys. processes detg. the vertical distribution of water condensate and vapor in the mid-latitude lower atm. of Mars. The calcns. deal with the basic processes of mass transport and the kinetics of $H_2O$ condensation and evapn. In the mid-latitude summer hemisphere, theor. values of the aerosol layer, formed of ice particles ~3 μm in diam., agree with data from observations, provided that the coeff. of turbulent diffusion as a function of time at 20-30 km altitude is ≤3 × $10^5$ $cm^2/s$. Ests. of the times of water-vapor condensation and also of the diffusion and pptn. of $H_2O$ condensate show that at >40 km altitude, the formation is unlikely of clouds with a perceptible optical thickness, from ice particles, in a dust-free atm. Formation, at the 50-70 km level, of the condensate layer obsd. during global dust storms is related to the increase in stratospheric temp. resulting from the increase in nontransparency toward solar radiation. At the summertime atm. temp. (averaged over time), the water vapor pressure above the $H_2O$ condensate layer can exceed the pressure of the satd. vapor by an order of magnitude, and the relative vol. content ($f_v$) of water vapor reaches $10^{-4}$. These values substantially exceed the $f_v$ values used in previous aeronomic models of Mars.

100: 10335k **High temperature phase equilibria in a solar-composition gas.** Saxena, S. K.; Eriksson, G. (Dep. Geol., Brooklyn Coll., Brooklyn, NY 11210 USA). *Geochim. Cosmochim. Acta* 1983, 47(11), 1865-74 (Eng). Using recent addns. to thermochem. data on minerals and information on their solid soln. behavior, new equil. phase diagrams were computed in a system of solar gas compn. (Si, Al, Mg, Ca, Fe, Ni, Ti, Na, K, C, H, O, S, N) in the pressure and temp. ranges of 1 to $10^{-6}$ bar and 1153-1773 K resp. These calcns. show that Fe-Ni alloy condenses before all silicates included here (except melilite) down to a pressure of 2 × $10^{-4}$ bar below which plagioclase and clinopyroxene condense first. Orthopyroxene condenses next followed by ilmenite. Pressure-temp. variation of the chem. compn. of melilite, clinopyroxene, orthopyroxene, metal alloy and plagioclase may be used for cosmothermometry and cosmobarometry for equil. assemblages. The major transition from the refractory oxides and melilite (the meteorite "inclusion assemblage") to an assemblage of Fe-Ni alloy, olivine, plagioclase, and pyroxenes ("planet-forming") takes place within a narrow interval of pressure and temp. Small fluctuations of either pressure or temp. across this narrow region result in drastic changes in types and modes of minerals, which may explain the wide mineralogical varieties of meteorites.

100: 10336m **The neon-21 production rate in stony meteorites estimated from beryllium-10 and other radionuclides.** Moniot, R. K.; Kruse, T. H.; Tuniz, C.; Savin, W.; Hall, G. S.; Milazzo, T.; Pal, D.; Herzog, G. F. (Dep. Phys., Rutgers, State Univ., New Brunswick, NJ USA). *Geochim. Cosmochim. Acta* **1983**, 47(11), 1887-95 (Eng). The $^{10}Be$ contents of 28 stony meteorites with known $^{21}Ne$ contents range 0.97-23 dpm/kg and give an av. $^{21}Ne$ prodn. rate ($P_{21}$) at std. temp. and pressure of (0.28 ± 0.02) × $10^{-8}$ $cm^3$/gctdMyr for shielding conditions corresponding to $^{22}Ne/^{21}Ne$ = 1.114 in an *H*-chondrite. The $P_{21}(^{10}Be)$ value agrees with that calcd. by others, based on $^{22}Na$, $^{81}Kr$, and $^{53}Mn$ but not on $^{26}Al$. Temporal variations in the cosmic ray flux do not explain the disagreement satisfactorily; major errors in the radionuclide half-lives are not indicated. The discrepancy is rooted in the data selection and the difficulties of making accurate corrections for shielding, chem. compn., and other sources of variability.

100: 10337n **Oxygen isotopes in impactites of the Logoiskii crater.** Glazovskaya, L. I.; Suvorova, V. A.; Parfenova, O. V. (Inst. Eksp. Mineral., Chernogolovka, USSR). *Dokl. Akad. Nauk SSSR* 1983, 272(4), 946-9 [Petrogr.] (Russ). Two types of impactites (zyuvites and tagamites) occur in the Logoiskii crater (~10 km diam.) in granite-gneisses and sedimentary rocks. The $\delta^{18}O$ value was detd. in the gneissic country rocks (11.7), zyuvite glass (8.8-10.7), tagamite glass (10.0), ignimbrite glass (7.0), and phenocrysts in ignimbrites (9.0‰). The content of bound water ($H_2O^+$) increases sharply from detrital grains of granite-gneisses in the matrix of zyuvites and tagamites to their glasses. The formation of impactites is attributed to shock metamorphism of melts. The glass of impactites is enriched in $^{16}O$ in relation to the county rocks.

100: 10338p **Heterogeneous interactions of the carbon, nitrogen, and sulfur cycles in the atmosphere: the role of aerosols and clouds.** Taylor, G. S.; Baker, M. B.; Charlson, R. J. (Dep. Civ. Eng., Univ. Washington, Seattle, WA 98195 USA). *SCOPE* **1983**, 21(Major Biogeochem. Cycles Their Interact.), 115-41 (Eng). Much of the title interaction occurs in the aq. phase due to the hygroscopic nature of compds. like $NH_4$, $HSO_4$, and the water soly. of trace gases, such as $SO_2$ and $NH_3$. To study the nature of these interactions, the equil. of the system $H_2O$ (liq.), $SO_2$, $NH_3$, $H_2SO_4$, and $CO_2$ were examd. Subsequently, oxidn. of S(IV) to S(VI) within the aq. phase is considered as a function of $NH_3$, $H_2O$ (liq.), and the source strength of S(IV). Results show that: (*1*) almost all S(IV) is in the gas phase when clouds are acidified below pH ≃5.5, whereas most $NH_3$ is present as $NH_4^+$; (2) the amt. of S(IV) available for oxidn. in the water phase is a strong function of pH, with more S(IV) available at higher pH; (*3*) the overall effect of increased $SO_2$ emissions could be a slowing of its oxidn. allowing geog. regions under the influence of acid pptn. to increase in area.

100: 10339q **Wind- and temperature-induced effects on mesospheric ion composition and inferred minor constituents.** Kopp, E.; Philbrick, C. R. (Phys. Inst., Univ. Bern, CH-3012 Bern, Switz.). *Eur. Space Agency, [Spec. Publ.] ESA SP* **1983**, ESA SP-183, ESA Symp. Eur. Rocket Balloon Programmes Relat. Res., 6th, 41-6 (Eng). The pos. ion compn., temp., and wind profiles were measured independently above Kiruna (Sweden). Ion chem. models of the *D*-region (ionosphere) predict a strong temp. dependence for the d. distribution of proton hydrates with different hydration order. Mesospheric temp. profiles derived from a steady-state ion chem. model were compared with measured scale-height temp. from the accelerometer falling-sphere expts. The effect of strong, horizonal, neutral-air transport and ion drag became evident in the total pos.-ion d. profiles, derived from the ion mass spectrometer measurements. The NO profiles, inferred from the ion compn. measurements, show a pronounced d. depletion at 97-98 km. The NO-depletion feature correlates with regions of high wind and may result from transport effects on the NO.

100: 10340h **Possible effects of stratospheric aerosol layers on the vertical transport of water.** Crescentini, L.; Fiocco, G. (Ist. Fis., Univ. Roma, Rome, Italy). *Nuovo Cimento Soc. Ital. Fis., [Sez.] C* 1983, 6C(3), 337-49 (Eng). A possible effect of stratospheric aerosols, composed of mixts. of water and $H_2SO_4$, is discussed on the vertical flux of water and the relative humidity at the base of the stratosphere. The existence of long-lived aerosol layers in the intertropical region is related to the vertical velocity profile in the ascending branch of the Hadley cell. Fluctuations of the temp. profile cause the water fraction to evap. at the bottom of the layer and to condense in the upper part, and affect the diffusion coeffs. Thus, aerosols within a range of radii grow and evap., resp., at the upper and lower levels of the layer: considering the change in water mass, a variable balance between the opposite effects of sedimentation and of the Hadley-cell updraft is detd. which causes the particles to oscillate along the vertical. At the bottom, water vapor is sepd. from the particles by mol. diffusion, which has little effect on the aerosol motion. Calcns. indicate that the mechanism does not work but that its effect is small, a redn. of humidity of a few percent typically, and could be ≤10% for unusually large amts. of $H_2SO_4$, such as those obtained after strong volcanic eruptions.

100: 10341j **Sodium clouds in the lower thermosphere.** Kirchhoff, V. W. J. H.; Takahashi, H. (Inst. Pesqui. Espaciais, BR Sao Jose dos Campos, Brazil). *Report* **1983**, INPE-2793-PRE/355, 21 pp. (Eng). Avail. NTIS. From *Sci. Tech. Aerosp. Rep.* 1983, 21(19), Abstr. No. N83-31136. Very large short lived enhancements obsd. in Na nightglow measurements are described. Only 9 such events have been obsd. in about 8 yr of regular nocturnal observations at 23° S. These enhancements are due to excess Na deposited in the upper atm. by meteoroid ablation.

100: 10342k **High spectral resolution lidar to measure optical scattering properties of atmospheric aerosols. 1. Theory and instrumentation.** Shipley, S. T.; Tracy, D. H.; Eloranta, E. W.; Trauger, J. T.; Sroga, J. T.; Roesler, F. L.; Weinman, J. A. (Univ. Wisconsin, Madison, WI 53706 USA). *Appl. Opt.* 1983, 22(23), 3716-24 (Eng). A high spectral resoln. lidar technique is described for measuring optical scattering properties of atm. aerosols. Light backscattered by the atm. from a narrow-band optically pumped oscillator-amplifier dye laser is sepd. into its Doppler broadened mol. and elastically scattered aerosol components by a 2-channel Fabry-Perot polyetalon interferometer. Aerosol optical properties, such as the backscatter ratio, optical depth, extinction cross section, scattering cross section, and the backscatter phase function, are derived from the 2-channel measurements.

100: 10343m **High spectral resolution lidar to measure optical scattering properties of atmospheric aerosols. 2. Calibration and data analysis.** Sroga, J. T.; Eloranta, E. W.; Shipley, S. T.; Roesler, F. L.; Tryon, P. J. (Univ. Wisconsin, Madison, WI 53706 USA). *Appl. Opt.* 1983, 22(23), 3725-32 (Eng). Calcn. and data anal. procedures developed for the high spectral resoln. lidar (HSRL), which measures optical properties of atm. aerosols by interferometrically sepg. the elastic aerosol backscatter from the Doppler broadened mol. contribution, are described. Data obtained during flight evaluation testing of the HSRL system are presented with ests. of uncertainties due to instrument calibration. HSRL measurements of the aerosol scattering cross section are compared with in situ integrating nephelometer measurements.

100: 10344n **Lidar differential absorption and scattering technique: theory.** Zuev, V. E.; Makushkin, Yu. S.; Marichev, V. N.; Mitsel, A. A.; Zuev, V. V. (Inst. Atmos. Opt., 634055 Tomsk, USSR). *Appl. Opt.* 1983, 22(23), 3733-41 (Eng). The results of theor. investigations of the differential absorption technique for sounding atm. $H_2O$ vapor are described. The sources of errors and their effect on the results of interpreting the sounding data are analyzed. The math. problems for inverting the lidar returns are considered.

100: 10345p **Laser sounding of atmospheric humidity: experiment.** Zuev, V. V.; Zuev, V. E.; Makushkin, Yu. S.; Marichev, V. N.; Mitsel, A. A. (Inst. Atmos. Opt., 634055 Tomsk, USSR). *Appl. Opt.* **1983**, 22(23), 3742-6 (Eng). The results are presented of studies of atm. humidity profiles by lidar on the basis of a tunable ruby laser. The lidar parameters are given, and the technique of lidar measurements of $H_2O$ vapor content in the atm. using the differential absorption technique is described. The results of reconstitution of humidity profiles for ≤17-km altitudes are given, based on laser sounding.

100: 10346q **Differential absorption lidar technique for measurement of the atmospheric pressure profile.** Korb, C. Laurence; Weng, Chi Y. (Goddard Space Flight Cent., NASA, Greenbelt, MD 20771 USA). *Appl. Opt.* **1983**, 22(23), 3759-70 (Eng). A new 2-wavelength lidar technique for remotely measuring the pressure profile using the trough absorption region between 2 strong lines in the O *A* band is described. The theory of integrated

100: **111950n Microwave integrated optical modulator.** Cross, Peter S.; Baumgartner, Richard A.; Kolner, Brian H. (Hewlett Packard Lab., Palo Alto, CA 94304 USA). *Appl. Phys. Lett.* **1984,** 44(5), 486-8 (Eng). A Ti-diffused $LiNbO_3$, traveling wave modulator was fabricated and tested at microwave frequencies. A Mach-Zehnder interferometer optical configuration and a coplanar waveguide elec. transmission line were used. For a 4-mm interaction length, the modulator has a 3-decibel bandwidth of 13 GHz and requires only 2 V to switch at $\lambda$ = 840 nm. The frequency response is measured directly using an ultrahigh speed photodiode, and the test setup therefore constitutes the highest bandwidth efficient electrooptical transmission system ever reported.

100: **111951p Microprocessor based data acquisition and control system for a balloon borne quadrupole mass spectrometer.** Nevejans, D.; Frederick, P.; Arijs, E. (Aeron. Inst., 1180 Brussels, Belg.). *Bull. Cl. Sci., Acad. R. Belg.* **1982,** 68(5), 314-32 (Eng). A microprocessor based data acquisition and control system was designed and constructed for use in balloon borne quadrupole mass spectrometers. It combines the capabilities of an ion counter/discriminator, a spectrum accumulator, an intelligent mass scan controller, a multiplexed analog to digital converter, a relay controller, a PCM encoder for serial digital output, multiple analog outputs and an interface to remote control signals. The system was flown successfully in the stratosphere.

100: **111952q Scientific programs for the Spacelab ES013 grille spectrometer.** Muller, C.; Laurent, J. (Inst. Aeron. Spat. Belg., B-1180 Brussels, Belg.). *Bull. Cl. Sci., Acad. R. Belg.* **1982,** 68(6), 431-42 (Eng). The ES013 IR spectrometer, jointly built by ONERA (France) and IASB (Belgium), will perform, in late 1983, earth limb observations, in absorption, during sunrise or sunset, and in emission from the Spacelab space station. The organization of spectral intervals and observation programs is presented as well as the reasons for the choices made for the 1st flight.

100: **111953r Gold absorbing film for a composite bolometer.** Dragovan, Mark; Moseley, S. Harvey (Enrico Fermi Inst., Univ. Chicago, Chicago, IL 60637 USA). *Appl. Opt.* **1984,** 23(5), 654-6 (Eng). Au absorbing films were effective for a mm bolometer operated at 1.5 K. At 1.5 K, Au is significantly better than Bi because it has a lower heat capacity for the absorbing film. Either Bi or Au is suitable at 0.3 K. Below 0.3 K a superconducting absorbing film can have a heat capacity low enough not to dominate the heat capacity of the detector and, therefore may give better performance than a nonsuperconducting absorbing film.

100: **111954s A translational energy spectrometer for investigation of electron capture-produced long-lived excited ions.** Berlinger, P.; Vanek, W.; Wutte, U.; Winter, H. (Inst. Allgemeine Phys., Tech. Univ. Wien, A-1040 Vienna, Austria). *Contrib. - Symp. At. Surf. Phys.* **1984,** 184-9 (Eng). Edited by Howorka, F.; Lindinger, W.; Maerk, T. D. Inst. Atomphys. Univ. Innsbruck: Innsbruck, Austria. A translational energy spectrometer equipped with a metastable ion detector was built and tested. The instrument and its applications for studying inelastic collisions of multiply charged ions with atoms in connection with primary and secondary long-lived highly excited ions are described.

100: **111955t Fiber optic electric field sensor investigations. II.** Mermelstein, Marc D. (Tracor, Inc., Rockville, MD USA). *Report* **1983,** TRACOR-T-83-RV-5260; Order No. AD-A129138, 24 pp. (Eng). Avail. NTIS. From *Gov. Rep. Announce. Index (U. S.)* **1983,** 83(19), 4753. In June 1981 the Office of Naval Research contracted with Tracor, Inc. to exptl. investigate the potential of interfacing piezoelec. polymers and single mode fibers for the development of an optical fiber elec. field sensor. The continuation and conclusion of that effect is reported herein. Expts. were performed whereby a single mode fiber was embedded in a piezoelec. active vinylidene fluoride-tetafluoroethylene copolymer film. Application of an external elec. field to the fiber copolymer composite induces strains in the polymer film that are transferred to the core of the optical fiber. The resultant phase shift in the propagating light is detected with a Mach-Zehnder interferometer and correlated to the amplitude of the applied elec. field.

100: **111956u Superconducting interferometers with two niobium-niobium oxide-lead/indium ($Nb-NbO_x-Pb/In$) 'window' junctions.** Cucolo, A. M.; Paterno, G. (Ist. Fis., Univ. Salerno, 84100 Salerno, Italy). *Cryogenics* **1983,** 23(12), 639-42 (Eng). The fabrication procedure and performance of planar interferometers with 2 $Nb-NbO_x-PbIn$ "window" Josephson junctions are presented. Threshold characteristics for different values of the parameter $\beta = LI_1/\phi_0$ are reported, where L is the loop inductance, $I_1$ the max. Josephson current of the interferometer and $\phi_0 = 2.07 \times 10^{-15}$ Wb is the flux quantum. For $\beta < 1$, the exptl. data are compared with the magnetic field dependence computed in terms of a simple theor. model.

100: **111957v The polarization ratio of crystal monochromators. A survey by the International Union of Crystallography Commission on crystallographic apparatus.** Jennings, L. D. (Army Mater. Mech. Res. Cent., Watertown, MA 02172 USA). *Acta Crystallogr., Sect. A: Found. Crystallogr.* **1984,** A40(1), 12-16 (Eng). A tabulation of some 40 measured values of the polarization ratio $K$ (a measure of the fractional polarization introduced into an x-ray beam by a crystal monochromator) is presented. The values may be represented through the parameter $n$ given by $K = \cos^n 2\theta_M$. The measured values of $n$ cluster around unit and the theor. rationale for this result is discussed. Possible explanations of outlying values are considered. A recommendation is made that the polarization ratio of an app. be measured using a direct method and whenever possible, but methods for estg. a value of $K$ are also given.

100: **111958w Nature of a change in the refractive index during titanium diffusion in lithium tantalate.** Shashkin, V. V. (Inst. Fiz. Poluprovodn., Novosibirsk, USSR). *Fiz. Tverd. Tela (Leningrad)* **1983,** 25(12), 3719-21 (Russ). Study of the mechanism of a change of $n$ in a Ti diffusion zone of $LiTaO_3$ which accounted for the effect of spontaneous polarization $P_3$ showed that the formation of a waveguide layer is primarily due to a shift of the phase transition crit. temp. $T_c$ and a change of mol. refraction. The latter effect amts. to ~76 and 32% for $\Delta n_1$ and $\Delta n_3$, resp. (the indexes 1 and 3 denote the ordinary and extraordinary beam, resp.), at the Ti concn. $\sim 9 \times 10^{20}$ $cm^{-3}$.

100: **111959x Source of optical radiation with a controlled spectrum.** Kosyachenko, L. A.; Shemyakin, V. A.; Pivovar, A. V.; Guts, V. V.; Sklyarchuk, V. M.; Starikov, D. I.; Kuzovaya, V. L. (USSR). *Opt.-Mekh. Prom-st.* **1983,** (12), 19-20 (Russ). A source of visible radiation with elec. controlled spectrum was developed which emits light with virtually any desired spectral compn. at 0.35-0.70 $\mu$m. The source employs 7 broadband $\alpha$-SiC LED's (Guts, V. V. et al., 1979), a set of interference $ZnS-MgF_2$ multilayer filters with the transmission halfwidth 0.001-0.04 $\mu$m, and a 2-lens optical mixer. The light beams emitted by the diodes after passing through the interference filters are transformed by short-focal-length lenses into parallel beams which are collected by long-focal-length lenses into a single point on an opaque glass surface. This design is characterized by a good quality optical mixing and a relatively low optical loss.

100: **111960r High-efficiency photodetector for ultraviolet radiation.** Alferov, Zh. I.; Gorelenok, A. T.; Danil'chenko, V. G.; Kamanin, A. V.; Korol'kov, V. I.; Mamutin, V. V.; Tabarov, T. S.; Shmidt, N. M. (Fiz.-Tekh. Inst. im. Ioffe, Leningrad, USSR). *Pis'ma Zh. Tekh. Fiz.* **1983,** 9(24), 1516-19 (Russ). A high-efficiency UV photodetector consists of a rare-earth-doped InP thin epitaxial layer on a InP:Fe substrate. The dark resistance of the layer is 3 × $10^6$-$10^8$ Ω and its current-potential characteristics remains linear up to the light field strengths $E$-8 × $10^3$-$10^4$ V-$cm^{-1}$. At $\lambda$ = 300 nm, the abs. sensitivity of the detector is ≥40 A-$W^{-1}$ at 50 V. The excellent UV sensitivity of the device was ascribed to low surface recombination rate in InP.

100: **111961s $TE_{01}$ mode propagation at $\lambda$ = 394 $\mu$m in smooth and corrugated circular metallic waveguides.** Marhic, M. E.; Spektor, D. (Dep. Electr. Eng. Computer Sci., Northwestern Univ., Evanston, IL 60201 USA). *Infrared Phys.* **1983,** 23(5), 281-8 (Eng). The attenuation of the $TE_{01}$ mode in straight circular metallic waveguides was measured at $\lambda$ = 394 $\mu$m. The radiation was produced by an optically-pumped HCOOH waveguide laser, with an inner diam. as small as 4 mm. The losses were measured in smooth waveguides as well as in guides with a fine helical corrugation on the inner wall, designed to reduce the detrimental $TE_{01} \leftrightarrow TM_{11}$ coupling. The losses in both types of guides were substantial, being, resp., of the order of 30 and 60% per m, for 4.8 mm inner diam. tubes; this is more than an order of magnitude larger than the theor. expected values.

100: **111962t Effect of fiber mismatch on nonlinear distortions of signals in optical communication links.** Klimov, I. I.; Morozov, V. N.; Shidlovskii, V. R. (Fiz. Inst. im. Lebedeva, Moscow, USSR). *Kvantovaya Elektron. (Moscow)* **1983,** 10(5), 1024-6 (Russ). The nonlinear distortion was studied of the signal in a mismatched coupler in fiber-optic communication links, which arise due to interference of fiber modes. The value of the nonlinear distortions can be higher than earlier obtained exptl. ests. because of strong wavelength deviation under modulation of the semiconductor laser radiation power. The distortion amplitude depends substantially on conditions of matching between the laser and the fiber.

100: **111963u Ion-assisted deposition of optical thin films: low energy vs high energy bombardment.** McNeil, John R.; Barron, Alan C.; Wilson, S. R.; Herrmann, W. C., Jr. (Dep. Electr. Comput. Eng., Univ. New Mexico, Albuquerque, NM 87131 USA). *Appl. Opt.* **1984,** 23(4), 552-9 (Eng). O ion-assisted deposition of $SiO_2$ and $TiO_2$ for optical coatings was investigated as a function of ion energy (30-500 eV) and c.d. (0-300 $\mu A/cm^2$) at the optic. Both low and high energy ion bombardment improve $SiO_2$ film stoichiometry, although slightly greater improvement is realized for the low energy case. For $TiO_2$ films, low energy bombardment improves stoichiometry, while high energy bombardment is clearly detrimental. A redn. in H content by a factor of 10 is obsd. in $SiO_2$ films deposited with high energy ion bombardment. Durable films are produced at low substrate temps. (50-100°). Film stress characteristics are discussed.

100: **111964v Dopant effect on transmission loss increase due to hydrogen permeation.** Kuwazuru, M.; Mochizuki, K.; Namihira, Y.; Iwamoto, Y. (KDD Res. Dev. Lab., Tokyo, Japan 153). *Electron. Lett.* **1984,** 20(3), 115-16 (Eng). The transmission loss increase due to H permeation in a $GeO_2$-doped silica fiber and a $P_2O_5$-doped silica fiber were investigated. The difference of the loss increase between them was obsd., and is due to the difference of the electronegativity between silica, Ge, and P.

100: **111965w Dual-mode optical-fiber interferometric sensor.** McMillan, J. L.; Robertson, S. C. (Hirst Res. Cent., GEC Res. Lab., Wembley, UK HA9 7PP). *Electron. Lett.* **1984,** 20(3), 136-7 (Eng). A dual-mode optical fiber can be used as an interferometric temp. sensor either by fringe counting or, more importantly, by anal. of the change in chromatic transmission spectrum with temp. Such a sensor has the geometric flexibility of an all-fiber interferometric sensor as well as the advantages of a wavelength modulating device, i.e. high accuracy, noise immunity, abs. measurement capability and no need for intensity calibration.

Nagle, John F. (Dep. Phys., Carnegie-Mellon Univ., Pittsburgh, PA 15213 USA). *Biochemistry* **1984**, 23(7), 1538-41 (Eng). A new subgel phase was demonstrated to occur in hydrated diimyristoylphosphatidylethanolamine (DMPE) by use of dilatometric and calorimetric techniques. The formation of the subgel phase took place very slowly at temps. near 0°, but it could still be obsd. at 25°. Once formed, the subgel phase melted ($\Delta H_h$ = 16.0 kcal/mol and $\Delta V$ = 0.085 mL/g) directly into the liq.-cryst. phase at a temp., $T_h$ = 56.3°, that was higher than the gel to liq.-cryst. transition temp., $T_m$ = 49.6°. Thus, the gel phase appeared to be metastable over its entire temp. range. In this regard, DMPE behaved differently from dipalmitoylphosphatidylcholine and distearoylphosphatidylcholine, but similarly to dilaurylphosphatidylethanolamine. This unusual long-lived metastability provides cells in addnl. option in detg. the properties of membranes.

100: 116760k **Interaction of synthetic signal sequence fragments with model membranes.** Nagaraj, R. (Cent. Cell. Mol. Biol., Hyderabad, 500 007 India). *FEBS Lett.* **1984**, 165(1), 79-82 (Eng). Peptide fragments corresponding to the signal sequence of chicken lysozyme, labeled with the fluorescent dansyl group were synthesized. The emission characteristics and fluorescence polarization of the dansyl group were used to study the interaction of signal sequence fragments with liposomes. The peptide fragments bound to liposomes and were assocd. with the hydrophobic core of the bilayer.

100: 116761m **Interaction of glycophorin with phosphatidylserine: a Fourier transform infrared investigation.** Mendelsohn, Richard; Dluhy, Richard A.; Crawford, Terry; Mantsch, Henry H. (Newark Coll. Arts Sci., Rutgers Univ., Newark, NJ 07102 USA). *Biochemistry* **1984**, 23(7), 1498-504 (Eng). Glycophorin (I), from human erythrocyte membranes, was isolated in pure form and reconstituted into unilamellar liposomes with bovine brain phosphatidylserine (PS). Fourier transform IR (FT-IR) spectroscopy was used to monitor the protein conformation as well as the effect of protein on lipid order and melting. I, at levels of 1 mol%, nearly abolished the gel-to-liq. crystal phase transition seen in pure PS vesicles between 8 and 16° by inducing significant disorder into the lipid gel phase. A transition of reduced magnitude remained between 14 and 22° in the lipid/protein complexes. Evidence was presented for specific interaction of I with the interfacial region of PS. In general, the effects on lipid melting produced by protein at the 1 mol% level were more pronounced than those noted in a previous study of I-phosphatidylcholine interactions. Two bands were obsd. for the protein amide I (C:O stretching) mode. A main feature at 1653 $cm^{-1}$ indicated that the bulk of the secondary structure was random coil or $\alpha$-helical. A weaker shoulder at 1675 $cm^{-1}$ suggested the occurrence of a small proportion of the $\beta$-sheet form. The results confirmed previous CD studies. FT-IR studies of a ternary complex of PS-dipalmitoylphosphatidylcholine-$d_{62}$ (DPPC-$d_{62}$)-I indicated that I preferentially interacts with the PS component. The melting of the DPPC-$d_{62}$ and PS components could be sep. monitored. DPPC-$d_{62}$ had a reduced transition width and increased melting temp. in the ternary system compared with the binary lipid mixt. Thus, the utility of deuterated phospholipids in FT-IR studies for monitoring the preferential partitioning of proteins in complex lipid environments is demonstrated.

100: 116762n **pH-Induced destabilization of phosphatidylethanolamine-containing liposomes: role of bilayer contact.** Ellens, Harma; Bentz, Joe; Szoka, Francis C. (Sch. Pharm., Univ. California, San Francisco, CA 94143 USA). *Biochemistry* **1984**, 23(7), 1532-8 (Eng). The mechanism of pH-induced destabilization of liposomes composed of phosphatidylethanolamine (PE) and a charged cholesteryl ester was studied by following the release of encapsulated aq. contents. The kinetics of release were measured continuously by using the water-sol. fluorophore, $\alpha$-aminonaphthalene-1,3,6-trisulfonic acid, in combination with the water-sol. quencher, $p$-xylylenebis(pyridinium) bromide. With this fluorescence assay, the release of contents from liposomes composed of PE and cholesteryl hemisuccinate (I) was a function of pH, the PE:I ratio, and the acyl chain compn. of PE. Leakage was very slow at pH 5.5 and increased dramatically with decreasing pH to 4.0. Replacing PE by phosphatidylcholine (PC) eliminated the effect of pH on leakage. Anal. of the kinetics of release by a mass-action model demonstrated that bilayer destabilization and leakage occur subsequent to aggregation. The requirement of bilayer contact for destabilization had been found previously for acidic phospholipid bilayers in the presence of divalent cation and for satd. PC bilayers below the isothermal phase transition temp. The PE-contg. bilayers examd. here satisfied the same requirement.

100: 116763p **Mechanism of the antioxidative action of $\alpha$-tocopherol in biological membranes.** Makarova, T. B. (USSR). *Deposited Doc.* **1982**, VINITI 5913-82, 56-61 (Russ). Avail. VINITI. The mechanism of $\alpha$-tocopherol (I) antioxidant activity in phosphatidylcholine lysosomes exposed to UV irradn. was studied by ESR at -170 to -150°. Two types of I radicals appear to be involved in the antioxidant activity. Apparently, the side-chain network of I inserts itself into the hydrophobic portion of the membrane, and is thus able to facilitate the removal of radicals from the hydrocarbon zone of the membrane.

100: 116764q **A nuclear magnetic resonance spectroscopic investigation of the headgroup motions of lysophospholipids in bilayers.** Wu, W.; Stephenson, F. A.; Mason, J. T.; Huang, C. (Sch. Med., Univ. Virginia, Charlottesville, VA 22908 USA). *Lipids* **1984**, 19(1), 68-71 (Eng). Fully hydrated lysophospholipids in both the gel and the liq. cryst. states exhibited neg. chem. shift anisotropies ($\Delta\sigma$) by $^{31}P$ NMR. The magnitude of $\Delta\sigma$ for monoacyl lysophospholipids is smaller than that of the corresponding diacyl phosphopholipids in bilayers by ~2-fold. Evidence is presented to suggest that the redn. in $\Delta\sigma$ can be attributed primarily to an addnl. motional averaging of the lysophospholipid headgroup, and the addnl. motion can best be explained by rapid rotational motions of the headgroup about the C(1)-C(2) glycerol bond.

100: 116765r **Phosphatidylcholine and cholesterol interactions in model membranes.** Guyer, W.; Bloch, K. (Dep. Chem., Harvard Univ., Cambridge, MA 02138 USA). *Chem. Phys. Lipids* **1983**, 33(4), 313-22 (Eng). Various phosphatidylcholines differing either in the stereochem. around their chiral center or in the position of a cis double bond along the acyl chains were synthesized to study crit. contact regions in the phospholipid mol. with adjacent cholesterol (I) in model membranes. Microviscosities calcd. from fluorescence depolarization of diphenylhexatriene and chain order from spin label studies were measured to monitor phys. membrane properties. The enhancing effect of I on the microviscosity of membranes contg. phosphatidylcholines with comparable acyl chain length was largest when the 2 acyl chains were satd. and smallest when both were unsatd. Membranes prepd. from phosphatidylcholines having a single cis double bond at different positions along the $sn$-2 acyl chain showed roughly the same changes of microviscosity or chain order on incorporation of I. No discrimination was evident in the interaction between I and enantiomeric phosphatidylcholines or between the enantiomeric phosphatidylcholine mols. themselves. Thus, the rigidifying effect of I in membranes does not depend on specific sites of interaction, and, with respect to phys. membrane properties, phosphatidylcholine behaves as an achiral mol.

100: 116766s **Electrostatic models of the gramicidin and the delayed rectifier potassium channel.** Jordan, Peter C. (Dep. Chem., Brandeis Univ., Waltham, MA 02254 USA). *Biophys. J.* **1984**, 45(1), 100-2 (Eng). The ion pore-forming component-membrane-$H_2O$ system is analyzed as an electrostatic problem involving a set of distinct dielec. phases. The 2 ion pore-forming components considered were gramicidin A and the delayed-rectifier K channel. Gramicidin A channel properties are clearly dependent on coulombic interactions and can be modelled as a uniform cylinder. In contrast, electrostatic considerations produce structural constraints on the phys. and elec. geometry of the delayed-rectifier K channel. In each model, the proposed pore geometry is discussed with respect to exptl. observations.

100: 116767t **Alamethicin. A rich model for channel behavior.** Hall, James E.; Vodyanoy, Igor; Balasubramanian, T. M.; Marshall, Garland R. (Dep. Physiol. Biophys., Univ. California, Irvine, CA 92717 USA). *Biophys. J.* **1984**, 45(1), 233-47 (Eng). Synthetic analogs of alamethicin were used to test various hypotheses of its mode of action in membrane channel formation. A channel structure is proposed in which the C-terminal residues bond together as a $\beta$-barrel, leaving the $\alpha$-helixes free to rotate under the influence of the elec. field and gate the channel. Though the no. of monomers/channel varies with exptl. conditions, the gating charge/monomer stays close to that expected from an $\alpha$-helical gate. The sign of the voltage which turns on a channel can be altered by varying the charge on the alamethicin analog. Channels are always slightly cation-selective even though formed by monomers with neg., pos., or zero formal charge. Channels are less stable in low ionic strength solns. than in high. Finally, alamethicin conductance parameters vary systematically with changes in membrane thickness. These results and others in the literature can be explained by a fairly detailed structural model. The model can be easily generalized to a form more suited to high-mol.-wt. single-peptide-chain proteins.

100: 116768u **Phospholipid-protein molecular interactions in relation to immunological processes. 1. Raman spectroscopic and calorimetric studies of phospholipid-polyglycine molecular interactions in model membranes.** Bertoluzza, A.; Bonora, S.; Fini, G.; Morelli, M. A.; Simoni, R. (Fac. Med. Chir., Univ. Bologna, Bologna, Italy). *J. Raman Spectrosc.* **1983**, 14(6), 395-400 (Eng). Dipalmitoylphosphatidylcholine (DPPC)-poly(Gly) and dimyristoylphosphatidylcholine (DMPC)-poly(Gly) systems were studied by Raman spectroscopy and differential scanning calorimetry. Both techniques indicate that the interactions are confined to the lipid-$H_2O$ interface, although some penetration of poly(Gly) into the lipid phase is obsd. in DMPC systems. Cholesterol does not change the surface character of the interaction of the lipids with poly(Gly). Because of the similarity of the behavior of the 2 lipids, DMPC can be used instead of DPPC in studies with artificial membranes. DMPC has the advantage of a lower $T_m$. Since lipid-$H_2O$ interactions define the conformational state of extrinsic proteins and consequently their biol. behavior, these surface interactions could be a 1st approach to understanding the mechanism of processes such as the immunol. response.

100: 116769v **Inverted micellar structures in bilayer membranes. Formation rates and half-lives.** Siegel, D. P. (Miami Val. Lab., Procter and Gamble Co., Cincinnati, OH 45247 USA). *Biophys. J.* **1984**, 45(2), 399-420 (Eng). Two sorts of inverted micellar structures have previously been proposed to explain morphol. and $^{31}P$ NMR observations of bilayer systems. These structures only form in systems with components that can adopt the inverse hexagonal $H_{II}$) phase. LIP (lipidic particles) are intrabilayer structures, whereas IMI (inverted micellar intermediates) are structures that form between apposed bilayers. Here, formation rates and half-lives of these structures are calcd. to det. which (or if either) of these proposed structures is a likely explanation of the data. Calcns. for the egg phosphatidylethanolamine and the $Ca^{+}$-cardiolipin systems show that IMI form orders of magnitude faster than LIP, which should form slowly, if at all. This result is probably true in general, and indicates that the lipidic particle seen in electron micrograph

100: 126984a **A note on spectrocolorimetric estimation of methdilazine hydrochloride N. F.** Emmanuel, J.; Naik, Pradeep M. (Goa Coll. Pharm., Goa, 403001 India). *Indian Drugs* **1983,** 21(3), 119-20 (Eng). *Methdilazine* (**I**) **[1982-37-2]** in tablets and

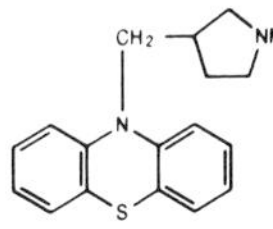

I

syrups was detd. by oxidn. with $K_2S_2O_8$ in the presence of *p*-aminobenzoic acid (stabilizer) to give a purple chromophore that was measured to 560 nm. Tablets were extd. with $H_2O$ for anal., and the syrup was extd. from a mixt. with satd. NaCl by using $CHCl_3$, evapg. the ext., and dissolving in dil. HCl. Recoveries were 101-102%.

100: 126985b **Fluorimetric and spectrophotometric determination of labetalol hydrochloride and its tablets.** Mohamed, Mohamed E. (Coll. Pharm., King Saud Univ., Riyadh, Saudi Arabia). *Pharmazie* **1983,** 38(11), 784 (Eng). *Labetalol* (**I**) **[36894-69-6]**

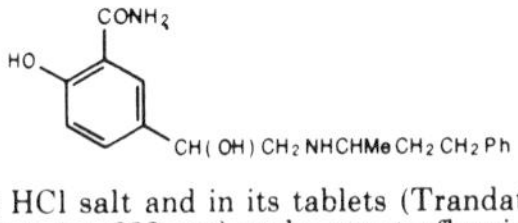

I

was detd. as its HCl salt and in its tablets (Trandate) by differential spectrophotometry (at 332 nm) or by spectrofluorimetry (at 425 nm with 301 nm as the excitation wavelength). Both methods were precise and accurate. In the fluorimetric method, the intensity of fluorescence was linearly related to concn. at 0.5-5 μg/mL only. With spectrophotometry, reproducible results were obtained at 10-80 μg I/mL.

100: 126986c **Application of high-performance liquid chromatographic chiral stationary phases to pharmaceutical analysis: structural and conformational effects in the direct enantiomeric resolution of α-methylarylacetic acid antiinflammatory agents.** Wainer, Irving W.; Doyle, Thomas D. (Div. Drug Chem., Food Drug Adm., Washington, DC 20204 USA). *J. Chromatogr.* **1984,** 284(1), 117-24 (Eng). Four α-methylarylacetic acids were directly resolved as enantiomeric amide derivs. on a high-performance liq. chromatog. chiral stationary phase consisting of covalently bound (*R*)-*N*-(3,5-dinitrobenzoyl)phenylglycine. Enantiomers of ibuprofen, naproxen, fenoprofen, and benoxaprofen all exhibited optimum sepn. ($\alpha$ = 1.12, 1.23, 1.12, and 1.10, resp.) as 1-naphthalenemethylamides. Derivs. of ibuprofen were studied in detail; measurable sepns. were obsd. for secondary and tertiary amides of diverse structural types, but not for the primary amide or for ester derivs. The results, which are consistent with an interaction model involving stacking of amide dipoles, with or without supplementary $\pi$-$\pi$ and H-bonding effects, appear to be strongly dependent on conformational requirements of both the solute and the chiral stationary phase.

100: 126987d **Colorimetric and polarographic methods of analysis for prenalterol hydrochloride.** Mohamed, Mohamed E.; Wahbi, Abdel Aziz M.; Gad Kariem, Elrasheed A. (Coll. Pharm., King Saud Univ., Riyadh, Saudi Arabia). *Anal. Lett.* **1983,** 16(B19), 1545-53 (Eng). *Prenalterol* (**I**) **[57526-81-5]** is detd. in bulk and in tablets

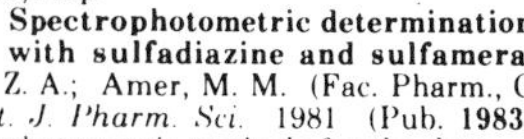

I

in pH 4.1 acetate buffer by colorimetry based on redn. of Fe(III) and subsequent measurement at 511 nm of the red color obtained by treatment of the resultant Fe(II) 1,10-orthophenanthroline, or by differential-pulse polarog. following nitrosation with 0.1M $NaNO_2$-dil. HCl. The differential polarogram was obtained under const. amplitude pulses of 50 mV superimposed on a linearly increasing DC-voltage ramp. The peak height (h) of the polarogram was measured at the peak potential of -0.2 V on the dropping Hg electrode vs. Ag-AgCl ref. electrode. The linearity ranges obsd. were 0.6-6.0 μg/mL and 2-12 μg/mL, resp. The mean recoveries were 99.5 and 100.5%, resp.

100: 126988e **Spectrophotometric determination of sulfathiazole in admixture with sulfadiazine and sulfamerazine.** Hassan, S. M.; El Sherif, Z. A.; Amer, M. M. (Fac. Pharm., Cairo Univ., Cairo, Egypt). *Egypt. J. Pharm. Sci.* 1981 (Pub. **1983**). 22(1-4), 39-46 (Eng). Spectrophotometric methods for the detn. of *sulfathiazole* (**I**)

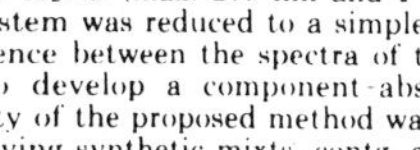

I

**[72-14-0]** and total pyrimidylsulfonamide content in mixt. contg. *sulfadiazine* **[68-35-9]** and *sulfamerazine* **[127-79-7]** are based on the soln. of simultaneous equations depending upon spectra in different media. The spectral max. of sulfadiazine and sulfamerazine coincide with each other in 0.1N HCl (max. 242 nm and **I** max. 280 nm) and in aq. 10% $H_2CO$ (max. 270 nm and **I** max. 290 nm), thus the 3-component system was reduced to a simple 2-component anal. In addn., the difference between the spectra of the compds. in both media was used to develop a component-absorbance difference method. The validity of the proposed method was demonstrated in a recovery study involving synthetic mixts. contg. authentic samples of the 3 sulfonamides. The spectra in 10% $H_2CO$ could be used for identification.

100: 126989f **The analysis of organic pharmaceuticals with N-bromosuccinimide. Part 2. Determination of pyrazolidinedione derivatives in pharmaceutical preparations.** Taha, A. M.; El-Zeany, B. A.; Amer, M. M.; El Sawy, O. A. (Fac. Pharm., Assiut Univ., Assiut, Egypt). *Egypt. J. Pharm. Sci.* 1981 (Pub. **1983**), 22(1-4), 47-53 (Eng). *Phenylbutazone* (**I**) **[50-33-9]** and *sulfinpyrazone*

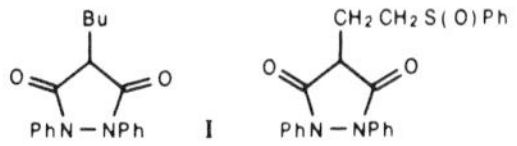

I II

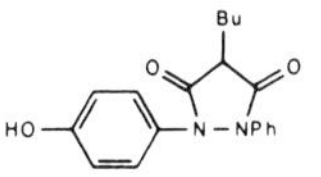

III

(**II**) **[57-96-5]** were detd. in tablets and *oxyphenbutazone* (**III**) **[129-20-4]** was detd. in tablets and suppositories by titrn. with *N-bromosuccinimide* **[128-08-5]** in HOAc medium with methyl orange indicator. The tablets were extd. with HOAc. Suppositories were dissolved in warm 5% NaOH, the fat was extd. with ether, and the aq. phase was dild. with HOAc. Recoveries were 97.7-102%.

100: 126990z **The application of 9-chloroacridine for the determination of some pharmaceutical substances.** Safwat, H. M.; Moussa, Baheia A. (Fac. Pharm., Cairo Univ., Cairo, Egypt). *Egypt. J. Pharm. Sci.* 1981 (Pub. **1983**). 22(1-4), 97-104 (Eng). Primary arom. amines, including *carbarsone* **[121-59-5]** and sulfonamides, can be detd. by dissolving in EtOH, mixing with *9-chloroacridine* **[1207-69-8]** in $CHCl_3$ and some HCl, and measuring the absorbance at 435 nm. Carbarsone is hydrolyzed with 60% $H_2SO_4$ before dissolving in EtOH, and the soln. is refluxed with the reagent soln. before measuring at 440 nm. This method avoids the alcoholysis of the reagent. Beer's law held for 0.25-1.5 mg/25 mL for carbarsone and 0.025-0.20 mg/25 mL for the other compds. Recoveries were 97.4-100% for 2 sulfonamides.

100: 126991a **Spectrophotometric determination of isoniazid.** Hassib, S. T.; Moreed, L. (Fac. Pharm., Cairo Univ., Cairo, Egypt). *Egypt. J. Pharm. Sci.* 1981 (Pub. **1983**). 22(1-4), 143-52 (Eng).

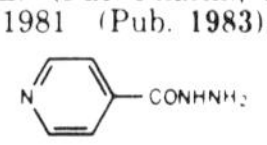

I

*Isoniazid* (**I**) **[54-85-3]** is detd. by heating with $ClCH_2CO_2H$, dissolving the melt (pyridinium salt) in $H_2O$, adding 0.1 N NaOH to form the pyridinium betaine, and measuring the absorbance at 389 nm in 80% EtOH soln. Beer's law holds for 5-30 μg/mL. Mean recoveries from tablets were 99.8-100.6%. Pyridoxine, nicotinamide, and nicotinic acid do not interfere.

100: 126992b **Spectrophotometric analysis of thioridazine hydrochloride.** Taha, A. M.; El-Rabbat, N. A.; El-Kommos, M. E.; Refaat, I. H. (Fac. Pharm., Univ. Assiut, Assiut, Egypt). *Egypt. J. Pharm. Sci.* 1981 (Pub. **1983**). 22(1-4), 163-9 (Eng). *Thioridazine*

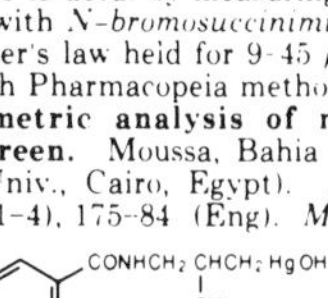

I

(**I**) **[50-52-2]** in tablets is detd. by measuring the absorbance of the violet product formed with *N-bromosuccinimide* **[128-08-5]** in 50% $H_2SO_4$. At 548 nm, Beer's law held for 9-45 μg/mL. Results agreed with those of the British Pharmacopeia method.

100: 126993c **Colorimetric analysis of mersalyl using titan yellow and methyl green.** Moussa, Bahia A.; El Kousy, Naglaa (Fac. Pharm., Cairo Univ., Cairo, Egypt). *Egypt. J. Pharm. Sci.* 1981 (Pub. **1983**). 22(1-4), 175-84 (Eng). *Mersalyl* (**I**) **[492-18-2]**,

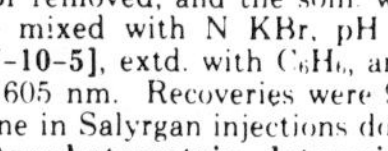

I

50 mg, was ignited in an O flask with $HNO_3$ as absorbent, 20% NaOH was added to pH 5, the soln. was dild. with $H_2O$, mixed with pH 8.3 borax buffer and 0.1% *titan yellow* **[1829-00-1]**, heated at 60-70° for 15 min, cooled, and the absorbance was measured at 510 nm. Alternately, 50 mg was ignited in an O flask contg. $H_2O$ and satd. aq. Br, excess Br removed, and the soln. was dild. with pH 2.2 HBr. Aliquots were mixed with N KBr, pH 2.2 HBr, and 0.5% *methyl green* **[54327-10-5]**, extd. with $C_6H_6$, and the absorbance of the ext. was detd. at 605 nm. Recoveries were 99.3-100.2% by both methods. Theophylline in Salyrgan injections does not interfere.

100: 126994d **Spectrophotometric determination of analgin with iodonitrotetrazolium chloride (INT).** Vasileva Aleksandrova, P.; Aleksandrov, A. (Bulg.). *Nauchni Tr. Plovdivski Univ.* **1982,** 20(3, Khim.), 41-51 (Bulg). *Analgin* (**I**) **[68-89-3]** was detd.

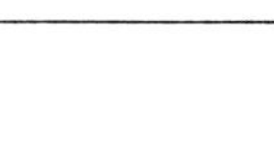

I

100: **185214x Comparative study of the ability of four amino= glycoside assay techniques to detect the inactivation of aminoglycosides by β-lactam antibiotics.** Pfaller, Michael A.; Granich, George G.; Valdes, Roland; Murray, Patrick R. (Sch. Med., Washington Univ., St. Louis, MO 63110 USA). *Diagn. Microbiol. Infect. Dis.* **1984,** 2(2), 93-100 (Eng). In vitro inactivation of aminoglycosids (*tobramycin* **[32986-56-4]**, *gentamicin* **[1403-66-3]**, and *amikacin* **[37517-28-5]**) by β-lactams (*cefazolin* **[25953-19-9]**, *cefotaxime* **[63527-52-6]**, *moxalactam* **[64952-97-2]**, *carbenicillin* **[4697-36-3]**, *piperacillin* **[61477-96-1]**, *mezlocillin* **[51481-65-3]**, and *azlocillin* **[37091-66-0]** was measured using the EMIT fluorescence polarization immunoassay (TDX), RIA, and bioassay. No significant inactivation of aminoglycosides was produced by high levels of the 3 cephalosporins as measured by EMIT, RIA, or bioassay. Inactivation of tobramycin and gentamicin by mezlocillin and azlocillin was comparable to that seen with piperacillin but less than that with carbenicillin. In general, the bioassay detected the greatest degree of aminoglycoside inactivation and the EMIT assay detected the least for all drug combinations. The TDX and RIA techniques were equiv. in their ability to detect aminoglycoside inactivation by β-lactam antibiotics.

100: **185215y The effect of nitrous oxide on the oxyhemoglobin dissociation curve.** Fournier, Louis; Major, Diane (Hosp. Cent., Laval Univ., Quebec, PQ Can.). *Can. Anaesth. Soc. J.* **1984,** 31(2), 173-7 (Eng). The influence of $N_2O$ on the oxyHb dissocn. curve (ODC) was studied using blood from 20 healthy patients. When the blood samples were exposed to 50% $N_2O$ during the detn. of the ODC, a left shift was obsd. and the $P_{50}$ was decreased by 1.06 kPa (8 mmHg). This shift cannot be explained by temp., pH, $PCO_2$, or 2,3-diphosphoglycerate effects. Following exposure of the blood to $N_2O$-free gases, the shift disappeared rapidly, and a normal $P_{50}$ (3.46 kPa) (26 mmHg) was obtained. In keeping with this reversibility, blood samples taken before and during 45 min. of $N_2O$-curare anesthesia showed identical dissocn. curves to those which had been obtained during the in vitro $N_2O$ exposure expt.

100: **185216z Direct infrared spectroscopic identification of drugs from thin-layer chromatograms.** George, Ute; Kolbe, A.; Lauermann, Ilse (Sekt. Chem., Martin Luther Univ., DDR-4020 Halle, Ger. Dem. Rep.). *Instrum. Anal. Toxikol., [Hauptteil Vortr. Symp.]* **1983,** 127-32 (Ger). Edited by Mueller, R. Klaus. Abt. Toxikol. Chem. Inst. Gerichtl. Med. Karl-Marx-Univ.: Leipzig, Ger. Dem. Rep. The use of IR-spectroscopy in the identification of drugs and their metabolites in biol. materials is discussed.

100: **185217a Determination of plasma heparin by polybrene neutralization.** Nash, Patricia V.; Bjornsson, Thorir D. (Med. Cent., Duke Univ., Durham, NC 27710 USA). *Anal. Biochem.* **1984,** 138(2), 319-23 (Eng). A method for detg. *heparin* **[9005-49-6]** activity (in unit/mL) in human plasma is described. The method is based on neutralization of heparin by Polybrene, a polymd. quaternary ammonium salt. It uses serial incubations of plasma with increasing amts. of Polybrene in conjunction with thrombin-induced coagulation times to define the amt. of Polybrene required to neutralize heparin in the sample. Ref. curves involve linear relationships between amts. of Polybrene required to neutralize known amts. of heparin. Coeffs. of variation for the assay vary 4-10% over the range 0.05-1.0 unit/mL.

100: **185218b Microdetermination of isoniazid and acetylisoniazid in plasma by high-performance liquid chromatography.** Lacroix, C.; Laine, G.; Goulle, J. P.; Nouveau, J. (Lab. Biochim., Cent. Hosp. Gen., 76083 Le Havre, Fr.). *J. Chromatogr.* **1984,** 307(1), 137-44 (Fr). A method for the anal. of *isoniazid* **[54-85-3]** and *acetylisoniazid* **[1078-38-2]** in plasma or serum is described. The method involves a simple, quick and economically optimized reversed-phase HPLC assay using a small sample size without solvent extn. step. Isoniazid was condensed with cinnamic aldehyde after trichloroacetic acid deproteinization. The method correlated well with a fluorimetric procedure. Precision, sensitivity, and accuracy were good. Common antituberculous drugs did not interfere.

100: **185219c Isomodal column switching high-performance liquid chromatographic technique for the analysis of ciglitazone and its metabolites in human serum.** Cox, J. W.; Pullen, R. H. (Upjohn Co., Kalamazoo, MI 49001 USA). *J. Chromatogr.* **1984,** 307(1), 155-71 (Eng). The application of isomodal column switching HPLC as an alternative to gradient elution was investigated for the anal. of *ciglitazone* (**I**) **[74772-77-3]**, a potential oral antidiabetic agent, and its monohydroxyl metabolites in human serum. An app. was designed to perform online fractionation of the serum ext. into nonpolar (drug) and polar (metabolite) fractions which were then automatically routed into individually optimized, isocratic, reversed-phase HPLC systems for simultaneous anal. Sample fractionation was performed with a reversed-phase guard column, and solvent routing was accomplished with microprocessor-controlled switching valves. Serum was extd. for anal. by a 1-step mode sequencing procedure using disposal bonded-phase columns, and quantitation was accomplished with spiked serum stds. Performance specifications of the method were defined for precision, accuracy linearity, and sensitivity. The column switching method was both expedient and reliable, and it may have general utility for the routine, quant. anal. of drug-metabolite mixts. that cannot be assayed by simple isocratic elution methods.

100: **185220w Haloperidol determination in serum and cerebrospinal fluid using gas-liquid chromatography with nitrogen-phosphorus detection: application to pharmacokinetic studies.** Abernethy, Darrell R.; Greenblatt, David J.; Ochs, Hermann R.; Willis, Christopher R.; Miller, Del D.; Shader, Richard I. (Sch. Med., Tufts Univ., Boston, MA 02111 USA). *J. Chromatogr.* **1984,** 307(1), 194-9 (Eng). A gas-liq. chromatog. method with N-P detection is described for the detn. of *haloperidol* **[52-86-8]** in blood serum and cerebrospinal fluid (CSF). The method utilized a 3% SP-2250 on Chromosorb WHP column. He (30 mL/min) was used as the carrier gas. Operating temps. were injection port 310°, column 270°, and the detector 275°. A chloro analog of haloperidol was used as the internal std. The drug was assayed following extn. with org. solvents. Retention times for haloperidol and McN-JR-1854 (the internal std.) were 3.5 and 7.0 min, resp. The correlation coeff. (10 std. curve over a 4 mo period) was 0.99 or greater. Day-to-day coeff. of variation in the slopes of the calibration curve was 7.8%. Sensitivity of detection was 0.5 ng/mL for a 2-mL sample. Pharmacokinetics data obtained in dogs by this method are given.

100: **185221x Simultaneous determination of lidocaine and its deethylated metabolites using gas-liquid chromatography with nitrogen-phosphorus detection.** Willis, Christopher R.; Greenblatt, David J.; Benjamin, David M.; Abernethy, Darrell R. (Div. Clin. Pharmacol., Tufts-New England Med. Cent., Boston, MA 02111 USA). *J. Chromatogr.* **1984,** 307(1), 200-5 (Eng). A gas-liq. chromatog. procedure utilizing a column packed with 3% SP-2250 on Supelcoport and a N-P detector is described for the simultaneous detn. of *lidocaine* **[137-58-6]** and its de-ethylated metabolite in human blood samples. The method does not require plasma sample clean-up or derivatization. He (30 mL/min) was used as the carrier gas. Operating temps. were; injection port 310°, column 200°, and detector 275°. The plasma samples were extd. with org. solvent and the sample was then chromatographed. The limit of sensitivity was approx. 0.05 μg/mL of lidocaine and its 2 metabolites. The extn. recovery was >90%. Coeff. of variations were 7.1, 16.5, and 13.6% for lidocaine, *monoethylglycineylidide* **[7728-40-7]**, and *glycinexylidide* **[18865-38-8]**, resp. Pharmacokinetics of these compds. in healthy humans were studied utilizing this method.

100: **185222y Silica capillary gas chromatographic determination of ibuprofen in serum.** Heikkinen, Liisa (Sch. Pharm., Univ. Helsinki, SF-00170 Helsinki, 17 Finland). *J. Chromatogr.* **1984,** 307(1), 206-9 (Eng). A gas chromatog. (fused-silica capillary column, OV-351) procedure for rapid estn. of *ibuprofen* **[15687-27-1]** at therapeutic levels and in cases of possible overdosage is described. Ibufenac was used as the internal std. The accuracy of the method was 98.1% with a relative std. deviation of 1.1%. Recovery was 70%.

100: **185223z Liquid chromatographic determination of amikacin in serum with spectrophotometric detection.** Kabra, Pokar M.; Bhatnager, Pradip K.; Nelson, Marla A. (Sch. Med., Univ. California, San Francisco, CA 94143 USA). *J. Chromatogr.* **1984,** 307(1), 224-9 (Eng). A liq. chromatog. method for the routine assay of *amikacin* **[37517-28-5]** in blood serum is described. A column packed with reversed-phase octyl packing material (Ultrasphere Octyl) was used; the mobile phase was MeCN-phosphate buffer (52:48). The flow rate was maintained at 2.0 mL/min at 50° and the column effluent was monitored at 340 nm (spectrophotometric detection). Kanamycin was used as the internal std. Amikacin was derivatized with trinitrobenzenesulfonic acid prior to extn. The limit of detection of amikacin in this method was <0.5 mg/L, when a 50-μL serum sample was used. The within-day coeff. of variation (CV) ranged from 3.5 to 6.9% and the day-to-day CV from 2.8 to 3.1%. This method is simple, precise, sensitive, and selective; the results correlate well with existing radioimmunoassay methods.

100: **185224a Bromperidol radioimmunoassay: human plasma levels.** Tischio, John; Hetyei, Nancy; Patrick, James (Res. Lab., Ortho Pharm. Corp., Raritan, NJ 08869 USA). *J. Pharm. Sci.* **1984,** 73(4), 546-8 (Eng). A sensitive RIA procedure is described for assay of *bromperidol* **[10457-90-6]** in human plasma after therapeutic drug administration. Antisera were generated in rabbits against a haloperidol-bovine serum albumin conjugate. Tritiated haloperidol was used as the radioligand. A single $Et_2O$ extn. of alkalinized plasma was used to sep. bromperidol from its more polar metabolites and to reduce assay variability encountered with a direct plasma assay. The lower limit of detection is ~0.5 ng/mL for parent drug in plasma. The assay exhibited within- and between-assay variabilities of ~9 and 14%, resp. A 103-106% recovery of bromperidol from quality control plasma samples was obsd. over the concn. range of 1-150 ng/mL. A correlation coeff. of 0.9999 with respect to measured vs. expected bromperidol content in the quality control plasma samples was exhibited. Cross-reactivity characteristics of the antisera indicated that dehydrobromperidol could significantly interfere (~25% cross-reactivity) with the RIA procedure. However, biotransformation studies have not suggested this compd. as a metabolite of bromperidol. Predose ($C_{min}$) plasma levels of bromperidol in schizophrenic patients maintained on drug therapy are also reported.

100: **185225b Simple radioligand binding assay for the deter= mination of urinary scopolamine.** Scheurlen, M.; Bittiger, H.; Ammann, Birgit (Hum. Pharmacol. Inst., Ciba-Geigy, Tuebingen, Fed. Rep. Ger.). *J. Pharm. Sci.* **1984,** 73(4), 561-3 (Eng). A sensitive radioligand binding assay is described for the detn. of *scopolamine* **[51-34-3]** in human urine. As a measure for the drug concn., the quant. displacement by scopolamine of [$^3$H]quinuclidinyl benzylate from rat brain receptors was used. The assay is sensitive to concns. asslow as 1.2 ng/mL. It can be performed easily and

Now that I have introduced you to, and made you use, *Chemical Abstracts* you can see that despite its size and some inevitable complexities, it is not too difficult to use, is it? More important though, I hope you can see just how powerful an index this is and how valuable a service it can be for you. Soon I shall be introducing you to on-line electronic searching of *Chemical Abstracts*: this form of searching is undoubtedly going to replace manual searching of the printed abstracts at an ever-increasing rate. However, those who have some familiarity with searching the printed version will find the transfer to electronic searching much easier.

### 3.2.4. Specialised Indexes and Abstracts

If *Chemical Abstracts* covers all that for you, why bother with anything else? Well, the first and most obvious reason might have struck you already: perhaps you did as I told you and headed off to your library, only to find that it doesn't have *Chemical Abstracts*, nor is it available in any library convenient to you. It is, as you might imagine, a very expensive service. Even if you can get access to a set of *Chemical Abstracts*, perhaps you want to search only a very narrow subject field, and a service produced by and for specialists in that field might well be more convenient, and might have other advantages too: its layout might be more helpful to you, it might be easier to browse through, or it might just be more up-to-date. So, if you find a specialised service within the field you are investigating, do use it, either as a supplement to or perhaps as a substitute for *Chemical Abstracts*.

Of course we have already encountered some specialist indexing services, haven't we? Do you remember those I mentioned in Section 2.5 and 2.6? They of course deal with special forms of literature (reports, theses etc): those we are looking at in this Section are oriented towards special subjects. So without further delay let's look at just a few specialist sources I think you might find useful. We start with what must be an obvious major source of information for analytical chemists:

> *Analytical Abstracts.* Nottingham: Royal Society of Chemistry (UK Chemical Information Service), 1954 to date. Monthly.

This has international coverage of all branches of analytical chemistry. Each issue is arranged in the following nine major sections: general analytical chemistry, inorganic chemistry, organic chemistry, biochemistry, pharmaceutical chemistry, food, agriculture, environmental chemistry, apparatus, and techniques. There are more specific sub-headings as required. There is a monthly subject index, and each annual volume has a cumulated subject and author index. The importance of this service to you must be obvious, so let's just take a closer look at it with a practical exercise for you.

**SAQ 3.2c**

Use volume 46 (1984) of *Analytical Abstracts* to find a couple of papers on atomic absorption spectroscopy.

(*a*) A review paper on the use of atomic absorption spectrophotometry in flow-injection analysis.

(*b*) A review paper on the applications of atomic absorption spectroscopy in the determination of anions and organic compounds.

When you have found your two references write each of them out in the form I have already shown you.

For the last time, do try to use the original publication. Use the copies on the following pages only if you must.

AABSAR 46(2) 77–159 (1984) February 1984

# Analytical Abstracts

A publication of The Royal Society of Chemistry

CONTENTS



Printed by Heffers Printers Ltd., Cambridge, England

Mar 1983, **38** (3), 488–490 (Russ.).—To determine ≃4 to 8 μg $ml^{-1}$ of pyridine (**I**) or ≃100 to 200 μg $ml^{-1}$ of piperidine (**II**), the sample is acidified with HCl to pH 1.7 and the absorbance is measured at 256 nm; **I** interferes with determination of **II**. M.P.

2H71. **Determination of hydantoins in condensate water from lignite gasification.** Olson, E. S.; Diehl, J. W.; Miller, D. J. (Grand Forks Energy Technol. Center, US Dept. Energy, Grand Forks, ND 58202, USA). *Anal. Chem.*, Jun 1983, **55** (7), 1111–1115.—The sample is treated (at its normal pH and at pH 2.0) with activated charcoal, which is filtered off and extracted with boiling ethanol. The residue from evaporation of the solvent is analysed on a column (25 m × 0.32 mm) coated with Superox-FA (AT-1000) and temp.-programmed from 60° to 85° at 30° $min^{-1}$, to 100° at 0.5° $min^{-1}$, to 200° at 1° $min^{-1}$ and finally to 260° at 1.5° $min^{-1}$. A f.i.d. is used and the carrier gas is H (2 ml $min^{-1}$). The compounds are characterized by electron-impact and chemical-ionization m.s. ($CH_4$ as reagent gas) after separation on a similar column. G.P.C.

## *See also Abstracts:*

2B1, 2B31, 2B45, 2B57, 2B60, 2B65, 2B66, 2B68, 2B76, 2C9, 2C31, 2C46, 2C50, 2C61, 2C64, 2C80, 2D48, 2J11, 2J37, 2J65, 2J73, 2J81, 2J87, 2J121.

# J. APPARATUS; TECHNIQUES

## General apparatus, automatic analysers, computer hardware

2J1. **Enrichment and trapping of volatile trace components by means of chromatographic solvent evaporation.** Roeraade, J.; Blomberg, S. (Dept. Anal. Chem., Royal Inst. Technol. Stockholm, 100 44 Stockholm, Sweden). *Chromatographia*, Jul 1983, **17** (7), 387–393.—To evaluate the cited technique, a sample (400 μl) of pentane (as model compound) containing trace hydrocarbon impurities was injected on to a glass column (40 m × 0.6 mm), and N (15 ml $min^{-1}$) was passed over the pentane film that irregularly coated the column, slowly evaporating the solvent. The vapour was re-condensed in a solid $CO_2$ trap, and fractions were injected [split injection (1 : 40)] on to a fused-silica column (50 m × 0.32 mm) coated with 5% of methylphenylsilicone. Compounds eluted before the pentane peak were concentrated in the first few fractions; those eluted after pentane were concentrated in the final fractions. Quantitative enrichment was achieved, with a solvent barrier in front of the evaporating sample. The effects of evaporation temp. and of solvent - solute polarity are discussed. P.J.W.

2J2. **Efficient novel device for solvent extraction.** Peleg, I.; Vromen, S. (Endocrinol. Lab., Rambam Med. Centre, Haifa, Israel). *Chem. Ind. (London)*, 1 Aug 1983, (15), 615–616.—A description is given of apparatus devised by Cais and Shimoni (*Ann. Clin. Biochem.*, 1981, **18**, 317) and now marketed under the name Mixxor (Lidex Corp., Technion City, Israel). The performance of the system is demonstrated by the extraction of oestradiol and testosterone from aq. soln. into ethyl ether; the efficiency of extraction is better than that attained with a separating-funnel. Use of a Mixxor device for phase separation in r.i.a. is particularly recommended. B.H.

2J3. **Application of synergic extraction to analytical chemistry.** Akaiwa, H.; Kawamoto, H. (Dept. Chem., Fac. Technol., Gunma Univ., Kiryu, Gunma 376, Japan). *Rev. Anal. Chem.*, 1982, **6** (1), 65–86.—Synergic extraction equilibria are discussed, the advantages of synergism are indicated, and applications of the technique in extractive spectrophotometry and in the separation of bivalent metals are reviewed. (46 references.) B.H.

2J4. **Flotation techniques in analytical chemistry.** Hiraide, M.; Mizuike, A. (Fac. Eng., Nagoya Univ., Chikusa-ku, Nagoya 464, Japan). *Rev. Anal. Chem.*, 1982, **6** (2), 151–168.—A review is presented, with 58 references. B.H.

2J5. **Flow electrolysis at a porous tubular electrode with internal stirring.** Wang, J.; Freiha, B. A. (Dept. Chem., New Mexico State Univ., Las Cruces, NM 88003, USA). *Anal. Chim. Acta*, 1 Jul 1983, **151** (1), 109–116.—The design and properties of a porous tubular electrode, in which effective mass transport is ensured by a stirrer rotating inside the tube, are described. High currents are possible owing to the high surface area and mass transport rates. The dependence of the limiting current and the degree of conversion on stirring and flow rate and on electrode length is described. A hydrodynamic modulation mode, based on measuring the current difference while switching the stirrer on and off, is used to correct for background. The cell was used for the determination of caffeic acid, paracetamol, dopamine and $Fe(CN)_6^{4-}$ in flow-injection and continuous-flow analyses, with detection limits at the nM level. R.C.M.

2J6. **Flow-injection methods and atomic-absorption spectrophotometry.** Tyson, J. F. (Dept. Chem., Loughborough Univ. Technol., Loughborough, LE11 3TU, UK). *Anal. Proc. (London)*, Sep 1983, **20** (9), 488–491.—The literature is reviewed. (19 references.) J.E.P.

## *See also Abstracts:*

2A18, 2C47, 2D161, 2H2, 2H6, 2H20, 2H22, 2H25, 2H30, 2H42, 2J11, 2J75, 2J83, 2J90, 2J108, 2J122, 2J124.

## Chromatography, ion exchange, electrophoresis

2J7. **Critical evaluation of quality criteria for the optimization of chromatographic multi-component separations.** Debets, H. J. G.; Bajema, B. L.; Doornbos, D. A. (Res. Group Optimization, Dept. Pharm. Anal. Chem., 9713 AW Groningen, Netherlands). *Anal. Chim. Acta*, 1 Jul 1983, **151** (1), 131–141.—Quality criteria for optimizing two-component and multi-component chromatographic separations are discussed. The relation of quality criteria to chromatographic resolution and the usefulness of criteria in expressing the quality of separation in reversed-phase h.p.l.c. are evaluated. Chromatograms were simulated, with assumption of Gaussian peak shapes according to a specified formula, by means of a computer. The published criteria expressing the quality of multi-component chromatographic separations are criticized. R.C.M.

2J8. **Micro high-performance liquid chromatography.** Ishii, D.; Takeuchi, T. (Dept. Appl. Chem., Fac. Eng., Nagoya Univ., Chikusa-ku, Nagoya 464, Japan). *Rev. Anal. Chem.*, 1982, **6** (2), 87–112.—A review is presented, with 39 references. B.H.

2J9. **Chromatographic separation of low-boiling liquids on type-Y zeolites modified with alkaline-earth-metal cations.** Banakh, O. S.; Laperashvili, L. Ya.; Stril'chuk, L. V. *Izv. Akad. Nauk Gruz. SSR, Ser. Khim.*, 1983, **9** (1), 33–36 (Russ.). *Ref. Zh., Khim.*, 19GD, 1983, (17), Abstr. No. 17G136.

2J10. **Elution characteristic of sodium nitrite in reversed-phase h.p.l.c. with aqueous methanol as the mobile phase.** Jinno, K. (Sch. Mater. Sci., Toyohashi Univ. Technol., Toyohashi 440, Japan). *Chromatographia*, Jul 1983, **17** (7), 367–369.—The behaviour of $NaNO_2$ in h.p.l.c. has been studied. Acetonitrile, methanol and aq. $NaNO_2$ (50 ppm soln.) were separately injected (in 0.5 to 5-μl amounts) on to a column (25 cm × 4.6 mm) of Jasco FineSIL C-18 (10 μm), at temp. from 20° to 70°, with methanol, aq. 50% methanol or $H_2O$ as mobile phase (1 ml $min^{-1}$) and detection at 210 nm. It was found that $NaNO_2$ gave the best estimate of void time of the column when injected at the cited concn., although a higher concn. might be preferable if methanol were the mobile phase. P.J.W.

2J11. **Applications of a microprocessor-controlled valve-switching unit for automated sample clean-up and trace enrichment in high-performance liquid chromatography.** Little, C. J.; Tompkins, D. J.; Stahel, O.; Frei, R. W.; Werkhoven-Goewie, C. E. (Kontron Electrolab Ltd., Ashford, Middx. TW15 1A4, UK). *J. Chromatogr.*, 15 Jul 1983, **264** (2), 183–196.—Methods of valve-switching are described and their application to automated

Auger-electron spectrometry, X-ray photoelectron spectrometry and secondary-ion m.s. are considered to be the most appropriate techniques for such analyses.
World Surf. Coat. Abstr.

6B7. **Chromatography of metal chelates. XII. Thin-layer chromatography (t.l.c.) and high-performance liquid chromatography (h.p.l.c.) of metal chelates of 1-hydroxypyridine-2-thione.** Steinbrech, B.; Koenig, K.-H. (Inst. Anorg. Chem., Univ. Frankfurt, 6000 Frankfurt/M. 50, W. Germany). *Fresenius' Z. Anal. Chem.*, Dec 1983, **316** (7), 685-688 (Ger.).—The cited chelates of several metals, in $CHCl_3$ soln., were subjected to t.l.c. on silica gel and to h.p.l.c. on LiChrosorb Si 100, with mixtures of $CH_2Cl_2$ and/or $CHCl_3$ with tetrahydrofuran as the mobile phases. The $R_F$ values and chromatographic behaviour showed that good separation of chelates of 14 metals (V, Cr, Mo, Fe, Co, Ni, Rh, Pd, Ir, Pt, Cu, Zn, Hg and Tl) was attainable by t.l.c. on Kieselgel $60F_{254}$; the method was unsuitable for separating chelates of Ti, Ru, Os, Nb, Pb, Bi, Mn and Te. With h.p.l.c. on a column (25 cm × 4.6 mm) of LiChrosorb Si 100 (10 µm), Fe, Zn, Hg and Tl could also not be separated satisfactorily, but otherwise this was an equally effective method (gradient elution being necessary for mixtures of components with greatly different polarity, *e.g.*, chelates of V, Cr, Mo, Co and Cu); detection was at 320 nm. (For Part XI see *Anal. Abstr.*, 1984, **46**, 5A7.) D.G.E.

6B8. **Chromatography of metal chelates. XIII. Chromatographic behaviour of metal chelates of 1-hydroxypyridine-2-thione in comparison with its methyl and bromo-derivatives.** Steinbrech, B.; Koenig, K.-H. (Inst. Anorg. Chem., Univ. Frankfurt, 6000 Frankfurt/M. 50, W. Germany). *Fresenius' Z. Anal. Chem.*, Dec 1983, **316** (7), 689-695 (Ger.).—In continuation of work in Part XII (see preceding abstract), the effects of a methyl group in the 3-, 4-, 5- or 6-position, of two methyl groups in the 4- and 6-positions and of Br in the 5-position on the t.l.c. and h.p.l.c. behaviour of the cited thionates of several metals were determined. The stability, polarity and retention characteristics of the chelates were modified by substitution. The $R_F$ values are reported for the chelates of 22 metals with four different mobile phases, and examples of successful h.p.l.c. separations are given. D.G.E.

6B9. **High-performance liquid chromatography of metal chelates.** O'Laughlin, J. W. (Dept. Chem., Univ. Missouri, Columbia, MO 65211, USA). *J. Liq. Chromatogr.*, Mar 1984, 7 (Suppl. 1), 127-204.—A review is presented, with 141 references, of the separation of chelates of β-diketones, Schiff bases derived therefrom, crown ethers, macrocyclic amines and organophosphorus compounds, *inter alia*, by h.p.l.c. The use of various detection systems is discussed. P.J.R.

6B10. **Effect of eluent electrolyte on chromatographic behaviour of ionic solutes on polyacrylamide gel.** Shibukawa, M.; Ohta, N. (Dept. Chem., Sch. Med., St. Marianna Univ., Miyamae, Kawasaki, Japan). *Bunseki Kagaku*, Oct 1983, **32** (10), 557-561 (Japan.).—Various inorganic ions and metal complexes were chromatographed on a column of Bio-Gel P-2 with aq. Na acetate, NaCl, $NaNO_3$, $NaClO_4$, LiCl and KCl as mobile phases. Based on the ion-partition mechanism proposed, equations were derived that related the eluent counter-ion and co-ion effects to the distribution coeff. of the test species. The experimental results agreed well with the predicted behaviour, supporting the proposed mechanism. The distribution coeff. of ionic solutes can therefore be determined by use of the derived equations.

6B11. **Influence of macro amounts of elements on preconcentration of micro amounts of elements by extraction chromatography.** Bol'shova, T. A.; Shapovalova, E. N. (M. V. Lomonosov Moscow State Univ., Moscow, USSR). *Zh. Anal. Khim.*, Aug 1983, **38** (8), 1489-1502 (Russ.).—A review is presented, with 98 references. M.P.

6B12. **Use of micro X-ray spectrometry to analyse solid micro-systems in composite electrochemical deposits.** Naumovich, N. V.; Rutman, M. E.; Stefanovich, A. A.; Chekan, V. A. *Poroshk. Metall. (Minsk)*, 1983, (7), 109-112 (Russ.). Ref. Zh., Khim., 19GD, 1983, (23), Abstr. No. 23G180.

6B13. **Spectroscopic determination of isotopic composition of minor components [*e.g.*, lithium] of small portions of powdered samples [*viz*, metals and rubidium chloride].** Drobyshev, A. I.; Turkin, Yu. I.; Yakimova, N. M. *Probl. Sovrem. Anal. Khim.*, 1983, (4), 96-100 (Russ.). Ref. Zh., Khim., 19GD, 1983, (23), Abstr. No. 23G56.

6B14. **Atomic-absorption spectroscopy as a tool for the determination of inorganic anions and organic compounds. I. A review.** Garcia-Vargas, M.; Milla, M.; Perez-Bustamante, J. A. (Dept. Anal. Chem., Fac. Sci., Univ. Cadiz, Cadiz, Spain). *Analyst (London)*, Dec 1983, **108** (1293), 1417-1449.—A review is presented of the use of a.a.s. in the determination of inorganic ions and organic compounds. Indirect methods, based on the formation of compounds in soln. or in the atomization cell and on the formation of volatile compounds in determination of halogens, and direct methods for non-metallic elements and organic compounds are included. (237 references.) G.A.B.

6B15. **Recent progress in mass spectrometry for inorganic analysis.** Adams, F. (Dept. Chem., Univ. Antwerp (U.I.A.), 2610 Wilrijk, Belgium). *Spectrochim. Acta, Part B*, Nov-Dec 1983, **38** (11-12), 1379-1393.—A review is presented that includes sections on spark-source and plasma-source m.s. and on laser micro-probe mass analysis. (53 references.) P.J.R.

6B16. **On the separation of caesium, strontium and cerium.** Toth, L. (Inst. Radiochem. and Phys., Veszprem Univ. Chem. Eng., 8201 Veszprem, Hungary). *Radiochem. Radioanal. Lett.*, 15 Nov 1983, **59** (4), 245-252.—A simple method is described for the enrichment and separation of $^{90}Sr$, $^{137}Cs$ and $^{144}Ce$ so as to allow their direct quantitative γ-ray spectroscopic determination. Dowex 50W-X8 resin at near-natural pH is used to concentrate the three cations from dil. soln., followed by elution with 0.5 to 2M-$HNO_3$. Further concentration and purification can be performed by chromatography with bis-(2-ethylhexyl) phosphate on Chromosorb as sorbent and a similar mobile phase. Final isolation of Sr and Ce can be achieved by chromatography on 2-thenoyltrifluoroacetone or quinolin-8-ol supported on Chromosorb. G.E.

6B17. **Study of the simultaneous determination of copper, lead and cadmium at a gold electrode, with the enhancement effect of iodide, by derivative anodic-stripping voltammetry.** Cui, C. (Inst. Environ. Chem., Acad. Sinica, Beijing, China). *Huaxue Xuebao*, Oct 1983, **41** (10), 927-933 (Ch.).—In $H_2SO_4$ - tetraethylammonium iodide medium, the peak heights of $Pb^{II}$ and $Cd^{II}$ are enhanced, and the peaks of $Bi^{III}$ and $Cu^{II}$ are well formed and completely resolved, at a gold electrode. The peak potentials for $Cu^{II}$, $Pb^{II}$ and $Cd^{II}$ are +0.25, −0.12 and −0.27 V, respectively. The detection limits for $Cu^{II}$, $Pb^{II}$ and $Cd^{II}$ are 0.2, 0.2 and 0.05 ppb, respectively. The process was studied by means of cyclic triangular-wave voltammetry. The enhanced wave was useful in determining these metals in tap-water and river water; recoveries of $Cu^{II}$, $Pb^{II}$ and $Cd^{II}$ averaged 96 to 100%, 94 to 110% and 100 to 110%, respectively. Z.J.

6B18. **Extraction - kinetic methods for determining some transition metals in special-purity substances.** Kreingol'd, S. U.; Shigina, E. D.; Loginova, E. V. (All-Union Sci.-Res. Inst. Chem. Reagents and Special-Purity Substances, Moscow, USSR). *Zh. Anal. Khim.*, Aug 1983, **38** (8), 1397-1403 (Russ.).—The selectivity of conventional kinetic determinations of traces of Cu, Fe, Ti, V or Nb in high-purity chemicals was considerably improved by extracting the metals as complexes into organic solvents (*e.g.* $CHCl_3$ or butanol) and carrying out the indicator reactions in mixed media after treating the extracts with 5 to 40% of ethanol and 0.2 to 15% of $H_2O$. Suitable extractants (and solvents) were: 2,2'-biquinolyl ($CHCl_3$) for Cu, 1,10-phenanthroline ($CHCl_3$) for $Fe^{II}$, catechol plus dioctylamine (butanol) for $V^V$ and benzoin α-oxime ($CHCl_3$) for $Nb^V$. The indicator reactions were: oxidation of *p*-phenetidine with $H_2O_2$ for Cu and Fe, oxidation of *o*-phenylenediamine with $H_2O_2$ (for Ti) or $KBrO_3$ (for V) and the oxidation of 2-aminophenol with $H_2O_2$ for Nb. The analysed materials were nitrates, carbonates or sulphates of Li, Na, K, Ca, Ba, Be, Mg, Al, Cu, Pb and Cd, nickel chloride and zirconium dioxide; the limits of detection were typically ≈10 ppb. M.P.

# Index to Volume 46
# 1984

ROYAL SOCIETY OF CHEMISTRY

## Introduction to the Index

### Subject Index

The alphabetically ordered subject index contains entries based on the most important subjects mentioned in the title and abstract. Letter-by-letter alphabetization is used for the bold subject headings, hyphenated words being treated as single entries. Repetition of words or phrases that occur in preceding entries is avoided by the use of indentation.

*E.g.*, **Waste water,** analysis of, by g.c., 6H63
detmn. of cadmium in, by a.a.s., 2H61
by anodic-stripping voltammetry, 10H65

The letter B following an abstract number indicates that the item refers to a book.

Organic compounds are indexed under the parent compounds. Esters and salts of organic acids are indexed under the names of their acid constituents.

### Author Index

The names of all authors are listed alphabetically, and prefixes are treated as part of the surname. Where two or more authors have the same surname and initials, no attempt at resolution has been made. For example, George Brown and Gillian Brown will both be included under the single entry "Brown, G."

The German ä, ö and ü and the Danish or Norwegian ø have been rendered as ae, oe, ue and oe, respectively. The suffix B to an abstract number indicates that the item refers to a book.

## Subject Index Guide

Abstracts are indexed under the most specific heading that can be assigned on the basis of the title and abstract text. Some headings have "*see also*" references to direct the user to more specific or closely related headings. "*See*" references indicate that the subject is indexed under another heading.

A



B



C

A

Now that you have examined the two major abstracting services in your field, I shall introduce you to two very specialised services. These demonstrate my point about their usefulness if they happen to cover the subject field in which you are interested. The first is:

> *Mass Spectrometry Bulletin.* Aldermaston: Mass Spectrometry Data Centre (AWRE), 1966 to date. Monthly.

This indexes references from all types of literature to mass spectrometry. It is arranged in subject sections. It has detailed indexes which include separate indexes of compounds, compounds and ions, elements, and materials. There are no full abstracts such as you would find in *Chemical Abstracts* or *Analytical Abstracts*, but there are some brief summaries to amplify titles of documents. The other service is:

> *Journal of Chromatography: Bibliography Section.* Amsterdam: Elsevier, 1968 to date. Quarterly.

This is a comprehensive listing of all relevant books and journal papers. It is arranged in subject sections, covering liquid column-chromatography, gas chromatography, planar chromatography, and electrophoresis. There are detailed sub-divisions to these general sections.

Before we leave this Section I must mention a couple more services and make one or two general points. You might come across the *Science Citation Index* and, since it uses a unique indexing method which will otherwise be quite unfamiliar to you, I shall say a few words about it here. The details of this are:

> *Science Citation Index.* Philadelphia: Institution for Scientific Information, 1961 to date. Every 2 months, annual and quinquennial cumulations.

The principle of citation indexing is based on the fact that almost all scientific papers refer to ('cite') earlier documents which provide support and precedent for or elaboration of what the author is writing (Want an example? Look at almost any scientific paper you choose, including the one I gave you as an exercise in the Introduction, which had a typical list of references at the end). Obviously by simply citing the earlier document the author of the paper is stating some kind of subject relationship between that document and his own paper. Also we may assume that papers which independently cite the same earlier document will stand, to some degree at least, in a subject relationship to one another. The principle of citation indexing is, therefore, to take each of these earlier cited documents as a subject indexing term and to group under it all those later papers which have cited it during the period being examined for indexing. Citation indexing is therefore an automatic process and can offer some advantages over traditional methods: one principal advantage is that it is multi-disciplinary and unexpected new leads can be opened up simply because an author has cited a particular earlier document in which you might be interested. Having introduced this service I shall not say more about it here. However, the publishers issue a useful leaflet outlining how you can search in *Science Citation Index*, and so I include a copy for illustration.

# *Science Citation Index*® *Sample Searches in Chemistry*

Area of interest: "leukotriene synthesis" Prominent known author: E. J. Corey

*(The index entries on this page are for illustration only.)*

***Source Index***—To find current articles by known authors, and to locate full bibliographic descriptions.

***Citation Index***—To find recent articles which cite earlier, known works.

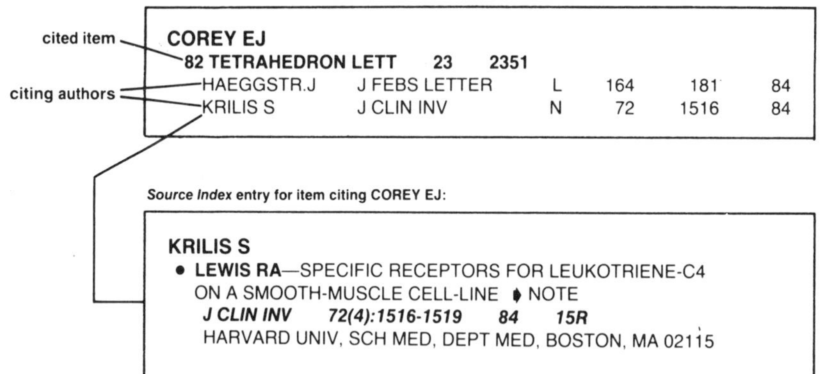


***Subject Index***—To find current articles which have words of interest in their titles.

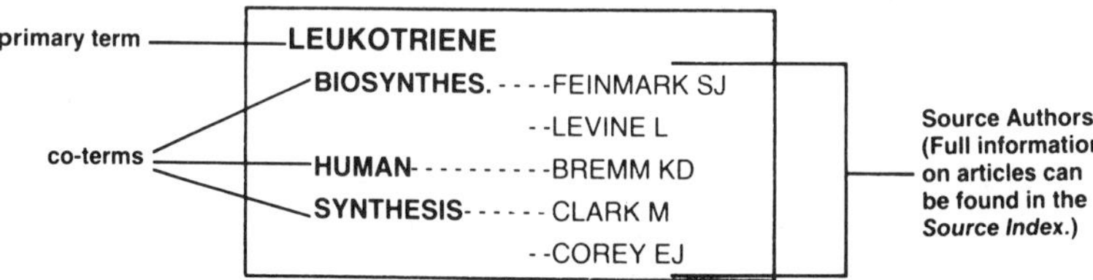


***Corporate Index***—To find articles of interest when a corporate or academic institution is known to publish work on the topic.

**MASSACHUSETTS**
**CAMBRIDGE**
• **HARVARD UNIV**
• *DEPT CHEM* . . . . . . . . . . . . . . . . . . . . . . . . . . . . . . . . . . . . . . . .

| | | | | |
|---|---|---|---|---|
| BRUNE WH | J PHYS CHEM | 87 | 4503 | 84 |
| COREY EJ | EXPERIENTIA N | 39 | 1084 | 84 |
| COREY EJ | TETRAHEDR L | 24 | 4883 | 84 |

For more information on the unique *Science Citation Index*, call the Institute for Scientific Information® at 1-800-523-1850, extension 1371.

Bibliographies, indexes, abstracts—they all have one major disadvantage: they take time to produce, publish, and distribute. In other words, the gap between work being completed and your reading about it is stretched still wider. If you notice a paper in a journal you have just received, you might be reading about work completed a year or more ago; even if you examine each copy of any abstracting or indexing service as soon as it arrives, that gap will have stretched to eighteen months, two years, or more. There are (as I think you must realise by now) far too many journals, even within analytical chemistry, for you to scan regularly (even if you can get access to a library which can afford them all!). So how can you reduce the risk of missing some vital paper until it appears in an indexing or abstracting service? About the best answer is to subscribe to our final service in this Section:

*Current Contents: Physical, Chemical and Earth Sciences.*
Philadelphia: Institute for Scientific Information, 1961 to date. Weekly.

Quite simply this reprints the contents pages of all the relevant journals received by the publishers (of *Science Citation Index*, you will note) during the week. Titles of individual papers are given in English only (so are translated when necessary). There are weekly indexes of keywords and authors. The chemistry section includes a section on analytical, inorganic, and nuclear chemistry. I have included a few pages, just as an illustration for you, so that you can see what this service is like. If you do need to keep as up-to-date as you can in a range of journals, you will find this a most convenient and valuable service.

Now let's just summarise what we have learned in this Section so far, by trying this question.

**SAQ 3.2d**

Here is a list of publications or services which can give you bibliographic information. List them in the order you would expect them to be published.

(*a*) Abstract Journals,
(*b*) Bibliographies,
(*c*) Current Awareness Services,
(*d*) Indexing Journals,
(*e*) Guides to Literature,
(*f*) Review Papers.

June 3, 1985 Volume 25 Number 22

# CURRENT CONTENTS®

## Physical, Chemical & Earth Sciences

(ISSN 0163-2574)

### *Citation Classics®* :

**Davis R E.** Predictability of sea surface temperature and sea level pressure anomalies over the North Pacific Ocean. *J. Phys. Oceanogr.* *6*:249-66, 1976.

### *Current Comments®*:

Drycleaning. Part 1. The Process and Its History: From Starch to Finish

CONTINUED **BER BUNSEN GES PHYS CHEM,89(3)85**



## Analytical, Inorganic & Nuclear Chemistry

**talanta**

Multilingual Journal—Abstracts in the Language of the Article

Pergamon Press Ltd., England

VOL. 32 NO. 4 APRIL 1985



CONTINUED

**CONTINUED** TALANTA, 32(4)85



AFZ94

# BUNSEKI KAGAKU

Articles in Japanese—Abstracts in English
(Section E—Articles and Abstracts in English)

Japan Soc. Anal. Chem.

VOL. 34 NO. 4 APRIL 1985

AGF18

JOURNAL OF
ELECTROANALYTICAL CHEMISTRY
AND INTERFACIAL ELECTROCHEMISTRY
Articles and Abstracts in English

Elsevier
Sequoia S.A.

VOL. 185 NO. 1 APRIL 10 1985

**TrAC**
**trends in analytical chemistry**
Articles and Abstracts in English
Elsevier Science Publishers

VOL. 4 NO. 4 APRIL 1985



AGF90

**JOURNAL OF LIQUID CHROMATOGRAPHY**
Abstracts in English
Marcel Dekker Inc.

VOL. 8 NO. 4 1985 (P,L)



CONTINUED

 PC&ES, V. 25, #22, June 3, 1985 81

### 3.2.5 Literature Searching

Right, now you know all the secrets and even have had a fair amount of practice in searching and using the analytical literature.

Π I have an information problem. What is the first thing I do?

No! I don't dash off to the library and dive into *Chemical Abstracts*. If we take that approach we shall waste an awful lot of time and build up a great deal of frustration. No, the correct answer to this question is that we stop and think (remember that advice?) and plan our first and subsequent moves logically.

Literature searching is a vital scientific technique and needs to be planned and performed as methodically as any other part of scientific research. By now you must realise that there are many different avenues to be followed in searching for information (and that not all roads necessarily lead to *Chemical Abstracts*). But if you approach and then set about any problem logically and consistently you will increase your chances of finding efficiently the information you need. Do you remember the diagram of literature production in Section 2.6.6? Well, when you start looking for information in a completely new subject you would normally start with the more general sources (such as textbooks and encyclopaedias), at the bottom of the diagram; as you need more specialised or detailed information you will progress towards the top, through handbooks and review papers, coming into the highly specialised primary literature (such as journal papers and reports) only later in your search when you already have a good background in the subject, or need very specialised information. I now give you a strategy of eight steps to help you to plan any literature search you might have to undertake. Don't treat these steps as rigid ones, each to be completed before going to the next: some at least will overlap with others (certainly the first three steps are really almost simultaneous); some will take only a few minutes to carry out, while others will take a few hours.

(*a*) **Define the topic**. Know exactly what you are looking for, and why. Decide exactly on the parameters you wish or need to set on your topic. Encyclopaedias and dictionaries can help you at this stage. If properly carried out, this can save you a lot of wasted time later on.

(*b*) **Define your sources.** Decide what kinds of documents are likely to be of most use for the particular problem you have defined: do you need a handbook, a standard, or a reference work, or must you carry out a search for a more detailed document? If in doubt do a quick search to test different sources against one another.

(*c*) **List your search terms**. Make a list of the terms under which material is likely to be indexed. Think of possible synonyms; remember that you might need to search under either or both of different American or British terms or spellings; and remember also that more general or more specific terms might be needed to carry out a successful comprehensive search. Obviously specialised dictionaries will be of use at this stage.

(*d*) **Test search.** Even at this stage don't start writing down references as you find them; perform a test search based on what you have decided up to this point. You might find too much material, or not enough; you might discover that the source you thought would be most use is not giving you that kind of information and so have to look at another. Redefine your search in the light of what you find at this test stage.

(*e*) **Conduct your search.** Do you see how much work you have had to do before you really start looking seriously for references? The preliminary work shouldn't have taken you very long, but by now you know just what you are looking for and where you are likely to find it; so steps 1–4 have saved you time at this stage. When you are consulting indexing or abstracting services, always start with the most recent volumes and work backwards in time, either until you have gathered enough references, or until some obvious stop date arrives (such as the year of discovery of a particular compound you might be investigating). When you have done that you can bring your search up-to-date by looking through the unbound most recent issues of the abstracting or indexing journals you are using.

(*f*) **Record the references.** Do you remember that exercise on bibliographic citations in the Introduction to this Unit? At this stage note all the relevant material you find, and if in any doubt take more information than you are likely to need subsequently. Remember that it is always much easier to edit material out at a later stage than to have to repeat a search or part of a search. Also bear in mind that your memory is always fallible and that you must write down everything to which you might need to refer.

(*g*) **Examine relevant material.** As you find your references sort them into a rough order of priority so that you can start looking at material as soon as possible. Check immediately on those which are available in your own library, and remember to allow plenty of time to request anything you might want from the inter-library loans system.

(*h*) **Evaluate your findings.** As you are finding references check them for relevance, first by reading abstracts of them, also by looking at useful-seeming documents you find. Keep a close watch for any likely review papers, remembering that they can do a lot of the finding and evaluation for you. At this stage *Science Citation Index* could be used as a cross-checking source.

The main message is obvious, isn't it? Before you take any of these steps or do anything to seek information, pause and think about what you are going to do, and why you are going to do it. Then while you are searching, make sure you record everything that might be of future value to you and leave nothing entrusted to your memory. All the time you are searching check that the results match what you hoped or expected to find and be prepared to modify your search if they don't.

As ever, we will finish with an exercise for you.

**SAQ 3.2e**

(*a*) Without referring back to the text, take the following steps in a literature search and arrange them in order:

Define the sources
Record references
Evaluate findings
Conduct the search
Test search
Define the topic
Examine relevant material
List the search terms

(*b*) Having listed the steps in order, look at the three categories of documents below and write them down beside your list of search steps to correspond with whereabouts in your search you would expect to be consulting them:

(*i*) Abstracts/Indexing services/reviews,
(*ii*) Journal papers/patents/reports,
(*iii*) Encyclopaedias/dictionaries/textbooks.

**Objectives**

When you have worked through this Section you should be able to;

- recognise the different types of published bibliographic reference services available;
- understand the functions and potential usefulness of bibliographics, literature guides, indexing and abstracting publications;
- recognise particular important services and find information by using them;
- carry out a manual (as opposed to electronic, covered in Part 4) literature search to find publications by using published sources.

# 4. Commercial On-line Services

## 4.1. INTRODUCTION TO ON-LINE SERVICES

### 4.1.1. What is On-line Searching?

By now you know about the many published sources of information for the analytical chemist—how to find them and how to exploit them. You also know (I hope) that looking for information is not a haphazard process: it requires thought, planning, and a systematic approach. You must also have realised that because of the vast amount of published literature, abstracting and indexing services can become large and complex, and that searching them thoroughly can prove a time-consuming exercise. Those of you who went and found a set of *Chemical Abstracts* to answer the questions in Section 3.2.3 will be well aware of the potential problems—remember you looked at only six months' output of literature. Most indexing and abstracting services nowadays, as well as many reference works, are compiled from computerised data bases. Many 'intermediaries' and 'end users' (or—translating from the currently fashionable jargon—librarians and analytical chemists) own or have access to microcomputers or computer terminals. Printing and publishing a constantly changing data base is a bulky, costly, and time-consuming business, especially if you also need to up-date it regularly. So the answer seems obvious, doesn't it? Why not put the customer's computer terminal in contact with the supplier's computer and let him search the data base directly? On-line retrieval isn't quite as simple as that, but there you have the basic system.

Π What are the advantages of on-line searching?

Well, we have already mentioned one: 'hard-copy' publication (ie printed books or journals) is bulky, time-consuming, and expensive. Instead of paying a high subscription to an abstract journal, of which you might consult only a small part anyway, it is likely to be more cost-effective to use an on-line version where you pay only for the time you spend with the data base and for the references you print. That is a fairly powerful argument, but I think the major advantage of on-line searching is simply that the computer can search in seconds material which it would take you hours, or days, to search manually. So your searches can be more flexible: you can add or subtract new terms and follow up new search avenues instantly. You can also take a print-out at any stage of your search, to save you having to write out laboriously lists of the references you find.

It sounds Utopian doesn't it? But are there disadvantages? Not many, but keep the following points in mind. Almost any terminal or microcomputer will do, but you still have to have the equipment, and either a modem or an acoustic coupler (these are special units to link a computer to the telephone system: the modem is more sophisticated and gives a clear line, but generally costs at least two hundred pounds; the acoustic coupler is simpler and portable, and costs less than one hundred pounds, and gives quite satisfactory results). Some of the access charges seem very steep (up to 2.50 dollars per minute for some data bases) and you have to pay for use of the telephone line (but not very much when you consider that a call to a local number can give you instant access to computers all over the world). If you are an inexperienced or an inefficient searcher—or if you get carried away with enthusiasm for the system (I have seen it happen!)—you can run up some pretty frightening bills. But if you use the system efficiently and sensibly you should not find on-line searching unduly expensive—especially if you calculate the cost to yourself of the alternative, viz—the amount of time and effort you would have to put into a manual search.

There are basically two kinds of data base you can search, and two types of organisation to give you on-line access for searching. The first kind of data base is the bibliographic: this is an abstracting or indexing journal of the kind we have already examined made available for searching as a computerised data base. When you search a bibliographic data base the information you retrieve will be in the form of a bibliographic reference to (and frequently an abstract of) a paper, report, or other publication. The other type of data base is the full-text or directory: here you can gain access directly on-line to the full original information you are seeking. We shall learn more about these in Parts 4.2 and 4.3. Initially, all my examples of on-line searches will be through bibliographic data bases. The two types of organisation are the original data base publishers and the computer-host service (only in very rare instances does the same organisation perform both functions): I shall explain this further in 4.1.2. Also at this stage it will be helpful for you to remember that in searching a data base the computer is searching both the assigned standard index terms and the terms as they occur in titles, abstracts, etc. ('free-text searching' this is called). One other on-line search service I can usefully mention to you here is SDI (no, not the star wars programme: this is a much longer established use of the acronym and stands for *Selective Dissemination of Information*). Any host service will allow you to create a search and then store the result in their computer: each time the particular data base you wish to search is updated your search will be matched against the new records and a print-out of any relevant records will be sent to you.

Before we look at on-line searching in more detail, let me just make two last points. You know that any literature search has to be planned carefully and conducted systematically: well, that advice for manual searching applies even more rigorously to computerised searching. You can't browse through an electronic data base in the same way you can through a printed book or journal: there are facilities to help you find your way through the data base on-line, but with computerised searching you really have to know exactly what you are doing at all times and to have planned your search in advance; but then you will realise by now that I consider anything that makes you stop, think, and plan to be distinctly advantageous! And what is my last point here? On-line searching has enormous power, tremendous potential for further development, and is expanding rapidly in all areas of chemical documentation. But amidst all this electronic euphoria, *don't forget the printed word*! I have shown you the part that books and other forms of the literature have to play in analytical chemistry, and although some of them may eventually be replaced almost entirely by electronic services, the pattern of their conversion and replacement will be uneven. Printed publications will remain important to you for many years—some of them will never be replaced—in parallel with the burgeoning electronic services we are examining here.

### 4.1.2. Commercial On-line Host Computers

How do we get access to these data bases? It isn't very difficult, nor is it terribly expensive (provided you know what you are doing). Almost any computer terminal (visual display unit or teleprinter—you will even be able to use a telex terminal soon) will do, or a personal computer or microcomputer. You have to be able to link this to the telephone network: a proper modem connection is available and recommended, but I have been working perfectly satisfactorily for several years with a simple, inexpensive, acoustic coupler (for which you just make your telephone connection then place the handset into the coupler which is already of course plugged in to your terminal). You dial your host computer, log-in to the system and start searching. Most major computer hosts offer direct lines for you to use for their service in most major countries, but probably the easiest, cheapest, and most efficient way to contact a host computer is through the International Packet Switch Service. You usually dial (in the UK at least) a fairly near exchange and, on connection, type in your identifying code (you must register to use the IPSS, and then you will be given a code); next, you are asked for an address, you then type in the code for the service you wish to access and are almost instantly connected to that host computer. Look at fig 4.1a and you can see this process and its world-wide applications: of the eight host services I shall mention to you, three are in the USA, four on the continent of Europe and one in the UK. Connection to each *via* the IPSS (or through the direct-link numbers they maintain in many countries in addition to IPSS) is instantaneous, and is charged at the same standard telecommunication rate (in other words you are not paying for a telephone call to California!). I should also mention here that there are networks similar to IPSS in different parts of the world (eg TYMNET and TELENET) offering the same facilities.

In a moment I shall show you what this process looks like on the screen or teleprinter of your terminal. But what is all this talk of computer hosts and data bases? What is the difference between them?

**Data Bases.** These are the services you are actually searching (eg *Chemical Abstracts*, *Chemical Engineering Abstracts*, *Kirk-Othmer*, etc): we shall examine some of these in more detail in Section 4.2 and 4.3. There are by now many hundreds of data bases on all

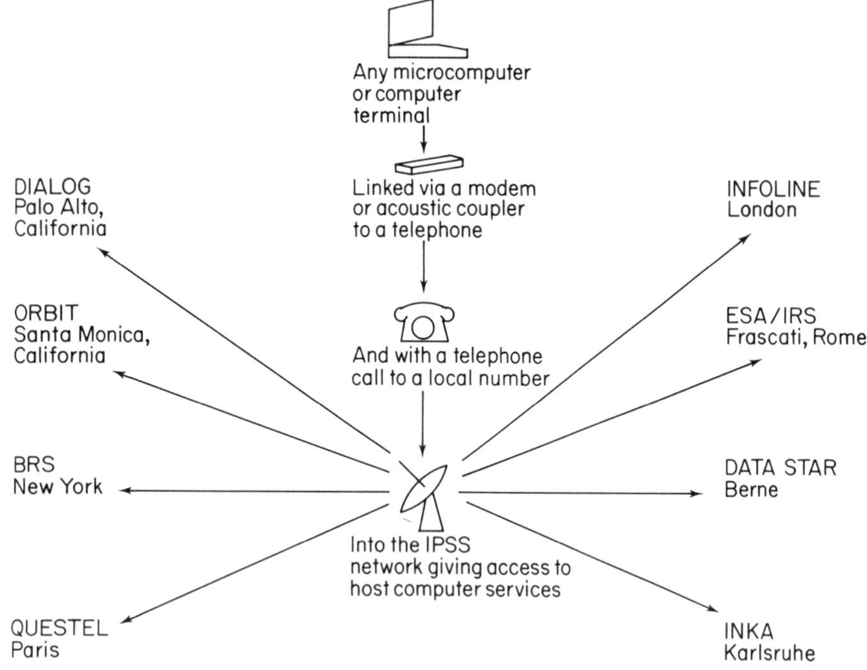


**Fig. 4.1a**

manner of subjects and it would not be economically feasible for many of them to offer public on-line searching, if each had to arrange its own individual access (although there are some independent data bases which can only be accessed directly). The development and rapid expansion of on-line searching has, therefore, been carried out by a few large suppliers, the computer host services.

**Computer Hosts.** These are the various services you dial which give you access to different numbers and ranges of individual data bases. Some offer access to very large numbers of data bases of many kinds on many subjects (the most prominent example of this is Dialog), while others specialise in particular subject areas with fewer data bases. Very few data bases are exclusive to one host, and several of the data bases we shall look at are offered by different host services (eg you can search *Chemical Abstracts* on the Dialog, Orbit, BRS, ESA/IRS, Questel, Datastar, or the Inka service). There are many more than the eight host services I have indicated in Fig. 4.1a; but these eight are the ones you are most likely to come across and to use frequently. Let me just tell you a little more about each of them.

BRS, Latham, New York. This offers over 70 data bases in the scientific/medical, business/financial, reference, education, and the social sciences/humanities field. It has particular significance for us in offering two major chemical full-text services (American Chemical Society Journals, and Kirk-Othmer—more of these in Section 4.3).

Data Star, Berne, a division of Radio-Suisse. This offers over 50 data bases, and specialises in the biomedical and the chemical area, but also covers the technical, business, news, and the reference field.

Dialog Information Services. Palo Alto, California. This is one of the founder hosts (operating since 1972), set up as a division of the Lockheed Missile and Space Company. It is one of the major hosts, and certainly the largest, offering over 220 data bases on very many subjects.

ESA/IRS, Frascati, Rome. This is another pioneer host service, set up by the European Space Agency. It offers over 70 data bases, and specialises in science and technology, but it also carries some business and reference data bases.

Inka On-line Service. Karlsruhe. This offers some 40 data bases, mainly in science and technology. It is the host system for STN which gives a special form of access to *Chemical Abstracts* (more of this later).

Pergamon Infoline, London. This is a specialist host, which offers some 35 data bases and specialises in chemistry and patents, although it also covers some business and other scientific and technical fields.

SDC/Orbit, Santa Monica, California. Another early host system, this offers over 60 data bases on a wide range of subjects.

Telesystemes Questel, Paris, a division of the French national telecommunications service. This offers some 50 data bases and specialises in chemistry, life sciences, patents, applied sciences, social sciences, and French data bases. It is the only European host for ISI's *Index Chemicus* (again, more about this later).

Now you know that when you want to search a data base you will do so by first accessing a host service. Before we go on to see how to construct and enter a search, I did promise to show you what a typical access call looks like on the screen. Have a look at Fig. 4.1b. I have numbered the lines in fives to make it easier for me to explain what is happening. I decided to access the *Chemical Abstracts* data base on the Data Star host system. As my institution is in the West of Scotland I dialled a Glasgow number and in line 1 received a coded reply indicating that I was connected to the IPSS. In line 2 I typed in my IPSS identification number (for obvious reasons I have blocked out all my real passwords and used fictional equivalent ones instead). At line 3 I am asked for the address of whatever service I want to access, and have entered the address for Data Star in line 4. In line 5 the system repeats the address to show that the connection has been made and in line 6 confirms that I am connected to Data Star I also now have to enter my identification authorising me to search on Data Star (each host gives you an identifying code for its system when you first register with it) and then the system asks me for my security password. I have entered this in line 11 and, now identified to Data Star, I receive any general messages about the system. In line 22 I am told that there are further system messages, but I decline these and in line 23 am asked what data base I want to search: I type in CHZZ, (Data Star code for *Chemical Abstracts*) and lines 24–26 confirm my connection. In line 27 I am asked to

```
GLA A001-4126040002
NABCDEF123XYZ
ADD?
A9228464110115
228464110115+COM

D A T A - S T A R , PLEASE ENTER YOUR USERID : ABCDEF

ENTER YOUR A-M-I-S PASSWORD

00000000
XXXXXXXX
********
GHIJKL

- FRIDAY 25. OCT. 1985: DUE TECHNICAL MAINTENANCE D-S WILL NOT BE

  AVAILABLE FROM 0630 TO 0930 HOURS LOCAL TIME.

- POLLUTION ABSTRACT (POLL) ONLINE TUESDAY 22. OCT. 1985. CONSULT

  BROADCAST MESSAGE FOR FREE TIME.

- A T T E N T I O N: PHARMAPROJECTS (PHAR) EACH ONLINE/OFFLINE CITATION

  COSTS SFR. 55.00 FOR NON HARD COPY SUBSCRIBERS.

- ICCF ONLINE NOW. FOR CHARGES/DETAILS SEE BROADCAST.

- SATURDAY 2. NOV. 1985 : DUE HARDWARE INSTALLATION D-S WILL NOT BE

  AVAILABLE FROM 0800 TO 1800 HOURS LOCAL TIME.

- ADDITIONAL TELEPAC-NUMBER 484110115 MAY ALSO BE USED.

ENTER YES IF BROADCAST MSG IS DESIRED_:

ENTER DATA BASE NAME_:  CHZZ

*SIGN-ON   10.10.52                     24.10.85

D-S/CHZZ/CHEM ABS 1967-V103,I14/1985        SESSION      69

COPYRIGHT BY AMERICAN CHEMICAL SOCIETY, COLUMBUS/OHIO, U.S.A.


D-S - SEARCH MODE  - ENTER QUERY

      1_:  ..off

*CONNECT TIME CHZZ:    0:00:56 HH:MM:SS    0.016 DEC HRS.   SESSION    69*

*SIGN-OFF   10.11.07                       24.10.85

CLR PAD    (00)  00:00:01:11 30 7
```

**Fig. 4.1b**

start searching, but on this occasion I have entered only the data base to demonstrate how to make the connection, so I type in the command '. . off' in line 28 to end my search. My clearance out of Data Star is confirmed in lines 29 and 30, and in line 31 I am told that the IPSS connection is also cleared. I don't know how long it took you to read that explanation, but from line 1 to line 31 the dialogue with the system took me about 90 seconds. That is how quick and how easy it is to access any of these host services.

### 4.1.3. On-line Search Techniques

We have accessed a host and entered the data base we want to search (we shall look at particular data bases in more detail in Section 4.2 and 4.3). What do we do now? Well, in fact we should already have done a significant amount of work before we ever sit down at our terminal. Do you remember paragraph 3.2.5? My remarks about deciding what we want and what terms to use are even more important in on-line searching: the computer actually charges you for wasting time! Even more than in manual searching, for a computer search you must plan your strategy carefully before you begin. You can adapt it as you search, of course, but careful pre-planning is absolutely essential if you are not going to waste on-line time (and money).

In mathematical terms on-line searching involves a combination of set theory and Boolean logic. Don't worry, it isn't as complicated as that might sound, and you will quickly get the hang of it. What this means is that you construct your search by grouping related terms in sets, and by combining these sets by using the terms ('logical operators' they are called technically) 'and', 'or', 'not', as necessary. I can't pretend that there are no complications to this, but these complications (and their solutions) will become obvious, and the basic process does remain fairly straightforward.

Let's look at an example. You are working in a food laboratory and have been studying water determination in ham and corned beef. Now you want to know what work has been done on water determination in other cooked meats. We are going to search *Food Science and Technology Abstracts*. Look at Fig. 4.1c in conjunction with my explanation.

Our first step is to work out a concise, but accurate and reasonably comprehensive definition of the topic, bearing in mind that we must consider synonyms. For the purposes of this example I have produced the definition you find at the head of Fig. 4.1c. This definition has already indicated some of the combinations of terms we shall use. So we start by creating Set 1. The term 'water' has given us 42,069 references, and a parallel term 'moisture' 17,770 references. By combining them with 'or' we create a set of the sum of these references; but where both terms occur in any one reference, that reference is recorded only once (the overlap area in the diagram). So our total for Set 1 is 54,008 references, not the 59,838 obtained as the gross sum of both terms. We do exactly the same to create Set 2 with the alternative terms 'determination' (20,342 references) and 'analysis' (24,137); our set here, using the 'or' combination, is 39,109 references. Now, for Set 3 we want to know where Set 1 and Set 2 overlap—the determination or analysis of water or moisture—so we combine our two sets with 'and'. The area of overlap where

SEARCH STRATEGY FOR TOPIC DEFINED AS:
The determination or analysis of water or moisture in cooked meats, excluding ham and corned beef.

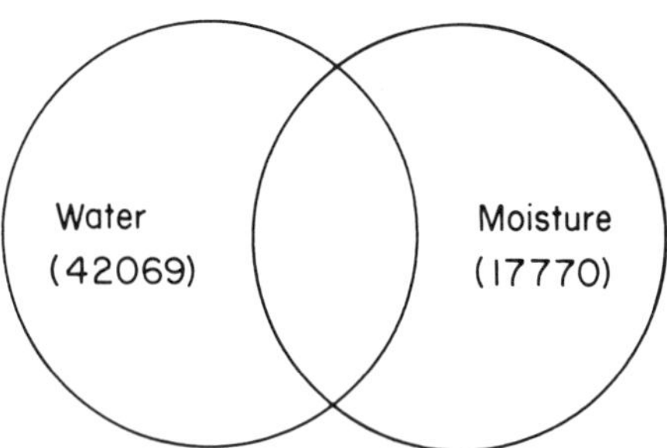


Set 1. Water or Moisture
The sum of the numbers of references under both terms, but with those which overlap entered only once.
= 54,008

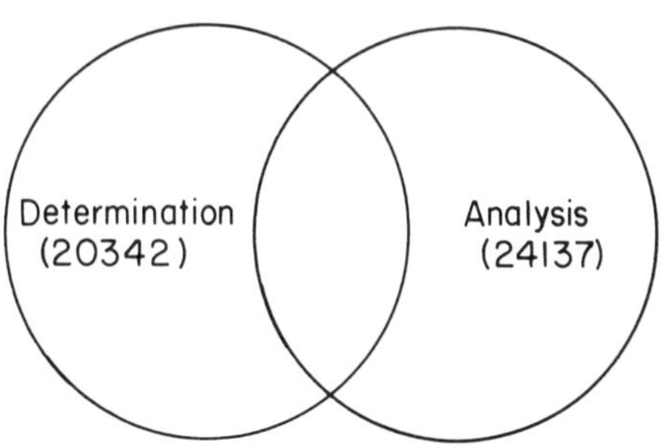


Set 2. Determination or Analysis
The sum of the numbers of references under both terms, but with those which overlap entered only once.
= 39,109

**Fig. 4.1c**

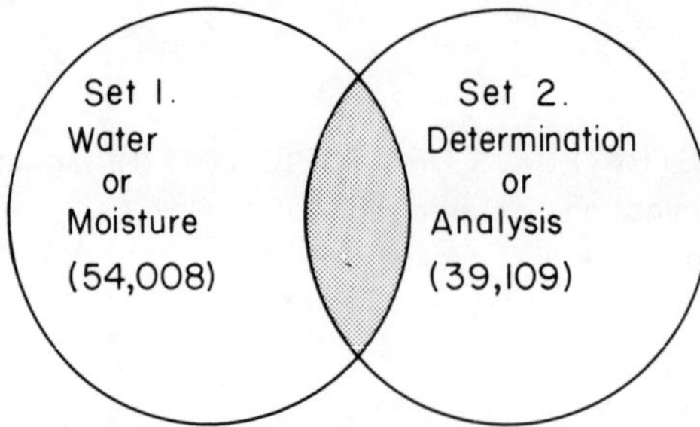


Set 3. (Water or Moisture) <u>and</u> (Determination or Analysis)
Only those references which match
from each set (i e the shaded overlap area only).
= 8340 refs.

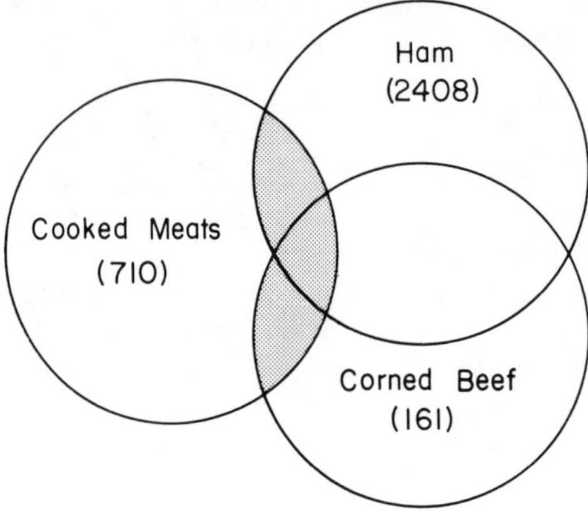


Set 4. Cooked Meats <u>not</u> (Ham or Corned Beef)
The whole of the 'Cooked Meats' set
excluding where it overlaps with 'Ham' or
'Corned Beef' (i e excluding the shaded area).
= 664 refs.

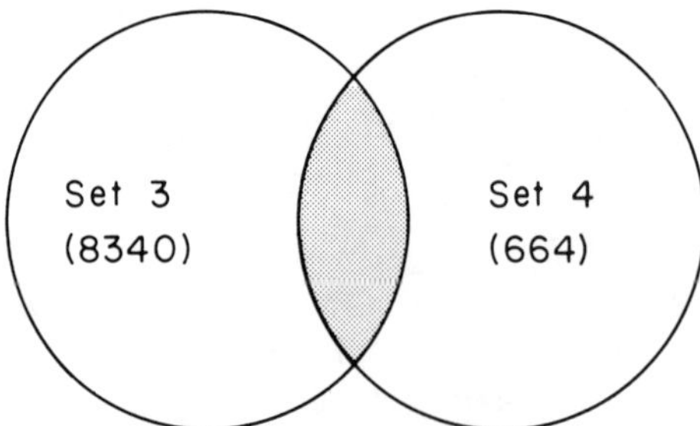


Set 5. Set 3 ([Water or Moisture] and [Determination or Analysis])
<u>and</u> Set 4 (Cooked Meats not [Ham or Corned Beef])
Only those references which match from each set
(i e the shaded overlap area only).
= 19 refs.

**Fig. 4.1c (cont.)**

the references in the two sets coincide gives us Set 3 of 8,340 references. In other words we now have a set of over 8,000 references on the analysis or determination of water or moisture in any substance covered by the data base. But we are concerned only with cooked meats, and excluding ham and corned beef; so we create Set 4 by finding all the references to cooked meats (710) but excluding the overlap areas with 'ham' or 'corned beef' by using the combining term 'not'. Our set then comes out at 664 references. By the way, if you look at the diagram carefully you might notice an apparent error at this stage of our search: this in fact represents one of the practical complications that can crop up in on-line searching and which experience teaches you to foresee, and to deal with. If you haven't spotted the problem, I am going to make you wait a little bit before I tell you what it is and how to deal with it—but I shall come back and explain it before the end of this Section. Now we have reached the final stage of our search, that of combining set 3 and Set 4 by 'and': the overlap area this time gives us a nice manageable set of 19 references. That is how you produce a typical search strategy—with just a little practice it becomes quite straightforward. Next we will look at the mechanics of entering our search into a terminal, but before we do, let's see you have a shot at compiling a strategy.

**SAQ 4.1a**

You have been working on the analysis of lead levels in soil samples by using the dithizone method. Now you wish to study in some detail atomic absorption and polarographic methods for the same application. To find out what has been published on these methods you are going to carry out an on-line search of *Chemical Abstracts*. Write down a definition of the subject you will search and which will be suitable for use as a search strategy. Write out your definition in the algebraic form: eg (a or b) and (c and d) not (e or f) etc, using the appropriate 'and', 'or', 'not' operators.

Finally in this Section let's look at how we actually enter our search at the terminal. This will not only show you how to perform a search but will also enable me to introduce one or two of the practical points I mentioned as affecting on-line searching. The first of these points is a general one. Throughout this search, and most of the others I show you, I shall enter the terms one at a time and create or combine sets step by step; this is to make clear to you the structure and procedure of our searches. But in fact, if and when you become an experienced on-line searcher you will find that you can enter a whole search on-line algebraically in one step: a search taking minutes will then be reduced to one taking seconds. But these short cuts are for when you are a more experienced searcher. For our purposes in this Unit we shall enter our searches step by step, slower though it is, to make sure you understand exactly what we are doing.

You will need to look at Fig. 4.1d in conjunction with my explanation: once again I have numbered the lines on the print-out to make it easier for you to follow. This time I have conducted our search on the ESA/IRS host in Frascati (Rome); in lines 1–19 I have accessed the host computer in the manner I showed you earlier. Now, one potential difficulty we come across immediately is that different host services use different command languages and different methods of prompting users. In Data Star we saw (albeit briefly) how the system prompts us in ordinary language (eg 'Enter data base name'); other hosts use similar systems, but ESA/IRS (and Dialog) simply use a question mark as the prompt and wait for the user to respond with a specific command. Obviously we are going to see the commands used by ESA/IRS (you can also search Dialog by using them): all the hosts issue prompt cards or booklets for use beside the terminals (as well as voluminous and highly detailed manuals), and I have included three of the briefest ones (ESA/IRS, Data Star and Infoline) as examples to show you the different kinds of commands used. These are Fig. 4.1e, 4.1f, and 4.1g. The basic search strategies are the same as I have already shown you, but the mechanics vary from host to host. Have a look through the different commands and instructions in Fig. 4.1e–g after we have worked through our sample search.

But now let's get back to Fig. 4.1d, and see how our search looks on screen or print-out. After logging in I receive the first '?' prompt at line 20 and have typed in my response 'B20': 'B' for 'begin' to tell the system I want file no. 20 (*Food Science and Technology Abstracts*). I am put in contact with the data base I have specified and at the prompt in line 26 I begin to search with 'swater': 's' for 'select' precedes any term on which I wish to search. The system response is in line 27. The '1' indicates Set 1: when working on-line each individual 'select' or 'combine' command creates a set, even if only one term is selected; these sets are, of course, numbered sequentially. '42069' is the number of references comprising that set and my search term 'water' is repeated to remind me what I asked the system to search. Set 2 'moisture' is created in the same way, then in line 30 I create Set 3 which is 'water or moisture' by combining ('c1or2') Set 1 or Set 2; in line 31 I am told that my Set 3 has 54,008 references. 'Determination or analysis' is created the same way and then in line 38 I combine my Set 3 with the newly-created Set 6 to give (in line 39) Set 7 of 8,340 references on: (water or moisture) and (determination or analysis). Now, at line 40 I want all the references on cooked meat (or cooked meats).

```
?****
GLA A001-4126040001
         NABCDEF123XYZ
ADD?
A92222620021

22226200230+COM
TERPAC H12B V122


Please enter your ESA-QUEST password
@@@@@@@@
######## Connection accepted in file032 11:20:32
         Port=027-E :  Quest-language selected

*-> Nouvelle version de PASCAL en ligne,
    (fichier 14) : voir ?PASCAL
*-> Special Quest Availability - 19 & 20
    October from 10.00 to 18.00
*-> NEW FILES ONLINE
    Files 67, 104, 113, 116
*-  NEW SERVICES - Aerospace Daily (72)
    available to all.  Trainee Program
    - see ?FIXBUDGET for details
? B20

---------25Oct85  11:21:14  User     ---
    0.54 AU 0.71 Minutes in File  32
    0.54 AU approx Total
File  20:FOOD SCIENCE:1969-85,12
SET ITEMS DESCRIPTION  (+=OR;*=AND;-=NOT)
--- ----- ------------------------------
? swater

   1 42069 WATER
? smoisture

  2 17770 MOISTURE
? clor2

  3 54002 1OR2
? sdetermination

  4 20342 DETERMINATION
? SANALYSIS

  5 24137 ANALYSIS
 ? c4or5

  6 39109 4OR5
c3and6
?
```

**Fig. 4.1d**

```
  7  8340 3AND6
? scooked(w)meat?

  8   710 COOKED(W)MEAT?
? sham

  9  2408 HAM
? scorned(w)beef

 10   161 CORNED(W)BEEF
?  c9or10

 11  2519 9OR10
? c8not11

 12   664 8NOT11
? c7and12

 13    19 7AND12
? t13/6/1-19


               TYPE 13/6/1-5
85-10-s0035 FSTA  85062132  JOURNAL
   Effects of feeding a live yeast culture on market turkey
performance and cooked meat characteristics

84-06-s1389 FSTA  84038076  JOURNAL
   Quantitative determination of melanoidins from meat
   Methode zur quantitativen Bestimmung von Fleischmelanoidinen

84-02-s0230 FSTA  84012156  ARTICLE
   Determination of ascorbic acid and ascorbates in fresh, frozen
and cooked meat

83-04-s0544 FSTA  83022204  JOURNAL
   Thoughts on analysis of polyphosphates in cooked meat products

81-06-r0378 FSTA  81033104  LECTURE
   Consumer and instrumental edibility measures for grouping of
fish species
   (In 'Advances in fish science and technology' see FSTA (1981)
13 6R300 .)


               TYPE 13/6/6-10
81-05-s0938 FSTA  81028080  JOURNAL
   Water binding determination in meat:  Improvement of the
centrifugation method
   Beitrag zur Bestimmung der Wasserbindung des Fleisches:
Untersuchungen zur Verbesserung der Zentrifugiermethode
```

**Fig. 4.1d (cont)**

80-06-s0966 FSTA 80035568 JOURNAL
Protein determination with the biuret method in solutions containing sodium dodecyl sulphate (SDS)
Proteinbestimmung mit der Biuret-Methode in Natrium Dodecylsulfat (SDS)-haltigen Losungen

79-11-s1799 FSTA 79065580 JOURNAL
Microbiological determination of vitamin B12 in heat-treated foods

79-11-s1702 FSTA 79065192 JOURNAL
Recovery of functional meat proteins from abattoir by-products

76-09-S1448 FSTA 76052596 JOURNAL
ROUTINE METHODS OF EXAMINATION FOR MEAT AND MEAT PRODUCTS. XI. DETERMINATION OF THE DRIP LOSS (DRAINED WEIGHT) OF JELLY FROM CANNED COOKED MEAT AND BRAWN. DETERMINATION OF THE o FREE FAT, PURE GELATIN AND JELLY
ROUTINE-UNTERSUCHUNGSMETHODEN FUER FLEISCH UND FLEISCHWAREN. XI. BESTIMMUNG DES ABTROPFGEWICHTES BEI GELEEHALTIGEN KONSERVEN AUS GEKOCHTEN FLEISCH UND BEI FLEISCHSUELZEN. ERMITTLUNG DES GEHALTES AN FREIEM FETT UND AN REINER GELATINE SOWIE DES GELEE-ANTEILES

TYPE 13/6/11-15

75-11-S1581 FSTA 75061408 JOURNAL
METHODS FOR ANALYZING PERCENTAGE LIPID OF GROUND BEEF AND BEEF-SOY BLENDS

74-04-S0457 FSTA 74019508 JOURNAL
DETERMINATION OF SHEAR FORCE VALUE OF MAJOR BEEF MUSCLES

73-08-S0855 FSTA 73042344 JOURNAL
PENTOSE CONTENTS OF MEAT AND MEAT PRODUCTS

72-03-S0340 FSTA 72016076 JOURNAL
WARMED-OVER FLAVOUR IN COOKED MEATS

71-08-S1000 FSTA 71043392 JOURNAL
A PRELIMINARY REPORT ON FACTORS AFFECTING THE DRYING REHYDRATION AND STABILITY OF FREEZE-DRIED CHICKEN MEAT

TYPE 13/6/16-19

71-03-S0356 FSTA 71015204 JOURNAL
PRODUCTION EFFICIENCY, GRADES AND YIELDS WITH THE LARGE WHITE TURKEY AS RELATED TO SEX AND AGE

71-03-S0232 FSTA 71014708 JOURNAL
BIOCHEMICAL AND QUALITY CHANGES IN POST-RIGOR OVINE MUSCLE AFTER THAWING AT LOW AND HIGH TEMPERATURE

70-07-S0537 FSTA 70029068 JOURNAL
EXTERNAL FAT COVER INFLUENCE ON RAW AND COOKED BEEF. I. FAT AND MOISTURE CONTENT

**Fig. 4.1d (cont)**

69-05-S0347 FSTA 69019688 JOURNAL
ANALYSIS OF PREPARED MEAT MAYONNAISE
ZUR BEURTEILUNG VON FLEISCHSALAT
? t13/4/6

TYPE 13/4/6
81-05-s0938 FSTA 81028080 JOURNAL
Water binding determination in meat: Improvement of the centrifugation method
Beitrag zur Bestimmung der Wasserbindung des Fleisches: Untersuchungen zur Verbesserung der Zentrifugiermethode
Hofmann, K.; Jolley, P.; Bluchel,E.
Bundesanstalt fur Fleischforschung, 8650 Kulmbach, Federal Republic of Germany
Fleischwirtschaft
Vol 60, no 9, p. 1717-1720, 5 Ref. 1980, In GERMAN; Summary in ENGLISH
Category Code : S (MEAT, POULTRY AND GAME)

Detn. of the water binding capacity (WBC) of meat by centrifugation was improved by preventing reabsorption of exuded meat juice by the meat (rehydration), by partially filling the centrifuge tubes with an absorbent material (gypsum). Results for raw and cooked meat were compared with those from Bouton's modified centrifugation method FSTA (1971) 3 8S994, and Grau and Hamm's filter paper method Zeitschrift fur Lebensmitteluntersuchung und Forschung, (1957) 105, 446. Tabulated data showed that Bouton's method gave higher values in raw but not in cooked meat (attributed to rehydration of the raw meat). Some methods for calculating numerical values of WBC are discussed; in this paper it is expressed as the ratio of bound water: total water content. As the same trend is observed when WBC is expressed as wt. of meat before and after centrifuging, this dispenses with detn. of total water content and considerably simplifies the detn. The effects of comminution, salting, 4 days' storage and sp. (beef and pork) are also discussed

Controlled Terms: Moisture content / meat, water binding capacity centrifugation detn. of / Centrifugation / * / Binding / * / Beef / water binding capacity centrifugation detn. of beef / Pork / water binding capacity centrifugation detn. of pork /
? logoff

----------25Oct85 11:33:10 User -----
15.92 AU 11.94 Minutes in File 20
0.12 AU 1 Online Print
16.04 AU approx Total

ESA-QUEST session terminated at 11:33:14

CLR PAD (00) 00:00:13:23 183 20

**Fig. 4.1d (cont)**

**esa** — Information Retrieval Service 

## QUICK REFERENCE LIST OF ESA-QUEST COMMANDS

| COMMAND | Abbreviation | FUNCTION | EXAMPLES |
|---|---|---|---|
| BEGIN<br>BEGIN N<br>(N = file number<br>= file name) | B<br>B | **Begin** Search<br>Begin Search with immediate entry in a file<br>(can be used for any file switch) | BEGIN B<br>BEGIN 3 B3 B METADEX<br>B PASCAL B CHEMABS |
| EXPAND | E | **Expand** Term from Basic Index<br>Expand Term with Prefix Code<br>Expand Term with Suffix Code<br>Expand Related Terms<br>Expand Related Tems over 99<br>Direct Expand of Related Terms<br>Expand after Select | E COMPUTER<br>E AU = EDWARDS<br>E COAL/TI<br>E E6<br>E R1/100<br>E COMPUTER/RT<br>E |
| SELECT | S | **Select** Term<br>Select Term with Prefix Code<br>Select Term with Suffix Code<br>Select from Expand List (single term)<br>Select from Expand List in Range (one set)<br>Select from Expand List in Ranges (separate sets)<br>Select Stem<br>Continuing long selection<br>Select using Contextual Logic:<br>—within a citation<br>—word and prefixed terms in a citation<br>—in the same field<br>—in the same sentence<br>—adjacent terms<br>—fragments of a word (file 2 only) | S COMPUTER<br>S CC = 1234<br>S STEEL? ?/CT<br>S E6<br>S E8,E15<br>S E6:E12<br>S COMPUT? S TAR? ?<br>S? S?15<br><br>S FISH (C) POND? ?<br>S MF = C6H1206 (C) GRAPE<br>S COLD (F) WATER?/TI,CT<br>S SPIN (S) WAVE? ?<br>S REMOTE (W) SENSING<br>S FLUORO (T) BENZENE |
| QUEST INDEX | Q | **Automatic display** of ESA-QUEST files subject coverage<br>Automatic specification of a group of files<br><br>Automatic temporary selection in a group of files (crossfile searching for finding quickly the most suitable file where to develop a search) | Q TOPIC?<br>Q TOPIC AGRICULTURE<br>Q TOPIC PHYSICS, 27<br>Q S DESERT? (C) IRRIGATION |
| COMBINE | C | **Combination of set numbers** | C*1 C1 and 2<br>C (1*2) + (3*4)<br>C1-3/OR C99*2 |
| TYPE<br><br>For viewing data on a sreen only, use DISPLAY e.g. D42/2/28<br>D- means, Display Back | T | **Type** ESA-QUEST Accession Number<br>Type Set Number<br>Type Forward<br>Type Certain item out of Set<br>Type Certain Range out of Set<br>Type in Formats (1,2,3,4,5,6,) | T 74N12345<br>T 6<br>T<br>T 42/2/28<br>T 42/2/28-41<br>T 99/6 T 74N12345/4 |
| DISPLAY SETS | DS | **Display Set History**<br>Display Specific Set in History<br>Display Specific Range in History | DS<br>DS3<br>DS3-10 DS7-99 |
| ZOOM | Z | **Automatic on the spot analysis** of set contents (statistics on indexing terms, titles, authors, corporate sources, codes, molecular formulae etc.)<br>ZOOM on indexing terms<br>in 50 citations of last set<br>in 120 citations of set 4<br>in 20 latest citations of set 4<br>in 70 latest citations of set 4<br>(with word by word analysis)<br>ZOOM on authors<br><br><br>ZOOM on classification codes<br>Online explanations (varies for each database) | <br><br><br><br>ZOOM or Z<br>Z 4 (120)<br>Z 4 (20) LATEST<br>Z 4 (70) LATEST W<br><br>Z AU<br>Z 4 (120) AU<br>Z LATEST AU<br>Z CC Z4 (120) CC<br>Z? |

NB. Refer to ESA-QUEST User Manual for detailed explanations.

IRS/OLS/February 1983

esrin, via galileo galilei, 00044 frascati italy ☎ (06) 94011 - telex 610637 esrin i

**Fig. 4.1e(*i*)**

| COMMAND | Abbreviation | FUNCTION | EXAMPLES |
|---|---|---|---|
| .DELETE | | **Delete** Last Set | .DELETE |
| | | Delete Range of consecutive sets including Last Set | .DELETE 26-31 |
| | | Delete All Sets | .DELETE ALL |
| END/SAVE | | **Search Save** (the computer allocates a Serial number: xxx) | END/SAVE |
| .EXECUTE xxx | .EX xxx | Search Execute (for a direct EXECUTION) | .EX 13BJ |
| .EXECUTE xxx/N | .EX xxx/N | Seach Execute up to line N ( " " ) | .EXECUTE 13BJ/25 |
| .EXECUTE STEPS xxx | .EXS xxx | Seach Execute Steps (for a direct EXECUTION by steps) | .EXS 13BJ |
| .EXECUTE STEPS xxx/N | .EXS xxx/N | Search Execute Steps up to line N ( " " ) | .EXS 13BJ/25 |
| .RECALL .xxx | | Search Recall (optional) | .RECALL 13BJ |
| .RELEASE xxx | | Search Release (preceded by RECALL Search No.) | .RELEASE 13BJ |
| END/SDI | | Search Save for SDI (applicable when the search begins with a BEGIN command) ( for cancellation .RECALL xxx them .RELEASE) | END/SDI |
| PAGE | P | **Page Forward** | P |
| | | Page Back | P- |
| PRINT | PR | **Offline print** of dispayed | PR |
| | | Offline print of set number (max. 50 items in format 2) | PR6 PR99 |
| | | Offline print item specifying a Format and no. references | PR6/4/1-250 |
| | | Offline print of certain Item or Range of items | PR6/4/37 PR6/4/20-45 |
| PRINT- | PR- | Delete last offline print order | PR- PR-; PR- |
| ORDER ISSUE | OI | **Online Ordering** of Original Documents (set/format/item) | OI 8/M/2-3 |
| | | Specifying a given supplier | OI S = UBUTEX 8/H/2-3 |
| | | Using ESA-QUEST accession number | OI 1802833/M |
| | | format: H hardcopy M microfiche | |
| ORDER DELETE | OD | Delete Original Document Order (refer to the order no.) | OD11 |
| ORDER STATUS FULL | OST | Full Description of all the Orders | OSTF OSTFW |
| ORDER TYPE | OT | Description of One Order | OT12/S |
| ORDER SHOW | OS | Description of one order on a screen | OS12/4 |
| LIMIT ALL/XX | LALL/XX | **Limit** All Successive Sets to given parameter/s | LALL/83 LALL/PA LALL/83/PA |
| LIMIT ALL | LALL | Cancel Limit All | LALL |
| LIMIT | L | Limit Set by Timespan | L9/80-83 L9/90-98 |
| | | Limit Set by File Update | L9/UP L9/UP-2 |
| | | Limit Set by Type of Document | L9/PA L9/FILM |
| N.B. CHECK ESA-QUEST MANUAL AND ONLINE TUTORIALS (?LIMIT n) FOR LIMIT POSSIBILITIES ON INDIVIDUAL FILES. | | Limit Set by Type of Term or Code | L9/MAJ L9/NEW |
| | | Limit Set by Subject Area | L9/B L9/A L9/ANI |
| | | Limit Set by Field | L9/CT L9/TI,AB |
| | | N.B. ELEMENTS CAN BE USED TOGETHER, RESPECTING THE FOLLOWING ORDER Lset Number/Time/Type of Document/Type of Code/Field/Language or Subject Area or Term | e.g. L9/80-82/TI L7/82/N/MAJ L7/98/PA/CT* |
| KEEP | K | **Transfer** of references in special set 99: | |
| | | Keep, Single item of given set | K 6/45 |
| | | Keep, Range of items in given set | K 6/45-52 |
| | | Keep ESA-QUEST Accession Number | K 74N12345 |
| END | | **End Search** | END |
| LOGOFF | | Terminate Terminal Sessione | LOGOFF |
| LOGOFF HOLD | | Temporary Interruption (max. 20 min.) | LOGOFF HOLD |
| LOGOFF BUT | | End of Session, start with another password without disconnecting | LOGOFF BUT |
| MESSAGE | M term. no./ | **Message** to a Terminal | M101/HELP ME... |
| .FILE N. (N = file no.) | | **File Switch** (without cancelling sets created before in other files) | .FILE 4 |
| ?XXXXXXX (X = code) | | **Online Consultation of standard Information** (on access, charges, files, search commands, ecc.) | ?GLOSSARY ?LIMIT8 ?ZOOM ?FILES ?SET ?PRIMORD ?QUESTIN |
| ..SET | | **Change standard COMPUTER-USER INTERFACE** | .SET PASSWORD CODE TO ABC |

**Fig. 4.1e(*ii*)**

# QUICK GUIDE To Most Useful System Commands

**HELP-DESK TELEPHONE NUMBERS**

| | |
|---|---|
| UK: in UK | 01-930 5503 |
| from elsewhere | +44-1-930 5503 |
| SWITZERLAND: | 031-659500 |
| GERMANY: | 089-492051 |
| FRANCE: | 6-079 18 00 |

## SAVING AND PURGING

| Commands and Format | Notes | Examples |
|---|---|---|
| **Saving search strategy** | | |
| ..SAVE<br>..SV | DS will prompt for search name | 3_: ..SAVE |
| ..SAVE *abcd* | Temporary save, until end of day<br>User chooses 4-character name | 3_: ..SV DEMO |
| ..SAVE PS (*abcd*) | Permanent save<br>User chooses 4-character name | 3_: ..SAVE PS(LAO1) |
| **Executing saved searches** | | |
| ..EXEC<br>..E | DS will prompt for search name | 1_: .. EXEC |
| ..E *abcd*<br>..EXEC PS (*abcd*) | | 1_: ..EXEC DEMO<br>1_: ..EXEC PS(LAO1) |
| **Automatic Newsletter/SDI** | | |
| ..SDI<br>..SDI *set no. paragraphs*/ID = *abcd*<br>*or* BIBL<br>*or* ALL | DS will prompt for rest of command<br>ID = profile identifier<br>DS executes SDI profiles automatically each time file is updated<br>Results mailed to user | 3_: .. SDI<br>3_: ..SDI 3 SO,AB/ID = TV GAMES; DR SMITH |
| **Cancelling commands** | | |
| ..PURGE<br>..PG | DS purges previous search statement | 7_: ..PURGE |
| ..PURGE *n* | DS purges selected statement(s) or range of statements or ALL<br>Useful to delete selected statements before saving strategy | 7_: ..PURGE 1<br>7_: ..PURGE 1,3<br>7_: ..PG 1-4<br>7_: ..PURGE ALL |
| ..PURGE PS(*abcd*)<br>..PURGE *Qxxxx* | DS purges permanent save name<br>DS purges offline print request number (at any time during same day) or SDI profile number (at any time) | 7_: ..PURGE PS(LAO1)<br>7_: ..PURGE Q0007 |
| **Displaying stored commands** | | |
| ..DISPLAY<br>..D | Same options as for ..PURGE | 7_: ..DISPLAY<br>7_: ..DISPLAY ALL |

## LIMITING

| Commands and Format | Notes | Examples |
|---|---|---|
| ..LIMIT/*set no. xx eq zz*<br>..L | DS will prompt for rest of command<br>XX = label of paragraph or field to be limited<br>eq = operator to be imposed<br>'EQ' or '=' (equals)<br>'GT' or '>' (greater than)<br>'LT' or '<' (less than)<br>zz = value | 5_: ..LIMIT<br>5_: ..LIMIT/4 LG = EN<br>5_: ..L/3 YR = 80 |

Please check individual database guides to see which fields may be LIMITed. Also note that ..LIMIT cannot be used for date limiting in certain databases (see below)

### Data Searching/Limiting

To restrict a search by date, use one of the following options

5_: 83.YR.
or 5_: YR=83

Exceptions:

| | |
|---|---|
| CEAB | ..LIMIT |
| HARF | ..LIMIT YR=83 83$2.AN. |
| HBRO | ..LIMIT |
| IEAB | 1983.SO. |
| NCMH | LIMIT |
| NTIS | 83.YR. Limit to journal year e.g. ..L/1 JNYR=83 |
| PREM | JULY.SO. |
| ULIT/UFOR | Limit to ED paragraph (6 characters) e.g. ..L/1 ED=83$4 |

**Fig. 4.1f(*i*)**

## DATA-STAR BASIC COMMANDS

| Commands and Format | Notes | Examples |
|---|---|---|
| **Selecting Files** | | |
| **4 character file label** | Find file label in Data-Star Newsletter | **ENTER DATABASE NAME: PTSP** |
| | or from NEWS database, | **ENTER DATABASE NAME: NEWS** |
| | decide which files have relevant data by using CROS database | **ENTER DATABASE NAME: CROS** |
| **Entering Single Terms** | Enter directly | **1_: TELEVISION** |
| **Controlled Vocabulary (Descriptors)** | Controlled Vocabulary phrases are bound by hyphens: other punctuation is stripped | **1_: LUNG-NEOPLASMS** |
| **Truncation** | | |
| **$** | Unlimited | **1_: TELEVISION$** |
| | If "more than 100 terms", re-enter a restricted truncation or continue by entering | |
| | 2_: ↙ | |
| **$n** | Restricted to n further characters | **1_: TELEVISION$1** |
| # | MEDLINE: # retrieves pre-exploded tree numbers | |
| | eg 1_:D586 # | |
| | EXCERPTA MEDICA: # retrieves pre-exploded classification codes | |
| **Root** | | |
| | ROOT Displays all terms from vocabulary file which begin | **1_: ROOT SMITH-A** |
| | with specified stem | **1_: ROOT LIFE** |
| **Operator** | (in reverse order of processing priority) | |
| **OR** | at least one term must be present | **1_: ENERGY OR POWER** |
| **NOT** | 1st term must be present, 2nd term must not be present | **1_: ENERGY NOT SOLAR** |
| **AND** | terms in same record | **1_: SOLAR AND ENERGY** |
| **SAME** | terms in same record & same paragraph | **1_: SOLAR SAME ENERGY** |
| **WITH** | terms in same record, same paragraph & same sentence | **1_: SOLAR WITH ENERGY** |
| **ADJ** | terms in same record, same paragraph, same sentence & adjacent in specified order | **1_: SOLAR ADJ ENERGY** |
| () | use parentheses to specify 1st processing priority | **1_: SOLAR ADJ (ENERGY OR POWER)** |
| **Paragraph Qualification** | | |
| *.pa.* | pa = 2 digit paragraph label; | **1_: TELEVISION.DE.** |
| | (print out a record online to see available paragraphs) | **1_: TELEVISION$1.TI,DE.** |
| | paragraph qualification can be made 'after-the-fact' using set number | **1_: LUNG-NEOPLASMS** |
| | | **2_: 1.MJ.** |
| **Printing Online** | | |
| **..PRINT** | Data-Star assumes prints to be from last set | **3_: ..PRINT** |
| **..P** | and will prompt for rest of command | |
| **..PRINT**/*paragraphs//* | 1st record from last set | **3_: ..PRINT/ALL//** |
| *or* **BIBL** | | |
| *or* **ALL** | | |
| **..PRINT set no. paragraphs/numbers** | specified documents from specified search statement | **3_: ..PRINT 1 TI/1,3** |
| **or BIBL or range** | | **3_: ..PRINT 1 TI,AB/1 – 5** |
| **or ALL or ALL** | | **3_: ..PRINT 1 BIBL/ALL** |
| **To interrupt printing press 'BREAK'** | | **_: BREAK** |
| **To continue with next record press "CARRIAGE RETURN"** | | **_: ↙** |
| **To temporarily halt printing press "CONTROL P"** | | |
| **TO RETURN TO SEARCHING, TYPE ..S** | | **_: ..S** |
| **Changing files** | | |
| **..CHANGE** | Data-Star will ask for file label, | **3_: ..CHANGE** |
| **..C** | or enter file label with 'change' | **3_: ..C/PTSP** |
| **Disconnect/logging off** | | |
| **..OFF** | | **5_: ..OFF** |
| **..O** | | |
| **..OFF CONT** | to retain sets until end of the day | **5_: ..OFF CONT** |

## OFFLINE PRINTING

| Commands and Format | Notes | Examples |
|---|---|---|
| **Offline Prints** | | |
| **..PRINTOFF** | Data-Star assumes prints to be from last set | **5_: ..PRINTOFF** |
| **..PO** | and prompts for rest of command | |
| **..PRINTOFF set no. paragraphs/number(s)/ID = abcd** | ID is identifying information to aid searcher | **5_: ..PRINTOFF 4 TI,AB/ALL/** |
| **or BIBL or range** | (not mailing address) | **ID = DEMO; DR JONES** |
| **or ALL or ALL** | Options: | |
| | Printout may be sorted by paragraph labels: SORT = xx | **5_: PO 4 ALL/ALL/** |
| | a record of search strategy may be requested: SS = Y | **ID = DEMO/SORT = AU/SS = Y/NP = Y** |
| | no split documents on page: NP = Y | |
| | one document only on page: NP = S | |
| **Merging Offline Print Output** | | |
| **..MERGE** *Qyyyy,Qzzzz* | to merge up to 6 sets of offline prints | **6_: ..MERGE Q0003,Q0004** |
| **..M** | (ID and SORT must be the same on each set) | |
| | Qyyyy, Qzzzz = offline print numbers | |

Note: **ALTERNATE COLOUR** indicates command abbreviations.

**Fig. 4.1f(*ii*)**

# User Cue Card

## Basic Commands

This cue card describes the use of the main commands on InfoLine. Full explanation of the use of each command can be found in the InfoLine User Guide System Reference Manual. If you have difficulty, please consult the System Reference Manual and the Database Reference Manual for the file you are using. Explanation is available online if you enter ? before the topic you want explained eg ?COMMANDS.

The InfoLine Help Desk will always be happy to help you

| | | |
|---|---|---|
| London | Telephone | 01 377 4957 |
| | Telex | 8814614 |
| Washington | Telephone | (800) 336 7575 |
| | | (703) 442 0900 |

## Commands

**FILE**

To select an initial file after LOGIN and to change file during a search session. Set numbering restarts after a FILE command.

**FILE PATS**

**SELECT**,SE,S,

Synonyms:
COMBINE,C
FIND,FD

To select a single term from the index of your choice

**SELECT TT=PLASTIC**
**S PACKAGE**
**S TT=PACK***
**S GALVANI* ING**

The InfoLine truncation symbol is *

To combine terms previously selected using normal Boolean operators

**S 1 A 2**

To enter a complete search statement with both index terms and Boolean operators

**S HORSE* A SADDLE**

**S AU=SMITH J AND PY=82**

**EXPAND,**EX,E

Synonyms:
NEIGHBOR,NBR

To look at a given section of the index, containing a number of alphabetically or numerically arranged terms

**EXPAND AIRCRAFT**

To look at a given section of the index, beginning with a common word stem

**E TT=POLYMER***

To look at a section of the index beginning with a given term

**E TT=POLYMER****

To select terms from an EXPAND, use the alphabetic identifiers down the left hand margin

**A,D,E-H**

**Fig. 4.1g(*i*)**

| | |
|---|---|
| **DISPLAY**,DI,D<br><br>Synonyms:<br>TYPE,TY,T | To display or print online all or selected parts of records retrieved during a search session in pre-defined or user-defined formats. Enter DISPLAY, set number, format and records required<br><br>**DISPLAY 5 F1/1-10**<br>**D 6 TI,IV,PD/ALL** |
| **PRINT**,PR | To request off-line prints of all or selected parts of records retrieved during a search session in a predetermined or user defined format. Enter PRINT, set number, format and records required<br><br>**PRINT 1 F4/ALL**<br>**PR 5 IV,AS/1-50** |
| **CPRINT**,CP | To cancel off-line prints requested during your search session but which have not yet been MAILed<br><br>**CPRINT** |
| **MAIL** | To despatch off-line prints requested by the PRINT command to a stored address or to a user specified address. Changing files or logging out causes an automatic MAIL command<br><br>**MAIL**<br>**MAIL J.JONES:ANYTOWN:ANYWHERE:THE WORLD** |
| **SCAN**,SC<br>Synonyms:<br>STRS | To execute a sequential search on a previously created document set. Enter SCAN, set number, field, and string to be scanned for<br><br>**SCAN 5 TI GREAT EXPECTATIONS** |
| **SORT** | To sort records from a retrieved document set in a specified order. Enter SORT, set number and field to be sorted<br><br>**SORT SET 10 AU** |
| **LOGOUT**<br>Synonyms:<br>LOGOFF/STOP/<br>END | To end session, produce an end of session summary accounting report and disconnect terminal |
| **LOGIC OPERATIONS** | The following logic operators, or their abbreviated forms, may be used to combine terms<br>**AND** **A**    **WITH** **W**<br>**AND NOT** **A N**    **WITHIN** **WN**<br>**OR** **O** |

A Space should always be inserted on either side of a logic operator.
WITH and WITHIN are both database dependent and require proximity and contextual relationships between terms. Please check database manuals before using these operators.

**Fig. 4.1g(*ii*)**

Compound terms have to be linked together and in ESA/IRS this is done by putting '(w)' between the two terms. In this case the search will produce only items where the term 'meat' immediately follows the term 'cooked'. Although we do not want to do so in this instance, we could specify that the two terms occur within a given number of words of each other, or within the same sentence or section ('field') of a record. The other thing to note is the question mark I have put after 'meat'. This question mark serves to truncate the term at the point indicated and takes that part of the word as the basis for any longer terms beginning with the specified stem—in this case the system would find both 'cooked meat' and 'cooked meats'. Obviously this can be very useful and allow you to search on a number of terms beginning with the same stem, all on a single truncated entry; but you must be careful in using it—experience will show you that unless you take great care, truncation can produce a host of unwanted irrelevant terms and entries ('noise' in computer jargon). Sometimes it is better just to combine singular and plural forms of your term with 'or' (as in SAQ 4.1a where I suggest the entry 'soil or soils'); but even here, stop and think to make sure that the singular and plural forms do have identical meanings (when you think about it, 'cooked meat' and 'cooked meats' have slightly different interpretations!). Anyway, we have our Set 8 of 710 references and next I specify two more sets for 'ham' and 'corned beef'. And now here is the complication I mentioned earlier. You might expect the term 'cooked meats' to include all references containing items dealing with 'ham' and 'corned beef'. But the numbers of references in our sets show that this cannot be so: the way we have done the search, excluding 'ham' and 'corned beef', is perfectly correct; but if we want to make sure that we have included all kinds of cooked meats we are going to have to specify them. Our entry might then have to look something like 'cooked meats or brawn or pork luncheon meat'—and any other cooked meats we care to think of. This is another point for you to watch when searching, but note also that in completing the search as we began it we still produce a reasonably useful set of references at the end. Anyway, having made a set (no. 11 at line 47) of 'ham or corned beef', at line 48 I specify that these terms are to be excluded from Set 8, and so create Set 12 of 664 references. My final stage is then to combine this with the previously constructed Set 7 and produce, at line 51, Set 13 of 19 references.

Now I want to see what these references are. The alternative methods of seeing the references are either to have them printed at my terminal (so using up valuable on-line time), or to have them printed off-line at the host computer and sent by post. This latter option is the cheapest way of having references delivered, but it does mean a delay of a few days before they are received. In this example I shall assume that there is some urgency and so I want to look at the references immediately. I can choose (within certain defined parameters) the format in which I want the references presented; initially I look at the headings and titles to see how relevant the references seem: this format is the quickest and shortest available on ESA/IRS. After the prompt at line 52 I have entered 't' (for 'type') '13' (the set number), '/6' (the code for the format I want), '/1–19' (which of the references I want—here all 19 of them). Then in lines 53–119 the computer prints out in groups of five all the references I have asked for. From these I can get an idea of which seem most relevant (so we can get some opportunity to browse—sometimes this stage reveals a list of irrelevant items so that we know we have to re-think our search). Having looked at them all I decide I want to see no. 6 in more detail. So at line 120 I repeat 't13', but this time with format '/4' and only reference no. '/6'. The format 4 specifies that the entire record is to be printed, and this is done on lines 122–153. Now we see a reasonably familiar form of abstract entry with title, authors, source citation, and abstract. Also printed are the additional indexing terms allocated by the data base system: these can be useful guides if we have to re-direct or completely re-cast a search. At this point I decide to end my search and so, at line 154, I enter the command 'logoff' which terminates my session on the host (if I wanted to move into another data base I would have repeated the 'b' command with the code number of the new data base I wanted).

That search took almost twelve minutes, partly because I entered terms individually and combined them step-by-step, and partly because I had references printed on-line at a very busy time of day. The same search could have been reduced to just a minute or two (especially if I decided not to have any references printed on-line). But the important point to remember is that the computer searched over 250,000 abstract references published since 1969; how long do you suppose it would have taken you to search all that manually with the same degree of precision?

Let's reinforce the point by having you try a similar exercise.

**SAQ 4.1b**

Below is a search already defined for you. Write out how you would enter this search on your terminal. Use the commands 's' for 'search' and 'c' for 'combine', and use the logical operators 'and', or 'not' as you think necessary. Join complex multi-word terms with the '(w)' operator if necessary, and use the '?' for truncating terms if you think you need to. Set your search out line-by-line and number each set you create. Stop at the final search command you would enter (ie don't bother asking for prints).

Your search is:

HPLC of fatty acids, excluding oleic acid.

**SAQ 4.1b**

We have had time to look at only a couple of searches and only a few practical points have cropped up. But I think you should now have a good idea of the potential power of on-line searching. Now we are going to look at some specific data bases, and I hope you will then get further practical insights into the nature of on-line searching.

**Objectives**

When you have worked through this Section you should be able to:

- appreciate the advantages (and disadvantages) of on-line over manual searching;
- understand the overall system of commercial on-line search networks;
- recognise the major commercial on-line host services likely to be of use to you;
- compile a logical search strategy for any on-line search you might wish to make;
- know how to input your search strategy into a typical major host computer.

## 4.2. ON-LINE BIBLIOGRAPHIC DATA BASES

### 4.2.1. Introduction

Bibliographic data bases—direct on-line access to abstracting or indexing services—were the first publicly available systems put on-line by the major host services; and they still form the most numerous type of publicly accessible data base. Their structure is much as you would expect now that you have worked through Section 3.2: at the end of your on-line search you find a bibliographic citation and frequently, but not always, an abstract of the item. You still have to find a copy of the document if you want to read the full text. In fact, many hosts now enable you to order off-prints or photocopies of the full text of items you have retrieved by entering a simple on-line command: but this is an expensive, if quick and convenient way of acquiring copies. Having worked through Section 3.1 of this Unit you should be able to find cheaper means of obtaining the documents you want.

I should mention here one major difference between searching on-line and manually which I have not yet mentioned: when you search a printed index to an abstracting journal you are generally doing so by using terms allocated to each item by the indexer in accordance with the rules and procedures adopted for the particular service. You can also search by these terms on-line (provided, of course, that you know what they are) by using special commands. But you can also search by 'free text' terms, as I have mentioned earlier: that is, searching by normally occurring terms in titles and abstracts. There are advantages and disadvantages in both kinds of searching, but for many bibliographic data bases, experience will, I think, show you that combination of both kinds of searching is the most advantageous method. When looking for any particular substance in *Chemical Abstracts* I usually search both its common name (or names) and its CA Registry Number to make sure I locate all the relevant references.

There is now an enormous range of bibliographic data bases available, and for the rest of this Section I shall just show them to you (with a couple more exercises and examples for your further experience). The scene is again dominated by *Chemical Abstracts* but once more, I do advise you also to bear the other data bases in mind. It can be very advantageous to search in a data base specialising in the area in which you are working; some peripheral subjects in, for example, certain biomedical fields are not covered as thoroughly by *Chemical Abstracts* as by more specialised services. Also remember that in on-line searching you pay only for what you use: you don't have to maintain subscriptions to be able to search in *Biological Abstracts*, or in the major medical data base *Medline*, whenever you find it necessary.

It is still too soon to see a very clear pattern of data base provision emerging, but as I indicated earlier, some hosts are beginning to specialise in certain subject areas. Data bases are being put up on-line to test the market, and some at least are likely to disappear, in time, due to lack of use. As in the provision of journals, however, the Royal Society of Chemistry is taking a lead in trying to supply a co-ordinated range of data bases: in addition to sponsoring *Chemical Abstracts* in the UK and parts of Europe, the Society's

UK Chemical Information Service is also responsible for *Chemical Business News* (from 1984), *Chemical Engineering Abstracts* (from 1970), *Chemical Hazards in Industry* (from 1984), *Current Biotechnology Abstracts* (from 1983), *Laboratory Hazards Bulletin* (from 1981), *and Mass Spectrometry Bulletin* (from 1966). The *Chemical Abstracts* Service also offers a fairly systematic coverage of chemical information; so, now let's look at various bibliographic data bases in a little more detail.

### 4.2.2. Chemical Abstracts

In the on-line field, as in printed services, chemical information tends to be dominated by *Chemical Abstracts*. But if I emphasise this service in this Section, as I feel I must, do remember my earlier warning about it not being the only service of value to analytical chemists. In on-line searching especially you have access to a wide range of services, and often a more specialised data base might cover your subject more thoroughly, or simply be more convenient or more immediately relevant to your information needs. When I refer to data bases I shall introduce them to you in a standardised format (the equivalent to a bibliographic citation of a printed service); you will need to note that on-line data bases often have slightly (and sometimes considerably) different titles from their printed equivalents. So, here goes with our first data base description:

> CA SEARCH/CAS ONLINE. Chemical Abstracts Service, American Chemical Society. The data base corresponds to the printed *Chemical Abstracts* (condensed version without abstracts for CA SEARCH, full version including abstracts for CAS ONLINE. Over 6,000,000 records on file. 1967 to date, up-dated fortnightly (400,000 new records per year).

As you can see, on-line *Chemical Abstracts* is available in two forms. The CA SEARCH version does not include the abstracts; this version was one of the first data bases to be made publicly available on-line and can now be searched on many hosts (including Dialog, DATA STAR, ESA/IRS, BRS, and Orbit). One thing you must watch out for when choosing a host on which to search a particular data base (where you have such a choice) is that there can be differences in the parts of the record you can search: on one host you might be able to search by document type (journal, report, patent, etc), but not on another host, or by journal title on one host but not on another. The second version of on-line *Chemical Abstracts* is CAS ONLINE. This is sponsored in the UK and parts of Europe by the Royal Society of Chemistry's UK Chemical Information Service, and is available in Europe on the STN host (connected with INKA) in Karlsruhe, West Germany. In this version you can retrieve abstracts on-line in addition to the bibliographic references; and you can search not only by subject terms in the way I have indicated so far, but also by molecular formulae and structure diagrams. This structural searching is highly sophisticated, but for our purposes I simply tell you that it exists: it is a fascinating subject, but rather advanced of what I am teaching you at the moment. If you do want to follow it up I would refer you initially to an article on the subject (P. Rhodes Chemical structures on-line. *Chemistry in Britain*, v.21, Jan 1985, 53–54, 58).

You know quite a bit about *Chemical Abstracts* and how to use it by now, don't you? Enough to know, certainly, that it is a very comprehensively indexed publication, both for general subjects and for chemical substances. Well, in addition to searching on free terms you can search on any CA index term as well, including CA Registry Numbers of substances (which are far easier than IUPAC names!). I have included for illustration (Fig. 4.2a) Dialog's brief description of its CA SEARCH files: look through this and you will get a more detailed picture of these files and how you can search them.

Another difference to note between hosts is that most of them split the file into segments by dates of issue, to make an enormous file more manageable, but ESA/IRS offers only a single file from 1967 to date (that is the one we searched in SAQ 4.1b. The CAS ONLINE version on STN also comprises a single file from 1967 to date. Now, before I round off this Section, let's just take a closer look at CAS ONLINE in an exercise for you.

**SAQ 4.2a**

Here are two searches for you to enter on-line. The STN commands for searching CAS ONLINE are the same as those used earlier for ESA/IRS. The first search is to be on the gas chromatographic analysis of myristic acid. You have already checked the *Chemical Abstracts Index Guide* (or some other directory of chemical substances which gives you the same kind of information) and found that the preferred name for myristic acid is tetradecanoic acid, with a CA Registry No. 544-63-8. Write out your search step-by-step as usual, with the relevant commands and operators. By the way, one difference you will note in STN is that your set numbers are prefixed 'L' and that you always use the 's' command (not 'c') to combine as well as to search (eg 's L1 and L2')

The second search seems much simpler even than the first, but what I want you to do this time is enter the whole search as a single entry, in algebraic form—eg 's (x or y) and (z or v) not w'. Use the usual commands and operators as necessary in your string. The subject to try is much the same as one we have already performed manually: analysis of the atmosphere of Mars and of Venus.

**SAQ 4.2a**

*Chemical Abstracts* is a detailed and complex service. You can check index terms on-line but it is better (and a lot cheaper) to do such checking before you approach your terminal (remember all my advice about planning your searches carefully). There are, however, two files to help to support the use of CA SEARCH and I just mention them to you here.

(*a*) CA CHEMICAL SUBSTANCE FILES. Several different hosts (Dialog, DATA STAR, Orbit, etc) offer dictionary files for chemical substances, giving their CA Registry Numbers, molecular formulae, CA substance index names, synonyms, ring data and some other chemical substance information. These files have different names in different hosts.

(*b*) CASSI. Chemical Abstracts Service, American Chemical Society. Corresponds to the printed *CAS Source Index*. 50,000 records on file. 1900 to date, up-dated quarterly.

This gives bibliographic data, library holdings, and information on publishers and suppliers for journals, serials, monographs, and conference proceedings which are customarily cited by their abbreviated titles. It has international coverage of the whole field of chemistry.

And finally, for commercial information there is a separate CA file:

> CHEMICAL INDUSTRY NOTES. Chemical Abstracts Service, American chemical Society, and Predicasts. Corresponds to a printed service with the same title. Over 350,000 records. 1974 to date, up-dated twice-weekly (2,000 new records per up-date).

399
308, 309, 320, 310, 311
# CA SEARCH
DIALOG® INFORMATION RETRIEVAL SERVICE

## FILE DESCRIPTION

CA SEARCH combines the condensed version of CHEMICAL ABSTRACTS with controlled vocabulary CA General Subject Index Headings and CAS Registry Numbers, each with its modifying phrase. Related general subject terminology from the ***CA Index Guide*** is also included. Chemical substances are represented by CAS Registry Numbers. Corresponding substance information may be searched in the DIALOG chemical substance files such as CHEMNAME™ (File 301).

All records from the 8th Collective Index Period forward are contained in File 399; Files 308-311 and 320 contain records from the individual Collective Index Periods as indicated in the table below.

## SUBJECT COVERAGE

The literature of chemistry and its applications is divided into the following principal areas:

- Applied Chemistry
- Biochemistry and Biology
- Chemical Engineering
- Classes of Substances
- Macromolecular Chemistry
- Organic and Inorganic Chemistry
- Physical and Analytical Chemistry
- Properties and Reactions

## SOURCES

The following sources are included in CA SEARCH: journal articles, patents, reviews, technical reports, monographs, conference and symposium proceedings, dissertations, and books.

## DIALOG FILE DATA

| File | Collective Index Period | Inclusive Dates | Accession Numbers | Update Frequency | File Size |
|---|---|---|---|---|---|
| 399 | 8th- | 1967- | 66000001- | Biweekly | 7,061,000* |
| 308 | 8th | 1967-1971 | 66000001-75157995 | Closed File | 1,314,665 |
| 309 | 9th | 1972-1976 | 76000001-85201798 | Closed File | 1,772,194 |
| 320 | 10th | 1977-1979 | 86000001-91222077 | Closed File | 1,275,366 |
| 310 | 10th | 1980-1981 | 92000001-95231484 | Closed File | 926,314 |
| 311 | 11th | 1982- | 96000001- | Biweekly (approximately 17,000 records per update) | 1,772,462* |

*As of November 1985, Volume 103, Issue 20.

## ORIGIN

CA SEARCH is produced by Chemical Abstracts Service. Questions concerning file content should be directed to:

Manager, User Education
Chemical Abstracts Service
P.O. Box 3012
Columbus, OH 43210

or

DIALOG Customer Services
DIALOG Information Services, Inc.
3460 Hillview Avenue
Palo Alto, CA 94304
Telephone: 800/3-DIALOG (334-2564)
Telex: 334499 (DIALOG)



(Revised January 1986) 311-1

**Fig. 4.2a(*i*)**

**FILES 308, 309, 320, 310, 311, 399**

# CA SEARCH
# DIALOG FILES 308, 309, 320, 310, 311, 399

## PATENT RECORD

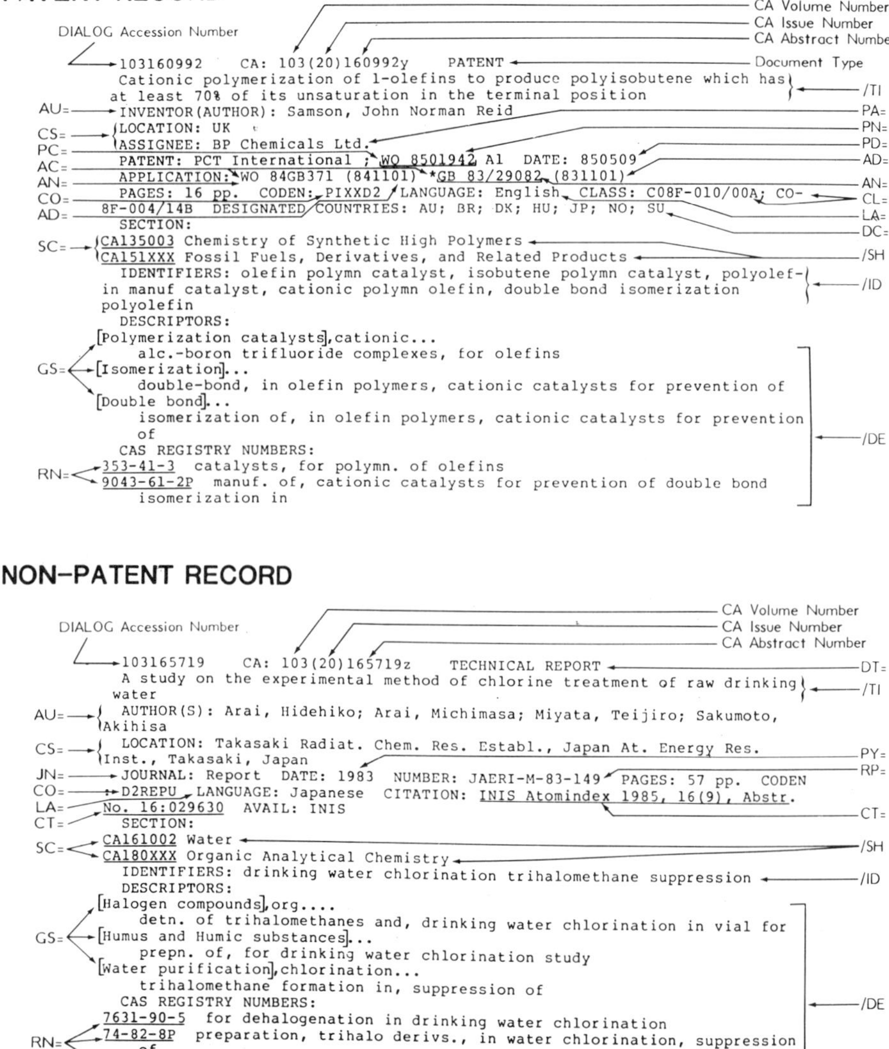

103160992 CA: 103(20)160992y PATENT
Cationic polymerization of 1-olefins to produce polyisobutene which has at least 70% of its unsaturation in the terminal position
INVENTOR(AUTHOR): Samson, John Norman Reid
LOCATION: UK
ASSIGNEE: BP Chemicals Ltd.
PATENT: PCT International ; WO 8501942 A1 DATE: 850509
APPLICATION: WO 84GB371 (841101) *GB 83/29082 (831101)
PAGES: 16 pp. CODEN: PIXXD2 LANGUAGE: English CLASS: C08F-010/00A; CO-8F-004/14B DESIGNATED COUNTRIES: AU; BR; DK; HU; JP; NO; SU
SECTION:
CA135003 Chemistry of Synthetic High Polymers
CA151XXX Fossil Fuels, Derivatives, and Related Products
IDENTIFIERS: olefin polymn catalyst, isobutene polymn catalyst, polyolefin manuf catalyst, cationic polymn olefin, double bond isomerization polyolefin
DESCRIPTORS:
[Polymerization catalysts],cationic...
alc.-boron trifluoride complexes, for olefins
[Isomerization]...
double-bond, in olefin polymers, cationic catalysts for prevention of
[Double bond]...
isomerization of, in olefin polymers, cationic catalysts for prevention of
CAS REGISTRY NUMBERS:
353-41-3 catalysts, for polymn. of olefins
9043-61-2P manuf. of, cationic catalysts for prevention of double bond isomerization in



## NON-PATENT RECORD

103165719 CA: 103(20)165719z TECHNICAL REPORT
A study on the experimental method of chlorine treatment of raw drinking water
AUTHOR(S): Arai, Hidehiko; Arai, Michimasa; Miyata, Teijiro; Sakumoto, Akihisa
LOCATION: Takasaki Radiat. Chem. Res. Establ., Japan At. Energy Res. Inst., Takasaki, Japan
JOURNAL: Report DATE: 1983 NUMBER: JAERI-M-83-149 PAGES: 57 pp. CODEN D2REPU LANGUAGE: Japanese CITATION: INIS Atomindex 1985, 16(9), Abstr. No. 16:029630 AVAIL: INIS
SECTION:
CA161002 Water
CA180XXX Organic Analytical Chemistry
IDENTIFIERS: drinking water chlorination trihalomethane suppression
DESCRIPTORS:
[Halogen compounds],org....
detn. of trihalomethanes and, drinking water chlorination in vial for
[Humus and Humic substances]...
prepn. of, for drinking water chlorination study
[Water purification],chlorination...
trihalomethane formation in, suppression of
CAS REGISTRY NUMBERS:
7631-90-5 for dehalogenation in drinking water chlorination
74-82-8P preparation, trihalo derivs., in water chlorination, suppression of
7697-37-2 uses and miscellaneous, for suppression of hydrolysis of halogenated orgs. in water chlorination



311-2 (Revised January 1986)

**Fig. 4.2a(*ii*)**

**FILES 308, 309, 320, 310, 311, 399**

# CA SEARCH
# DIALOG FILES 308, 309, 320, 310, 311, 399

## SEARCH OPTIONS

### BASIC INDEX[1]

| SUFFIX+ | FIELD NAME | INDEXING | SELECT EXAMPLES |
|---|---|---|---|
| /DE | Descriptor[2] | Word & Phrase[3] | S DOUBLE(W)BOND/DE<br>S HALOGEN COMPOUNDS,ORG?/DE<br>S RN=353-41-3(S)CATALYSTS/DE |
| /ID | Identifier[4] | Word | S DRINKING(W)WATER/ID |
| /SH | CA Section Title | Word & Phrase | S HIGH(W)POLYMERS/SH<br>S ORGANIC ANALYTICAL?/SH |
| /TI | Title[5] | Word | S DRINKING(W)WATER(F)TREATMENT/TI |

+If no suffix is specified **all** Basic Index fields are searched.
[1]Any term in the Basic Index may be limited to a full term using /FF, e.g., S OLEFIN/FF.
[2]Also /DF.
[3]Phrase indexing is for the General Subject Index Headings only.
[4]Also /IF.
[5]Also /TF.

### ADDITIONAL INDEXES

| PREFIX | FIELD NAME | INDEXING | SELECT EXAMPLES |
|---|---|---|---|
| AC= | Application Country | Phrase | S AC=GB |
| AD= | Application Date | Phrase | S AD=841101 |
| AN= | Application Number[6] | Phrase | S AN=WO 84GB371<br>S AN='GB 83/29082' |
| AU= | Author or Inventor | Phrase | S AU=ARAI, H? |
| CL= | Patent Classification Number | Phrase | S CL='C08F-010/00A' |
| CO= | CODEN | Phrase | S CO=PIXXD2 |
| CS= | Corporate Source | Word | S CS=(ENERGY(W)RES?) |
| CT= | Citation[7] | Word | S CT=(INIS(W)ATOMINDEX) |
| DC= | Designated Country[7] | Phrase | S DC=BR |
| DR= | Designated Region[7] | Phrase | S DR=CH |
| DT= | Document Type[8] | Phrase | S DT=TECHNICAL REPORT |
| EC= | Elected Country[7] | Phrase | S EC=KR |
| ER= | Elected Region[7] | Phrase | S ER=BE |
| GS= | General Subject Index Heading | Phrase | S GS=HALOGEN COMPOUNDS |
| JA= | Journal Announcement | Phrase | S JA=CA10319 |
| JN= | Journal Name | Phrase | S JN=CHEM. BER. |
| LA= | Language | Phrase | S LA=JAPANESE |
| PA= | Patent Assignee | Word & Phrase | S PA=(BP(W)CHEMICALS)<br>S PA=BP CHEMICALS LTD. |
| PC= | Patent Country | Phrase | S PC=WO |
| PD= | Patent Date | Phrase | S PD=850509 |
| PN= | Patent Number | Phrase | S PN=WO 8501942 |
| PU= | Publisher | Phrase | S PU=MCGRAW-HILL |
| PY= | Publication Year | Phrase | S PY=1985 |
| RN= | CAS Registry Number | Phrase | S RN=9043-61-2P |
| RP= | Report Number | Phrase | S RP='AAEC/E579' |
| SC= | CA Section Code | Phrase | S SC=CA135003 |
| SN= | International Standard Serial Number (ISSN)[9] | Phrase | S SN=0013-466X |
| UD= | Update[10] | Phrase | S UD=10320 |

[6]From 1972 forward.
[7]Searchable from January 1982 forward.
[8]The following four entries may be SELECTed using DT=: BOOK, CONFERENCE PROCEEDING, DISSERTATION, TECHNICAL REPORT. To limit to patents use /PAT.
[9]From January 1978 forward.
[10]Files 311 and 399 only.

(Revised January 1986) 311-3

**Fig. 4.2a(*iii*)**

**FILES 308, 309, 320, 310, 311, 399**

# CA SEARCH
# DIALOG FILES 308, 309, 320, 310, 311, 399

## LIMITING

Sets or terms may be limited by Basic Index suffixes, i.e., /DE, /DF, /ID, /IF, /SH, /TF, /TI, (e.g., S S7/TI), as well as by the features listed below:

| SUFFIX | FIELD NAME | EXAMPLES |
|---|---|---|
| NONE | Publication Year | S S5/1985 |
| NONE | DIALOG Accession Number | S S2/102003673-103999999 |
| /ENG | English Language | S S8/ENG |
| /NONENG | Non-English Language | S S3/NONENG |
| /PAT | Patent Records | S S9/PAT |
| /NPT | Non-Patent Records | S S9/NPT |

## SORTING

| SORTABLE FIELDS | EXAMPLES |
|---|---|
| Online (SORT) and offline (PRINT).<br>AU, CO, CS, PY. | SORT 6/ALL/AU<br>PRINT 6/5/1-59/PY,D |

## MAPPING

| MAP FIELDS | EXAMPLES |
|---|---|
| AN, CL, PA, PN, RN. | MAPRN TEMP STEPS S5 |

## FORMAT OPTIONS

| NUMBER | RECORD CONTENT |
|---|---|
| Format 1 | DIALOG Accession Number (CA Volume and Abstract Number) |
| Format 2 | Bibliographic Citation and Keyword Phrase(s) |
| Format 3 | Bibliographic Citation |
| Format 4 | Full Record With Tagged Fields |
| Format 5 | Full Record |
| Format 6 | Title and Journal Reference |
| Format 8* | Title, Keyword Phrase(s) and Descriptors (no accession number) |
| Format 9* | Title Only (no accession number) |

*No online TYPE or DISPLAY charge.

## DIRECT RECORD ACCESS

| FIELD NAME | EXAMPLES | | |
|---|---|---|---|
| DIALOG Accession Number | TYPE 103007837/8 | DISPLAY 103007837/5 | PRINT 103007837/6 |

311-4 (Revised January 1986)

**Fig. 4.2a(*iv*)**

This service extracts articles from business journals world-wide, covering the chemical process industry, including the pharmaceutical, petroleum, paper and pulp, agricultural, and food industries.

### 4.2.3. Other Principal Chemical Bibliographic Data Bases

The *Chemical Abstracts* files are of obvious importance, but they are by no means the only ones likely to be of use to you. Here are brief introductions to another six data bases likely to be of importance to you (with practical exercises for you on a couple of them).

> MASS SPECTROMETRY BULLETIN. Mass Spectrometry Data Centre, Royal Society of Chemistry. This data base corresponds to a printed version with the same title. Over 120,000 records on file, 1966 to date, up-dated monthly (900 new records per month).

This service indexes records from over 400 primary journals, abstracting journals, report lists, book, and conference proceedings, covering literature on mass spectrometry. It includes: instrument design and techniques, isotopic analysis, chemical analysis, organic chemistry, atomic and molecular processes, surface phenomena and solid state studies, thermodynamics, and reaction kinetics.

> INTERNATIONAL PHARMACEUTICAL ABSTRACTS. American Society of Hospital Pharmacists. Corresponds to a printed version with the same title. 90,000 records. 1970 to date, up-dated monthly (1,200 new records per month).

This indexes the primary pharmacy and drug-related literature, scanning over 600 journals world-wide. It covers all phases of the development and use of drugs, and of pharmacy practice and education (and of course, includes analysis). At this point, let's just have a bit more search practice.

**SAQ 4.2b**

> This time we will search on Dialog, by using just the same commands and operators we are used to with ESA/IRS. Write out a search to locate references on the analysis of Isoniazid in tablets.

**SAQ 4.2b**

Now, on with more data bases.

> *FSTA*. International Food Information Service. Corresponds to the printed *Food Science and Technology Abstracts.* 250,000 references. 1969 to date, up-dated monthly (1,500 references per month).

This covers the entire field of food science, including microbiology, food hygiene, etc. It scans 1,500 journals world-wide, as well as patents, books, reports, and standards. We have already looked at a search on this data base, in para 4.1.3.

> LABORATORY HAZARDS BULLETIN. Royal Society of Chemistry. 15,000 records. 1981 to date, up-dated monthly.

This data base gives information on hazards likely to be encountered by the chemical and biochemical laboratory research worker, which may affect health and safety. It includes hazardous chemicals and reactions, new safety precautions and legislation, and biological hazards. Data are abstracted from periodicals and new books; also included are audiovisual materials, conferences, courses, and new equipment concerned with laboratory safety. Once again, let's have some practice shall we?

**SAQ 4.2c**

My main reason for this question, apart from giving you a bit more practice, is to show you a search on another host, DATA STAR. This time you don't have to enter the 's' command; the truncation symbol is '$' and if you want to join two words consecutively or in a compound term you use 'adj' between them (instead of '(w)' we have used in other hosts). Write out a search on *Laboratory Hazards Bulletin* for references to the problems of the safe disposal of dioxin.

Back to the data bases.

CHEMICAL ENGINEERING ABSTRACTS. Royal Society of Chemistry. This data base corresponds to a printed version with the same title. Over 60,000 records. 1970 to date, up-dated monthly (5,000 new records per year).

Theoretical and practical material is covered in this service, with a bias towards the practical, and on all aspects of chemical engineering. The service covers over 100 of the world's major primary chemical and process engineering journals. And now here is a final bit of practice for you.

**SAQ 4.2d**

> We stay in the DATA STAR host. Write out a search on *Chemical Engineering Abstracts* for references to the use of X-ray fluorescence techniques in chemical engineering.

The final data base in this paragraph is:

> INDEX CHEMICUS ONLINE. Institute for Scientific Information. Corresponds in part to the printed *Current Abstracts of Chemistry* and *Index Chemicus.* Over 350,000 records. 1962 to date, up-dated monthly (20,000 new references per year).

This service from the publishers of *Science Citation Index* (remember it?) indexes articles from about 110 journals in which new organic compounds have been reported. The compounds can be retrieved by molecular formula and structure searching.

Now you have seen the principal chemical bibliographic data bases, but there are still more services you need to be aware of, so let's go straight on into the last paragraph of this Section.

### 4.2.4. Other Relevant Bibliographic Data Bases

Throughout this Unit I have been trying to avoid presenting you simply with annotated lists. But at last I admit defeat! I simply want to make sure now that you are aware of the range of data bases available to you: I have chosen another ten for this purpose and, in the time and space available, can do no more than list and describe them. Do note the range of topics they cover, from related subject fields (life sciences) to wide-ranging or highly specialised data bases. And remember that all of them are bibliographic data bases similar in structure to those we have looked at in some detail.

> BIOSIS. BioSciences Information Service. Corresponds to the printed *Biological Abstracts.* 4,000,000 records. 1969 to date, up-dated twice-monthly (14,000 records per up-date).

This is the biologist's answer to *Chemical Abstracts*! But there could be occasions when you might find it useful. It covers the worldwide literature on every aspect of all life-science subjects, including biochemistry, pharmacology, toxicology, etc.

> CHEMICAL BUSINESS NEWS. Royal Society of Chemistry. 10,000 records. 1984 to date, up-dated weekly (over 500 new records per week).

This carries abstracts and brief related factual items on trends and current affairs in the chemical industry and its end-markets. It covers companies producing and using chemicals, chemicals production and use, plant capacities, economics, and trends of the chemical industry.

> CHEMICAL HAZARDS IN INDUSTRY. Royal Society of Chemistry. 2,000 records. 1984 to date, up-dated monthly (200 new records per month).

This service gives bibliographic data from journals, books, audiovisual materials, and trade literature on hazards caused by chemicals likely to be encountered in the chemical industry. It covers their toxic or hazardous effects, safety practices, equipment, legislation, and waste management.

> CHEMICAL REACTIONS DOCUMENTATION SERVICE. Derwent Publications. 65,000 records. 1942 to date, up-dated quarterly (750 reactions per up-date).

Bibliographic references and mini-abstracts are offered covering novel organic chemical reactions, reagents, synthetic methods, and similar data. This service is not automatically available, you have first to take out a subscription.

> CURRENT BIOTECHNOLOGY ABSTRACTS. Royal Society of Chemistry. 10,000 records. 1983 to date, up-dated monthly.

This abstracting service covers the scientific, technical, and technocommercial literature of biotechnology, and includes techniques and applications in the chemical and related industries.

> INPADOC. Internal Patent Documentation Centre. Over 10,000,000 records. 1968 to date, up-dated weekly.

If you are seeking a specific patent or set of patents, this service gives up-to-date on-line access to the patent documents issued by 51 national or regional patent offices. Patents are also, of course, included with other document types in many of the other abstracting services we have already considered.

> PAPERCHEM. Institute of Paper Chemistry. Corresponds to the Institute's printed *Abstract Bulletin*. Over 165,000 records. 1968 to date, up-dated monthly (1,100 new records per month).

A specialist service, this covers the world-wide literature on pulp and paper technology and related subjects, including the chemistry of cellulose, hemicellulose, carbohydrates, lignin, and extractives. It screens some 1,000 journals in over 20 languages as well as the patent gazettes of major countries.

> RAPRA. Rubber and Plastics Research Association. Corresponds to the printed *RAPRA Abstracts*. Over 200,000 records. 1972 to date, up-dated every 2 weeks (25,000 new records per year).

Another specialist service, this covers all the commercial and technical aspects of polymer materials, processing, and products.

> RINGDOC. Derwent Publications. Corresponds to the printed *RINGDOC abstract Journal*. 900,000 records. 1964 to date, up-dated monthly (4,000 new records per month).

Another Derwent service requiring a subscription, this covers the scientific journal literature on pharmaceuticals, including all aspects of drugs (analysis, biochemistry, chemistry, endocrinology, microbiology, pharmacology, etc).

> SCISEARCH. Institute for Scientific Information. Corresponds to the printed *Science Citation Index* with additional records from *Current Contents*. Over 2,000,000 records. 1974 to date, up-dated twice-monthly (21,000 new records per up-date).

As you might remember from our brief look at *Science Citation Index* earlier in this Unit, this is a multi-disciplinary index to the literature of science and technology. The data base can be searched either by citations or key terms (as well as by authors, corporate sources, etc).

So there you have nearly twenty bibliographic data bases which might be of use to you within analytical chemistry (or within specific aspects or applications of analytical chemistry). Now we take our final step into full-text electronic services.

**Objectives**

When you have worked through this part of the Unit you should be able to:

- appreciate the major features of bibliographic data bases;
- recognise the major data bases relevant to analytical chemistry;
- recognise some of the wide range of other relevant data bases;
- appreciate a few more of the practical aspects of on-line searching.

## 4.3. ON-LINE FULL-TEXT AND DIRECTORY DATA BASES

### 4.3.1. Introduction

We have examined printed publications of many kinds, and then printed abstracting and indexing services, to help us to trace these original publications. We have then seen how quickly and conveniently these abstracting and indexing services can be searched on-line through a computer terminal. So why not go the whole way and instead of accessing references to documents, go directly on-line to a data base comprising the documents themselves? That is the final logical step in the process, and exactly the one we shall investigate in this final Section of our Unit.

It sounds deceptively simple just to call up on-line on your computer terminal the text of any document you want, but there are practical complications: we are just at the beginning of this particular communications revolution and have a very long way to go before we are anywhere near the gigantic and super-efficient electronic document stores depicted in science fiction. But we have begun. To quote a once-popular TV programme, 'we have the technology': optical discs combined with new-generation computers will be able to combine vast documentary storage potential with swift and accurate retrieval. But there are many technical and, above all, economic and commercial issues to be resolved first. In the rest of this Section I shall show you the relevant chemical services already available. They fall into three categories: encyclopaedias, data-books/bases on chemical substances, and journal papers. But compared with the bibliographic data bases there aren't yet many of them. I anticipate that for some time yet the provision of on-line full-text services will expand in the same way that it has so far developed—by offering new services based on particular published works or groups of publications. The encyclopaedia, which is time-consuming to compile and costly to produce (and buy) and which risks running quickly out of date in parts, seems an obvious candidate for conversion into on-line access. The text can be up-dated as regularly as the publisher wishes (or the readers, or subject matter, demand), and such amendments will be immediately available to users, without having to re-print the whole work or publish awkward supplements. Similarly, data-books can be up-dated regularly, and they have the added benefit for conversion into electronic data bases of already being in the form of highly structured files. Journal-paper data bases ('the electronic journal') raise particular points. We shall deal with these, and other points relating to encyclopaedias and reference works, later in this Section as we examine particular services.

Staying with the general points for a moment, are there any disadvantages in full-text services for publishers or users? Well, publishers have to use a lot of storage and provide sophisticated retrieval software for potentially enormous data bases; the commercial interaction between printed and data base services is often uncertain, so publishers have generally put on-line only services for which they know there will be considerable demand. For his part, the user has to remember that print-outs from a full-text data base can be long (and expensive): once again (as always you must by now realise!) you do need to know what you are doing before you start searching. But those who are reasonably familiar with the printed service will be able to search the on-line version more efficiently and more profitably than those who are not. And as you build up experience you will have no difficulty in using other services which will become available in the future.

How do you search a full-text data base? And are there significant differences from searching a bibliographic data base? No, not really at the search stage—the search process is almost identical. The significant differences arise when you come to look at or print what you have retrieved in your search. The subject searching is just the same as that we have already practised, *with one difference.* In joining terms to search bibliographic data bases we have used the operator 'and': this operator should be avoided when searching full-text data bases. Just think, you might be joining terms from opposite ends of a very long document, in which case they have no contextual relationship with one another. On all the hosts you can use special commands to specify that the terms you are joining must be within a certain distance of one another. All our examples in this Section will be searches on either the Data Star or the BRS host, which, happily, use identical search languages. When joining terms within a full-text data base you should always use one of the following operators *instead of* 'and':

| | |
|---|---|
| x SAME y | terms in the same record and same paragraph; |
| x WITH y | terms in the same record, paragraph and sentences; |
| x ADJ y | terms adjacent to one another in the specified order. |

The other hosts have the same kind of operators, for example in ESA/IRS the equivalents are (in the same order as the examples above): '(F)', '(S)', '(W)'. It is at the 'print' or 'display' stage that we come up against the major difference between bibliographic and full-text data bases. Complete a search which has produced a set of 20 references, then type in '.. p all/doc=all', without thinking and you could spend a very long time watching your terminal, and trying to calculate how much it will all cost! The examples we shall use on BRS or Data Star allow us a print called 'oc' which shows us whereabouts in a document our search terms have been found. So, by combining this with the 'ti' (title) format we can print out for any set a list of titles and occurrence tables for the documents we have retrieved. Then we can select one or two of the most relevant documents and have 'hits' printed (on-or off-line)—that format will print for us only those paragraphs of the item in which the terms have been found.

Let's look at an example. Do you remember the Kirk-Othmer *Encyclopaedia of Chemical Technology*? We are getting a little ahead of ourselves here as I shall tell you a bit more about the on-line version in the next paragraph, but we will just take a second look, this time on-line, at a search we did some time ago. It is the search in which I had you looking for information on the analysis of suspect paintings and finding a very useful article in the supplement. In order to show you how you can browse on-line I shall, for once, assume that I haven't done much homework and take a fairly leisurely approach to my search. You will need to look at Fig. 4.3a in conjunction with my explanation.

I start by selecting 'painting or paintings' (I could just as easily have truncated it as 'painting$') and find 60 hits. Next I decide to look at 'forged or forger$' (truncated to give me 'forgery' or 'forgeries') and find 21 hits. On combining these first two sets (by using the 'with' operator, not the 'and') only one reference is found, so I asked to see what is the title of the entry ('.. p 3 ti/doc=all'). I decide this looks interesting, but there might be quite a few relevant paragraphs, so I ask to see the occurrence table ('4_: .. p 3 ti,oc/doc=1'). By the way, when specifying what documents in the set I want to see it doesn't matter whether I say 'doc=all' or 'doc=1' as, of course, there is only one anyway. The occurrence table tells me in fact that my terms are found in just a single paragraph of the text (para. 7), so I decide to read this on-line. This time I type in '.. p 3 ti,hits/doc=1' so that the only paragraph printed will be the one where the search terms have been found ('hits'). I found the resulting paragraph very interesting, but it wasn't really what I was after. So I think again, and this time search for 'pigment adj analysis' which produces only a single hit. I ask to see the title of the article this hit is in, and the occurrence table, and find it to be another paragraph in the same article on fine-art examination and conservation, this time para. 73. So, again, I ask to see just the text of that paragraph and find a useful, concise explanation of the subject.

So there you see a fairly easy and successful search of a full-text data base. It isn't always so straightforward of course, and I must remind you again of two points you must always bear in mind when searching a full-text data base. Always refine your search to give yourself precise sets with as few references as are consistent with your search; and always check how many paragraphs in each reference carry hits from your search before you ask for the text to be printed. Even an occurrence table can take sometime to print if there are lots of hits: just imagine a set of over 50 references each with hits in a dozen or more paragraphs! As I have told you so often, do always think about what you want and what you are doing; with the experience you are gaining here (and I hope either now or subsequently actually at a terminal) you will quickly become a successful searcher of

```
D-S/KIRK/VOL24,SUPPLEMENT 25            SESSION    90

COPYRIGHT BY JOHN WILEY & SONS, INC. NEW YORK, NY 10158, USA

D-S - SEARCH MODE  - ENTER QUERY

    1-:  painting or paintings

    RESULT          60

    2-:  forged or forger$

    RESULT          21

    3-:  1 with 2

    RESULT           1

    4-:  ..p 3 ti/doc=all

      1

TI Fine art Examination and Conservation.

          END OF DOCUMENTS

D-S - SEARCH MODE  - ENTER QUERY

    4-:  ..p 3 ti,oc/doc=1

      1

OC PARAGRAPH

    TX   (7)

TI Fine Art Examination and Conservation.

R0601 * END OF DOCUMENTS IN LIST
```

**Fig. 4.3a**

```
D-S - SEARCH MODE  - ENTER QUERY

    4-:  ..p 3 ti, hits/doc=1

     1

TI Fine Art Examination and Conservation.

TX 7 OF 216

   Several interesting cases of well-known forgeries are discussed in
   ref. 4.  Probably the most notorious case is that of the forger Han
   van Meegeren, who faked works by Vermeer and de Hoogh.   Several of
   his works were acquired by reputable Dutch museums, and one was sold
   to Hermann Goering during the German occupation of the Netherlands.
   When, after the war, van Meegeren was arrested and charged with
   collaboration, he claimed his innocence, protesting that this
   painting and many others, were not originals, but rather forgeries
   by his hand.   These claims were hotly contested, and he had to
   prove, in captivity, that he could indeed fake Vermeer's style so
   well as to deceive the stylistic experts.   The technical examination
   which proved these works to be forgeries was the work of Paul
   Coremans (5).   Later, **210 Pb dating confirmed the recent date of
   manufacture (6).

RO601 * END OF DOCUMENTS IN LIST

D-S - SEARCH MODE  - ENTER QUERY

    4-:  pigment adj analysis

    RESULT        1

    5-:  ..p 4 ti,oc/doc=1

     1

OC PARAGRAPH

   TX (73)

TI Fine Art Examination and Conservation.

RO601 * END OF DOCUMENTS IN LIST
```

**Fig. 4.3a (cont)**

```
D-S - SEARCH MODE  - ENTER QUERY

    4-:  ..p 3 ti,hits/doc=1


     1

TI Fine Art Examination and Conservation

TX 7 OF 216

   Pigment Analysis.  Identification of pigments present on paintings
   of unquestioned attribution provides reference information regarding
   the use of certain pigments in given periods or schools.   In the
   study of paintings with uncertain provenance, the pigments
   identified in it may either negate the possibility of the proposed
   attribution or lend more credibility to it.  Most pigment analyses
   are done in one of three ways:  microscopy and michrochemical tests,
   x-ray diffraction-powder analysis, or energy-dispersive x-ray
   fluorescence spectrometry.  Although the latter technique does not
   directly identify the pigment but rather the chemical elements
   present, this information often allows deduction of the identity of
   the pigment.   The technique can be used nondestructively, without
   any need for sample removal (see Nondestructive testing).  Thus,
   many areas on a painting can be analyzed, giving better overall
   information.


RO601 *  END OF DOCUMENTS IN LIST
```

**Fig. 4.3a (cont)**

both full-text and bibliographic data bases. Now, finally, let's look at a few full-text data bases in some more detail.

### 4.3.2. Kirk-Othmer

I'm sure you remember the Kirk-Othmer *Encyclopaedia of Chemical Technology*, don't you? If you don't, look back at paragraph 2.2.6 now. I don't intend to repeat here what I have already told you about its coverage. Well, the whole of the third edition is available for you to search as an on-line data base. We shall begin with one of our standard data base descriptions for this service.

> KIRK-OTHMER ENCYCLOPAEDIA OF CHEMICAL TECHNOLOGY. John Wiley and Sons. Corresponds to the printed encyclopaedia, 3rd edn, 1978 and supplements. Up-dated concurrently with each new printed volume.

The full text of all 25 volumes (and supplements) is available for searching. The on-line data base also carries informative abstracts of each article, and these are only in the on-line version. This data base, at the time of writing this Unit, is available only on the BRS and the Data Star hosts. The full-text on-line service allows you precise retrieval and display of extensive tabular data, index terms, and cited references, as well as retrieval by searching for terms in the text. In para 4.3.1 we have already considered particular features of searching full-text encyclopaedic data bases, and we have looked at a search on Kirk-Othmer itself. So I am now going to help you to take another, closer, look at this service (and remind you of on-line search techniques) with another exercise.

**SAQ 4.3a**

You want information on the use of lithium perchlorate as an oxidant in rocket fuels. Write out a search on Data Star, by using the commands and operators I have shown you, to find some relevant data in the Kirk-Othmer data base. When you have completed the search phase, write out two commands to scan and then print a relevant and individual reference.

I reached a set with just one reference in my demonstration of this search, and reached the stage (at prompt 7) of asking for a print showing the occurrences in that reference. Or trying to ask for a print! So far I have shown you everything working smoothly; but I ought to show you also what happens when you make a mistake. I have actually made two errors in typing my 'print' command (in all honesty, neither of them was deliberate!): I have forgotten a set number and mis-spelled 'ti'. The serious error is the lack of a set number, but the system simply tells me what I have done wrong and asks me to try again. Getting it right this time, I discover there is one relevant paragraph in an article on 'lithium and lithium compounds'. I look at the relevant paragraph (when I remember to leave a space between the 'p' and the set number!) and decide that although it tells me a little I still need more information. So, what do I do now? I am pretty sure there must be another relevant article somewhere in either Set 4 or Set 5. I decide to print the titles of Set 5 (there are only eight references anyway) and I notice document no. 5, 'explosives and propellants—propellants', which looks very likely to be what I am after. So next I ask to see the relevant occurrences and find that they are only in a descriptor term and a table. But the table might have some useful data, so, once again, I ask for a print of the hits, and a table of the properties of some common inorganic oxidisers is printed for me, including of course data for lithium perchlorate. If I still wanted more information I would either have to look at one or two of the other articles in my set on lithium perchlorate, or consider ways of widening the search slightly (perhaps looking more generally at oxidants for rocket fuels).

Anyway, there you have an electronic encyclopaedia on chemistry: with a little practice, and a lot of thought and planning, you will find how to make the best use of it, and see how easy it can be to extract relevant information. Now, let's move on and look at a few more data bases.

### 4.3.3. Directories of Chemical Substances

By now you are (or should be) pretty familiar with on-line searching, and have extended your practice into full-text data bases. Of the full-text services encyclopaedias such as Kirk-Othmer are about the most complex kinds, so you shouldn't find much difficulty with the services we shall study in this paragraph. Let's start by re-visiting another old friend.

> MERCK INDEX. Merck and Co. Corresponds to the printed work of the same title. 10,000 records. Current data only, up-dated every six months.

You remember this, don't you? It has concise descriptions of chemicals and drugs. It was pretty easy to use as a book and is, if anything, even easier on-line. There is one useful general point to make about these directories of substances: all the entries are concise and, usually, none of them is of any great length; so you don't have to be quite so cautious about scanning and printing relevant sections.

```
      2_:   duphaston
       RESULT              1 DOCUMENT
      3_:   ..p 2 all/doc=all
       1
AN MONOGRAPH NUMBER:  3460
ST STATUS:  10th Edition.
TI TITLE:  Dydrogesterone
RN REG. NO: 152-62-5
ON CAS NAME: Pregna-4,6-diene-3,20-dione.
   10*alpha*-pregna-4,6-diene-3.20-dione.
SY SYNONYMS: 6-dehydro-retro-progesterone.  10*alpha*-isoprognenone.
DC DRUG CODE NUMBER:  None.
TN TRADE NAME: Prodel. Retrone.  Duphaston (Philips-Roxane).
   Gestatron  (Leo).  Gynorest (Mead Johnson).
MF MOLECULAR FORMULA:  C21H2002,
MW MOLECULAR WEIGHT:  mol wt 312.44.
PC PERCENT COMPOSITION:  C 80.73*percent*, H 9.03*per cent*, 0
   10.24*percent*.
RF LITERATURE REFERENCE:  Prepn: Westerhof, Reerink, Rec. Trav. Chim.
   79, 771 (1960): Rappoldt, Westerhof, ibid. 80, 43 (1961).
MP MELTING POINT: mp 169-170*degrees*.
LD LETHAL DOSE: None.
PD PHYSICAL DATA: Crystals from acetone + hexane, mp 169-170*degrees*.
   (*alpha*)D25 -484.5*degrees* (chloroform). uv max: 286.5 nm
   (*epsilon* 26,400).
DT DERIVATIVE TYPE:  None.
DF DERIVATIVE MOLECULAR FORMULA: None.

DD DERIVATIVE DATA:  None.
US USE:  None.
TC THERAPEUTIC CATEGORY:  Progestogen.
TV THERAPEUTIC CATEGORY (VET) : None.

END OF DOCUMENTS IN LIST
BRS SEARCH MODE - ENTER QUERY
```

**Fig. 4.3b**

As you will know, the major use of the *Merck Index* (and similar reference works) is to look up data on specific substances. In the printed version you start at the index and turn to the relevant entry under whatever name has been chosen for that entry; on-line you can enter whatever name you know and (provided it is accepted as a synonym in the text) go straight to the relevant entry. Look at Fig. 4.3b.

I wanted data on a drug I knew as Duphaston, so I entered that name and was told it occurred in only one entry. I took the risk of looking at the entire entry and found it to be the relevant one with all the data I expected; the main title of the entry and preferred name of the drug is, in fact, Dydrogesterone. The data fields are all labelled for us, all of them being either self-evident or spelled out for us anyway. You should notice that *Chemical Abstracts* Registry Numbers and preferred names are quoted: you can search the data base by using these as well as the common or official names. If you do search by the CA Registry Number, you enter your command with just the registry number, followed by 'r.n.' (for example, '152-62-5.r.n.'). Just take another quick look for yourself by looking up a substance.

**SAQ 4.3b**

We have already come across myristic acid: we know that *Chemical Abstracts* calls it tetradecanoic acid and has given it a Registry Number 544-63-8. Write down two alternative commands for looking up myristic acid on-line in the Merck Database. One way of seeking it will give you five references; write down the commands you would enter, first to scan all five quickly, then to look at one in more detail (you can make up set and document numbers, or just use 'n', as you prefer).

Our next two services are both based on well-known printed works, and in fact both are reference books I would have included in Section 2 of the Unit if space had not been so tight. The first service is in many respects similar to the *Merck Index*. It is:

> MARTINDALE ONLINE. Pharmaceutical Society of Great Britain. Corresponds to the printed *Extra Pharmacopeia* (28th edn). 5,000 compounds. Current data. Up-dated every six months.

This service covers the nomenclature, properties, effects, uses, dosage, and preparations of compounds for drugs and ancillary substances. I have a short example, mainly to compare this service with the *Merck Index*. You will find a typical entry less comprehensive than in Merck, but including slightly different types of data. In Fig. 4.3c I have looked for a cough mixture called Furadantin. The entry for this turns out to be under the title Nitrofurantoin Sodium. 'RF' indicates the type of record (alternatives might have been 'drug definition and description' or 'uses') and 'LE' tells us the length of the on-line record. The date of publication ('ED') and heading ('HD') lead us into the preparation paragraph. The entry is completed by a list of the indexing descriptor terms ('DE').

The data bases we have examined in detail so far are all offered by the BRS or the Data Star hosts (which use identical search systems). To complete this paragraph here are notes on just three more data bases of chemical substances, from other hosts.

> HEILBRON. Chapman and Hall. Corresponds to the printed *Dictionary of Organic Compounds* (5th edn) and *Dictionary of Organometallic Compounds.* 70,000 entries giving 150,000 chemical substances. Current data, up-dated every six months.

This data base offered by Dialog (although not actually live on-line yet as I write this) is intended for chemical identification and to give the physical and chemical properties, uses, hazard and reference data for the more important chemical substances. Our last two services deal with commercial data on chemical products. The first is:

> PESTICIDE DATABANK. British Crop Protection Council and Commonwealth Agricultural Bureaux. 3,000 products. Current data, up-dated quarterly.

Here you will find information on products and related chemicals, including their properties, uses, toxicology, formulations, and analysis. This data base is offered by Pergamon Infoline, as is our last example:

> FINE CHEMICALS DIRECTORY. Fraser Williams (Scientific Systems). Corresponds to a printed directory with the same title. Over 170,000 records. Up-dated monthly (4,500 new records per month).

This is a comprehensive catalogue of commercially available research chemicals, including organics, inorganics, biochemicals, dyes, and stains. It includes catalogue entries from many different suppliers.

```
D-S/MART/MARTINDALE ONLINE              SESSION     91

COPYRIGHT BY THE PHARMACEUTICAL SOCIETY OF G.B. LONDON GB.


D-S - SEARCH MODE  - ENTER QUERY

    1-:  furadantin

    RESULT          1


    2-:  ..p 1 all/doc=1

       1

an 5652/n8-f.

TI Nitrofurantoin Sodium

RF PREPARATIONS.

LE 452 Characters.

ED Jan. 1983.

HD PROPRIETARY PREPARATIONS.

PR Furadantin

   Manufacturer or source:  Norwich-Eaton, UK
   Nitrofurantoin, available as Suspension containing 25 mg in each 5 ml
   (dilution not recommended) and as scored Tablets of 50 and 100 mg.
   (Also available as Furadantin in Austral., Ger.,  Ital., Norw.,
   S.Afr., Switz., USA).

DE nitrofurantoin, prop, suspension, oral-dosage-form, tablet, pharmm,
   diluting.


RO601 * END OF DOCUMENTS IN LIST
```

**Fig. 4.3c**

It is likely that between my writing this paragraph and your reading it there will be at least one or two more similar data bases available (certain in fact: as I write this I discover that a new data base from the Royal Society of Chemistry, Agrochemicals, is about to be made available on Data Star). But with the knowledge and experience you have gained in this Section of the Unit you will be able to exploit satisfactorily any such new data bases as you come across them. And now, as our final step, we proceed to look at a prototype of the electronic journal.

### 4.3.4. American Chemical Society Journals

All those journal papers are being written and published, then subsequently indexed and abstracted in on-line data bases. With the spread of microcomputers into laboratories, offices, and homes, and with the growth of electronic networks, bulletin boards, and national or international communications facilities (like this IPSS I have been using), what are the prospects for eventually replacing the printed journal with an on-line electronic version? Well, from the few surveys I have seen published it seems that scientists generally expect and wish the printed publications to continue. Once again, the publishers have to examine the commercial implications of changing to electronic publishing. There are other considerations as well; for example, how will the refereeing system operate when papers can be transmitted electronically almost as soon as they have been written? The implications of the 'electronic journal' are widespread and fundamental. On the credit side of electronic-journal publishing are the tremendous increase in searching power and sophistication and the improvement in the efficiency of communications. But these are thoughts for the future, really. We are still just at the threshold of this revolution.

The American Chemical Society, in conjunction with the BRS host service, has made available the first journal data base in the field of chemistry:

> AMERICAN CHEMICAL SOCIETY PRIMARY JOURNAL DATA BASE. American Chemical society. Parallel availability with the printed journals. Over 30,000 articles. 1980 to date, up-dated twice monthly.

The full text of papers appearing in 18 primary ACS journals is made available on-line in this data base, the on-line availability being simultaneous with the printed publication. Each record contains abstracts, complete reference and footnote listings, captions and CAS Registry No. references in addition to the complete text of the article. I have listed for you the titles of the eighteen journals included in this data base.

JOURNAL TITLES INCLUDED IN THE ACS PRIMARY JOURNAL DATA BASE

*Accounts of Chemical Research*
*Analytical Chemistry*
*Biochemistry*
*Chemical Reviews*
*Environmental Science and Technology*
*Industrial and Engineering Chemistry Fundamentals*
*Industrial and Engineering Chemistry Process Design and Development*
*Industrial and Engineering Product Research and Development*
*Inorganic Chemistry*
*Journal of Agricultural and Food Chemistry*
*Journal of Chemical and Engineering Data*
*Journal of Chemical Information and computer Sciences*
*Journal of Medicinal Chemistry*
*Journal of Organic Chemistry*
*Journal of Physical Chemistry*
*Journal of the American Chemical Society*
*Macromolecules*
*Organometallics*

You know how to search now, so to look at this data base a little more closely I shall let you do the initial work, with our final exercise.

**SAQ 4.3c**

You are interested in Fourier-transform infra-red spectroscopy and, in particular, are looking for documents concerning its use for the analysis of aerosols. First of all, construct a search on BRS (by using just the same commands as we have used throughout Section 4.3) to find relevant documents.

Having found a small set of just three documents, write out the commands you would use, first to compare all three, then to concentrate on one you choose as seeming particularly relevant.

**SAQ 4.3c**

**SAQ 4.3d**

Having completed the previous question you want to look at wider applications of Fourier-transform infrared spectroscopy. Write down the command you would use to scan ten of the general references. Then, again, select one of these and write out the command you would use to look at it more closely.

Finally, having completed that stage, can you think of a further way of searching to find some more material which might prove useful?

**SAQ 4.3d**

Now you have had a look at this data base, if you want to consider its practical uses and implications further I would recommend you to a journal paper which discusses them: S.W Terrant, *et al*. Online searching: full text of American Chemical Society primary journals. *J. Chem. Inf. Comput. Sci*., v.24 (1984), 230–235. You could always try finding a copy on-line! The all-embracing single data base carrying all chemical papers is a distant dream (or perhaps nightmare): indeed, there are good grounds for arguing that it can never happen. If the ACS Journal Database establishes a place for itself in the market, no doubt other publishers will follow the lead. Eventually a pattern of electronic publishing will emerge, but, in looking to that future let me emphasise again that printed or paper copies will always be with us as well in some form or another.

So, there we have come to the end of this Unit with an uncertain glimpse of the future; or, I hope, another beginning for you. If you have done no more than work with the materials in this Unit you will now have a good knowledge and fair experience of analytical literature in all its aspects. I hope you have already started applying this knowledge and these skills for your own purposes; but, in any case, remember that I have helped to give you a key to a wider world of chemical information. I wish you success in applying your skills in the fields to which your interests or circumstances lead you.

**Objectives**

When you have worked through this part of the Unit you should be able to:

- appreciate the main features of full-text data bases;
- understand the developments likely to occur in the expansion of full-text data bases;
- recognise those full-text data bases most relevant to analytical chemistry;
- search successfully full-text as well as bibliographic data bases.

# Self Assessment Questions and Responses

**SAQ 1.7a**

I have given you, on the next page, a copy of the title page of a book on rock analysis, and one of the first and the last page of a journal paper on polycyclic aromatic hydrocarbon separation (with the title pages of the journal in which it was published). The book has 379 pages. Write out a bibliographic citation for each, by using the full form of citation I have indicated in my text.

**Response**

Your two entries should have looked like this:

Jeffery, P. G. and Hutchison, D. *Chemical methods of rock analysis*. 3rd ed. Oxford: Pergamon, 1981. 379pp.

Lucke, Richard B. *et al*. Integrated, multiple-stage chromatographic method for the separation and identification of polycyclic aromatic hydrocarbons in complex coal liquids. *Analytical Chemistry*, v.57 (March 1985), 633–639.

Whether you got those exactly right or not, read my comments (remember what I said about useful supplementary advice being given in these answers).

The authors in both cases are pretty straightforward, but I hope you remembered to reduce those of the journal paper to the first-named (Lucke) and *et al*. Many books and most journal papers tell you where the authors work, but you don't need to quote that in any citation. Both titles are also pretty straightforward, but there is no way round quoting in full a long (but explicitly informative) journal-paper title. The edition of the book is stated clearly and must be included in your citation. For the rest of the details of the book: if you are quoting place of publication give only the main (or first-named) location. The name of the publisher can be further shortened by omitting the word 'Press' (unless doing so might cause ambiguity). I have written out the title of the journal in full, but if you adopted the abbreviation given at the head of the page of the paper (*Anal. Chem.*) that was perfectly correct. You can also just use the year of publication if you wish. And do you notice that at the head of the first page of the paper you are given a citation to that paper (although the order of the date and volume number is reversed from my example)?

To finish off, let me just draw a couple of points to your attention. Look at the last page of the journal paper and you will see a list of references breaking most of the rules I have just given you! Then turn to Sections 2.1 and 2.2 of this Unit and you will see that I have referred to reference books always by their titles, giving their authors or editors as the second item. Different journals have different preferred formats for bibliographic references, and particular circumstances can dictate different layouts. But if you always compile your references in the manner I have shown you, you will have clear citations in a consistent, standard, and acceptable format and which you can later adapt if necessary to a different arrangement.

If you are unsure about some of these details, don't let it worry you for the present. You will see examples of an awful lot of bibliographic citations, and be given more practice in compiling them, before we finish this Unit!

***********************************

**SAQ 2.1a**

This is the first SAQ I have set you in this Section, so let me make a very important general point right at the start. I cannot be certain that you will have convenient access to all, or any, of the books I refer to in this Unit; so, you are being given copies of at least a few of the pages of each major work to enable you to answer the SAQ if you do not have access to the necessary books. But, if you do have convenient access in a local public, college, or works library to any of the books on which I set you questions, then *ignore the copies and use the books themselves* to answer the questions; you will derive much more benefit from this Unit if you are able to do so. And please bear with me when I remind you briefly of this advice at every relevant question. The important thing with these questions is to make you look things up for yourself in the books concerned, and you will do that more profitably in the material themselves; the copies will be useful substitutes, *but only if you cannot get access to the books.* ⟶

**SAQ 2.1a (cont.)**

So, let's start by having you look up some British Standards. For these questions you will need a copy of the *BSI Catalogue*: use the copies only if you cannot get hold of a copy.

(*a*) Imagine you have to select a technique and apparatus for a gas chromatographic analysis to a British Standard. What two standards would you need to refer to? Just write down their numbers and dates of publication.

(*b*) Imagine you are preparing chemical samples for distribution to and use in chemical laboratories in schools in your area. You need to check the preparation of potassium nitrate. As your work has always to be to the British standard, what standard specification do you need to check? Again just write down the serial number and date of publication.

(*c*) Suppose you are going to carry out an analysis for lead by anodic stripping voltammetry to British standards, by using hydrogen peroxide as a reagent. What British standard would you need to check for a specification for the reagent and method of test? Just give the serial number and date of the specification. I must warn you that unless you are already familiar with the standard concerned you might find this particular question very tricky. If you cannot find the right standard after three or four attempts, don't get upset but go straight to my response: there is a moral to this question!

**Response**

(*a*) BS 4587 : 1970 (*Recommendations for the selection of apparatus and techniques for the analysis of gases by gas chromatography*) and BS 5443 : 1977 (*Recommendations for a standard layout for methods of chemical analysis by gas chromatography*). I think you will have found these quite easily in the subject index, although if you checked under 'chromatography' you will have found only a reference to BS 3282 which is not what you require to answer this question. If you did start at 'chromatography' I hope your initial lack of success prompted you to look under 'gas chromatography' where you would find the reference you need. I hope also that you checked each reference, in the main chronological section of the catalogue. *Never* quote information directly from an index but always check back to the source: there might be significant differences or qualifications in the source material which will not be apparent in the index.

(*b*) BS 5936 : 1981 (*Specification for chemicals used in schools*). There is no index entry under 'potassium nitrate', so where do you look? If a direct approach under a specific term is unsuccessful then try more general terms (or perhaps synonyms, but that does not apply in this case). Here the clue in the question was *chemicals for use in schools*: look under 'chemicals' and there is an entry for 'Chemicals used in schools', and, similarly, an entry under 'School equipment'. I shall have quite a lot to say about indexes, especially in the SAQ responses in this Section and the next. There is an enormous range of indexes of different kinds and, more importantly, of different qualities, and there is no magic formula for always finding the information you want at the first attempt: you will gain experience and learn some principles from working through this Unit but much will depend on common sense, ingenuity, and sometimes luck. But familiarity with particular types of publication and with specific sources, such as I am trying to give you in this Unit, will give you a head start.

(*c*) BS 6376. (*Reagents for chemical analysis*) Part 1: 1983. (*Methods of test*), Part 2: 1984 (*Specifications*). Did you manage to find that? Unless you already knew of the existence of this standard I think you probably found it difficult; and if you failed to find it, don't worry, but make sure you read this response thoroughly and have a look at the *BSI Catalogue* (or the photocopies) before you move on. You start at 'lead' in the subject index and think you are home and dry with BS3908 (at the sub-heading 'sampling and analysis'). But when you check back to the catalogue you discover that in the fifteen parts anodic stripping voltammetry is a method not covered. No likely index entries at 'anodic' or 'voltammetry', so where now? 'Analysis' gets you nowhere so you probably turn in desperation (if you haven't given up already!) to 'chemical'. There is in fact an entry in the subject index under 'chemicals' at the sub-heading 'reagents, analytical grade'; similarly, there is an index entry at 'reagents', but I wonder how many of you found it, or realised that it would refer you to the information you need. And I bet you thought this was going to be easy when you answered part (*a*)! This might seem like an unfair trick question, but it does help to make a couple of vital points right at the start: indexes are not always easy to use and they do vary considerably from publication to publication and even within the same index sometimes; that applies both to the quality of the index and the depth to which material is indexed. And in the end the information you needed is to be had in the standard I quoted, even though it took some finding. If a publication, or series of publications, is likely to be of long-term use to you it is well worth your time getting to know its layout, strengths and weaknesses at a fairly early stage (but more of that in Section 2.2).

***********************************

**SAQ 2.1b**

By the time you have worked through my SAQs you will have found yourself working, in imagination at least, in an extraordinary variety of laboratories and industries! But apart from giving you a wide range of questions to try, and making them as interesting as I can for you, there is a more serious purpose too: I want you to realise just how easily you can find information on subjects which might be completely unfamiliar to you.

Here a couple of straightforward questions from the *Annual Book of ASTM Standards* find you concerned with metals analysis and gas chromatography (nothing very esoteric here, but wait for some later questions!).

(*a*) You find yourself working in the metal industry. You have to determine the amount of silicon in a bronze sand-casting ingot. How many methods are recommended by ASTM? What weight of sample is needed?

(*b*) You are going to use a flame-photometric detector in a gas chromatography analysis. Find an ASTM standard which will tell you how to go about your task.

Use the copies only if you have to: try and consult copies of the books if you can.

**Response**

The first thing to notice is that there is an index to each volume as well as the general index, Vol. 00. It is possible (even likely) that if you are working in a particular industry you might have available only the volume or volumes of ASTM standards immediately relevant to your work. But you can still find your standards even without the general index. I am going to assume here that you used the general index, Vol. 00, to find your information: the comments about the index are just the same whether you have used the general index or the volume indexes.

The second point is that each volume begins with a list of the standards it contains. If you are going to use one particular volume a lot, it might be useful to go through this list of contents and get to know just what standards are in the volume, but I must emphasise that this is *not* the way to look, for a specific standard (unless of course you have failed to find anything in the index and are getting desperate!).

(*a*) Not very difficult if you checked the index under 'bronze': there is a sub-heading 'ingots' and 'chemical analysis' taking you to Standard E 54 in volume 03.05: no page references, but as the standards are arranged in alphabetical/numerical order of their codes you have no problem in turning to E 54. If you started from the other end and checked 'silicon' you will have found what seems to be the entry you need, headed 'silicon' content'. But E 54 is not in fact listed there, even though it does give you a method for silicon-content analysis: like the BSI index this one relies on terms which appear only in the titles of standards; they try to make the titles as descriptive and specific as possible, but not everything is included. If you did start at 'silicon', I hope when you were unsuccessful you thought to try looking under 'bronze'. Anyway, you should have found two methods recommended, each requiring 5 g of sample.

(*b*) Again, no problem if you looked directly for 'flame-photometric detectors'. If you tried 'gas chromatography' you will have found yourself referred to 'chromatographic techniques/procedures' with a sub-heading 'gas chromatography'. Check there and you find yourself directed to E 840, as you were under 'flame-photometric detectors'.

One last point: did you notice in the general index how all the standards on particular topics are brought together no matter what volumes they appear in? That could be very useful if you need to do an exhaustive survey of methods or techniques with a wide range of applications, or if you find yourself working in a field not directly connected with the subject in which you are used to working. Use the general index and you are unlikely to miss any important relevant ASTM standard, whatever the volume in which it is published.

***********************************

**SAQ 2.1c**

You have to determine the percentage of free fatty acid in a sample of soap. What reagent do you use in the principal recommended method and what weight of sample is recommended? Are there any alternative methods you could use?

If you cannot get hold of a copy of the book, use the copies provided, although these will make an easy question extremely simple. If you do use the book, remember that I am setting this question some time before the new edition is published. You might find yourself using this new edition, so watch out; some at least, if not all, the details in my response are likely to differ from those in the book you may be using.

**Response**

I always advise you to use the indexes in any reference book you consult, although for this example it is quite possible that you would realise there is a chapter on soap and turn straight to it. Whatever approach you used, this is a pretty easy work in which to find information. Check the index under 'fatty acids' and you find a sub-heading 'free' with a reference to pp. 178 and 180 (soaps). Check under 'soaps' and you are taken to the relevant chapter, although you still have to find the particular entry you need for fatty acids. And if you checked 'soap' without looking at the next word ('soaps') you might have a bit of difficulty finding the right page. Always remember that you might need to check under different forms of words in indexes and to note whether both singular and plural forms of words are used, as in this example, or only one form (normally the plural is preferred, but by no means invariably).

Either way, through the index or the contents pages, you should have found your way quite easily to p. 178 where you find that you need 100 ml of ethanol and 5 g of sample. Turn to p. 180 and you find an appendix with two alternative methods: the barium chloride and the glycerol method.

**********************************

**SAQ 2.1d**

This is a nice easy work to use, so this question shouldn't detain us long. Quite simply, find:

(*a*) a method for determining starch in peanut butter;

(*b*) a method for determining salt in nuts.

Try and have a look at the book if you can: this question becomes more-or-less a demonstration if you work from the copies. But the copies are there to be used if you cannot get hold of the book itself. And in either case I shall be able to illustrate the points I want to make in my response.

**Response**

As you can see, there is a clear and detailed contents page, where you might have noticed chapter 27 (Nuts and Nut Products). But much better always to use the index, where you find clear references to 27.012 and 27.016 for sodium chloride in nuts, and starch in peanut butter respectively. Note that you are not referred to pages but to chapter and paragraph numbers: just as easy to find and more precise. And the index has been compiled by using terms found within each chapter, so you find these entries without difficulty whether you look under 'peanut butter' or 'starch' (*a*). For (*b*) you find an entry under 'nuts', but the sub-heading is not 'salt', but 'sodium chloride'; similarly you find an index entry under 'sodium chloride', but at 'salt' you need to follow the cross-reference which says 'Salt, see also Chlorides; Sodium chloride'.

But no real problems here are there? Wouldn't it be nice if every reference work were as easy to use as this!

***********************************

**SAQ 2.1e**

Here you are working in a medical laboratory (I told you we would get around in this Unit didn't I). You are going to analyse some Isoniazid, using a sample taken from a batch prepared for injection. Where do you find a method for analysis? Is there any difference in acidity between the compound itself and its injection solution? You expect you will have to carry out an infra-red analysis of Isoniazid before long. Is there a published spectrum and what method would you use to prepare the sample?

Once again, this question is going to be pretty easy if you have to rely on the copies, so do try and use the actual books if you can.

**Response**

Not a very difficult question to answer, was it? But there are just a couple of points to watch.

As long as you realise that the index to both volumes is at the end of volume 2, and that the page numbers run consecutively from volume 1 through volume 2, then you will find your information easily. The index takes you to p244 (vol. 1) for the monograph on Isoniazid and the method of analysis. A second entry in the index under 'Isoniazid injection' takes you to p. 627 (in vol. 2) where you note that the acidity of the injection solution is adjusted to pH 5.6 to 6.0 instead of the 6.0 to 8.0 quoted in the monographs.

Question satisfactorily answered? *No*! Remember this work is updated by supplements: there might just be something else to find. Check the index in each supplement ('addendum' they call it) in turn and, sure enough, in the 1983 addendum you find a reference to page p. 238, where there is a new monograph on Isoniazid to replace the one you already found in the main volume. There may not be many significant differences between the two, but even a minor alteration may be critical, and you might have found yourself in trouble if you relied on an older entry, not realising it had been superseded.

And, finally, not much problem with the ir spectrum is there? The index leads you to S128 for the spectrum and another entry under 'Isoniazid injection' refers you to the same spectrum, and to a method (M186) to which you turn and find how to prepare a sample. Note that in this volume you are referred to serial numbers for both the spectra and the methods, not to page numbers. Although this time you will find newer data, I hope you would also remember to check the updating supplements to this part of the work as well.

***********************************

**SAQ 2.2a**

You are about to be interviewed for a job which will entail the analysis of natural gas. It will be prudent for you to know about the methods involved and so you go to a library and find a copy of Wilson and Wilson's *Comprehensive Analytical Chemistry*. (Don't look at the copies yet!)

First you must select the relevant volume. Note the volume number you would select as the most likely source of information on methods of analysis of natural gas, by using the list of volumes above. This is not so easy a task as it would be if you had access to a set of the volumes, but try it anyway—if you get really stuck the copies give the game away.

Next turn to the copies (or better still to the books themselves if you have access to a set) and find the relevant section. Note where you look and the terms under which you search.

Finally, note what is practically the only reliable method for the detailed analysis of natural gas.

**Response**

Finding the right volume is probably the hardest part of this task; you should have found your way to Volume XIII, Analysis of Complex Hydrocarbon Mixtures. You might have chosen another volume if you approached the problem from the viewpoint of technique rather than that of application; this is a major difficulty when using a multi-volume work which has no general index, and there is no easy answer other than trial-and-error searches through likely volumes, or familiarising yourself with the work.

Having reached Volume XIII you find that it is itself in two parts. By now you should have checked the index to each volume: if you look at the first page of the index to Part A you are told that there is a comprehensive index to both parts at the end of part B. Checking this you will find an entry under 'Natural gas' directing you to pp 766–769; here you do find the information you are seeking and the statement on p. 766 that 'Gas chromatography ... For a detailed analysis of natural gas it is practically the only reliable method (ASTM Method D 1945)'. At the end of the section is a list of references so that you can find further information if you need it.

You might have continued to use the contents pages to find your information: at the end of this process you would have a good knowledge of the contents of the work, but it would have taken you a lot of unnecessary reading and potentially valuable time to find the right section, before you even start to read the information you want; and it is possible that by this method you would not even find the right section.

If you checked the index under another term, probably 'Gas', you would similarly have had a lot of checking to do before finding the correct section (unless you already knew that you were looking for gas chromatography, but even then there are a lot of index entries for that term).

So always start by looking in an index under the most specific term you are seeking: only if that is unsuccessful should you start thinking of and looking up alternative, often broader but perhaps synonymous, terms. You might also need to remember where a book is published. American books use some terminology quite different from that in British books, even for some everyday things. You will find an example of this shortly.

************************************

**SAQ 2.2b**

> You are working for a food company when, as a result of reorganisation following a merger, you are told that you are being transferred from solid-food analysis to work on the analysis of fizzy drinks. This is subject you have never worked on before, so you need to find out about the techniques likely to be involved. You find a set of the *Encyclopedia of Industrial Chemical Analysis* in your works library and start looking for information.
>
> Use the copies I have provided, or better still use a set of the encyclopaedias if you can, and look up a useful article. Note where you look, what terms you would look up, and then look at the relevant article.
>
> After reading this article you may decide you want to refresh your memory about techniques of precipitation in gravimetric methods. Again, note where you look and what terms you look up.

**Response**

The first point with this work is to realise that it is two works in one, as I described in the text. You are looking first for an application so you should go to the index Volume 20 and see what you find. Here your troubles start: no entries under 'drinks', 'fizzy drinks', 'mineral waters', 'Cola' or the like (although you will eventually find all these words in the text). If you realise that carbonation is involved then you will find a reference under that term directing you to volume 8 where you discover that the relevant article is under 'carbonated beverages'. It took some finding didn't it? Or perhaps you are used to American terminology and found the article fairly easily. Some indexes do your thinking for you, but this is an example (in this instance at least, in other examples you would have no difficulty) of an index where you have had to consider fairly carefully the terms you need to look up, and your difficulties are compounded by American usage.

By the way, did you notice that this index is arranged letter-by-letter and not word-by-word? What is the difference? Consider the two following columns:

| (*a*) | (*b*) |
|---|---|
| Carbon | Carbon |
| Carbon black | Carbonate |
| Carbon dioxide | Carbon black |
| Carbon tetrachloride | Carbon dioxide |

| Carbonate | Carbonised |
|---|---|
| Carbonised | Carbon tetrachloride |

(*a*) is arranged word-by-word and (*b*) letter-by-letter. It dosn't make much difference in a short list like this, but, apply it to a lengthy, complex list of terms (like the index page in your photocopies) and you will, I hope, realise that here is another point to be careful about when you are consulting an index. This arrangement is entirely at the discretion of the compiler or publisher; you will find many examples of each kind of arrangement and even, I am afraid, in some badly edited works, the occasional example of both arrangements in the same index.

Next, for gravimetric methods, you are looking for a technique. If you look at index Volume 20 believing this to be a general index to the entire encyclopaedia (I did when I first used it, disregarding all my own good advice about checking first!) you will find nothing: you need to be in Section 1 (Volumes 1–3) and they have their own index (see what I mean about it being two works in one). But once you get so far you will find life much easier: clear, full, index entries under 'precipitation' and under 'gravimetric', even with a cross reference from 'precipitation' to 'gravimetric'. No problems here, I hope, about finding all the relevant information you need (and more). I have copied only a small part of the very long article about gravimetric methods to show you what it is like.

***********************************

**SAQ 2.2c**

You are working in a forensic laboratory when your director suddenly tells you that in a day or two he wants you to help in an analysis of a suspect painting. You have never worked with paintings before and point this out to your director; he just tells you that it will be an ir and pigment-analysis task.

You decide that you must know more about the work involved before you start and have access to a copy of the Kirk-Othmer *Encyclopedia of Chemical Technology*.

Use the copies provided (or better still, if you have access to a complete set of the 3rd edition, use it and ignore the copies) and find the information you would need. Just note how you first approached the work and the terms you looked up and where.

Of course it won't be difficult to find if you use the photocopies, but the list of contents gives you a indication of what articles are, and are not, included.

**Response**

*Did you look for a likely term directly in the text?* In an A-to-Z encyclopaedia this seems a useful approach. But in this instance you would have been unsuccessful: you could look under 'Art' or 'Paintings' and find nothing: you will find articles under 'Pigments' and 'Infrared', but these do not help you. Browsing in an encyclopaedia is always fascinating and usually instructive, but if you want relevant information quickly always use the general index (if there is one). In this example it is absolutely essential to use the index because the article you need is in fact in the supplement and not in the main body of the work.

*Did you look up likely terms in the index first?* At the expense of the random fascination of browsing in an encyclopaedia, this is always the quickest and surest way of finding the information you need. Only the term 'Pigment analysis' fails in the index. Any of the others suggested by your requirement will lead you into the relevant article: you can find it under 'Art forgeries', 'Fine art examination and conservation', 'Infrared reflectography' (at the expense of browsing through various terms at 'Infrared', and noting that you do not find any relevant term at 'IR'), or 'Paintings'.

Once you have found your relevant article, then browse through it first to see if you need to read it all or just relevant sections. If, after reading the article, you want more information still, there is a list of references at the end.

***********************************

**SAQ 2.2d**

You want a fairly quick review of the methods of analysis and determination of glucose: you would also like a note of any analysis in which glucose is used as a reagent.

Look in the Meites *Handbook of Analytical Chemistry* if you can, but if you do not have easy access to this then use the copies provided instead).

Note the page numbers where you find relevant information and look the references up in the text. How many methods are given for analysis of aldoses (eg glucose) and which one requires the longest heating? What methods can you use to determine glucose? What substance can be determined by using glucose as a reagent?

**Response**

If you look at the contents page of this book and glance through the text (especially of the complete book and not just the photocopies), you will quickly realise that a good index is essential. And, happily, this mine of information does have an excellent index complete with its own explanatory introduction. If you did not read this introduction to the index when you looked for information then do so now.

You should have no problems once you have looked up 'Glucose': the entries are systematically arranged for you under 'Analysis', 'Determination', 'Properties', 'Reagent', and 'Separation', as explained in the introduction to the index.

On p. 6–81 you find four methods for aldoses (and one for hexoses), with method no Su 2.5 needing 20 min heating. For alternative methods of determination you are referred to p. 5–21 for potentiometric redox titrations of carbohydrates, or to p. 13–81 where photometric methods to determine some carbohydrates are given. At p. 6–187 you find that glucose can be used as a reagent in the determination of tryptophan.

*********************************

**SAQ 2.2e**

The *Merck Index* is primarily intended for quick reference to data on a comprehensive range of subjects: so here are just a few sample quick reference questions to answer from it and get an idea of its contents and layout. Again, if you can use the book itself, do so; if not, use the copies provided. Jot down brief answers to *all* the questions before reading my response.

You are given a tablet marked 'Valadol'. What is it?

What are the chemical constituents of raspberry?

You are given a sample labelled as a stable aqueous solution of curare. What concentration would you expect it to have? We all know about curare as a poison used on arrows and darts, but what therapeutic use does it have?

What is the formula of, and what are the alternative names for Valium? ⟶

**SAQ 2.2e (cont.)**

What substance which occurs in nutmeg butter is represented by the formula $C_{14}H_{28}O_2$?

What is the Ruzicka large-ring synthesis?

**Response**

If you went directly to the text for Valadol you will have drawn a blank: the name index refers you to entry 39 (*not* p. 39, note) where you find that it is a proprietary brand of Acetaminophen.

Much as I like eating raspberries it wasn't until I began setting these questions that I discovered their main chemical constituents to be sugar, and malic and citric acid. As a natural substance, 'raspberry' will be found without difficulty either from the index or directly in the text under its common name.

Curare likewise can be found both in the index and the main text under its common name. Towards the end of an excellently concise summary of the properties of this substance (including references to some fascinating literature if your interest is caught by South American poisons—a hint to any budding Sherlock Holmes using this Unit) you are told that a stable aqueous solution should be standardised to contain 20 units per ml, and that curare has a therapeutic use as a skeletal muscle relaxant.

You won't find the name 'Valium' in the main text but the index takes you to entry No 2967, where you discover a formula of $C_{16}H_{13}ClN_2O$, a main name of diazepam and a variety of alternative names from Ansiolin to Vivol.

Once you have found the formula index (hope you didn't waste time trying to identify the formula from another book first) you will have no difficulty in identifying $C_{14}H_{28}O_2$ as most likely to be either Ethyl laurate or Myristic acid: reference to entries 3768 and 6138 tells you that Myristic acid occurs in nutmeg butter.

Check the index under 'Ruzicka', and you find nothing: you must either have realised already, or have to find out from the contents page, that there is a section on organic name reactions. There, in its correct alphabetical name sequence you find a brief description of the Ruzicka large-ring synthesis.

Did you find those questions pretty easy? Of course, one of the main advantages, and a main purpose of a quick reference work is to find concise answers easily to fairly precise questions. If you need more detail in your answers then you could move on from what you have learned in this book either to one of the more comprehensive reference books we have already looked at (or one of the more specialised ones we shall look at in the next Section), or to any of the references to various publications given in the text (eg the monograph on curare).

**********************************

**SAQ 2.2f**

Right, let's see how much of my advice you remember.

You want to buy a reference book (the subject is immaterial to the principle of this question). You are in a well-stocked bookshop and have narrowed your choice down to three titles, A, B, and C.

A is the 5th edition, 1984, of an American book published by Wiley–Interscience and originally issued in 1960. It has 600 pages and costs thirty pounds (or fifty dollars).

B is the first edition of a work published in 1978 by the then Chemical Society. It has 300 pages and costs twelve pounds (or twenty dollars).

C is the 2nd English edition, dated 1980, of a work first published in German in 1970. The book is published in Amsterdam by Elsevier, has 400 pages and costs thirty pounds (or fifty dollars).

First of all, note in *your own* order of preference, *four* criteria you would use (cost, length, etc.) to judge these works. Then make a final decision and note which book—A, B, or C—you would buy.

**Response**

There are various criteria you could use, as you may recall from the text you have been working through, many of which I am sure you could think of yourself without my assistance. The judgments are yours: there is no absolutely correct answer to this question. The following are criteria you might consider.

(*a*) Date—how up-to-date is the information in the book?

(*b*) Publisher—is the publisher well-known for good scientific books? (I hasten to point out that in this example all three are highly reputable publishers with excellent international reputations).

(*c*) Country of origin—is it vital that the book refers to British or U.S practice exclusively, or does it make no difference?

(*d*) Edition—has the work been regularly updated in new editions, suggesting it has become accepted as a useful (even standard) work? On the other hand, if it is republished annually, can you afford to buy a new edition each year?

(*e*) Price—does the book seem value for money, especially if compared with a rival volume?

(*f*) Treatment—does the book present information in a mainly tabular or textual format? Which do you prefer? Which seems more suited to the subject of the work?

(*g*) Authors—have you heard of them? Does the book state where they work and does that indicate a high qualification for their producing this book? Do you have a prejudice for or against academic as opposed to industrial scientists?

(*h*) Length—perhaps this should be considered with (*e*) and (*f*), but a longer book might either contain more detailed information, or cover a wider subject area, or be verbose and repetitious.

If you have selected any four of those criteria you are well on the way to making an informed decision. It is slightly unfair of me to ask you to rank your criteria in order because several, and ultimately all, must be considered in conjunction. But I think my list is more-or-less in the order I would rank them (remember I am only talking about reference books here: the criteria might be different or in a different order for different kinds of books. I certainly wouldn't choose a book of poetry this way!).

By the way, I hope you would remember to read the book's preface or introduction, and take with a very large pinch of salt whatever the publisher's advertising manager has to say on the dust jacket. And, finally, what were the first two steps you took before you even went into the bookshop?

(*a*) You defined clearly in your own mind exactly what kind of book you were looking for, and why you wanted it.

(*b*) You asked for advice from someone—a more experienced colleague or superior perhaps, or a librarian.

Right, now I will put my own head on the block and tell you that assuming I had the money, I would have selected A: it is right up-to-date, regularly revised (but not so frequently as to cost me a fortune!), produced by a very reputable scientific publisher (as are all three), and has either greater detail or wider coverage than B and C for the money.

If you chose differently, don't worry: any such decision is ultimately subjective. As long as you had good reasons in your own mind for making whatever decision you did, all is well.

************************************

**SAQ 2.3a**

An easy one to start with, but there's a useful point to be made in the responses. If you don't already know, what will happen if you mix hydrogen peroxide with acetone?

You are going to work with ketone peroxides and, in particular, with a chemical compound $C_9H_{18}O_6$ whose IUPAC name you have checked and discovered to be

3,3,6,6,9,9-Hexamethyl-1,2,4,5,7,8-Hexaoxacyclononane.

What hazards must you beware of? Before you try to look up this compound I must warn you that we have unwittingly stumbled into a doubtful area of chemical terminology. To avoid over-complications at this stage I will just advise you that in the *Handbook of Reactive Chemical Hazards* you will find this compound named

3,3,6,6,9,9-Hexamethyl-1,2,4,5,7,8-Hexaoxaonane.

This is just another example of the kind of complication that can beset you when you come to look for published information.

Remember to use the book if you can, and use the copies only if you have to.

**Response**

You might already have known that hydrogen peroxide and acetone form an explosive mixture. But even if you did know, it would do you no harm to check on the conditions of their reaction more closely if you are expecting to work with this combination (or with either substance for that matter).

By now you should know to check indexes first for information: there is more than one index to this book, so you have to make sure you go to the right one for whatever type of information you are seeking. Here you can look up either 'acetone' or 'hydrogen peroxide' in the Index of Chemical Names and Synonyms; you will find an entry under each name. If you turn to the entry for acetone you find a cross-reference under 'oxidants' to the entry for hydrogen peroxide, where, at pp. 961–962 you find the information you need.

When you look up 'ketone peroxides' you are seeking a generic class of substances, so you must look in the Index of Class, Group, and Topic Titles. A straightforward reference to p. 91 takes you directly to the information you need, advising you that extreme caution is required in handling ketone peroxides in high concentrations. You are also given a list of specific chemical substances for further details, including

3,3,6,6,9,9-Hexamethyl-1,2,4,5,7,8-Hexaoxaonane.

Back to the Index of Chemical Names and Synonyms (arranged alphabetically, so we look under 'h-' for 'hexamethyl', ignoring the numerals for main filing purposes). Again a straightforward reference, to p. 680, where you find a brief entry describing the substance's alarmingly powerful explosive properties, together with references to the literature and further cross-references to related substances.

************************************

**SAQ 2.3b**

You have some urea-formaldehyde resin adhesive named 'Beetle', and that is all you know about it.

What would it be used for and who manufactured it?

Is it marketed in any form other than a resin?

How long could you store the resin?

To what British Standard would it need to conform?

You need to know if there is a firm near Cambridge which manufactures urea-formaldehyde adhesives: can you find the name of one?

Use the copies if you must, but try and use the book if you can.

**Response**

There is no index, so you have to rely on a good layout and a detailed contents page to help you find the information you need in this book. Well, to start with you should notice in the list of contents that there is a chapter on adhesives, materials and properties (chapter 4); and that in chapter 10 (Adhesives and Trade Sources) there is a trade names list. At p. 77 you find a section on urea-formaldehyde adhesives: this suggests that your 'Beetle' adhesive will be used for woodwork, but it does not tell you who made it. Don't lose this page; you will need to come back to it in a moment.

Turning to the trade names list in chapter 10, you find at p. 300 an entry under 'Beetle', confirming that it is used for furniture assembly; it also answers another of your questions by telling you that it is available both as a liquid and a powder. This entry refers you to code number 43 in the list of manufacturers. If you used the adhesive products directory, in chapter 9 you will have found an entry for 'Beetle' under urea-formaldehyde, giving you further technical data for this specific product, and referring you to code number 43 for the manufacturer. You will find this code number on p 318 where you discover the manufacturer to be B.I.P. Chemicals International Ltd.

Turning back to p. 77 you find that urea-formaldehyde resins have a shelf-life at 20 ° C of three to six months; at the end of the entry there is a reference to BS 1204, 1979 and BS 1203, 1979. Of course, you could have found these standards by using the BSI Catalogue, but do you remember some of the difficulties you encountered there? It is very convenient to have such information all gathered together for you here in the one specialised volume.

For a list of manufacturers of urea-formaldehyde adhesives you now need to turn forward again to chapter 10, Adhesives and Trade Sources. In Table 10.2, on p. 315, you find under 'urea-formaldehyde' a list of numbers referring you to the manufacturers listed in Fig. 10.4 (where you found code number 43 a few minutes ago). You look these numbers up and discover that only one manufacturer, CIBA-Geigy, is located near Cambridge.

So, we have completed our examination of reference books with a work you have to get to know if you want to make the best use of it, particularly as it lacks an index. But I think these questions will have shown you how much useful information you can find in specialised reference books such as this.

***********************************

**SAQ 2.3c**

In a table accompanying this question I have listed all the reference works I have mentioned or discussed in Section 2.1 to 2.3.

Keep this list; it will be a useful memory-jogger for you. Also, I shall be referring to it again for a different purpose when we reach Section 3.1.

Now, imagine you are in a well-stocked library. To help to create the illusion I have listed the books in the order in which my library shelves them, using the Universal Decimal Classification we encountered earlier (more explanations in Section 3.1). I have listed below ten questions, or types of information. For each one I want you to note the serial number from my list (ie 1–29, not the classification number), of the book you would go to first to try and find the information required.

There is not necessarily an absolutely correct answer to each question as the information may be found in more than one book; also, you will find that I have not referred to every book in the list.

Just note where you would go *first* for:

(*a*) a description of a process for the desulphurisation of coal;

(*b*) the title of a scientific reference book on wines and spirits;

(*c*) a list of the uses of different enzymes;

(*d*) the formula and properties of lead stannate;

(*e*) an explanation of biorheology;

(*f*) a method of identifying aspirin;

(*g*) the hazards presented by butyric acid;

(*h*) an outline description of organic dyes; ⟶

**SAQ 2.3c (cont.)**

(*i*) methods for the extraction of juices from fruit;

(*j*) a method of assay for the ammonia content of ammonium chloride

**Response**

Remember, I am going to tell you where I would look first (or the book in which I actually found this question). I will also try and indicate where else you might have looked. If you disagree with my answer, then go to a library and check the books for yourself to find out whether or not you are correct: the exercise will be worth your while no matter whether you eventually find yourself right or wrong.

To save time, and to make you refer back to the list (it will help you to remember titles by keeping on looking them up), I shall myself refer to the serial numbers in the replies. And remember that if I don't mention a work you have chosen you are not necessarily wrong (but check back to the text of the Unit if you are in any doubt); but I do think you would find the specified information most easily in the works I choose.

(*a*) 22 gives the best up-to-date description, but you might also have tried 23 or 24 or, possibly 5.

(*b*) You ought to have said 4; you might just find something in the bibliographics of various other works, but that would be at the unnecessary expense of a lot of time searching.

(*c*) 16 has a useful list, but you should also find the information in 5 or 18.

(*d*) 7 gives you this quickly and easily, although you could also find it in several reference works, notably 6 and 8. If you go to any of the major encyclopaedias for this kind of information (you will also find it in 5 and 22 after some searching) it is rather like using a sledgehammer to crack a nut.

(*e*) 5 gives the best introductory article. If you weren't sure what biorheology was then you should have looked it up first in a dictionary: I haven't given you any examples of general scientific and technical dictionaries here, but there are many good ones available in most libraries. The main point about asking this particular question is that biorheology is not a chemical concept, so you need to start with a more general scientific encyclopaedia. Remember one of the things I promised you about this Unit? You will be able to find information on subjects even if they are quite incomprehensible to you at first!

(*f*) 19 is the obvious and best source, although you could also refer to 11, 12, 13 or 15.

(*g*) The obvious sources are 9 or 10, although in fact 7 also tells you briefly what the hazards are.

(*h*) 22 and 24 are the best first sources, although you might also find something in 5.

(*i*) I set this question from 25, but you could just as correctly have tried 22 or 23.

(*j*) 23 covers this in some detail, but you might also have looked in 11 or 12, or even 13 or 14. If you had to follow a specific standard procedure then that would have been made plain to you, for you to seek the relevant specific standard.

Do you get the message? There are several different possible sources for the information needed to answer most of these questions, although some sources are, of course, more likely than others for each type of question. The other point is that if your choice of reference books is limited to a very few titles, you should still be able to extract quite a lot of useful information from them.

And finally, remember: if you have access to a well-stocked library, there are far more than 29 reference books available of potential value to the analytical chemist!

**********************************

**SAQ 2.4a**

Of the types of journal contents listed below, put a cross beside those you would expect to find in a research journal, and a tick besides those you would expect to find in a monthly news journal. Some will appear in both types of journal, of course, so put both a cross and a tick beside those.

Book reviews

Advertisements

Feature articles

Equipment reviews

Research papers

Conference reports →

**SAQ 2.4a (cont.)**

Review papers

Letters to the editor

Company and personnel reviews

**Response**

*Book reviews*: these could appear in both kinds of journal. Those in news journals are likely to be shorter than those in research journals.

*Advertisements*: most in news journals; but some research journals do carry a few advertisements, especially from scientific publishers.

*Feature articles*: only in news-type journals.

*Equipment reviews*: only in news journals.

*Research papers*: almost entirely in research journals, of course. But one or two American journals do mix research papers with feature articles and news items.

*Conference reports*: some research journals cover these, if haphazardly. News journals would be expected to cover them more widely, although it is surprising how many conferences seem to slip past unreported (mind you it is surprising just how many conferences take place!).

*Review papers*: with a few exceptions, as noted under 'Research papers', you will find these only in research journals. But more about review papers in Section 2.5.

*Letters to the editor*: in either type of journal, although their character and numbers will differ usually between a news and a research journal (more and generally shorter in the news journal). By the way, don't confuse letters to the editor, often arising out of something published in a recent issue, with the letters journals I mentioned in the text. Letters in letters journals are in fact very brief research communications (more-or-less unrefereed mini-papers).

*Company and personnel reviews*: only in the news journal.

I know you can, if you try, find exceptions to all those answers, but the general points are valid.

***********************************

**SAQ 2.4b**

You want to keep yourself up-to-date with what is happening in chemistry at large and in analytical chemistry in particular.

From the list below note the numbers of two journals you would choose to read regularly to try to fulfil these aims.

1. *The Analyst*
2. *Analytical Chemistry*
3. *Analytical Proceedings*
4. *Chemical Communications*
5. *Chemical Engineering Progress*
6. *Chemistry in Britain*
7. *Talanta*

**Response**

My choice (bearing in mind that I live and work in the United Kingdom) would be 6 and 3 respectively for general and analytical chemistry. If you chose differently, don't worry, but read the following notes and go back to the text to check details if you think you need to.

Or—dare I repeat myself? *Go to a library and look at copies of all the relevant journals you can find.*

1. It will take you a lot of hard reading to keep up-to-date by using this. It is a research journal, not a news monthly.

2. Not my first choice, but only because I preferred one published in the United Kingdom. If you chose this one knowing it to be an American publication then your choice was perfectly reasonable.

3. My choice to keep up-to-date with analytical chemistry in the United Kingdom; outside the U.K. I probably would have preferred no. 2.

4. These are pre-announcements of research findings: very interesting in themseleves, but to keep you up-to-date with news you need a more balanced selection of articles and news reviews.

5. If you want news and articles on engineering aspects of chemistry in the USA this would be the right choice; but it does not quite fit the requirements I specified in my question.

6. This would be my choice for news and general feature articles on chemistry as a whole.

7. This publishes only research papers and so is not suitable for keeping up-to-date with news.

***********************************

**SAQ 2.5a**

Look at the following uses for reviews and list them in order of likely importance to you as an analytical chemist in relation to any specialised topic in which you might be interested.

(*a*) To find lists of papers.

(*b*) To introduce an unfamiliar subject.

(*c*) To give you a compact set of results from a wide range of papers.

(*d*) To evaluate published papers in comparison with one another.

(*e*) To identify areas of the subject still needing research.

**Response**

To some extent the answer to this question is subjective, but I would suggest the following as the most likely order of usefulness to you:

(*d*)
(*c*)
(*a*)
(*b*)
(*e*)

Don't worry if you disagree with me, but read the following observations anyway.

(*d*) I think that in common with most users of reviews you should find the evaluative and comparative functions of reviews their most significant feature.

(*c*) Reviews can save you an awful lot of reading time. After all, that is one of the reasons you use textbooks and handbooks. So this is an important reason for using reviews.

(*a*) The lists of papers are very useful. I have put this third, mainly because I think both (*d*) and (*c*) are more important, but also because you can get lists of papers by other means, as you will see in section 3.2.

(*b*) Yes, a review can do this; but I think that initially you would be better served by an entry in an encyclopaedia or a handbook.

(*e*) A valid reason for using reviews when you are at the research stage; but I suspect that it is unlikely to apply to most of you just yet.

***********************************

**SAQ 2.5b**

> You want to know about research on new instrumentation techniques in mass spectrometry over the last few years. Write down the titles of three sources of reviews where you would look for likely papers and to give yourself the best coverage you can. Then, when you have looked at these three sources, write down where you would turn next to look for further or more up-to-date reviews. ⟶

**SAQ 2.5b (cont.)**

You will be able to answer this question from the titles quoted in the text of this Section (in other words you can ignore the titles in the three lists).

**Response**

Our first choice is fairly obvious, isn't it? In one of the Royal Society of Chemistry's Specialist Periodical Reports you have a title on your very subject, *Mass Spectrometry*. You would expect to find material on mass spectrometric instrumentation techniques here (and, sure enough, you will in fact find exactly such an article in vol. 6, 1981).

The next two sources aren't so obvious, perhaps. But I would try two regular publications which might bring us more up-to-date, namely: Fundamental Reviews (*Analytical Chemistry*) and CRC Critical Reviews—*Analytical Chemistry*. I think these are the most likely as they deal specifically with analytical chemistry. But if you chose perhaps *Chemical Society Reviews* or *Chemical Reviews*, then the principle of using a regular journal source to up-date an annual or biennial publication is the same; but at the same time you are much less certain to find a specific subject such as mass spectrometric instrumentation, in what are more research oriented journals covering the whole field of chemistry.

And that final step? I hope you would think to turn to either, or both of *CA Reviews Index* and *Index to Scientific Reviews*, to see if anything relevant has been published in less obvious sources.

*********************************

**SAQ 2.6a**

Look at the following list of literature types. Then write them down in the chronological order in which you would expect each type to be produced in the publication of any research project (eg if you think the first step in publishing new research results would be to write a reference book, put that at the top of your list). Some of the types will, in practice, overlap at least slightly, but try the exercise anyway I shall now make one or two important points arising out of it.

⟶

**SAQ 2.6a (cont.)**

Journal papers
Textbooks
Patents
Reports and conference papers
Handbooks, standards, and reference books
Review papers

**Response**

The order I suggest is as follows:

1. *Reports and conference papers*. I put these together because either or both are likely to be produced in the course of a research project, and more-or-less as soon as it is completed. In fact, a conference paper might also be produced some time after the completion of a project as well.

2. *Patents*. Anything of potential (or actual) commercial value discovered during the project will normally be patented immediately.

3. *Journal papers*. Don't forget this is the standard form of communication of scientific research results. But it will not normally be written until the project is completed or at least some significant results are available.

4. *Review papers*. These obviously have to follow journal papers. Here is the first step in assimilating new results into the corpus of the traditional scientific reference works and textbooks.

5. *Handbooks, standards, and reference books*. Useful results will normally be assimilated into these at a fairly early opportunity. The precise order in which these might be published will be determined normally by commercial, *not* scientific considerations.

6. *Textbooks*. Finally (and often after some considerable time unless there has been a major revolutionary innovation), the new results will be assimilated into scientific knowledge and be presented in textbooks.

You might care to ponder on the implications of that pattern for your own learning process. Now at least you have another good reason for studying this Unit! But back to the text for final explanations.

***********************************

**SAQ 3.1a**

> Specify any *two* types of library to which you could claim automatic right of access to consult their stock.

**Response**

(*a*) The first obvious choice is the public library: you can use a large municipal reference library (often with a specialist science and technology department) or borrow books from a lending library where you live or where you work.

(*b*) If there is a library in your organisation you will, of course, be able to use it. And it will be conveniently located.

Those are the two I would expect you to have nominated. But you might also have opted for (*c*) or (*d*).

(*c*) The library of a college or polytechnic, if you are following a course at the institution. If you are not, you might still be allowed to use the library if you ask. But remember that unless you have some formal connection with the institution you have no automatic right of access to any college, polytechnic, or university library.

(*d*) You can use the Science Reference Library in London without formality; or apply for a reader's ticket at the National Library in Cardiff or that in Edinburgh, although I think you are rarely, if ever, likely to need to use either of these.

***********************************

**SAQ 3.1b**

> (*a*) First of all note any two forms in which documents might be given you in a library, apart from books and journals.
>
> (*b*) Then note any three other kinds of materials you might find kept in a library, apart from books, journals, and whatever you gave as your answer to (*a*).

**Response**

(*a*) The most likely are photocopies, computer print-outs, or microforms; microforms may include microfilm, microfiche, or ultrafiche (which contain several hundred pages per fiche rather than the standard 60 to 100). Any other forms really belong in the answer to (*b*).

(*b*) Take your pick from:

Colour slides and film strips,
Overhead transparencies,
Videocassettes and cine films,
Audio tapes (and cassettes) and gramophone records,
Sheet music,
Computer programs or files (tapes, discs, etc),
Teletext data services,
Maps,
Photographs.

Some or all (or none, of course) of these may be available in many libraries nowadays. And on the horizon we have the videodisc and other forms of new communication technology.

So the overall moral is: don't just use the library for traditional documents. There is much more to it than that. Look on the library as a source of information, which might be found in many types of material including electronic data bases. And don't think of the library just as it appears to you—it is the gateway to a far wider world of information and documents.

**********************************

**SAQ 3.1c**

I have given you in the next few pages some copies of parts of the Dewey Decimal Classification schedules and indexes. No, I am not going to be strict this time—it is not a book you are likely to consult much at all (you probably will find yourself using the resulting classification numbers frequently); but if you are answering this question in a library and they have a copy of the 19th edition, ask whether you can use it for a few minutes to answer this question (and have a general look at it while you have the opportunity). ⟶

**SAQ 3.1c (cont.)**

Anyway, use the alphabetical index first and find classification numbers for the following:

(*a*) Spectroscopic analysis,

(*b*) Analysis of drugs,

(*c*) Chemical technology of soft drinks.

In each case check back to the schedules from the index to make sure you have the correct number. And ignore in the schedules any references to tables or to standard subdivisions: I am trying to keep this fairly simple for you, not teaching you to become librarians!

**Response**

(*a*) This is a pretty good index to use, isn't it? You are led from 'spectroscopic' to 'spectrochemical' and there find a few alternative numbers each briefly defined. Checking the schedules you find the general number to be 543.085 8, or a more specific one within qualitative chemistry of 544.6 or quantitative chemistry of 545.83. Then there are the numbers for organic analysis. This is part of the answer to the problem I raised in the text, and also demonstrates one of the weaknesses of a rigid classification system: there have to be alternative locations for the one subject to allow for the different contexts in which it might appear.

(*b*) Check under 'analysis', 'drugs', or 'pharmaceuticals' and you are led either directly or at one remove to 615.190 1. And here you are in another area of the scheme I haven't mentioned: pharmaceuticals are in the medical section.

(*c*) You must have noticed by now that this is an American publication. The common term 'soft drinks' leads you to the terminologically exact but hardly everyday term 'non-alcoholic beverages'. The 640-schedules relate to 'home economics and family living', so you will have guessed (or discovered in the schedules) that this is not the number you are looking for. But under the rather cryptic sub-heading 'comm. proc.' you find a reference to 663.6, and I hope you will have realised that 660 is chemical technology; so this is likely to be the number you seek, confirmed by checking the schedules.

***********************************

**SAQ 3.1d**

Look at the following three classification numbers and write down the classification scheme to which each belongs:

(*a*) 543.089 4

(*b*) QD 330

(*c*) 663.14.031.3:543.053

**Response**

(*a*) This is from the Dewey Decimal Classification and has straightforward decimal numbers. You might think it was from the Universal Decimal Classification, but in that case it would have had the decimal point after the 9 as well as after 543. This number signifies liquid chromatography.

(*b*) This is obviously a Library of Congress Classification number by its mixture of letters and numerals. This number signifies aromatic chemistry.

(*c*) This is clearly a Universal Decimal Classification number. You can tell by the colon, even if you didn't notice the way the first number was sub-divided by decimal points. This number as a whole signifies sample-collection apparatus (543.053) for nitrogen compounds as microbiological raw materials (663.14.031.3).

***********************************

**SAQ 3.1e**

Write down the name words you would look up in the catalogue to find works by the following authors or for which the following bodies are responsible:

(*a*) Knut Schmidt-Nielsen,

(*b*) The Chemical Society's Continuing Education Committee,

(*c*) Richard T. O'Connell, ⟶

**SAQ 3.1e (cont.)**

(*d*) Anthony Hartley De Borde,

(*e*) The United Kingdom Department of Industry's Microprocessor Applications Project,

(*f*) Glasgow University's Department of Chemistry.

**Response**

(*a*) You would find the entry under SCHMIDT-NIELSEN, Knut, in nearly all library catalogues. If you looked up NIELSEN, Knut Schmidt-, you ought to find a reference telling you to look under the form of the name I have indicated.

(*b*) You should look first under the name of the main body, then there will normally be sub-headings under that for any subordinate committees, thus: Chemical Society. Continuing Education Committee. You should not find any entries directly under Continuing Education Committee as it is not usual practice to put even references from such obviously subordinate bodies. Another point, by the way: I know the Chemical Society is now the Royal Society of Chemistry, but this book was published when it was still the Chemical Society. You might be given a reference between the two names whenever an organisation changes its name, but you might also have to remember to look in both places for yourself. Obviously libraries can't go changing all their headings every time some organisation changes its name.

(*c*) You will find the entry under O'CONNELL, Richard T. It will be alphabetical as though the apostrophe did not exist (ie as if spelled OCONNELL), otherwise all the entries O'... would fall together in a separate sequence.

(*d*) There would be an entry under DE BORDE, Anthony Hartley. There might be a reference from BORDE, Anthony Hartley De, but the cataloguer might just as easily have assumed that you would know to look directly under DE BORDE. Going back to the point about alphabetisation, in most catalogues surnames prefixed De ... will together in a sequence, thus:

DE BONO
DE BORDE
DE VERE
DEAN
DELAWARE

But do you recall my mentioning earlier (in Section 2.2) the difference between word-by-word and letter-by-letter filing? The order of this sequence could also be:

DEAN
DE BONO
DE BORDE
DELAWARE
DE VERE

This second list is letter-by-letter filing, the first sequence was word-by-word. No great problem in this example, but imagine the implications in a card catalogue with thousands of entries.

(*e*) I expect you would look directly under Department of Industry. If so, you would normally (but not always—some libraries would catalogue this directly under 'Department') find a reference saying 'See Great Britain. Department of Industry'. As the names of ministries and departments can be identical in different countries, most libraries enter government documents under the name of the country first. If you looked under Microprocessor Applications Project you should also have found a reference to the full main heading: Great Britain. Department of Industry. Microprocessor Applications Project. Libraries do not usually use more than one sub-heading, but do make exceptions for some government bodies like this. They should never use more than two sub-headings.

(*f*) 'Department of Chemistry' could be a subordinate department to many organisations, so I hope you wouldn't have looked under that. My point in putting this particular question is that the institution is in fact likely to be found under 'University of Glasgow'. There should be a reference from 'Glasgow', but not certainly, and you might have to think for yourself to look under 'University' if you find nothing under 'Glasgow'. The reason for entering it under University of Glasgow is quite simply that that is the official name of the institution.

************************************

**SAQ 3.2a**

When I set you some of the first of these practical SAQs I promised you that we would find ourselves in many varied laboratories and types of employment. Well, let's go the whole hog and find ourselves analysing extra-terrestrial materials. ⟶

**SAQ 3.2a (cont.)**

You will be working eventually on the analysis of samples returned from the planet Mars. You decide that initially you want a fairly broad introduction to the obviously special problems involved, and so you turn to *Current Technology Index*. Quite simply, go to the 1983 annual volume and see if you can find a likely paper. When you find one write down its details as a standard bibliographic reference, as I showed you in the Introduction (ie Author, title of paper, title of journal, volume, date, and pages).

I haven't had to say this for a little while, have I? But here we go again: *Current Technology Index* is widely available in libraries of all kinds in the United Kingdom, so try and use a copy of the book. Use the copies on the next few pages only if you are quite unable to find the book itself.

**Response**

I expect most of you would look first under 'Mars'. If you started at 'space', 'orbiting' or 'samples' you will have found nothing and had to try and think of other terms. Anyway, at 'Mars' you find reference to a heading you are unlikely to have thought of, namely 'Astronautics flights: Mars'. Turn to that heading and you will have found a very specific entry headed 'Astronautics flights. Mars: samples: analysis: laboratories: earth orbit' with a reference to a paper which, I hope, you will have transcribed more-or-less as:

> Hesselberg, P. Orbiting quarantine facility: the Antaeus Report. *Spaceflight*, v.25 (Jan 1983), 26–27.

If your reference is significantly different from mine (that is more than just punctuation or capitalisation) go back to Section 1.7 and refresh your memory.

Being an unfamiliar work I expect you would have at least glanced at what entries there are under 'analytical chemistry'. In fact at 'Analysis, checmical' you find very few entries but a lot of useful references to more specific headings.

So this isn't very difficult to use is it? You have found a useful-looking item in a non-chemical source journal. Of course if you want more detailed chemical information you will have to use more specialised sources. I shall soon be giving you practice in those.

************************************

**SAQ 3.2b**

I have tried to make my descriptions and explanations of *Chemical Abstracts* as clear and simple as possible. But given the enormous amount of material indexed there are bound to be complexities. These are best confronted by actually using the service, so here are a few questions to make you use the different indexes I described. All the questions are set from one volume (Vol. 100) only, and so remember that you are covering only six-months chemical literature. And let me remind you to try to use the books themselves if you can get access to a set of *Chemical Abstracts*: believe me, you will get a much greater sense of achievement by finding these needles in what will at first seem a dauntingly large haystack, than if you rely on the copies on the following pages. I must also warn you that, due to difficulties in printing this Unit, the copies of *Chemical Abstracts* we have given you here look simpler than the real thing. You are only seeing about two thirds of a single column each time—there are in fact three columns to each page of the indexes and two columns per page in the abstracts section in the original publication. But sets of *Chemical Abstracts* are not all that widespread, so use the photocopies if you must; they will serve to demonstrate my points and give you useful initial experience. In each case refer to the abstracts and write your references out as a full bibliographic citation, as I showed you in Section 1.7 (but note that you will have to edit out, in doing so, some of the data provided with the abstracts and references).

We are going to start with the easiest search, by authors, and then proceed into gradually more complex subject searches.

(*a*) You have been told that Richard A. Dluhy has published papers on applications of Fourier-transform infrared spectroscopy. Find two papers, of which he is one of the authors, on this subject, by using the author index to volume 100 of *Chemical Abstracts*. I should warn you that reference 98632p is not one of the two papers you are seeking. ⟶

**SAQ 3.2b (cont.)**

(*b*) Let's go back to our space question. Having read the paper you found in *Current Technology Index*, you want more detailed information. In particular you want information on the water content of the atmosphere of Mars. Use the General Subject Index to volume 100 of *Chemical Abstracts* to find a paper on this subject. And meanwhile back on Earth you are also involved stratospheric analysis. You have been told of a new balloon-carried quadrupole mass spectrometer linked to a microprocessor and developed in Europe. Use the same General Subject Index to find a paper describing this.

(*c*) Do you remember a question about Isoniazid earlier in this Unit? Following up what you learned in the *British Pharmacopoeia* you want to know whether more recent work has been done on its determination in tablets. Use the Chemical Substance Index to volume 100 of *Chemical Abstracts* to find a relevant paper.

(*d*) To cross-check on this and related substances, look in the Chemical formula Index and note just the reference number and the CA Registry Number (you don't have to look the full reference up this time) of a paper on Isoniazid mixed with another compound. You need to find the formula first so you could check it quickly and easily in either the *British Pharmacopoeia* or *Merck Index* (or look back to the photocopy with SAQ 2.1.6 if you are stuck).

**Response**

(*a*) There is no intermediary to the Author Index so you turn straight to Part 1 (A–Lea) and look up Dluhy, Richard A. You see the last reference under his name, written in conjunction with three other authors, is obviously one of the two you need. Turn to the abstracts in numerical order and 2450d gives you a reference which you should have described more-or-less like this:

Dluhy, Richard A. *et al*. Fourier-transform infrared spectroscopic studies of the effect of calcium ions on phosphatidylserine. *Biochemistry*, v.22 (1983), 6318–6325.

But what about a second paper? Did you notice that the first entry under 'Dluhy, Richard A.' was a cross-reference to Mendelsohn, Richard? (he is also one of the co-authors of the paper you have already found). Turn to Part 2 (Leb–Z) of the author index, look under Mendelsohn, Richard, and you find that R.A Dluhy and the others are co-authors of the second paper you are looking for. Check back to abstract 116761m and you can produce your second reference:

> Mendelsohn, Richard *et al*. Interaction of glycophorin with phosphatidylserine: a Fourier-transform infrared investigation. *Biochemistry*, v.23 (1984), 1498–1504.

(*b*) I hope you remembered that before you ever use the General Subject or Chemical Substance Indexes you must first consult the Index Guide. I would think we should generally expect to look under 'Mars' for our reference, and the first thing we learn from the Index Guide is that we must in fact look under 'Planets–Mars'. You will reach the same entry if you start under 'atmosphere' in the Index Guide, as the note there tells you to look under the names of celestial bodies concerned if you are looking for extraterrestrial atmospheres. So, into the General Subject Index under 'Planets–Mars' and among four entries under 'atmosphere' you find one on 'water in, modelling of at 10334j. Turn to the abstract and you can produce the reference:

> Kulikov, Yu. N and Rykhletskii, M.V. Modeling of vertical water distribution in the atmosphere of Mars. *Astron. Vestn.*, v.17 (1983), 144–152.

Do you notice the annotation '(Russ)' in the entry? This tells you that the original paper is in Russian, but at least you have an English abstract to read before you decide whether to set about finding a translation.

You follow much the same procedure for the second part of this question. The Index Guide gives you no relevant entries at 'quadrupole', but you see that the term 'mass spectrometer' is used. Check there in the general subject index and you find a sub-heading 'quadrupole'; scanning the entries under that you quickly find 'microprocessor based data acquisition and control system for balloon-borne, 111951p'. Turn to the abstract and you can produce the following reference:

> Nevejans, D. *et al.* Microprocessor based data acquisition and control system for a balloon borne quadrupole mass spectrometer. *Bull. C1. Sci. Acad. R. Belg.*, v.68 (1982), 314–332.

Despite its provenance you can tell from the annotation '(Eng.)' that the full paper is in English. By the way, I expect you have noticed that for our bibliographic references we have adopted the acceptable abbreviations of the titles of journals as used by *Chemical Abstracts.*

(*c*) First of all you must find under what name *Chemical Abstracts* indexes Isoniazid. Starting as ever in the Index Guide you discover that the IUPAC name is used: 4-Pyridinecarboxylic acid, hydrazides, hydrazide (with a CA Registry No. 54-85-3). Turn to the Chemical Substance Index Part 4 (O–P) and look under Pyridinecarboxylic acid, then the various sub-headings, and you come to 4-Pyridinecarboxylic acid, hydrazides, with the further sub-heading 'hydrazide (isoniazid) (54-85-3)', so that you know you have the substance you are seeking. As usual the individual entries are arranged alphabetically by the abbreviations of their titles. Scanning these you find six entries beginning 'detn. of', the last of which refers to tablets, by spectrophotometry, with the reference 126991a. Of course you know by now to check the source abstracts and produce your references:

Hassib, S. T. and Moreed, L. Spectrophotometric determination of isoniazid. *Egypt. J. Pharm. Sci.*, v.22 (1981), 143–152.

Here you have a method clearly outlined in the abstract. And do you notice that the work was done in Egypt? It illustrates my earlier point about seeking truly international coverage of the literature.

(*d*) Finally, having discovered the formula of Isoniazid by the Hill System to be $C_6H_7N_3O$, you consult the Formula Index Part 1 (A–C15). At this formula you find a number of related compounds, but scanning down the list you find 4-Pyridinecarboxylic acid with the sub-heading hydrazide (and the Registry No. for isoniazid, 54-85-3). This in fact refers you back to the Chemical Substance Index you have already checked. But the next entry is for the hydrazide mixed with *N*-[4-[[(aminothioxomethyl)hydrazono]methyl]phenyl]acetamide. This has a registry number 62682-46-6 and refers you to abstract 271x.

Let me just sum up the implications from this SAQ. Do you now see how you can use *Chemical Abstracts* to locate specific work in the mass of chemical literature, in any language and from anywhere in the world? But don't forget that in this question you have searched only six months' literature: to perform a thorough search you would have to start at the most recent volumes and work back through individual volumes until you were satisfied that you had gathered all the references you need. But it wasn't too difficult was it? And just see what precise information you have found out of all that literature abstracted.

***********************************

**SAQ 3.2c**

Use volume 46 (1984) of *Analytical Abstracts* to find a couple of papers on atomic absorption spectroscopy.

(*a*) A review paper on the use of atomic absorption spectrophotometry in flow-injection analysis. ⟶

**SAQ 3.2c (cont.)**

> (*b*) A review paper on the applications of atomic absorption spectroscopy in the determination of anions and organic compounds.
>
> When you have found your two references write each of them out in the form I have already shown you.
>
> For the last time, do try to use the original publication. Use the copies on the following pages only if you must.

**Response**

Of course we start at the annual index. This isn't always the easiest index to use—you do have to think for some of your entries (not that that is necessarily a bad thing!).

(*a*) Check 'infra-red' or 'absorption' and you find nothing at all, not even a cross-reference. But keep looking and under 'spectrophotometry' you find the sub-heading 'atomic-absorption'. You then have to scan quickly through the individual entries under this until you find one headed 'in flow-injection analysis, review'. This is obviously what you want. It refers you to '2J6' which means you must turn to issue No. 2 (ie February 1984), Section J, item 6; there you find your entry which you should have transcribed as:

Tyson, J. F. Flow-injection methods and atomic-absorption spectrophotometry. *Anal. Proc.*, v.20 (Sept 1983), 488–491.

If you started in the index at 'Flow-injection analysis' you will, again, have had to scan the entries until you find one 'with a.a.s. detection' and then a sub-heading 'review', and of course the same reference to 2J6.

(*b*) Much easier this time wasn't it? In fact, if you were scanning the entries under 'Spectrophotometry, atomic-absorption' for the answer to (*a*) you probably found this one first. A straightforward reference 'applications of, review' leads you to 6B14 which, on exactly the same principle as for (*a*) you find and produce a reference:

Garcia-Vargas, M. *et al.* Atomic-absorption spectroscopy as a tool for the determination of inorganic anions and organic compounds, I: a review. *Analyst*, v.108 (Dec 1983), 1417–1449.

If you started looking under 'spectroscopy' you will have found nothing; but I think you will have realised that the heading used is 'spectrophotometry'.

Don't forget that I have just set you these as examples. If you were doing a real search you would be checking more volumes of *Analytical Abstracts* than just the one I have used, and perhaps looking more closely at some of the other references listed. But when you find a review paper you can also check on the references listed there and thus save yourself a lot of time.

**********************************

**SAQ 3.2d**

Here is a list of publications or services which can give you bibliographic information. List them in the order you would expect them to be published.

(*a*) Abstract Journals,

(*b*) Bibliographies,

(*c*) Current Awareness Services,

(*d*) Indexing Journals,

(*e*) Guides to Literature,

(*f*) Review Papers.

**Response**

The correct order is:

(*c*)
(*d*)
(*a*)
(*f*)
(*b*)
(*e*)

Even if you got that absolutely right, read my notes below as they help to summarise this Section.

**Current Awareness Services**. These are the obvious first step in the process. We have looked only at *Current Contents* as the most useful and up-to-date published service available. In fact many current-awareness services are now offered as electronic alerting services tailored (at a price) to your individual requirements. The Chemical Abstracts Service offers *CA Selects*, alerting bulletins on a wide range of specialist chemical topics (including several titles in chromatography, a dozen in spectroscopy and few in analytical chemistry); but these are not as up-to-date as *Current Contents* since they rely on data generated for the full *Chemical Abstracts.*

**Indexing Journals** will normally be the next stage as they simply take the details of journal papers (and any other publications included in their coverage) and list them with whatever subject indexing entries they use. But simply producing these index entries introduces some delay into the process.

**Abstract Journals** must take longer than indexing journals if only for the time taken to write an abstract of a paper. But remember that a very specialised regular abstracting publication might still be produced faster than a large, comprehensive indexing service.

**Review Papers**. Although I haven't covered them in the text of this Section, I hope you hadn't forgotten these as sources of bibliographic information. They also have a very definite place in this process, as they list all the papers discussed. But obviously it will take some time for a fairly large amount of literature to be read and assimilated, and the paper written up.

**Bibliographies** might, if you are lucky, be produced quite quickly. But usually you should consider them as unlikely to be as up-to-date as the other sources I have listed ahead of them.

**Guides to the Literature** are generally to be considered as ordinary books, with all the inherent delays that can mean. But a good one (such as any of those I listed here or in the Introduction to the Unit) can remain useful for some years.

*********************************

**SAQ 3.2e**

(*a*) Without referring back to the text, take the following steps in a literature search and arrange them in order:

Define the sources
Record references
Evaluate findings
Conduct the search
Test search
Define the topic →

**SAQ 3.2e (cont.)**

Examine relevant material
List the search terms

(*b*) Having listed the steps in order, look at the three categories of documents below and write them down beside your list of search steps to correspond with whereabouts in your search you would expect to be consulting them:

(*i*) Abstracts/Indexing services/reviews,
(*ii*) Journal papers/patents/reports,
(*iii*) Encyclopaedias/dictionaries/textbooks.

**Response**

The order and correspondences are:

| | |
|---|---|
| 1. Define the topic | |
| 2. Define the sources | Encyclopaedias/dictionaries/ textbooks. |
| 3. List search terms. | |
| 4. Test search | Abstracts/indexing services/ reviews |
| 5. Conduct the search | |
| 6. Record references | |
| 7. Examine relevant material | Journal papers/patents/reports |
| 8. Evaluate the findings | |

If your answer doesn't match mine then re-read the text of this Section and, if you are uncertain, read Section 2.6.6 again as well.

***********************************

**SAQ 4.1a**

> You have been working on the analysis of lead levels in soil samples by using the dithizone method. Now you wish to study in some detail atomic absorption and polarographic methods for the same application. To find out what has been published on these methods you are going to carry out an on-line search of *Chemical Abstracts*. Write down a definition of the subject you will search and which will be suitable for use as a search strategy. Write out your definition in the algebraic form: eg (a or b) and (c and d) not (e or f) etc, by using the appropriate 'and', 'or', 'not' operators.

**Response**

I hope you have a strategy looking something like this:

> (Analysis or determination) and Soil and (Atomic absorption or Polarographic) not Dithizone

Don't worry if your strategy looks a bit different, or includes some other terms, but just read through my comments here anyway.

'(Analysis or determination)' is fairly obvious I think. 'Soil' is also an obvious key term; in fact with more experience you will find that the best term to use would be 'Soil or Soils' to include both singular and plural forms (you have to specify everything like that to the computer), but I don't expect you to have thought of such subtleties at this stage! You might have used 'Soil samples' or 'Soil and Samples'—it is quite logical and correct if you did, although I think experience will show you that the term 'sample' is not strictly necessary here. 'Atomic absorption' or 'Polarographic' is also pretty obvious—you don't strictly need any qualifying terms like 'methods', although you might at some stage in your search decide to specify these techniques more precisely. The last term 'not Dithizone' is also pretty obvious in this context.

Don't stop there: look at your own work, or at applications arising out of any other Units you are studying, and think out a few typical strategies to search for information on these topics.

***********************************

**SAQ 4.1b**

Below is a search already defined for you. Write out how you would enter this search on your terminal. Use the commands 's' for 'search' and 'c' for 'combine', and use the logical operators 'and', or 'not' as you think necessary. Join complex multi-word terms with the '(w)' operator if necessary, and use the '?' for truncating terms if you think you need to. Set your search out line-by-line and number each set you create. Stop at the final search command you would enter (ie don't bother asking for prints).

Your search is:

HPLC of fatty acids, excluding oleic acid.

**Response**

The following is how I would expect you to have entered the search. Don't worry if you have the spacings wrong at this stage (but remember it does matter when you are live on-line), and I shall be surprised (but pleased) if many of you thought of entering the alternative forms of the term for HPLC. I have checked this search on-line in *Chemical Abstracts* (1967–1985 on the ESA/IRS host), and just to show you how it comes out I have put in brackets the numbers of references produced at each set.

1 sHPLC
(5735)

2 shigh(w)performance(w)liquid(w)chromatograph?
(11776)

3 clor2
(14215)

4 sfatty(w)acid?
(63353)

5 soleic(w)acid
(2785)

6 c4not5
(62446)

7 c3and6
(194)

Set 1 of the common abbreviation 'HPLC' produces over 5,000 references, but the full form finds nearly 12,000: note that I have truncated 'chromatograph-' so that our search will include 'chromatography' or 'chromatographic'. The two forms combined at Set 3 give us over 14,000 references. 'Fatty(w) acid? (the truncation gives us both singular and plural entries) is followed by oleic acid (singular form only this time) which we then exclude, reducing Set 4 of over 63,000 references to Set 6—almost 1,000 references fewer. See how the computer manipulates these enormous sets with no delay. Finally we combine Set 3 and 6 to produce a more manageable (but still rather daunting) set of 194 references.

That has searched the whole of *Chemical Abstracts* from 1967 to date at one go; it took me less than four minutes computer time (and I could have done it even faster!). Look at a collection of printed *Chemical Abstracts* for the same period and try to guess how long it would take you to do the same search manually (never mind copying out the 194 references!). Suddenly one of the major benefits of on-line searching becomes very obvious, doesn't it?

**********************************

**SAQ 4.2a**

Here are two searches for you to enter on-line. The STN commands for searching CAS ONLINE are the same as those used earlier for ESA/IRS. The first search is to be on the gas chromatographic analysis of myristic acid. You have already checked the *Chemical Abstracts Index Guide* (or some other directory of chemical substances which gives you the same kind of information) and found that the preferred name for myristic acid is tetradecanoic acid, with a CA Registry No. 544-63-8. Write out your search step-by-step as usual, with the relevant commands and operators. By the way, one difference you will note in STN is that your set numbers are prefixed 'L' and that you always use the 's' command (not 'c') to combine as well as to search (eg 's L1 and L2')

The second search seems much simpler even than the first, but what I want you to do this time is enter the whole search as a single entry, in algebraic form—eg 's (x or y) and (z or v) not w'. Use the usual commands and operators as necessary in your string. The search to try is much the same as one we have already performed manually: analysis of the atmosphere of Mars and of Venus.

AN CA103(15):120340c

TI Gas chromatography-mass spectrometry study of propolis. Analysis of phenolic acids and sugars
AU Maciejewicz, Wieslawa; Daniewski, Marek; Mielniczuk, Zbigniew
CS Zakl. Chem. Nieorg. Anal., Akad. Med.
LO Lublin 20-081, Pol.
SO Chem. Anal. (Warsaw), 29(4), 421-7
SC 12-1 (Nonmammalian Biochemistry)
DT J
CO CANWAJ
IS 0009-2223
PY 1984
LA Eng
AB The H20-sol. fraction of propolis, collected in southern Poland by Apis mellifera carcina, was analyzed. The fraction was composed mainly of D-glucose, D-ribofuranose, D-glucitol, talose, D-fructose, D-glucose, saccharose, cinnamic acid, 3,4-dimethoxycinnamic acid, 2-amino-3-methoxybenzoic acid, isoferulic acid, and caffeic acid. A gas chromatog.-mass spectrometric anal. of the volatile fraction (BzOH, phthalic acid, pentadecanoic acid, palmitic acid and undecanoic acid and their esters).
KW propolis compn carbohydrate phenol Apis
IT Propolis
(carbohydrates and org. acids of, of bee)
IT Sesquiterpenes and Sesquiterpenoids
Carbohydrates and Sugars, biological studies
Carboxylic acids, biological studies
(of propolis, of bee)
IT Honeybee
(A. mellifera carnica, carbohydrates and org. acids of propolis of )
IT 57-10-3D, esters 88-99-3D, esters 112-37-8D, esters 1002-84-2D, esters
(of propolis of bee)
IT 50-70-4, biological studies 50-99-7, biological studies 57-11-4, biological studies 57-48-7, biological studies 57-50-1, biological studies 65-85-0, biological studies 120-51-4 121-33-5 331-39-5 537-73-5 544-63-8, biological studies 613-83-2 621-82-9, BIOLOGICAL STUDIES 2316-26-9 3177-80-8 3843-74-1 4205-23-6 30077-17-9
(of propolis, of bee)

ANSWER 2

AN CA103(14) : 110013j
TI Gas chromatographic analysis of the injections of polyphase liposome 139 (antitumor agents)
AU Sun, Yuging; Zhao Liping; Gu, Xueqiu
CS Dep. Anal. Chem., Shenyang Coll. Pharm.
LO Shenyang, Peop. Rep. China
SO Shenyang Yaoxueyuan Xuebao, 1(3), 228-33

**Fig. 4.2b**

```
SC 64-2 (Pharmaceutical Analysis)

DT J
CO SYXUE3
PY 1984
LA Ch
AB Anal. of an antitumor polyphase liposome 139 injection sample
   (contg. mainly oleic acid [112-80-1] and other polyunsatd.
   fatty acids) and a com. oleic acid sample by gas
   chromatog. is described.  A sample was heated to sep. the oily
   layer from the water layer.  The oily layer was sapond. with
   0.5N NaOH-MeOH, and the fatty acids liberated were esterified
   with MeOH in the presence of BF3.Et2O for measurement by gas
   chromatog.  Reproducibility with a relative std. deviation of
    3% was obsd. (for oleic acid).  Myristic acid [544-63-8],
   palmitic acid [57-10-3], hexadecenoic acid [28039-99-8],
   heptadecenoic acid [28040-00-8], steric acid [57-11-4].
   Linoleic acid [60-33-3], octadecatrienoic acid [27213-43-0]
   and eicosadienoic acid [32839-28-4] also was detected in com.
   cleic acid samples.
KW oleic acid detn liposome injection;  gas chromatog polyphase
   liposome injection;  polyunsatd fatty acid liposome antitumor
IT Neoplasm inhibitors
      (fatty acids in polyphase liposome injections, anal. of, by
       chromatog.)
IT Liposome
       (polyphase, injections, anal. of fatty acid mixt. in, by
        gas chromat.)
IT Fatty acids, analysis
       (polyunsatd., detn. of, in antitumor polyphase liposome
        139 injection by gas chromatog.)
IT 57-10-3, analysis 57-11-4, analysis 60-33-3, analysis 112-80-1,
   analysis   544-63-8,  analysis   27213-43-0   28039-99-8  28040-00-8
   32839-28-4
```

**Fig. 4.2b (cont)**

```
FIG.4.2C

=> s atomospher? and analysis and (Mars or Venus)
       36137 ATMOSPHER?
      427085 ANALYSIS
        1157 MARS
        1161 VENUS
L7        39 ATMOSPHERE? AND ANALYSIS AND (MARS OR VENUS)

=> D L7 1-2 ALL

L7  ANSWER 1 OF 39

AN  CA102(4):36033b
TI  Possibility of spectroscopic detection of oxygen in the lower
    atmosphere of Venus
AU  Krasnopol'skii, V. A.
LO  USSR
SO  Kosm. Issled., 22(5), 812-15
SC  73-9 (Optical, Electron, and Mass Spectrometry and Other Related
    Properties)
DT  J
CO  KOISAW
IS  0023-4206
PY  1984
LA  Russ
AB  Theor. evaluation of the means of spectroscopic detection of 02 in
    the lower atm. of Venus showed that a detection threshold of 1 ppb
    can be attained using a diffraction spectrometer with the resolving
    power of 1 .ANG., assuming a const. relative 02 content throughout
    all atm. This detection threshold is equiv. to that of 3 and 30 ppb
    for 02 present only at heights of >25 km and only in a cloud layer,
    resp.
KW  Venus atm oxygen detection spectrometer; analysis oxygen planet Venus
    spectrometer
IT  Spectrometers
       (for oxygen detection in Venus lower atm.)
IT  Planets
       (Venus, detection of oxygen in lower atm. of, possibilities for
       spectroscopic)
IT  7782-44-7, occurrence
       (in Venus lower atm., spectroscopic detection of, possibilities
       for)

L7  ANSWER 2 of 39

AN  CA100(24):195485a
TI  Gas-chromatographic analysis of the chemical composition of the
    atmosphere of Venus
AU  Gel'man, B. G.;  Ivanova, S. G.;  Lamonov, N. I.;  Levchuk, B. V.;
    Mel'nikov, V. V.;  Okhotnikov, B. P.;  Rotin, V. A.;  Khokhlov, V. N.;
    Yakutina, S. I.
LO  USSR
```

**Fig. 4.2c**

```
SO  Analiz Neorgan. Gazov. Sb. Plenar, Dokl. 1 Vses. Konf. po Analizu
    Neorgan, Gazov, Leningrad, 27-29 Sent., 1983, L. 157-68
    From:  Ref. Zh., Khim. 1984, Abst. No 5G220
SC  53-9 (Mineralogical and Geological Chemistry)
DT  J
PY  1983
LA  Russ
AB  Title only translated.
KW  Venus atm compn
IT  Planets
       (Venus, atm. of, compn. of, gas chromatog. anal. of)

=> D L7 3-5 ALL

L7  ANSWER 3 of 39

AN  CA100(8):54913b
TI  Isotopic composition of noble gases in the atmoshphere of Mars:
    additional analysis of data
AU  Levskii, L. K.
CS  Leningr. Gos. Univ.
LO  Leningrad, USSR
SO  Geokhimiya, (12), 1778-80
SC  53-9 (Mineralogical and Geological Chemistry)
DT  J
CO  GEOKAQ
IS  0016-7525
PY  1983
LA  Russ
AB  The concns. were detd. of 36Ar, 132Xe, K, and I, resp., in the atm.
    of Mars:  (0.08-0.25) .times. 10-8 cm3/g, (0.08-0.25) .times. 10-10
    cm3/g, 100-800 ppm, and 0.003-0.15 ppm.
KW  Mars atm argon isotope
IT  Planets
       (Mars, atm. of, argon and xenon isotopes in)
IT  7440-37-1D, isotopes, occurrence  7440-63-3D, isotopes, occurrence
       (in Martian atm.)

L7  ANSWER 4 OF 39

AN  CA98(26):219418z
TI  Gas-chromatographic analysis of the atmospheric chemical composition
    of Venus by apparatus deployed by the Venera-13 and Venera-14 robot
    landers
AU  Mukhin, L. M.; Gel'man, B. G.; Lomanov, N. I.;  Mel'nikov, V. V.;
    Nenarokov, D. F.; Okhotnikov, B. P.;  Rotin, V. A.;  Khokhlov, V. N.
LO  USSR
SO  Kosm. Issled., 21(2), 225-30
SC  53-9 (Mineralogical and Geological Chemistry)
DT  J
CO  KOISAW
IS  0023-4206
PY  1983
LA  Russ
```

**Fig. 4.2c (cont)**

AB Results are presented on the gas-chromatog. estn. of H2 (2.5 .+-. 1) .times. 10-3, O2 (1.8 -+. 0.4) .times. 10.3, Kr (7 .+-. 3) .times. 10-5, H2O (7 .+-. 3) .times. 10-2, H2S (8 .+-. 4) .times. 10-3, COS (4 .+-. 2) .times. 10-3, and SF6 (N2O) (2 .+-. 1) .times. 10-5 (.apprx.1) % in the 29-58 km altitude range in the cloud layer of Venus.
KW Venus atm gas chromat. Venera
IT Planets
(Venus, atm. of, gases in)
IT 463-58-1 1333-74-0, occurrence 2551-62-4 7439-90-9, occurrence 7732-18-5, occurrence 7782-44-7, occurrence 7783-06-4, occurrence 10024-47-2
(in atm., gas chromatog. detn. of, of Venus)

L7 ANSWER 5 OF 39

AN CA97(20) : 166598d
TI Gas-chromatographical analysis of chemical composition of the atmosphere of Venus done by Venera 13 and Venera 14 probes
AU Mukhin, L. M.; Gel'man, B. G.; Lamonov, N. I.; Mel'nikov, V. V.; Nenarokov, D. F.; Okhotnikov, B. P.; Rotin, V. A.; Khoklov, V. N.
CS Inst. Kosmicheskikh Issled.
LO Moscow, USSR
SO Pis'ma Astron. Zh., 8(7), 399-403
SC 53-9 (Mineralogical and Geological Chemistry)
DT J
CO PAZHDA
PY 1982
LA Russ
AB Results are presented of the gas-chromatog. estn. of H2 (2.5 .times. 10-3), O2 (1.8 .times. 10-3), Kr (7 .times. 10-5), H2O (7 .times. 10-2), H2S (8 .times. 10-3), COS (4 .times. 10-3), and SF6 (2 .times. 10-5%) in the Venusian atm. at altitudes <50 km.
KW Venus atm gas chromatog Venera
IT Planets
(Venus, atm. of, gas-chromtog. anal. of, by Veneral landers)
IT 1333-74-0, occurrence 2551-62-4 7732-18-5, occurrence 7702-44-7 , occurrence 7783-06-4, occurrence
(in Venusian atm., gas-chromatog. estn. of)

**Fig. 4.2c (cont)**

**Response**

Your first search should have proceeded more or less as follows. Again, I have inserted the numbers of references retrieved, when I actually checked this search on-line.

| | | |
|---|---|---|
| 1 | s myristic(w) acid | (280) |
| 2 | s tetradecanoic(w) acid | (95) |
| 3 | s 544-63-8 | (3,291) |
| 4 | s L1 or L2 or L3 | (3,431) |
| 5 | s gas(w)chromatograph? | (25,390) |
| 6 | s L4 and L5 | (108) |

The main point to notice here is that the Registry No. produced far more references than just a free-text search on the names. The rest of the search was pretty straightforward: note that I truncated 'chromatograph-' as far along the stem as I could to avoid false references. To show you the CAS ONLINE format I have printed out the first three complete entries (including their abstracts and the lists of the index terms assigned to them) in Fig. 4.2b.

For the answer to the second search turn to Fig. 4.2c where you can see how this search looked on-line. I entered the statement in a single line as:

's atmospher? and analysis and (Mars or Venus)'

The system responded by itemising my terms and then combining them as I had instructed into a set of 39 references. ('L7' by the way because I had done some other searches before this one). The whole search took scarcely 30 seconds. I asked to have the first two references printed in full on-line and, as one of them didn't have any abstracts to show, I then asked for the next three as well ('D L7 3-5 ALL'): these continue on the second page. Normally I would have asked for all 39 references to be printed off-line and posted to me, so that my on-line time for this search would have been just a minute or so (searching *Chemical Abstracts* from 1967 to date!).

***********************************

**SAQ 4.2b**

> This time we will search on Dialog, by using just the same commands and operators we are used to with ESA/IRS. Write out a search to locate references on the analysis of Isoniazid in tablets.

```
FIG 4.2d

File 74: International Pharmaceutical Abs. - 70-85/Oct
(Copr. ASHP 1985)

        Set Items Description

? sanalysis
         1 12617 ANALYSIS
? sisoniazid
         2   401 ISONIAZID
? cland2
         3    70 1AND2
? stablet?
         4  5218 TABLET?
? c3and4
         5    25 3and4
? T5/8/1-10
5/8/1
   106766
   Hydrazine levels in formulations of hydralazine, isoniazid, and
 phenelzine over a 2 year period
   Descriptors:  Hydrazine - contaminants;  Hydralazine -  contamination
Isoniazid - contamination;  Phenelzine - contamination;  Stability -
hydrazine;  Hypotensive agents - hydralazine;  Antituberculars - isoniazid;
Antidepressants - phenelzine;  Chromatography - hydrazine;  Toxicity -
hydrazine;  Carcinogens - hydrazine
   CAS Registry No:  302-01-2; 86-54-4;  54-85-3; 51-71-8
   Section Headings:  Drug Analysis-(14);  Drug Stability-(10)

5/8/2
   105490
   Identification of isoniazid in pharmaceutical drug forms and biological
objects
   Descriptors:  Isoniazid-photometry;  Nitroblue tetrazolium - reactions;
Antituberculars - isoniazid;  Dyes - nitroblue tetrazolium;  Powders -
isoniazid;  Tablets - isoniazid;  Reactions - isoniazid
   CAS Registry No: 54-85-3;  298-83-9
   Section Headings:  Drug Analysis-(14)

5/8/3
   102948
   Control of impurities in isoniazid tablets
   Descriptors: Isoniazid - analysis; 1-Isonicotinoyl-2-lactosyl hydrazone -
analysis;     Antituberculars     -     isoniazid;     Contamination -
1-isonicotinoyl-  2-lactosyl hydrazone;  Stability - isoniazid
   CAS Registry No:  54-85-3
   Section Headings:  Drug Analysis-(14);  Drug Stability-(10)
```

**Fig. 4.2d**

5/8/4
099327
Assay of hydrazine in isoniazid and its formulations by difference spectrophotometry
Descriptors: Isoniazid - spectrometry, ultraviolet; Hydrazine - spectrometry, ultraviolet; Antituberculars - ioniazid; Hydrolysis - isoniazid; Stability - isoniazid
CAS Registry No: 54-85-3; 302-01-2
Section Headings: Drug Analysis-(14)

5/8/5
098117
New method for the determination of isoniazid in nonaqueous medium
Descriptors: Isoniazid - titrimetry; Isonicotinic acid - titrimetry; Antituberculars - isoniazid
CAS Registry No: 54-85-3; 55-22-1
Section Headings: Drug Analysis-(14)

5/8/6
096144
Colorimetric methods for the direct determination of p-aminosalicylic acid in the presence of isoniazid
Trade Names: p-Aminosalicylic acid
Descriptors: Aminosalicylic acid - colorimetry; Isoniazid - combination, aminosalicylic acid; Antituberculars - aminosalicylic acid
CAS Registry No: 65-49-6; 54-85-3
Section Headings: Drug Analysis-(14)

5/8/7
086031
Determination of isoniazid in the presence of combination drugs in dosage forms
Descriptors: Isoniazid - colorimetry; Antituberculars - isoniazid
CAS Registry No: 54-85-3
Section Headings: Drug Analytis-(14)

5/8/8
086018
High pressure liquid chromatographic assay of thiacetazone and isoniazid tablets
Trade Names: Thiacetazone
Descriptors: Isoniazid - combination, amithiozone; Chromatography, liquid - amithiozone, combination, isoniazid; Antituberculars - isoniazid, combination, amithiozone
CAS Registry No: 54-85-3; 104-06-3
Section Headings: Drug Analysis-(14)

**Fig. 4.2d (cont)**

5/8/9
084863
Spectrophotometric determination of some antitubercular drugs
Descriptors: Aminosalicylic acid - combination, isoniazid; Antituberculars - aminosalicylic acid, combination, isoniazid; Colorimetry - aminosalicylic acid, combination, isoniazid
CAS Registry No: 65-49-6; 54-85-3
Section Headings: Drug Analysis-(14)

5/8/10
083380
Spectrophotometric determination of norethynodrel and mestranol in contraceptive tablets
Descriptors: Mestranol - combination, norethynodrel; Colorimetry - mestranol, combination, norethynodrel; Contraceptives, oral - mestranol, combination, norethynodrel
CAS Registry No: 72-33-3; 68-23-5
Section Headings: Drug Analysis-(14)

?t5/7/8
5/7/8
086018 19-07017
High pressure liquid chromatographic assay of thiacetazone and isoniazid tablets
Rao, G. R.; Banerjee, S. K.; Mohan, K. R.
Quality Control Labs., IDPL, Hyderabad 500 037, India
Indian J. Pharm. Sci. 43:154-156 (Jul-Aug) 1981
Coden: IJSIDW
Languages: English
(3 References)
An HPLC method for the determination of thiacetazone (amithionzone; I) and isoniazid (II) in combination tablets and solutions, using acetonitrile, methanol and 0.01M ammonium acetate as the mobile phase, is described. Satisfactory separation was attained by the method and the accuracy of the method, based on tablet analysis, was 99.9-101.1% for I and 100.5-101.9% for II.
Paul R. Webster

**Fig. 4.2d (cont)**

**Response**

To show you a search on Dialog I have included a photocopy of my on-line search as Fig. 4.2d. The search itself is pretty straightforward. I have shown you how to print out the references in order to take a quick look at all 25 of the final set, and then at the end I have selected one reference (no. 8) as seeming particularly relevant, and have had it printed out in full.

***********************************

**SAQ 4.2c**

My main reason for this question, apart from giving you a bit more practice, is to show you a search on another host, DATA STAR. This time you don't have to enter the 's' command; the truncation symbol is '$' and if you want to join two words consecutively or in a compound term you use 'adj' between them (instead of '(w)' we have used in other hosts). Write out a search on *Laboratory Hazards Bulletin* for references to the problems of the safe disposal of dioxin.

**Response**

Again I have given you a photocopy (Fig. 4.2e) of my on-line search. Points to note are that on a smaller more specialised data base I have had to enter only two terms, and that I truncated 'dioxin-' to give me both singular and plural terms. DATA STAR always assumes you are searching unless you tell it otherwise (that is why you don't have to enter 's' or 'c' commands). At set 4 I tell the system I want it to print ('.. p') Set 3, showing me the titles only ('.. p 3 ti') of all the documents I found. When I have looked at these I have then, asked it to print out the sixth reference and give me the author, title, source, and abstract ('.. p 3 au,ti,so,ab/doc=6'); in fact it turns out that this item was published anonymously and so has no author.

***********************************

```
*CONNECT TIME LHBU:    0:02:37 HH:MM:SS   0.044 DEC HRS.  SESSION   7@:CHIN
*SIGN-ON   15.33.16                      29.10.85

D-H/CHIN/1984 - VOL02,ISS10/1985             SESSION   78

COPYRIGHT BY THE ROYAL SOCIETY OF CHEMISTRY, NOTTINGHAM, UK.


D-S - SEARCH MODE  - ENTER QUERY

    1_:  dioxin$

    RESULT     54

    2_:  disposal

    RESULT     114


    3_:  1 and 2

    RESULT        9


    4_:  ..p 3 ti/doc=all

        1

TI Re-Chem to raise incineration charges, phase out waste imports.


        2

TI All dioxin-containing wastes now regulated.


        3

TI Dangerous substances report.


        4

TI Incinerators 'did not cause birth defects'.


        5

TI New Studies urged by Re-Chem inquiries.
```

**Fig. 4.2e**

```
      6

TI Getting rid of dioxins safely.

      7

TI Re-Chem factory cleared.

      8

TI Bhopal fallout spurs a flap over waste hazards in Australia.

      9

TI Dioxin fears in Pontypool.

RO601 * END OF DOCUMENTS IN LIST


D-S - SEARCH MODE  - ENTER QUERY
    4_:  ..p 3 au,ti,so,ab/doc=6
      6

TI Getting rid of dioxins safely.

SO Chem-Engr-London, no: 409, p: 21, Dec 1984, ISSN 0302-0797.

AB Cremer & Warner (140 Buckingham Palace Rd., London SW1W 9SQ, UK, tel.
   01-730-0777) have developed an alternative method to incineration for
   disposal of dioxin.  The electrochemical process is described.  Rather
   than up-grading of incinerators, it can be used at source.

              END OF DOCUMENT


_:..OFF
```

**Fig. 4.2e (cont)**

```
ENTER DATA BASE NAME_:  CEAB

SIGN-ON   15.07.23                        9.10.85

D-S/CEAB/1971 - VOL4,ISS10/1985            SESSION   75

COPYRIGHT BY CHEMICAL ENGINEERING ABSTRACTS

D-S - SEARCH MODE - ENTER QUERY

    1_:  x-ray adj fluoresc$

    RESULT      12

:..p 1 ti/doc=all

      1

TI Infrared measurement of coatings and protective lacquers.

      2

TI Online analysis of light elements in slurries.

      3

TI Taking the uncertainty out of bulk-solids analysis.

      4

TI The residence time distributions of liquor and lime mud flows in the
   recausticizing process.

      5

TI Analytical instruments in process management.

      6

TI X-ray analysis in talc production.
```

**Fig. 4.2f**

```
        7

TI USING ON-STREAM X-RAY FLUORESCENCE FOR SLURRY COMPOSITION ANALYSIS.

        8

TI ADVANCES IN THE ANALYSIS OF AIR CONTAMINANTS.  A CRITICAL REVIEW.

        9

TI New health monitors:  Small but accurate.

        10

TI Multicomponent  On-stream  Analyzers  for  Process Monitoring and
   Control

        11

TI Morphology and Chemical Composition of the Houston Aerosol.

        12

TI EXPERIENCE GAINED IN THE EARLY STAGES OF OPERATION OF THE NEW 3000
   TPD KILN PLANT AT THE SCHELKLINGEN, WEST GERMANY/BETRIEB UND ERSTE
   ERFAHRUNGEN MIT DER NEUEN.

   2_:  ..p 1 au,ti,so,ab/doc=1-4

        1

TI Infrared measurement of coatings and protective lacquers.

AU Benson-I-B, Hindle-P-H.

SO Chart-Mech-Eng,  vol: 32, no: 2, p: 31-33, Feb 1985,  ISSN 0009-191X.

AB Organic coatings and protective lacquers can be measured,
   continuously online, by using an infrared absorption gauge specially
   developed for these applications.  The method is particularly suited
   to monitoring the thickness of extremely thin coatings on a variety
   of metal sheets, foils and metallized films.  A recent innovation has
   enabled coatings on highly reflective, moving, surfaces to be
   measured reliably.  This method complements x-ray fluorescence and
   nucleonic techniques.
```

**Fig. 4.2f (cont)**

```
      2

TI Online analysis of light elements in slurries.

AU Smith-L.

SO ProcessEng, vol: 65, no: 8, p: 29, Aug 1984, ISSN 0370-1859

AB A system based on x-ray fluorescence was developed for the online
   analysis of clay slurries.  The system prepares a dry filter cake from
   the slurry sample before analyzing for any 2 elements simultaneously.
   Operation of the unit is continuous, fully automatic, and controlled
   by microcomputer.  An almost instantaneous analysis is produced,
   compared with a delay of about 36 h previously.

      3

TI Taking the uncertainty out of bulk-solid analysis.

AU Parkinson-G, Short-H, Gomez-B, Pedersen-A.

SO Chem-Eng-Int-Ed, vol: 91, no: 16, p: 23, 25. 29,  6 Aug 1984
```

**Fig. 4.2f**

**SAQ 4.2d**

We stay in the DATA STAR host. Write out a search on *Chemical Engineering Abstracts* for references to the use of X-ray fluorescence techniques in chemical engineering.

**Response**

Look at Fig. 4.2f for my on-line search. There is a slight trick here, which is one of the reasons why I set you this question. The whole data base covers chemical engineering, so you don't need to include that term in your search. By selecting a very specific topic, such as X-ray fluorescence techniques, you are generally going to be given at once a well-defined set (it would be different in a search on *Chemical Abstracts* though, where there are likely to be very many references on the technique as applied over the whole field of chemistry). And so you do find here immediately a very limited set: just by entering 'X-ray adj fluoresc$' (I don't think the truncation is strictly necessary here, but it does cast the net slightly wider); we find only 12 references in our set. Once again I look at the titles of all twelve then ask for a fuller print-out of a few of our references. If we had wanted to be really wide ranging in our search and produce a much larger set of references, then we could, of course, have used an 'or' combination and searched alternative terms such as 'X-ray emission spectrometry', 'X-ray spectrometric analysis' etc.

**********************************

**SAQ 4.3a**

You want information on the use of lithium perchlorate as an oxidant in rocket fuels. Write out a search on Data Star, by using the commands and operators I have shown you, to find some relevant data in the Kirk-Othmer data base. When you have completed the search phase, write out two commands to scan and then print a relevant and individual reference.

**Response**

Once again I have done this search on-line and given you a copy of the results (Fig. 4.3d). First of all, let's do the search. Because I have used this data base before and also know the terminology involved, I have chosen three terms to pair with 'rocket$' (truncated for wider coverage): 'fuels', 'propellants' and 'motors'. I doubt whether many of you used more than the term 'rocket fuel', although I hope you truncated 'fuels'. If you used the operator 'adj' where I have used 'with', then you are quite correct: my search will be slightly wider than yours. 'Lithium adj perchlorate' is straightforward; I chose to combine it with my Set 4 by the 'same' operator (following my earlier advice) and reached a set with just one reference. At this point I would have expected you to complete your answer with two print commands:

.. p ti,oc/doc=all

.. p ti,hits/doc=1

The first would show you the occurrences for all the references retrieved, and the second would print out the relevant paragraphs only from your chosen most suitable reference.

At this point the theory and practice diverge slightly. Keep hold of Fig. 4.3d, but turn back to the main text where I am going to point out a few practical lessons that have arisen in the rest of this search.

**********************************

```
D-D/KIRK/VOL24,SUPPLEMENT 25          SESSION      94
COPYRIGHT BY JOHN WILEY 7 SONS, INC. NEW YORK, NY 10150, USA.


D-S - SEARCH MODE  - ENTER QUERY

    1_:  rocket$ with fuel$

    RESULT      28


    2_:  rocket$ with proellant$

    RESULT      29


    3_:  rocket$ with motor$

    RESULT       14


    4_:  1 or 2 or 3

    RESULT      49


    5_:  lithium adj perchlorate

    RESULT      8


    6_:  4 same 5

    RESULT      1


    7_:  ..p toi,oc/doc=1

E0503 STATEMENT NUMBER MISSING.
```

**Fig. 4.3d**

```
D-S - SEARCH MODE  - ENTER QUERY

    7_:  ..p 6 ti,oc/doc=1

       1

OC PARAGRAPH

    TX (87)

TI Lithium and Lithium Compounds

R0601 * END OF DOCUMENTS IN LIST


D-S - SEARCH MODE  - ENTER QUERY

    7_:  ..p6 ti,hits/doc=1

E1002 UNRECOGNISED COMMAND - REENTER OR RETURN TO CONTINUE WITH  LAST COM

MAND

       1

TI Lithium and Lithium Compounds.

TX 87 of 128.

   LiC104 *RT ARROW* LiCl + 2 021f mixed with a catalyst, eg, boron,

   manganese dioxide, iron, etc, decomposition begins at lower

   temperatures;  such mixtures have been described as a pyrochemical

   source of pure oxygen (103).  Lithium perchlorate also has been used

   as a rocket-fuel oxidizer (see Propellants).

R0601 * END OF DOCUMENTS IN LIST
```

**Fig. 4.3d (cont)**

```
D-S - SEARCH MODE  - ENTER QUERY

    7_:  ..p 5 ti/doc=all

      1

TI Semiconductors, Organic.


      2

TI Propylene Oxide.


      3

TI Oxygen-Generation Systems.


      4

TI Lithium and Lithium Compounds.


      5

TI Explosives and Propellants--Propellants.


      6

TI Drying Agents


      7

TI Chlorine Oxygen Acids and Salts_Perchloric Acid and Perchlorates.


      8

TI Batteries and Electric Cells, Primary--Introduction.

R0601 * END OF DOCUMENTS IN LIST

D-S - SEARCH MODE  - ENTER QUERY

    7_:  ..p 5 ti,oc/doc=5

      5
```

**Fig. 4.3d (cont)**

```
OC PARAGRAPH

    DE (45)

    TA (22)

TI Explosives and Propellants--Propellants.

                    END OF DOCUMENT

_:..p 5 ti,hits/doc=5

      5

TI Explosives and Propellants--Propellants.

DE 45 OF 101. Lithium perchlorate y7791-03-91, propellant oxidizer

   #9:637

TA (22)

   Table 5. Properties of Some Common Inorganic Oxidizers

   COLUMN HEADINGS (7):

   (1)  Oxidizer

   (2)  Available oxygen

   (3)  Melting point,  DEGREES C

   (4)  Density, g/cm**3

   (5)  Heat of formation, kJ/mol**a

   (6)  Heat capacity, J/(mol.K)**a

   (7)  Moles of gas per 100 g**b

   Table 5 - Line 1.

   (1)  potassium perchlorate**c

   (2)  46.0

   (4)  2.53

   (5)  -433.4

   (6)  112.5

   (7)  0
```

**Fig. 4.3d (cont)**

Table 5 - Line 2.

(1) ammonium perchlorate

(2) 34.0

(3) dec

(4) 1.95

(5) -290.3

(6) 128.0

(7) 2.55

Table 5 - Line 3.

(1) ammonium nitrate**c

(2) 20.0

(3) 169

(4) 1.72

(5) -365.2

(6) 137.2

(7) 3.75

Table 5 - Line 4.

(1) Lithium perchlorate

(2) 60.6

(3) 236

(4) 2.43

(5) -368.6

(6) 104.6**d

(7) 0.

END OF DOCUMENT

**Fig. 4.3d (cont)**

**SAQ 4.3b**

We have already come across myristic acid: we know that *Chemical Abstracts* calls it tetradecanoic acid and has given it a Registry Number 544-63-8. Write down two alternative commands for looking up myristic acid on-line in the Merck Database. One way of seeking it will give you five references; write down the commands you would enter, first to scan all five quickly, then to look at one in more detail (you can make up set and document numbers, or just use 'n', as you prefer).

**Response**

You could have used any two of these three commands:

myristic adj acid
tetradecanoic adj acid
544-63-8.rn.

The 'myristic adj acid' command gives you the five documents, the other two each give you the same single document. You don't need to look at the occurrence tables for a database like this, so you just want to look at the titles; you would enter:

.. p 1 ti/doc=all

the titles are then revealed as:

1 Menhaden oil
2 Myristic acid
3 Myrophene
4 Orris root oil
5 Reticulin (the Protein)

You notice that reference no. 2 actually carries the title 'myristic acid'; so, to look at the whole entry your final command will be:

.. p 1 all/doc=2

***********************************

```
* COPYRIGHT BY THE AMERICAN CHEMICAL SOCIETY,
* 1155 16TH STREET, N.W., WASHINGTON, D.C. 20036
* SING ON       8:15:21          11/06/85
CFTX 1980 - NOV 1985
BRS SEARCH MODE - ENTER QUERY
       1_:  fourier adj transform adj infrared adj spectroscopy
        RESULT          143 DOCUMENTS
       2_:  aerosol or aerosols
        RESULT          630 DOCUMENTS
       3_:  1 with 2
        RESULT            3 DOCUMENTS
       4_:  ..p 3 ti,oc/doc=all
         1
TI QUANTITATIVE ANALYSIS OF NITRATE ION IN AMBIENT AEROSOLS BY
   FOURIER-TRANSFORM INFRARED SPECTROSCOPY.
OC PARAGRAPH      SENTENCE  NS-WORD
    TI     (1)        1         6
    TI     (1)        1        10

         2
TI INFRARED SPECTROMETRY.
OC PARAGRAPH      SENTENCE  NS-WORD
    RF  (276)         2         7
    RF  (276)         2        11

         3
TI OZONE-CYCLOHEXENE REACTION IN AIR: QUANTITATIVE ANLYSIS OF
   PARTICULATE PRODUCTS AND THE REACTION MECHANISM.
OC PARAGRAPH      SENTENCE  NS-WORD
    TX     (2)        4         8
    TX     (2)        4        11

END OF DOCUMENTS IN LIST_:    ..p 3 ab/doc=3
       3
AB BOTH GASEOUS AND PARTICULATE PRODUCTS OF THE CYCLOHEXENE-OZONE
   REACTION WERE ANALYZED. MAJOR GASEOUS PRODUCTS WERE ALDEHYDES THAT
   CONSIST OF  ADIPALDEHYDE (CHO(CH2)4CHO), GLUTARALDEHYDE
   (CHO(CH2)3CHO), AND PENTANAL (CH3(CH2)3CHO). THE SUM OF THE PRIMARY
   YIELDS OF ALDEHYDES REACHES AS HIGH AS 50%. IN ADDITION TO
   ALDEHYDES, FORMIC ACID, CO, AND CO2 WERE PRODUCED, BUT FORMALDEHYDE
   WAS NOT DETECTED. MAIN PARTICULATE PRODUCTS WERE ADIPALDEHYDE,
   6-OXOHEXANOIC ACID (CHO(CH2)4COOH), ADIPIC ACID (HOOC(CH2)4COOH),
   GLUTARALDEHYDE, 5-OXOPENTANOIC ACID (CHO(CH2)3COOH), AND GLUTARIC
   ACID (HOOC(CH2)3COOH). ALL THESE COMPOUNDS WERE ANALYZED
   QUANTITATIVELY, AND THE FRACTION OF INITIAL CYCLOHEXENE CONVERTED TO
   AEROSOL ORGANIC CARBON WAS ESTIMATED TO BE 13. +-  3% AS THE VALUE
   EXTRAPOLATED TO A PPM CONCENTRATION RANGE OF REACTANTS. ALTHOUGH THE
   REACTION MECHANISM IS IN GENERAL EXPLAINABLE IN TERMS OF THE CRIEGEE
   MECHANISM, THE REACTION PATHWAY TO FORM FORMIC ACID IS QUITE UNIQUE
   IN THIS REACTION SYSTEM. THE ENTIRE MECHANISM WAS DISCUSSED ON THE
   BASIS OF THE QUANTITATIVE PRODUCT ANALYSIS DATA.
END OF DOCUMENT
```

**Fig. 4.3e**

```
END OF DOCUMENTS IN LIST_:  ..p 3 au,ti,so/doc=3
        3
SO ENVIRONMENTAL SCIENCE AND TECHNOLOGY,VOL. 019, NO. 10, 1985, P 935.
TI OZONE-CYCLOHEXENE REACTION IN AIR: QUANTITATIVE ANALYSIS OF
   PARTICULATE PRODUCTS AND THE REACTION MECHANISM.
AU (1) HATAKEYAMA, S. (2) TANONAKA, T. (3) WENG, J. (4) BANDOW, H. (5)
   TAKAGI, H. (6) AKIMOTO, H.
END OF DOCUMENT

END OF DOCUMENTS IN LIST_:  ..p 3 hits/doc=3
        3
TX PARAGRAPH 2 OF 36.  INTRODUCTION.  IN THE PRESENT STUDY, GAS-PHASE
   REACTIONS OF CYCLOHEXENE AND OZONE WERE STUDIED IN DETAIL IN ORDER TO
   REVEAL THE REACTION MECHANISM.  QUANTITATIVE ANALYSES OF GASEOUS
   PRODUCTS BY FOURIER TRANSFORM INFRARED SPECTROSCOPY (FT-IR) AND OF
   AEROSOL PRODUCTS BY GC AND GAS CHROMATOGRAPHY/MASS SPECTROMETRY
   (GC/MS)  ALL BASED ON AUTHENTIC SAMPLES WERE MADE FOR THE FIRST TIME,
   AND THEIR FORMATION MECHANISM WAS ASSESSED.

END OF DOCUMENT
```

**Fig. 4.3e (cont)**

**SAQ 4.3c**

You are interested in Fourier-transform infra-red spectroscopy and, in particular, are looking for documents concerning its use for the analysis of aerosols. First of all, construct a search on BRS (by using just the same commands as we have used throughout Section 4.3) to find relevant documents.

Having found a small set of just three documents, write out the commands you would use, first to compare all three, then to concentrate on one you choose as seeming particularly relevant.

**Response**

Look at Fig. 4.3e and you can see how I conducted this search. I entered the whole phrase (fourier adj transform adj spectroscopy') as a single search command, and got Set 1 of 143 documents. The terms 'aerosol' or 'aerosols' (you could just as easily have truncated this as 'aerosol$' if you prefer) produced 630 hits; combining the two sets by 'with' to ensure the relevance of the sets of terms to one another, I reached a set of three documents. At prompt 4 I take the next obvious step of looking at the titles and occurrences of all three documents. The occurrence tables look slightly different in this data base: they tell you the field where the hits were found (eg Ti title, TX text, etc), and the number of the relevant paragraph, sentence, and words. I decide that document 3 looks interesting and ask to see the abstract of it ('.. p 3 ab/doc=3'). Next I ask to see its bibliographic details, in case I need to record or cite the paper, and, finally, have the relevant 'hits' paragraph printed.

***********************************

**SAQ 4.3d**

> Having completed the previous question you want to look at wider applications of Fourier-transform infrared spectroscopy. Write down the command you would use to scan ten of the general references. Then, again, select one of these and write out the command you would use to look at it more closely.
>
> Finally, having completed that stage, can you think of a further way of searching to find some more material which might prove useful?

**Response**

We have already Set 1 of our previous search giving us 143 documents on Fourier-transform infrared spectroscopy. To look at just the titles of the most recent ten of these we enter the command:

.. p 1 ti/doc=1-10

Look at Fig. 4.3f and you will see how these titles are printed out.

I decided that document no. 1 looked particularly interesting, so my next step was to check on the occurrences ('.. p 1 oc/doc=1'): hits were recorded in the document's title, in two paragraphs of its text and in two footnotes. I next asked for the bibliographic details of the paper, and, finally for a print of the hits.

The last search I did, in Fig. 4.3g, and my answer to the final part of the question, was to see whether the author of the paper I had already found to be highly relevant (J.L Koenig) has published any other papers. This search command 'koenig adj j.au.' recorded nine hits. I had the titles of these displayed, and thereafter could have selected any of these documents for a more detailed examination in the manner I have already shown you.

```
          1
TI FOURIER TRANSFORM INFRARED SPECTROSCOPY OF CHEMICAL SYSTEMS

          2
TI CONFORMATIONAL ENERGIES OF STEREOREGULAR POLY(METHYL METHACRYLATE) BY
   FOURIER TRANSFORM INFRARED SPECTROSCOPY.

          3
TI SYNTHESIS HYPERFINE INTERACTIONS, AND LATTICE DYNAMICS OF THE
   INTERCALCULATION COMPOUNDS FEOCL: (CH3O)3P:1/6 AND FEOCL:(CH3CH2)3P:1/6.

          4
TI ACHOLEPLASMA LAIDLAWII MEMBRANES:  A FOURIER TRANSFORM INFRARED STUDY
   OF THE INFLUENCE OF PROTEIN ON LIPID ORGANIZATION AND DYNAMICS.

          5
TI DIRECT OBSERVATION OF SUBSTRATE DISTORTION BY TRIOSEPHOSPHATE
   ISOMERASE USING FOURIER TRANSFORM INFRARED SPECTROSCOPY.

          6
TI CHARACTERIZATION  OF THE PRETRANSITION IN
   1,2-DIPALMITOYL-SN-GLYCERO-3-PHOSPHOCHOLINE BY FOURIER TRANSFORM
   INFRARED SPECTROSCOPY.

          7
TI A FT IR SPECTROSCOPIC STUDY OF THE OZONE-ETHENE REACTION MECHANISM IN
   O2-RICH MIXTURES.

          8
TI KINETICS OF THE CLO + NO2 + M REACTION.

          9
TI MECHANISM OF THE GAS-PHASE REACTIONS OF C3H6 AND NO3 RADICALS.

         10
TI OXIDATION OF SULFUR DIOXIDE BY METHYLPEROXY RADICALS.

END OF DOCUMENTS

ENTER DOCUMENT SELECTION._:  ..p 1 oc/doc=1

          1
OC PARAGRAPH     SENTENCE  NS-WORD
      T1   (1)      1         4
      TX   (1)      3         9
      TX  (25)      3         4
      FN  (16)      1        17
      FN  (28)      1        13

END OF DOCUMENT
```

**Fig. 4.3f**

```
ENTER DOCUMENT SELECTION._:  ..pl au,ti,so/doc=1
C001 ENTER COMMAND IN CORRECT FORMAT._:  ..p 1 au,ti,so/doc=1

          1
SO ACCOUNTS OF CHEMICAL RESEARCH, VOL. 014, NO 6, 1981, P 171-178.
TI FOURIER TRANSFORM INFRARED SPECTROSCOPY OF CHEMICAL SYSTEMS
AU (1) KOENIG, J.L.
END OF DOCUMENT

ENTER DOCUMENT SELECTION._:  ..p 1 hits/doc=1

          1
TI FOURIER TRANSFORM INFRARED SPECTROSCOPY OF CHEMICAL SYSTEMS.
TX PARAGRAPH 1 OF 33.  INFRARED SPECTROSCOPY IS RECOGNIZED AS ONE OF OUR
   MOST POWERFUL PHYSICAL TOOLS FOR INVESTIGATING CHEMICAL SYSTEMS.  THE
   INTRODUCTION OF INTERFEROMETRIC TECHNIQUES LEAD TO THE DEVELOPMENT OF
   FOURIER TRANSFORM INFRARED SPECTROSCOPY (FT-IR).  THE IMPACT OF THESE
   IMPROVED INSTRUMENTS HAS BEEN DRAMATIC AND HAS PENETRATED NEARLY
   EVERY ASPECT OF CHEMICAL RESEARCH FROM ANALYZING THE ENVIRONMENT ON
   OTHER PLANETS TO CELLS IN THE BLOOD STREAM.  IT BEHOVES CHEMISTS,
   THEREFORE, TO DEVELOP SOME INSIGHT INTO THE NATURE OF THIS EVOLVING
   TECHNIQUE IN ORDER TO APPRECIATE THE LARGE AMOUNT OF ACTIVITY IN THIS
   AREA.  THIS ACCOUNT DESCRIBES THE TECHNIQUE OF FT-IR AND SELECTED
   CHEMICAL APPLICATIONS.

TX PARAGRAPH 25 OF 33.  FT-IR APPLICATIONS TO POLYMERS.  FOURIER
   TRANSFORM INFRARED SPECTROSCOPY IS HAVING A MAJOR IMPACT ON THE STUDY
   OF POLYMER SYSTEMS.  THE ADVANTAGES OF FT-IR ARISE PRIMARILY FROM THE
   HIGHER SIGNAL-TO-NOISE RATIO OBTAINABLE FROM THE HIGHER ENERGY
   THROUGHPUT OF THE SYSTEM AND THE CAPABILITY OF SIGNAL AVERAGING OF
   THE SPECTRA TO ACHIEVE A FURTHER INCREASE IN THE SIGNAL-TO-NOISE
   RATIO.  ADDITIONALLY, THE FULL BENEFIT OF THE HIGHER SIGNAL-TO-NOISE
   RATIO CAN BE REALIZED THROUGH THE USE OF THE DATA-PROCESSING
   POTENTIAL ARISING FROM THE AVAILABILITY OF THE MINICOMPUTER.

FN FN 16 OF 36.  M.K. ANTOON, K.M. STARKEY AND J. L. KOENIG,
   "APPLICATION OF FOURIER TRANSFORM INFRARED SPECTROSCOPY TO QUALITY
   CONTROL OF THE EPOXY MATRIX".  ASTM COMPOSITE MATERIALS  (FIFTH
```

**Fig. 4.3f (cont)**

```
ENTER DOCUMENT SELECTION._:   ..s
BRS SEARCH MODE - ENTER QUERY
       5_:  koenig adj j.au.
        RESULT          9 DOCUMENTS

       6_:  ..p 5 ti/doc=all
          1
TI FOURIER TRANSFORM INFRARED SPECTROSCOPY OF CHEMICAL SYSTEMS.

          2
TI HIGH-RESOLUTION CARBON-13 NUCLEAR MAGNETIC RESONANCE STUDY OF
   CONJUGATION IN SOLID POLYIMIDES.

          3
TI SPECTROSCOPIC STUDIES OF POLY:N,
   N'-BIS(PHENOXYPHENYL)PYROMELLITIMIDE:.   1. STRUCTURES OF THE
   POLYIMIDE AND THREE MODEL COMPOUNDS.

          4
TI SPECTROSCOPIC STUDIES OF POLY:N,
   N'-BIS(PHENOXYPHENYL)PYROMELLITIMIDE:.   2. STRUCTURAL CHANGES OF
   POLYIMIDE UPON YIELDING.

          5
TI FACTOR ANALYSIS FOR SEPARATION OF PURE COMPONENT SPECTRA FROM MIXTURE
   SPECTRA.

          6
TI CHARACTERIZATION OF SILANE-TREATED GLASS FIBERS BY DIFFUSE
   REFLECTANCE FOURIER TRANSFORM SPECTROMETRY.

          7
TI MAGIC ANGLE SPINNING CARBON-13 NUCLEAR MAGNETIC RESONANCE OF
   ACRYLIC-MELAMINE COATINGS.

          8
TI STUDIES OF CHAIN FOLDING IN SOLUTION-CRYSTALLIZED POLY(ETHYLENE
   TEREPHTHALATE).

          9
TI STUDY OF THE SOLID-STATE CROSSPOLYMERIZATION OF
   POLY(1,11-DODECADIYNE) THROUGH MAGIC-ANGLE CARBON-13 NMR.

END OF DOCUMENTS IN LIST_:  ..off
```

**Fig. 4.3g**

# Units of Measurement

For historic reasons a number of different units of measurement have evolved to express quantity of the same thing. In the 1960s, many international scientific bodies recommended the standardisation of names and symbols and the adoption universally of a coherent set of units—the SI units (Système Internationale d'Unités)—based on the definition of five basic units: metre (m); kilogram (kg); second (s); ampere (A); mole (mol); and candela (cd).

The earlier literature references and some of the older text books, naturally use the older units. Even now many practicing scientists have not adopted the SI unit as their working unit. It is therefore necessary to know of the older units and be able to interconvert with SI units.

In this series of texts SI units are used as standard practice. However in areas of activity where their use has not become general practice, eg biologically based laboratories, the earlier defined units are used. This is explained in the study guide to each unit.

Table 1 shows some symbols and abbreviations commonly used in analytical chemistry. Table 2 shows some of the alternative methods for expressing the values of physical quantities and the relationship to the value in SI units.

More details and definition of other units may be found in the *Manual of Symbols and Terminology for Physicochemical Quantities and Units*, Whiffen, 1979, Pergamon Press.

**Table 1** *Symbols and Abbreviations Commonly used in Analytical Chemistry*

| | |
|---|---|
| Å | Angstrom |
| $A_r$(X) | relative atomic mass of X |
| A | ampere |
| *E* or *U* | energy |
| *G* | Gibbs free energy (function) |
| *H* | enthalpy |
| J | joule |
| K | kelvin (273.15 + *t* °C) |
| *K* | equilibrium constant (with subscripts p, c, therm etc.) |
| $K_a$,$K_b$ | acid and base ionisation constants |
| $M_r$(X) | relative molecular mass of X |
| N | newton (SI unit of force) |
| *P* | total pressure |
| *s* | standard deviation |
| *T* | temperature/K |
| *V* | volume |
| V | volt (J $A^{-1}$ $s^{-1}$) |
| *a*, *a*(A) | activity, activity of A |
| *c* | concentration/ mol $dm^{-3}$ |
| e | electron |
| g | gramme |
| *i* | current |
| s | second |
| *t* | temperature / °C |
| | |
| bp | boiling point |
| fp | freezing point |
| mp | melting point |
| $\approx$ | approximately equal to |
| $<$ | less than |
| $>$ | greater than |
| e, exp(*x*) | exponential of *x* |
| ln *x* | natural logarithm of *x*; ln *x* = 2.303 log *x* |
| log *x* | common logarithm of *x* to base 10 |

**Table 2** *Alternative Methods of Expressing Various Physical Quantities*

1. **Mass (SI unit : kg)**

$$g = 10^{-3}\ kg$$
$$mg = 10^{-3}\ g = 10^{-6}\ kg$$
$$\mu g = 10^{-6}\ g = 10^{-9}\ kg$$

2. **Length (SI unit : m)**

$$cm = 10^{-2}\ m$$
$$Å = 10^{-10}\ m$$
$$nm = 10^{-9}\ m = 10Å$$
$$pm = 10^{-12}\ m = 10^{-2}\ Å$$

3. **Volume (SI unit : $m^3$)**

$$l = dm^3 = 10^{-3}\ m^3$$
$$ml = cm^3 = 10^{-6}\ m^3$$
$$\mu l = 10^{-3}\ cm^3$$

4. **Concentration (SI units : mol $m^{-3}$)**

$$M = mol\ l^{-1} = mol\ dm^{-3} = 10^3\ mol\ m^{-3}$$
$$mg\ l^{-1} = \mu g\ cm^{-3} = ppm = 10^{-3}\ g\ dm^{-3}$$
$$\mu g\ g^{-1} = ppm = 10^{-6}\ g\ g^{-1}$$
$$ng\ cm^{-3} = 10^{-6}\ g\ dm^{-3}$$
$$ng\ dm^{-3} = pg\ cm^{-3}$$
$$pg\ g^{-1} = ppb = 10^{-12}\ g\ g^{-1}$$
$$mg\% = 10^{-2}\ g\ dm^{-3}$$
$$\mu g\% = 10^{-5}\ g\ dm^{-3}$$

5. **Pressure (SI unit : N $m^{-2}$ = kg $m^{-1}$ $s^{-2}$)**

$$Pa = Nm^{-2}$$
$$atmos = 101\ 325\ N\ m^{-2}$$
$$bar = 10^5\ N\ m^{-2}$$
$$torr = mmHg = 133.322\ N\ m^{-2}$$

6. **Energy (SI unit : J = kg $m^2$ $s^{-2}$)**

$$cal = 4.184\ J$$
$$erg = 10^{-7}\ J$$
$$eV = 1.602 \times 10^{-19}\ J$$

**Table 3** *Prefixes for SI Units*

| Fraction | Prefix | Symbol |
|---|---|---|
| $10^{-1}$ | deci | d |
| $10^{-2}$ | centi | c |
| $10^{-3}$ | milli | m |
| $10^{-6}$ | micro | $\mu$ |
| $10^{-9}$ | nano | n |
| $10^{-12}$ | pico | p |
| $10^{-15}$ | femto | f |
| $10^{-18}$ | atto | a |

| Multiple | Prefix | Symbol |
|---|---|---|
| 10 | deka | da |
| $10^{2}$ | hecto | h |
| $10^{3}$ | kilo | k |
| $10^{6}$ | mega | M |
| $10^{9}$ | giga | G |
| $10^{12}$ | tera | T |
| $10^{15}$ | peta | P |
| $10^{18}$ | exa | E |

**Table 4** *Recommended Values of Physical Constants*

| Physical constant | Symbol | Value |
|---|---|---|
| acceleration due to gravity | $g$ | $9.81\ m\ s^{-2}$ |
| Avogadro constant | $N_A$ | $6.022\ 05 \times 10^{23}\ mol^{-1}$ |
| Boltzmann constant | $k$ | $1.380\ 66 \times 10^{-23}\ J\ K^{-1}$ |
| charge to mass ratio | $e/m$ | $1.758\ 796 \times 10^{11}\ C\ kg^{-1}$ |
| electronic charge | $e$ | $1.602\ 19 \times 10^{-19}\ C$ |
| Faraday constant | $F$ | $9.648\ 46 \times 10^{4}\ C\ mol^{-1}$ |
| gas constant | $R$ | $8.314\ J\ K^{-1}\ mol^{-1}$ |
| 'ice-point' temperature | $T_{ice}$ | 273.150 K exactly |
| molar volume of ideal gas (stp) | $V_m$ | $2.241\ 38 \times 10^{-2}\ m^3\ mol^{-1}$ |
| permittivity of a vacuum | $\epsilon_o$ | $8.854\ 188 \times 10^{-12}\ kg^{-1}\ m^{-3}\ s^4\ A^2\ (F\ m^{-1})$ |
| Planck constant | $h$ | $6.626\ 2 \times 10^{-34}\ J\ s$ |
| standard atmosphere pressure | $p$ | $101\ 325\ N\ m^{-2}$ exactly |
| atomic mass unit | $m_u$ | $1.660\ 566 \times 10^{-27}\ kg$ |
| speed of light in a vacuum | $c$ | $2.997\ 925 \times 10^{8}\ m\ s^{-1}$ |